BUSINESS MATHEMATICS WORKTEXT

Second Edition

BUSINESS MATHEMATICS WORKTEXT

Second Edition

Esther H. Highland
City University of New York

Charles Peselnick
DeVry Institute of Technology

PRENTICE HALL
Upper Saddle River, New Jersey   Columbus, Ohio

Library of Congress Cataloging-in-Publication Data

Highland, Esther Harris.
    Business mathematics worktext / Esther H. Highland, Charles Peselnick. — 2nd ed.
      p.   cm.
    Includes index.
    ISBN 0-13-040103-X (pbk).
    1. Business mathematics.   I. Peselnick, Charles.   II. Title.
HF5691.H48 2000
650'.01'51—dc21   99-41217
                                                        CIP

Editor: Stephen Helba
Production Editor: Louise N. Sette
Production Supervision: TSI Publishing Ramsey
Design Coordinator: Karrie Converse-Jones
Cover Designer: Rod Harris
Cover Art: ©Rod Harris
Production Manager: Pat Tonneman
Marketing Manager: Shannon Simonsen

This book was set in Times Roman by TSI Graphics and was printed and bound by The Banta Company. The cover was printed by Phoenix Color Corp.

© 2000, 1998 by Prentice-Hall, Inc.
Pearson Education
Upper Saddle River, New Jersey 07458

All rights reserved. No part of this book may be reproduced, in any form or by any means, without permission in writing from the publisher.

Printed in the United States of America

10 9 8 7 6 5 4 3 2 1

ISBN   0-13-040103-X

Prentice-Hall International (UK) Limited, *London*
Prentice-Hall of Australia Pty. Limited, *Sydney*
Prentice-Hall of Canada, Inc., *Toronto*
Prentice-Hall Hispanoamericana, S.A., *Mexico*
Prentice-Hall of India, Private Limited, *New Delhi*
Prentice-Hall of Japan, Inc., *Tokyo*
Simon & Schuster Singapore Pte. Ltd.
Editora Prentice-Hall do Brasil, Ltda, *Rio de Janeiro*

Preface

The second edition of *Business Mathematics Worktext* provides practical, up-to-date coverage of the wide variety of mathematical topics that students need to learn to be successful in today's business world. Calculations frequently necessary for employees in accounting, finance, statistics, taxation, insurance, real estate, and other areas of business are discussed and illustrated. This edition uses the same approach and emphasis that made the last worktext edition and the first five hardback editions successful teaching tools. The emphasis remains on a careful and logical approach to solving real-world business problems.

Current tax laws, business procedures, business trends, and the most recent government forms and tables have been incorporated. Text, problems, examples, and illustrations are up to date and reflect the latest prices and market conditions. Throughout the book, "Calculator Hints" sections show how to apply calculator operations to section examples. Much care has been taken to provide a complete discussion of each topic, so the student can understand why they are learning it and where it applies.

In preparing to write this text, the authors, not only focused on content, but on the way that the content is used in a business environment. Today, and even more so in the future, individuals will perform business mathematics calculations on a computer as part of constructing a computer spreadsheet. In fact, many employers now require that their employees have the capacity to use spreadsheet software. These changes have led us to include computer spreadsheet exercises emphasizing the most popular package—Microsoft Excel—for this edition of *Business Mathematics Worktext*.

FEATURES OF THIS EDITION

- Significant formulas presented within each chapter are conveniently summarized at the end of the chapter.

- Important terms are listed by chapter in the glossary at the end of the book. This provides a valuable reference and summary for the student.

- Examples used throughout the text cover every type of homework problem the student may encounter. Step-by-step solutions and explanations are provided.

- Material is presented in a concise, readable format. The text provides the necessary amount of content to give a thorough description of each topic and its function in the business world.

- Each chapter begins with learning objectives, which tell the student what skills will be acquired upon successful completion of the chapter.

- Space is provided below each problem for students to complete calculations. The pages are perforated so that homework and test pages may be removed and given to the instructor.

- "Calculator Hints" Sections throughout the text use calculator-key notation to show students how to efficiently set up and solve many types of problems using a variety of calculator features. Written explanations accompany the illustrations.

- Important formulas are outlined to separate them from the main body of text. Important terms are in **boldface** type for emphasis.

- Reviews of arithmetic, algebra, ratios, equations, percents, and approximation are included in Chapters 1 and 2 and Appendix A for students who have been away from math for a long time or whose skills need refreshing. Problems in these sections have a business orientation to maintain continuity with the remainder of the text.

- Computer Spreadsheet Exercises related to chapter topics are at the end of each chapter beginning with Chapter 2. Microsoft Excel spreadsheet software is emphasized, but other software can be used. The exercises enable students to integrate business, computers, and computer spreadsheet software as they might do in a real-world business situation. Many powerful spreadsheet functions are explained and utilized. Problems vary in degree of difficulty. Answers to *all* computer spreadsheet exercises appear in the *Instructor's Manual*.

- There is in-depth coverage of depreciation methods in Chapter 5. The section on MACRS reflects the latest tax laws.

- The payroll section in Chapter 6 incorporates the most recent changes in FICA and income tax withholding rates. Social security and medicare taxes are presented separately, and both the wage bracket and percentage methods of federal income tax withholding are explained.

- Interest rates on loans and investments in Chapters 7–11 reflect current market conditions.

- The Chapter 12 graphs contain real and recent data gathered by the federal government and business trade publications.

- Chapter 13 contains a section on reading the stock and bond tables of *The Wall Street Journal*.

- At the end of each chapter there is a review test called *Test Yourself,* which can be used as preparation for an exam or as a general review. The test is a summary of the topics presented in the chapter and is keyed to various chapter sections for reference. Worked-out solutions for all *Test Yourself* problems are given at the back of the book.

- Over 700 additional problems appear at the end of the chapters. The problems are arranged in order of increasing difficulty, and are similar in style and difficulty to the homework presented within the chapter. Answers to the odd-numbered problems appear in the back of book. Worked-out solutions to all of the end-of-chapter problems appear in the *Instructor's Manual*.

- Key tables are summarized in Appendix C at the back of the book for convenient student reference. For testing purposes, the instructor may give a student access to the tables without providing access to the main text.

- An appendix on the metric system is included for those who may use business mathematics in the global marketplace.

- Step-by-step solutions, including formulas, are provided for *all* problems in the *Try These Problems* sections are provided in the back of the book.

SUPPLEMENTS

- The *Instructor's Manual* contains step-by-step solutions to all problems in the text. Complete answers to the computer spreadsheet exercises are also provided. The manual includes some suggestions for teaching each chapter.

- An author-written, copy-ready *Test Bank* contains six tests per chapter. Of these tests, three are free response and three are multiple choice. The questions are consistent with those presented in the homework sets and end-of-chapter problems.

- The Prentice Hall Custom Test File creates and prints custom tests, and prints math symbols, equations, and graphs.

The content of *Business Mathematics Worktext* is appropriate for a two-semester course or selected topics may be used for a one-term course.

We would like to express our thanks to the many users who gave useful comments and valuable suggestions for improving this book. We also acknowledge and greatly appreciate the contributions of the following reviewers:

- James Abbott, *Broome Community College*
- Karen M. Durling, *Beal College*
- Ronald Goldfarb, *Middlesex County College*
- Frances L. Haldar, *North Central Community College*
- Thomas J. Kearns, *Northern Kentucky University*
- Daniel Kraska, *North Central Community College*
- Theodore Lai, *Hudson County Community College*
- Diane Morris, *Belmont Technical College*
- Young H. Rhie, *Springfield College*
- Hazel Struby, *Macon Technical College*

Special thanks and appreciation goes to Vicki Price, John McManamon, Doug Nottingham, and Bob Conklin for their helpful suggestions and comments, and to Ravi Iyer for assisting with some of the graphs in Chapter 12.

We would also like to thank Stephen Helba at Prentice Hall. His focus and commitment made sure that this project was completed in a most successful manner.

A great deal of time and effort has been taken to make the material presented clear, precise, and error free. Should you require clarification, find potential errors, or wish to make suggestions or comments, please contact *cpeselnick@devrycols.edu*.

Charles Peselnick

Esther H. Highland

Contents

PREFACE v

1   Essential Mathematics   1

    1.1   Order of Operations, 1
    1.2   Rounding of Decimals, 3
    1.3   Algebra Concepts, 7
    1.4   Ratio, 27
    1.5   Checking Answers, 31
           New Rules in This Chapter, 32
           End-of-Chapter Problems, 35
           Test Yourself, 39

2   Percents and Their Applications   41

    2.1   Changing Percents to Decimals and Fractions, 41
    2.2   Changing Decimals and Fractions to Percents, 49
    2.3   The Basic Equation, 53
    2.4   Checking Problems by Approximation, 54
    2.5   Applications of the Basic Equation, 59
    2.6   Measuring Change, 65
           New Rules in This Chapter, 68
           New Formulas in This Chapter, 68
           End-of-Chapter Problems, 71
           Test Yourself, 77
           Computer Spreadsheet Exercises, 79

3   Pricing a Product   81

    3.1   The Basic Equation, 81
    3.2   Markup on Cost Versus Markup on Selling Price, 83
    3.3   Percent Markup, 89
    3.4   Pricing Perishable Products, 97
    3.5   Markdowns, 101
    3.6   Breakeven Point, 104

New Formulas in This Chapter, 106
End-of-Chapter Problems, 111
Test Yourself, 121
Computer Spreadsheet Exercises, 123

4 Commercial Discounts   125

4.1 Trade Discounts, 125
4.2 Computing Discount and Net Price, 126
4.3 Computing List Price, 129
4.4 Series Discounts, 133
4.5 Single Equivalent Discount, 137
4.6 Cash Discounts, 143
4.7 Ordinary Dating and Other Credit Terms, 145
4.8 Partial Payments, 153
New Formulas in This Chapter, 154
End-of-Chapter Problems, 157
Test Yourself, 167
Computer Spreadsheet Exercises, 171

5 Accounting Topics: Depreciation and Valuation of Inventory   173

5.1 Depreciation, 173
5.2 Valuation of Inventory, 193
New Formulas in This Chapter, 197
End-of-Chapter Problems, 201
Test Yourself, 209
Computer Spreadsheet Exercises, 211

6 Accounting Topics: Partnership Profits and Payroll   213

6.1 Division of Profits in a Partnership, 213
6.2 Payroll, 225
End-of-Chapter Problems, 243
Test Yourself, 253
Computer Spreadsheet Exercises, 257

7 Borrowing Money: Interest   261

7.1 Simple Interest, 262
7.2 Exact Time between Dates, 269
7.3 Ordinary and Exact Interest: Time in Days, 275
7.4 Principal: Present Value, 281
7.5 Promissory Notes, 283
New Formulas in This Chapter, 290
End-of-Chapter Problems, 293
Test Yourself, 301
Computer Spreadsheet Exercises, 303

8 Borrowing Money: Bank Discounts   305

8.1 Simple Interest Versus Simple Discount, 305
8.2 Discounting a Promissory Note, 313
8.3 Maturity Value of a Discounted Loan, 321
8.4 Borrowing to Anticipate an Invoice, 323
8.5 Comparing Simple Interest and Simple Discount, 326

New Formulas in This Chapter, 326
End-of Chapter Problems, 329
Test Yourself, 337
Computer Spreadsheet Exercises, 341

9 Consumer Credit 343

9.1 Credit Plans, 343
9.2 Determining the Periodic Payment, 346
9.3 United States Rule, 351
9.4 Charge Accounts, 352
9.5 Annual Percentage Rate, 359
9.6 Rule of 78, 365
New Formulas in This Chapter, 366
End-of Chapter Problems, 369
Test Yourself, 377
Computer Spreadsheet Exercises, 381

10 Compound Interest 383

10.1 Simple Versus Compound Interest, 383
10.2 Finding Compound Amount by Table, 386
10.3 Using a Formula to Find Compound Amount, 399
10.4 Rate Changes during Period, 403
10.5 Additional Deposit Made during Period, 403
10.6 When *n* Is Larger than Last Entry in Table, 404
10.7 Effective Rate, 405
10.8 Present Value, 409
New Formulas in This Chapter, 411
End-of Chapter Problems, 415
Test Yourself, 423
Computer Spreadsheet Exercises, 425

11 Annuities 427

11.1 What Is an Annuity?, 427
11.2 Amount of an Annuity, 428
11.3 Present Value of an Annuity, 435
11.4 Finding Rent: Present Value of Annuity Known, 443
11.5 Finding Rent: Amount of Annuity Known, 447
11.6 Review of Compound Interest and Annuities, 451
New Formulas in This Chapter, 452
End-of Chapter Problems, 455
Test Yourself, 463
Computer Spreadsheet Exercises, 465

12 Understanding Statistics 469

12.1 Presentation of Numerical Facts, 470
12.2 Averages, 487
12.3 Dispersion, 491
12.4 Sampling, 497
12.5 Index Numbers, 497
New Formulas in This Chapter, 499
End-of Chapter Problems, 503
Test Yourself, 509
Computer Spreadsheet Exercises, 513

13   Investing   517

13.1   Reading a Financial Report, 517
13.2   Financial Ratios, 520
13.3   Stocks, 525
13.4   Bonds, 530
13.5   Stock Quotations, 535
13.6   Bond Quotations, 537
       New Formulas in This Chapter, 539
       End-of Chapter Problems, 543
       Test Yourself, 549
       Computer Spreadsheet Exercises, 553

APPENDIXES

A   Review of Arithmetic   555

Fractions, 555
Decimals, 566
New Rules in This Appendix, 572
Exercises, 575

B   The Metric System   579

Metric Units of Measure, 579
Exercises, 585
Problems, 587

C   Tables   563

Table C-1: Income Tax Withholding—Wage Bracket Tables, 590
Table C-2: Income Tax Withholding—Percentage Method Tables, 594
Table C-3: The Number of Each Day of the Year, 596
Table C-4: Annual Percentage Rate Table for Monthly Payment Plans, 597
Table C-5: Amount of 1 at Compound Interest, 602
Table C-6: Present Value of 1 at Compound Interest, 618
Table C-7: Amount of an Annuity of 1 Per Period, 634
Table C-8: Present Value of Annuity of 1 Per Period, 642
Table C-9: Periodic Rent of Annuity Whose Present Value is 1, 650
Table C-10: ACRS Cost Recovery Percentages, 658

ANSWERS AND SOLUTIONS TO TRY THESE PROBLEMS AND TEST YOURSELF PROBLEMS   659

Chapter 1, 659   Chapter 8, 678
Chapter 2, 661   Chapter 9, 681
Chapter 3, 662   Chapter 10, 685
Chapter 4, 665   Chapter 11, 688
Chapter 5, 667   Chapter 12, 690
Chapter 6, 672   Chapter 13, 693
Chapter 7, 675

ANSWERS TO ODD-NUMBERED END-OF-CHAPTER PROBLEMS   697

    Chapter 1, 697   Chapter 9, 704
    Chapter 2, 697   Chapter 10, 706
    Chapter 3, 698   Chapter 11, 706
    Chapter 4, 699   Chapter 12, 708
    Chapter 5, 699   Chapter 13, 708
    Chapter 6, 702   Appendix A, 709
    Chapter 7, 703   Appendix B, 709
    Chapter 8, 703

GLOSSARY   711

INDEX   717

1 ■ Essential Mathematics

OBJECTIVES

Upon completion of this chapter, the student should be able to:
1. *Calculate the answer to a series of similar or mixed operations.*
2. *Simplify expressions containing parentheses or similar terms.*
3. *Round off numbers to a specified number of decimal places.*
4. *Identify the terms, coefficients, and constants in algebraic expressions.*
5. *Solve equations by using the four arithmetic operations.*
6. *Express numbers in ratios.*
7. *Check answers by estimation.*

Numbers are used in a wide variety of business situations. Therefore, the ability to evaluate numerical information and knowledge of the fundamentals of both arithmetic and algebra are essential for success in the business world.

1.1 ORDER OF OPERATIONS

Business problems usually involve several steps and may include more than one mathematical operation. The order in which the operations are performed is not important if only addition, subtraction, or multiplication must be done. However, the order will affect the answer in a series of divisions or mixed operations.

Consider the following series of additions, subtractions, and multiplications, respectively. In each series, changing the order in which the numbers are written and used has no effect on the answer. Only one variation for each operation is illustrated, but any other could have been used.

Addition: $4 + 12 + 7 + 5 = 28$

$7 + 4 + 5 + 12 = 28$

Subtraction: $100 - 20 - 10 - 3 = 67$

$100 - 10 - 3 - 20 = 67$

The initial 100 is not part of the subtraction series: its understood sign is +.

Multiplication: $3 \times 2 \times 8 \times 5 = 240$

$8 \times 3 \times 5 \times 2 = 240$

For a series of divisions or mixed operations there are definite rules to follow.

1

Division: A series of divisions should be taken in the order written. Thus,

$$100 \div 10 \div 2 = 10 \div 2 = 5$$

This result would be different and incorrect if the second division were done first:

$$100 \div 5 = 20 \qquad \text{This is incorrect.}$$

Mixed Operation: In a series of mixed operations, multiplication and division are done before addition and subtraction. Consider, for example,

$$100 + \underbrace{20 \times 2}_{} - \underbrace{8 \div 4}_{}$$
$$= 100 + \quad 40 \quad - \quad 2$$
$$= 138$$

If the operations were done in the order written, the answer would be different and incorrect.

> **Rule 1-1.** A series of *only* addition, *only* subtraction, or *only* multiplication may be performed in any order. A series of *only* division must be done in the order written.

> **Rule 1-2.** In a series of *mixed* operations, multiplication and division are done first in the order written from left to right, then addition and subtraction are done in the order written from left to right.

■ **Example 1**

1. Evaluate $24 \div 6 \div 2$

   $24 \div 6 \div 2 = 4 \div 2$

   $= 2$

   Divisions should be done in the order written.

2. Evaluate $7 + 1 - 8 \times 4 \div 16$

   $7 + 1 - \underbrace{8 \times 4}_{} \div 16$

   $= 7 + 1 - \quad 32 \quad \div 16$

   $= 7 + 1 - \quad\quad 2$

   $= 6$

   Multiplication and division are done first.

■

Parentheses

Parentheses are used to enclose one or more numbers that are to be considered as one quantity. Sometimes the quantity enclosed in parentheses is multiplied by a number in front of it. If no number precedes the parentheses, it is understood that the quantity inside is to be multiplied by 1.

Although parentheses may be removed by multiplying each number inside by the number outside, it will be simplest in the problems in this text to complete all operations within the parentheses first.

If parentheses enclose a set of mixed operations, the procedure described in Rule 1-2 should be followed.

Note: Appendix A gives a complete review of arithmetic operations, including fractions and decimals.

ROUNDING OF DECIMALS **3**

■ **Example 2**   Evaluate the following expressions.

1. $12 - (9 - 3)$

$$12 - (9 - 3) = 12 - (6)$$
$$= 12 - 1 \times 6$$
$$= 12 - 6$$
$$12 - (9 - 3) = 6$$

Complete arithmetic within parentheses first.

2. $6(5 - 2)$

$$6(5 - 2) = 6(3)$$
$$6(5 - 2) = 18$$

Complete arithmetic within parentheses first.

3. $100\left(1 + 0.09 \times \dfrac{1}{2}\right)$

$$100\left(1 + 0.09 \times \dfrac{1}{2}\right) = 100(1 + 0.045)$$
$$= 100(1.045)$$
$$100\left(1 + 0.09 \times \dfrac{1}{2}\right) = 104.5$$

$0.09 \times \dfrac{1}{2} = 0.045.$

Multiplication is done before addition.

4. $400\left(1 - 0.12 \times \dfrac{5}{12}\right)$

$$400\left(1 - 0.12 \times \dfrac{5}{12}\right) = 400(1 - 0.05)$$
$$= 400(0.95)$$
$$400\left(1 - 0.12 \times \dfrac{5}{12}\right) = 380$$

$0.12 \times \dfrac{5}{12} = 0.05.$

Multiplication is done before subtraction.   ■

1.2 ROUNDING OF DECIMALS

Many problems in business mathematics result in numbers with several decimal places or, when division is involved, an indefinite number of decimal places. Answers to business problems are frequently rounded to two decimal places for dollars and cents, but this is only one case of a general rule.

> **Rule 1-3.** To round a decimal, determine the desired number of decimal places and consider only the value of the next decimal place. If this value is 5 or more, increase the previous digit by 1. If it is less than 5, leave the previous digit unchanged. Then drop all digits after the desired number of decimal places.

■ **Example 1**   1. Round $12.7936 to two decimal places (nearest hundredth).

$$\$12.7\ 9\ 3\ |\ 6$$
$$\uparrow$$

Consider only the third decimal place. Three is less than 5, and therefore 9 remains unchanged.

$$\$12.79$$

Drop remaining digits.

2. Round 5.183 to the nearest tenth (one decimal place).

        5.1 8 | 3   *Consider only the second decimal place. Eight is more than 5. Increase 1 to 2.*

        5.2   *Drop remaining digits.*

3. Round $59.69495 to the nearest cent (two decimal places).

        $59.6 9 4 | 9 5   *Consider only the third decimal place. Four is less than 5. Nine remains unchanged.*

        $59.69   *Drop remaining digits.* ∎

Accuracy with Calculator Solutions

Hand-held calculators can help you solve the problems in this text. The calculator makes it easy to get accurate results by carrying a large number of decimal places during a calculation. Many different models are available, but this section will use only those features that are common to most calculators. If your calculator does not show the indicated answers to the following problem, check your instruction book.

Beginning with Chapter 2, the text contains CALCULATOR HINTS that will help you use your calculator most efficiently.

■ **Example 2**

Use a calculator to evaluate $400 \left(1 - 0.11 \times \dfrac{7}{12}\right)$.

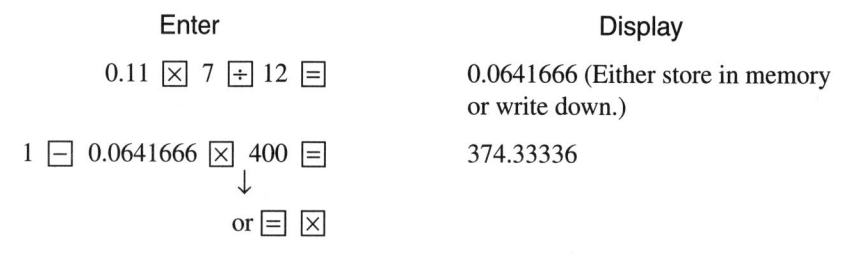

Enter	Display
0.11 ⊠ 7 ⊞ 12 ⊟	0.0641666 (Either store in memory or write down.)
1 ⊟ 0.0641666 ⊠ 400 ⊟ ↓ or ⊟ ⊠	374.33336

Rounded to two decimal places, the final answer is 374.33.

If you record and enter fewer than six decimal places for the first answer, the final result will be different.

    Using five decimal places (0.06416),   the answer is 374.34.

    With four decimal places (0.0641),   the answer is 374.36.

    With three places (0.064),   the answer is 374.40.

It is therefore advisable to keep and use at least six decimal places for intermediate answers if you are not using memory. If you can store the value in memory, all the decimal places will be retained.

If your calculator does not have a memory and you are writing down six decimal places, the order of calculations in this type of problem should be $0.11 \times 7 \div 12$ rather than $7 \div 12 \times 0.11$. The division results in an inaccurate answer because of the unending decimal that must be truncated at some point. If possible, this answer should not be used in further calculations. ∎

Chapter 1

TRY THESE PROBLEMS (Set I)

Evaluate the following:

1. $60 \div 5 \div 2$
2. $4 + 12 \times 2 - 16 \div 8$
3. $8 + 5 - 18 \div 3$

4. $25 - (10 - 2)$
5. $5(8 - 2)$
6. $100\left(1 + 0.04 \times \dfrac{1}{2}\right)$

7. $20\left(1 + \dfrac{12}{100} \times \dfrac{3}{4}\right)$
8. $120\left(1 - 0.12 \times \dfrac{7}{12}\right)$
9. $5\left(1 - \dfrac{15}{100} \times \dfrac{1}{3}\right)$

10. $\left(1 + \dfrac{10}{100} \times \dfrac{1}{2}\right)$
11. $\left(1 - \dfrac{16}{100} \times \dfrac{3}{12}\right)$

Round to thousandths:

12. 7.72841
13. 0.6258
14. 5.0496

Round to the nearest cent:

15. $10.0584
16. $0.326
17. $243.3459
18. $61.995

Round to one decimal place:

19. 3.179%
20. 0.4687%
21. 8.071%
22. 25.99%

Round to tenths, hundredths, and thousandths:

23. 56.2989
24. 8.8958
25. 15.00649
26. 123.5228

ALGEBRA CONCEPTS 7

1.3 ALGEBRA CONCEPTS

Algebra is a tool that will help you think through and solve business mathematics problems. It is used to express a problem in its simplest form, discarding all nonessential words and expressing the essential part in numbers and symbols. Most of the symbols and rules of algebra have been familiar since grade school arithmetic. The symbols +, −, and = still mean "add," "subtract," and "is equal to." Division in algebra is generally indicated by use of the fraction form, as in $\frac{2}{3}$, which means 2 divided by 3. Muliplication may be shown in several ways in addition to the familiar ×. The quantity 3 × 4 may also be written 3 · 4 or 3(4).

In arithmetic, only definite numbers are used: the digits 0 through 9 and combinations of them. These numbers never change their meaning or value and are called **constants.** The constant 3 remains the same whether we are talking about 3 men, 3 apples, or 3 inches. In algebra we also use another kind of number to represent a quantity for which the description is known but not the value. These **general numbers** are usually represented by letters of the alphabet and are called **variables.** Thus, we may decide that C stands for cost, but its value will depend on whether the problem is about the cost of a chair, an automobile, or some other item. The laws of arithmetic apply to algebra as well, for example,

$$\text{In arithmetic:} \quad 6 + 2 = 2 + 6$$

$$\text{In algebra:} \quad a + b = b + a$$

Note that in the arithmetic illustration we have made a statement only about two specific, definite numbers. But the algebra illustration makes a statement about two general numbers and is true for whatever values are given to a and b.

When constants, general numbers, and operations signs (+, −, ×, ÷) are combined, they form an **algebraic expression.** Each algebraic expression represents one quantity, although it may be made up of more than one number or symbol.

In arithmetic: *4 × 2 is the quantity 8.*

In algebra: *$a \times b$ would also represent the quantity 8 if we first agree that one of the letters is equal to 2 and the other to 4.*

Again, the arithmetic statement applies only to 4 and 2; the algebraic statement applies to any combination of numbers. The value of the expression $a \times b$ will depend on the values assigned to the general numbers a and b. This can be stated in algebra by

$$a \times b = c$$

If, for example, $a = 6$ and $b = 8$, then $c = 48$.

Note: $a \times b$ may be written as ab. No multiplication sign is needed when the expression combines letters or a number and letters. Multiplication is indicated when no sign is used.

The simple algebraic statement $ab = c$ has many familiar applications:

- Rate × time = distance ($R \times T = D$). The letters are changed to represent particular quantities, but the meaning is the same.

- Hourly wage rate × number of hours worked = total pay ($R \times H = P$).

- Tax rate × assessed value of a house = amount of taxes ($R \times A = T$).

- In a "$\frac{1}{3}$ off sale," $\frac{1}{3}$ selling price = deduction from selling price $\left(\frac{1}{3}S = D, \text{ or } \frac{S}{3} = D\right)$.

8 CHAPTER 1   ESSENTIAL MATHEMATICS

Substitution

The algebraic expression $4a + bc - ac$ represents a quantity of unknown value until we assign values to each of the letters. If, in a particular problem $a = 7$, $b = 2$, and $c = 5$, then the value of the expression can be found by putting the numbers in place of the corresponding letters; that is, we **substitute** to find the value.

$$a = 7 \qquad 4a + bc - ac =$$
$$b = 2 \qquad 4 \times 7 + 2 \times 5 - 7 \times 5 =$$
$$c = 5 \qquad 28 + 10 - 35 =$$
$$\qquad\qquad\qquad 38 - 35 = 3$$

If $a = 7$, $b = 2$, and $c = 5$, then $4a + bc - ac = 3$.

Note: As in arithmetic, multiplication and division are done before addition and subtraction.

Terms

The **terms** of an algebraic expression are the parts that are separated by plus and minus signs. In the expression just evaluated there are three terms, $4a$, bc, and $-ac$. The number 4 in the term $4a$ is called the **coefficient** of a. Where there is no numerical coefficient written, as in the term bc, the coefficient is understood to be 1.

■ **Example 1**

$2R + 5$ is an algebraic expression with two terms, $2R$ and 5. Its value depends on the value given to R. If $R = 10$, the value of the expression is found by substitution.

$$R = 10 \qquad 2R + 5 =$$
$$\qquad\qquad 2(10) + 5 =$$
$$\qquad\qquad 20 + 5 = 25$$

If $R = 10$, then $2R + 5 = 25$. ■

■ **Example 2**

$3xc + 5y - xy$ is an algebraic expression with three terms, $3xc$, $5y$, and $-xy$. It represents one quantity whose value depends on the values given to the letters. The numerical coefficient of the first term is 3, of the second 5, and of the third -1. If $x = 1$, $c = 3$, and $y = 2$, the value of the expression is found by substitution.

$$x = 1 \qquad 3xc + 5y - xy =$$
$$c = 3 \qquad 3 \cdot 1 \cdot 3 + 5 \cdot 2 - 1 \cdot 2 =$$
$$y = 2 \qquad 9 + 10 - 2 =$$
$$\qquad\qquad\qquad 19 - 2 = 17$$

If $x = 1$, $c = 3$, $y = 2$, then $3xc + 5y - xy = 17$. ■

■ **Example 3**

$\dfrac{3x}{a} + k$ is an algebraic expression with two terms, $\dfrac{3x}{a}$ and k. $\dfrac{3x}{a}$ may be written $3\left(\dfrac{x}{a}\right)$ and the coefficient of $\dfrac{x}{a}$ is 3. The coefficient of k is 1. If the values of x, a, and k are equal to 5, 15, and 4, respectively, then the value of the expression can be found by substitution.

$$x = 5 \qquad \dfrac{3x}{a} + k =$$
$$a = 15 \qquad \dfrac{3(5)}{15} + 4 =$$
$$k = 4 \qquad \dfrac{15}{15} + 4 =$$
$$\qquad\qquad 1 + 4 = 5$$

If $x = 5$, $a = 15$, and $k = 4$, then $\dfrac{3x}{a} + k = 5$. ■

Chapter 1

TRY THESE PROBLEMS (Set II)

1. In the algebraic expression $10Z - 6A + 12$:

   the number of terms is _____;
   the coefficient of Z is _____;
   the constant is _____.

2. In the algebraic expression $\frac{3}{4}T + 5P - 6$:

   the number of terms is _____;
   the coefficient of T is _____;
   the coefficient of P is _____;
   the constant is _____.

3. In the algebraic expression $\frac{3}{5}B - Y + S - 6$:

   the number of terms is _____;
   the coefficient of B is _____;
   the coefficient of Y is _____;
   the coefficient of S is _____.

4. Evaluate $R + 0.6R$ if $R = \$25$.

5. Evaluate $T - \frac{2}{3}T$ if $T = 45$.

6. Evaluate $3B + \frac{3}{2}A - 2$ if $B = 3$ and $A = 12$.

7. Evaluate $ac - c + 0.2a$ if $a = 3$ and $c = 8$.

8. Evaluate $\frac{5}{4}P - 3PT + 22$ if $P = 3$ and $T = 2$.

9. Find the value of $X + 2Y + \frac{3}{2}Z - 11$ if $X = 3$, $Y = 4$, and $Z = 9$.

10. Evaluate $\frac{5}{2}br + s$ if $b = 5$, $r = 6$, and $s = 1\frac{1}{2}$.

    the coefficient of br is _____;
    the coefficient of s is _____.

Combining Terms

In arithmetic, definite numbers may be combined to give one total:

$$5 + 3 + 2 = 10$$

But if the 5 represents 5 pounds of sugar, the 3 represents 3 chickens, and the 2 represents 2 cans of peas, the total of 10 would have no meaning. In order to add these items, they would have to be changed to the same denomination or unit, and in the supermarket this is done by adding the cost of such items in dollars and cents. (This is similar to changing fractions to a common denominator before they are added.)

In algebra, the same rule applies. The terms of an algebraic expression may be added only if they are **similar,** that is, if they contain the exact same letters. $4z$ and $6z$ may be added, just as we add 4 oranges and 6 oranges. The results are $10z$ and 10 oranges, respectively. However $3a$ and $4b$ may not be combined, since the letters in the terms are different.

> **Rule 1-4.** To add or subtract similar terms, combine their coefficients.

■ **Example 4**

Simplify the following algebraic expressions:

1. $2ab - ab$ — *Both terms contain ab and only ab and may therefore be combined.*

   $2ab - ab = ab$ — *Combine coefficients: $2 - 1 = 1$.*

2. $4r + 0.5r$

   $4r + 0.5r = 4.5r$ — *Combine coefficients: $4 + 0.5 = 4.5$.*

3. $S - \frac{2}{3}S$

   $S - \frac{2}{3}S = \frac{1}{3}S$ or $\frac{S}{3}$ — *Combine coefficients: $1 - \frac{2}{3} = \frac{1}{3}$.*

4. $2k - k + 2$

   $2k - k + 2 = k + 2$ — *Only terms with k may be combined. Combine coefficients: $2 - 1 = 1$.*

5. $4x + y - 2x - y + 4$

   $4x - 2x + y - y + 4 =$ — *Combine two x terms and two y terms. Term 4 is a constant and is not of the same kind as any of the others.*

   $2x + 0 + 4 =$

   $2x + 4$ — *Combine coefficients of x: $4 - 2 = 2$. Combine coefficients of y: $1 - 1 = 0$.* ■

Chapter 1

TRY THESE PROBLEMS (Set III)

Simplify the following expressions by combining similar terms.

1. $12T - 4T$

2. $\dfrac{5Z}{2} - Z$

3. $\dfrac{5A}{3} - A$

4. $4Y - \dfrac{1}{2}Y$

5. $9C - \dfrac{2}{3}C$

6. $7P - \dfrac{2P}{3}$

7. $R + 0.4R - 1$

8. $2W - 0.2W$

9. $7b + 2 - b - 3 + c$

10. $x + 0.5x + y$

11. $4t - 5y - 4t - 4y$

12. $8k + \dfrac{k}{3} - 12$

13. $7Z - \dfrac{Y}{2} + 2Z + Y - Z$

14. $A + 5 - \dfrac{3A}{4} - 2 + B$

The Equation

Perhaps the most important tool in algebra is the **equation,** a statement that two algebraic expressions are equal. The two expressions are connected by an equals sign (=).

The idea of two equal quantities is very common. For example, if we asked a cashier for change for a twenty dollar bill, we might exchange it for a ten dollar bill and two five dollar bills. We have set up a simple equation:

$$\text{Twenty dollars} = \text{ten dollars} + \text{five dollars} + \text{five dollars}$$

$$20 = 10 + 5 + 5 \text{ (in dollars)}$$

If the request is for change for a fifty dollar bill instead of change for a twenty, we would increase the left side of the equation from 20 to 50. If that is all we do, the equals sign is no longer correct; no one would take the right side of the equation as change for fifty dollars. If we add 30 to one side of the equation, we must add 30 to the other side to keep the two sides equal.

$$20 + \boxed{30} = 10 + 5 + 5 + \boxed{30}$$

$$50 = 50$$

Similarly, if the problem was to find change for a ten dollar bill and we subtract 10 from the left side of the equation, then we must also reduce the right side by 10.

$$20 - \boxed{10} = 10 + 5 + 5 - \boxed{10}$$

$$10 = 10$$

The same rule applies to multiplication and division. Dividing both sides of the equation by 4 yields

$$\frac{20}{4} = \frac{10 + 5 + 5}{4}$$

$$5 = \frac{20}{4}$$

$$5 = 5$$

And multiplying both sides by 2 yields

$$2(20) = 2(10 + 5 + 5)$$

$$40 = 2(20)$$

$$40 = 40$$

> **Rule 1-5.** Both sides of an equation may be increased, decreased, multiplied, or divided by the same number or by equal numbers; the results will be equal. (Division by zero is excluded.)

To expand on the money-changing problem, suppose the request is now for change of a one hundred dollar bill. However, the cashier only has 4 tens and 3 fives in the cash drawer. How much additional money is needed? The problem can be stated as follows: The value of one hundred dollars equals the value of 4 tens, 3 fives, and some additional amount needed to complete the exchange. In algebra, the situation can be represented by a simple equation.

$$100 = 4(10) + 3(5) + X, \text{ where } X \text{ represents the unknown amount.}$$

If we carry out the indicated multiplication and combine like terms on each side of the equation, we have

$$100 = 4(10) + 3(5) + X$$
$$100 = 40 + 15 + X$$
$$100 = 55 + X$$

Only one value of the letter X will make this equation true. To find this value, *we solve the equation.* This means changing its form so that X will stand by itself on one side of the equation. In this equation, 55 must be subtracted from the right side. Thus, remembering that we may subtract the same number from both sides of the equation without destroying the equality, we have

$100 = 55 + X$	
$100 - 55 = 55 + X - 55$	*Subtract 55 from both sides.*
$45 = X$	*Combine similar terms.*
$X = 45$	*This is the customary way of writing the solution, but either form is correct.*

This means that the cashier still needs 45 dollars to complete the change for the one hundred dollar bill.

Addition, multiplication, and division may be used in the same way to solve an equation. Addition and subtraction are inverse, or opposite, operations. Division and multiplication are also inverse operations.

■ **Example 5**

1. If a number is subtracted from the unknown, use addition to solve.

$Z - 7 = 30$	
$Z - 7 + 7 = 30 + 7$	*Add 7 to both sides of the equation.*
$Z = 37$	*Combine like terms.*

2. If the unknown is divided by a number, use multiplication to solve.

$\dfrac{W}{5} = 9$	$\dfrac{W}{5}$ *may also be written as* $\dfrac{1}{5}W.$
$\dfrac{W}{5} \times 5 = 9 \times 5$	*Multiply both sides of the equation by 5.*
$W = 45$	

3. If the unknown is multiplied by a number, use division to solve.

$7Y = 40$	
$\dfrac{7Y}{7} = \dfrac{40}{7}$	*Divide both sides of the equation by 7, the coefficient of y.*
$Y = 5\dfrac{5}{7}$	

4. Division is used to solve the following equation because p is multiplied by $\frac{3}{4}$.

$$\frac{3p}{4} = 6 \qquad \qquad \frac{3}{4}p \text{ may also be written as } \frac{3p}{4}.$$

$$\frac{\cancel{4}}{\cancel{3}} \times \frac{\cancel{3}p}{\cancel{4}} = 6 \times \frac{4}{3} \qquad \qquad \textit{Divide both sides of the equation by } \frac{3}{4}, \textit{ the coefficient of p. Dividing by } \frac{3}{4} \textit{ is the same as multiplying by } \frac{4}{3}. \left(\frac{4}{3} \textit{ is the reciprocal of } \frac{3}{4}.\right)$$

$$p = 8$$

To check the answers, substitute the value obtained in the solution into the original equation.

1. $Z - 7 = 30$
$37 - 7 \stackrel{?}{=} 30$
$30 \stackrel{?}{=} 30$

2. $\dfrac{W}{5} = 9$
$\dfrac{45}{5} \stackrel{?}{=} 9$
$9 \stackrel{\checkmark}{=} 9$

3. $7Y = 40$
$7\left(5\dfrac{5}{7}\right) \stackrel{?}{=} 40$
$\cancel{7}\left(\dfrac{40}{\cancel{7}}\right) \stackrel{?}{=} 40$
$40 \stackrel{\checkmark}{=} 40$

4. $\dfrac{3}{4}p = 6$
$\dfrac{3}{4}(8) \stackrel{?}{=} 6$
$6 \stackrel{\checkmark}{=} 6$ ∎

Chapter 1

TRY THESE PROBLEMS (Set IV)

Solve for the unknown and check.

1. $R + 7 = 15$

2. $B - 13 = 6$

3. $\dfrac{A}{4} = 2.5$

4. $3S = 42$

5. $\dfrac{2L}{5} = 24$

6. $3Y = 24.12$

7. $5A = \dfrac{3}{8}$

8. $0.7R = 35$

9. $\dfrac{2P}{3} = \$54$

10. $\dfrac{Y}{\frac{1}{3}} = 5$

11. $13 = \dfrac{1}{2}Q$

12. $\dfrac{2}{3} = 3H$

ALGEBRA CONCEPTS   21

It is frequently necessary to use more than one operation to solve an equation.

■ **Example 6** Solve the following equations:

1. $2Y + 6 = 18$

$$2Y + 6 - 6 = 18 - 6$$ *Subtract 6 from both sides of the equation to get the term with the unknown by itself.*

$$2Y = 12$$

$$\frac{2Y}{2} = \frac{12}{2}$$ *Divide both sides by the coefficient of the unknown.*

$$Y = 6$$

Check: $2Y + 6 = 18$

$$2(6) + 6 \stackrel{?}{=} 18$$ *Substitute 6 for Y.*

$$12 + 6 \stackrel{?}{=} 18$$

$$18 \stackrel{\checkmark}{=} 18$$

2. $\frac{X}{4} - 5 = 13$

$$\frac{X}{4} - 5 + 5 = 13 + 5$$ *Add 5 to both sides of the equation.*

$$\frac{X}{4} = 18$$ *Combine similar terms.*

$$\frac{X}{4} \times 4 = 18 \times 4$$ *Multiply both sides by 4.*

$$X = 72$$

Check: $\frac{X}{4} - 5 = 13$

$$\frac{72}{4} - 5 \stackrel{?}{=} 13$$ *Substitute 72 for X.*

$$18 - 5 \stackrel{?}{=} 13$$

$$13 \stackrel{\checkmark}{=} 13$$

Note: The addition or subtraction of a term from both sides of the equation is the algebraic basis for the equation's solution. The effect is to shift the term to the other side of the equation and change the sign.

3. $10K + 12 = 8K + 26$

$$10K - 8K = 26 - 12$$ *Collect all terms containing K on one side and all constants on the other.*

$$2K = 14$$ *Combine similar terms.*

22 CHAPTER 1 ESSENTIAL MATHEMATICS

$$\frac{\cancel{2}K}{\cancel{2}} = \frac{14}{2}$$

Divide both sides by 2, the coefficient of K.

$$K = 7$$

Check: $\quad 10K + 12 = 8K + 26$

$$10(7) + 12 \stackrel{?}{=} 8(7) + 26$$

Substitute 7 for K.

$$70 + 12 \stackrel{?}{=} 56 + 26$$

$$82 \stackrel{\checkmark}{=} 82$$

4. $\dfrac{3X}{4} + X = 12 + X$

$$\frac{3X}{4} + X - X = 12$$

Collect similar terms by subtracting X from both sides of the equation.

$$\frac{3X}{4} = 12$$

Combine similar terms by adding coefficients:
$\dfrac{3}{4} + 1 - 1 = \dfrac{3}{4}.$

$$\frac{\cancel{3}X}{\cancel{4}} \times \frac{\cancel{4}}{\cancel{3}} = \cancel{12}^{4} \times \frac{4}{\cancel{3}}$$

Multiply both sides by $\dfrac{4}{3}$, the reciprocal of $\dfrac{3}{4}$.

$$X = 16$$

Check: $\quad \dfrac{3X}{4} + X = 12 + X$

$$\frac{3(16)}{4} + 16 \stackrel{?}{=} 12 + 16$$

Substitute 16 for X.

$$\frac{48}{4} + 16 \stackrel{?}{=} 12 + 16$$

$$12 + 16 \stackrel{?}{=} 12 + 16$$

$$28 \stackrel{\checkmark}{=} 28$$

5. $\dfrac{9T}{2} - 15 = T\left(1 + 4 \times \dfrac{1}{2}\right)$

$$\frac{9T}{2} - 15 = T\left(1 + \cancel{4}^{2} \times \frac{1}{\cancel{2}}\right)$$

Complete arithmetic within parentheses.

$$= T(1 + 2)$$

$$\frac{9T}{2} - 15 = 3T$$

T(3) may be written as 3T.

$\dfrac{9T}{2} - 3T = 15$	*Collect similar terms by adding 15 and subtracting 3T on both sides of the equation.*
$\dfrac{3}{2}T = 15$	*Subtract similar terms by subtracting coefficients: $\dfrac{9}{2} - 3 = \dfrac{9}{2} - \dfrac{6}{2} = \dfrac{3}{2}$.*
$\dfrac{\cancel{2}}{\cancel{3}} \times \dfrac{\cancel{3}T}{\cancel{2}} = 15 \times \dfrac{2}{3}$	*Multiply both sides by $\dfrac{2}{3}$.*
$T = 10$	

Check:

$$\dfrac{9T}{2} - 15 = T\left(1 + 4 \times \dfrac{1}{2}\right)$$

$$\dfrac{9(10)}{2} - 15 \stackrel{?}{=} 10\left(1 + 4 \times \dfrac{1}{2}\right)$$

$$\dfrac{9(\cancel{10})^{5}}{\cancel{2}} - 15 \stackrel{?}{=} 10\left(1 + \cancel{4}^{2} \times \dfrac{1}{\cancel{2}}\right)$$

$$45 - 15 \stackrel{?}{=} 10(3)$$

$$30 \stackrel{\checkmark}{=} 30$$

Substitute 10 for T.

■

Chapter 1

TRY THESE PROBLEMS (Set V)

Solve for the unknown and check.

1. $5y - 5 = 45$

2. $3d + 5 = 28$

3. $7A - 4A = 81$

4. $6d = 5d + 10$

5. $\dfrac{4}{3}Y - Y = 8$

6. $\dfrac{2}{3}T + 6 = 13$

7. $\dfrac{Y}{9} + 30 = 40$

8. $5y - y + 4y = 24$

9. $12z - 7 = 14 + 5z$

10. $6R = 4R + 21$

11. $40 + \dfrac{2c}{3} = c$

12. $X + 1.5X = 50$

13. $5(t - 1) = 2t + 25$

14. $\dfrac{a}{3} + 6 = 10$

15. $2T - 5 = T\left(1 - \dfrac{2}{3}\right)$

16. $9R - 10 = R\left(1 + 9 \times \dfrac{2}{3}\right)$

1.4 RATIO

Many statements expressed with numbers have only limited meaning unless the numbers are compared with another quantity. For example, a computer operator offered $18.00 an hour may know that this is not enough to pay her expenses, but she doesn't know whether it's a reasonable salary unless she can compare it to other salaries for the same work. If her friend is earning $20.00 an hour, then her offer may seem low. But if the average salary for this work in her area is only $16.00 an hour, her offer is good. Similarly, if a student achieves an 80 on an exam, he does not really know how well he did unless he can compare his grade with those of the rest of the class.

A very useful method of comparison is a **ratio,** which may be defined as a comparison between two similar numbers. For example, if a college class has a total of 40 students, of which 15 are women and 25 are men, then the ratio of women to men is 15 to 25, also expressed as 15:25, or $\frac{15}{25}$. The ratio of men to total students is $\frac{25}{40}$. (Note that the number *to which* the comparison is made is always the denominator of the fraction.) Since the ratio, expressed in this way, is a fraction, it may be used as a fraction in calculations.

Like any other fraction, a ratio may be less than, equal to, or greater than 1. If a second class has 55 students including 15 women:

- The ratio of women in the first class to women in the second class is $\frac{15}{15}$, or 1.

- The ratio of the number of students in the second class to the number in the first class is $\frac{55}{40}$, which is more than 1.

In the preceding illustration, the ratio of women to men, $\frac{15}{25}$, may be reduced to $\frac{3}{5}$ and can be interpreted as:

- There are 3 women for every 5 men in the class, or
- There are three-fifths as many women as men in the class.

The second ratio, the number of men to the total, $\frac{25}{40}$, may be reduced to $\frac{5}{8}$ and would indicate that:

- Five-eighths of the class is men, or
- Out of every 8 students, 5 are men.

■ **Example 1**

Newtown Country Club recently purchased a new U.S. flag. The flag was 9.5 feet long and 5 feet wide. What is the ratio of the length to the width?

$$\text{Ratio} = \frac{\text{Length}}{\text{Width}}$$

Quantity to which the comparison is made is the denominator.

$$= \frac{9.5 \text{ feet}}{5 \text{ feet}}$$

$$\text{Ratio} = \frac{1.9}{1} \text{ or } 1.9 \text{ to } 1 \text{ or } 1.9\!:\!1$$

Reduce the fraction by dividing numerator and denominator by 5 feet.

This ratio is more meaningful if we approximate it by observing that it is very close to 2:1. This means that the length of the flag is almost twice the width. ■

■ **Example 2**

1. During one summer selling period, General Motors, Ford, and Chrysler sold a total of 153,000 new cars. Of these, GM sales were 97,000. What is the ratio of GM sales to the total?

28 CHAPTER 1   ESSENTIAL MATHEMATICS

$$\text{Ratio} = \frac{\text{GM sales}}{\text{Total sales}}$$

Quantity to which the comparison is made is the denominator.

$$= \frac{97{,}000}{153{,}000}$$

$$\text{Ratio} = \frac{97}{153} \cong \frac{100}{150} = \frac{2}{3}$$

$\cong$ *means "is approximately equal to."*

This means that out of every three cars sold by GM, Ford, and Chrysler in this period, approximately two were sold by General Motors.

2. If the ratio in question was the number of GM cars sold to the number of autos sold by Ford and Chrysler, the ratio would be:

$$\text{Ratio} = \frac{\text{GM sales}}{\text{Ford and Chrysler sales}}$$

$$= \frac{97{,}000}{(153{,}000 - 97{,}000)}$$

$$\text{Ratio} = \frac{97{,}000}{56{,}000} \cong \frac{100{,}000}{60{,}000} = \frac{5}{3}$$

This ratio indicates that five GM cars were sold for every three cars sold by Ford and Chrysler during this period. ■

■ **Example 3**   A retail store bought a lamp for $50 and sold it for $80. The markup (difference between cost and selling price) was $30. The following ratios may be useful to the store:

1. Cost to selling price $= \dfrac{50}{80} = \dfrac{5}{8}$.

   Cost was $\dfrac{5}{8}$ of selling price.

2. Markup to selling price $= \dfrac{30}{80} = \dfrac{3}{8}$.

   Markup was $\dfrac{3}{8}$ of selling price.

3. Markup to cost $= \dfrac{30}{50} = \dfrac{3}{5}$.

   Markup was $\dfrac{3}{5}$ of cost. ■

Chapter 1

TRY THESE PROBLEMS (Set VI)

For Problems 1–6, write the ratios as fractions and reduce to lowest terms.

1. 60 feet to 90 feet
2. 8 yards to 8 feet
3. 12 minutes to $1\frac{1}{2}$ hours
4. 45 dimes to 3 dollars

5. A student took 13 tests during the semester and passed 11.

   a. What is the ratio of tests passed to tests taken?

   b. What is the ratio of tests failed to tests passed?

6. A family has a weekly take-home pay of $840. One hundred and sixty dollars a week is spent on food, $50 a week on gasoline, and $240 a week on rent.

   a. What is the ratio of food expenditures to gasoline expenditures?

   b. What is the ratio of the amount spent on rent to food expenditures?

   c. What is the ratio of gasoline expenditures to take-home pay?

7. A survey of Big City showed 9,265 people employed and 564 unemployed.

   a. What was the ratio of unemployed workers to the total surveyed?

   b. What was the ratio of unemployed workers to employed workers?

   c. Interpret the ratios in parts a and b by approximation.

8. A company had total sales of $12.4 million in one year. Of this total, $5.7 million was from sales to foreign countries.

   a. What was the ratio of exports to domestic sales?

   b. If the dollar value of the company's domestic sales was three times the value of its exports, what would the ratio be?

   c. What was the ratio of exports to total sales? How would you interpret this ratio?

1.5 CHECKING ANSWERS

Few of us, students or teachers, can do much arithmetic and algebra without making a mistake in calculations, in placing the decimal point, and the like. But what we can do is check every answer to see if it makes sense, if it is a possible answer to the problem.

For example, if the problem gives the cost of an item that a merchant is reselling, and you are asked to find the selling price, the answer should be checked by such questions as:

- Is the selling price larger than the cost? It should be unless the merchant is losing money.

- Is the selling price about the same as the cost? It should be higher than cost but not too much. If an article costs $10, the selling price might be around $15 to $20. An answer that is much less or much more should be checked.

Other examples:

- If the answer to a problem is the cost of a house, the answer should usually be thousands of dollars. The possible range is large, but an answer in hundreds or billions is obviously wrong.

- If the problem is to determine the finance charge on a charge account, the charge should be small compared to the amount owed.

- If an answer represents the price of a new car, check whether a new car can really be bought for that amount.

This method for checking answers can be summarized by the following question. From experience, is the size of the answer reasonable for the conditions given in the problem?

A second method of checking answers is approximating solutions without actually doing all the work:

- If you are changing $\frac{4}{7}$ to a decimal, you can, without calculation, determine that the fraction's value is close to $\frac{4}{8}$, which equals $\frac{1}{2}$ or 0.5.

- If you are changing $\frac{44}{31}$ to a decimal, the answer must be greater than 1 because the numerator is greater than the denominator.

- To approximate the value of 2.3×6.98, round each number to the nearest whole number. The answer should be about $14 = 2 \times 7$. This will check the placement of the decimal point.

- To approximate the value of $\frac{1.2}{8.7}$, round to whole numbers. The fraction is then approximately equal to $\frac{1}{9}$, which is close to $\frac{1}{10}$ or 0.1.

■ **Example 1**

1. Without calculating, approximate the value of $\dfrac{3.7 \times 1.9}{4.3} - 1$.

$$\frac{3.7 \times 1.9}{4.3} - 1 \cong \frac{4 \times 2}{4} - 1$$

$$\frac{4 \times 2}{4} - 1 = \frac{8}{4} - 1$$

$$\frac{3.7 \times 1.9}{4.3} - 1 \cong 1$$

Actual value found by calculation is 0.63.

2. The balance in your checkbook at the beginning of June is $341.19. During the month you made no deposits, but issued checks for $27.50 and $34.98. About how much money do you have in the account after these checks are paid by your bank?

   a. First, the amount left must be smaller than the starting balance because there have been no deposits (additions), and two checks have been issued (subtractions).

   b. The two checks total about $60. (One is a little less than 30, one somewhat more.)

   c. The starting balance is near $340. The remaining balance approximates $340 − $60 or $280. (The accurate balance is $278.71.)

3. A coat is advertised as priced as $79.98, but during a special sale it will be sold at $\frac{1}{3}$ off its original price. About how much will you have to pay for the coat?

   a. Cost is about $80.

   b. $\frac{1}{3}$ of $80 is slightly more than $25. ($25 is $\frac{1}{3}$ of $75.)

   c. Therefore, the new selling price is about $80 − $25 or $55. (The accurate price is $53.32.) ∎

Now go to Try These Problems (Set VII) to practice checking answers.

NEW RULES IN THIS CHAPTER

- **Rule 1-1.** A series of individual additions, subtractions, or multiplications may be performed in any order. A series of divisions must be done in the order written.

- **Rule 1-2.** In a series of mixed operations, multiplication and division are done first in the order written from left to right, followed by addition and subtraction in the order written from left to right.

- **Rule 1-3.** To round a decimal, determine the desired number of decimal places and consider only the value of the next decimal place. If this value is 5 or more, increase the previous digit by 1. If it is less than 5, leave the previous digit unchanged. Then drop all digits after the desired number of decimal places.

- **Rule 1-4.** To add or subtract similar terms, add or subtract their coefficients.

- **Rule 1-5.** Both sides of an equation may be increased, decreased, multiplied, or divided by the same number or equal numbers; the results will be equal. (Division by zero is excluded.)

Chapter 1

TRY THESE PROBLEMS (Set VII)

Approximate by rounding to integers:

1. 2.6×8.3
2. $5.2 + 4.9 \times 3.1$
3. $16.848 - 5.24 \times 4.9$
4. $\dfrac{24.319}{6.25}$
5. $\dfrac{4.1 \times 6.8}{6.6 - 0.38}$

6. The value of $\frac{5}{7}$ is given as 0.0714. Is this answer possible? Explain.

In Problems 7 and 8, three choices are given for the value of the fractions. Choose the one that is possible but not necessarily correct:

7. $\dfrac{18.39}{5.8}$ **a.** 4.02   **b.** 3.17   **c.** 30.8

8. $\dfrac{4.2 \times 8.1}{5.1 - 0.9}$ **a.** 8.1   **b.** 0.81   **c.** 0.13

9. You start the week with $35 in cash. During the week a friend repays a $6.40 loan, you buy a calculator for $10.80 and make three other purchases for about one dollar each. At the end of the week you check and find you have $22.69 left. Is this balance approximately correct? Explain.

10. A compact disc is advertised as $\frac{1}{3}$ off its original price of $17.98. About how much money would you need to buy the album?

Chapter 1

END-OF-CHAPTER PROBLEMS

Round to tenths, hundredths, and thousandths.

1. 5.1697
2. 0.3268
3. 12.2498
4. 0.0921
5. 79.94496

Round to the nearest cent.

6. $46.2267
7. $1,712.691
8. $56.655
9. $12.397
10. $78.9975

Write the following ratios in fraction form reduced to lowest terms.

11. 65 to 70
12. 8 to 72
13. 16 to 96
14. 39 to 19

15. 92 to 8
16. 1.3 to 5.2
17. $\frac{1}{7}$ to $\frac{1}{11}$
18. 14 to (80 − 25)

19. (11.5 + 1.5) to (9 + 30)
20. $1\frac{3}{5}$ to $3\frac{2}{5}$

Evaluate each of the following.

21. $25\left(1 + 0.1 \times \frac{1}{4}\right)$
22. $300\left(1 - 0.1 \times \frac{1}{4}\right)$
23. $34\left(1 + 0.12 \times \frac{18}{360}\right)$

24. $300\left(1 - \frac{18}{100} \times \frac{1}{3}\right)$
25. $30 \times 3.82 + 4.7 \times 4$
26. $35 \times 2.3 - 15 + 1.75$

27. $500 \times 0.09 \times \frac{12}{360}$
28. $1{,}095 \div 5 \times \frac{90}{365}$

29. $240 \times 0.18 \times \frac{5}{12}$
30. $600 \times 0.14 \times t - 200$

Simplify the following algebraic expressions by combining similar terms and removing parentheses.

31. $5A + A$
32. $\frac{R}{4} + R$
33. $x + 0.7x + 5 - 0.02x$

34. $7T - \frac{1}{2}T + W - \frac{W}{2}$
35. $Z - 8Y + \frac{Z}{4} + 3$
36. $\frac{3S}{5} - R + S - R$

37. $9(4C + 0.2C)$

38. $6S\left(2 - \frac{1}{4}\right) + 12$

39. $5(N + 0.4N) - H + H(2 + 0.5)$

40. $(R - 0.4R) + S(4 + 0.4)$

Solve the following equations for the unknown and check.

41. $Y + 10 = 13$

42. $X - 8 = 11$

43. $14R - R = 52$

44. $Z + 4Z = 17$

45. $15S = 75$

46. $\frac{2A}{3} = 32$

47. $\frac{C}{3} - 5 = 0$

48. $\frac{P}{2} = 15$

49. $S + 0.32S = 7.92$

50. $3W - 0.5 = 0.5W + 37$

51. $C + 0.5C = 345$

52. $70 + 0.6S = S + 20$

53. $26 + 0.35S = S$

54. $1{,}215 = P\left(1 + 0.08 \times \frac{90}{360}\right)$

55. $351.21 = P\left(1 + 0.12 \times \frac{54}{360}\right)$

56. $3.90 = 260 \times r \times \frac{1}{4}$

57. $14.14 = 120 \times r \times \frac{7}{12}$

58. $1{,}500.46 = S\left(1 - 0.09 \times \frac{20}{360}\right)$

59. $1{,}237.50 = S\left(1 - 0.15 \times \frac{24}{360}\right)$

Approximate by rounding to integers.

60. $\frac{5.78 - 2.03}{3.5} + 2$

61. $\frac{3.3\,(4.92)}{5.06}$

62. $2.8\,(0.91679)$

63. $5.9\,(1.3) - \frac{5.8}{3.2}$

In Problems 64 through 69, three choices are given for the answer. *Without working out the problem,* approximate the answer and indicate which choice is possible but not necessarily correct.

64. $2.13 \times 11.97 =$

    a. 24.3   **b.** 18.7   **c.** 2.43

65. $\frac{52.6}{24.92} =$

    a. 20.7   **b.** 0.207   **c.** 2.07

Chapter 1

END-OF-CHAPTER PROBLEMS

66. $\dfrac{11.98 \times 4}{8.1} =$

   a. 6.7   **b.** 5.9   **c.** 0.59

67. 22.1 (2.917) − 0.0278 (20.3) =

   a. 6.5   **b.** 65   **c.** 43

68. A retailer bought a portable CD player for $54. He added $5.78 for expenses and $8.91 for net profit. At what price (to the nearest whole dollar) did he sell the radio?

   a. $69   **b.** $54   **c.** $63

69. Your checking account balance is $326.28 at the beginning of the semester. You issue checks for books ($76.89) and for college fees ($35). You also deposit two small checks, each for less than $25. Your balance after all these transactions have been completed is:

   a. under $200   **b.** between $200 and $300   **c.** over $300

70. A store advertised a $\tfrac{1}{3}$ off ticket price sale of all its summer clothing. If you chose a suit with a ticket price of $249.99 and a pair of shoes priced at $37.98, approximately how much money would you need to buy both items?

71. The owner of a small store owes suppliers payments of $562.50, $95.78, and $201. There is only $120 in cash. Approximate how much additional money he needs to cover his bills.

72. Alliance Technology has 160 people on their office staff, and 200 people are employed in the programming and consulting divisions.

   a. What is the ratio of office staff to programmers and consultants?

   b. What is the ratio of office staff to all employees?

73. The dairy section of a supermarket occupies 2,400 square feet of space in a store that has a total area of 36,000 square feet. What is the ratio of dairy space to all other space?

74. A budget includes an allowance of $55 a week for food and $15 a week for entertainment out of a total allowance of $275. What is the ratio of food costs to the total? Interpret your answer.

75. A builder sold a house for $160,000. The land cost $39,300, construction costs and other expenses were $58,100, and the remainder was profit. Approximate all ratios with meaningful fractions.

  a. What is the ratio of land costs to all other costs?

  b. What is the ratio of profit to selling price?

  c. What is the ratio of total cost to selling price?

  d. How can you check your answers to parts b and c?

76. State University has an entering class of 6,500 students. Of these, 2,100 commute to school, 3,000 can be accommodated in dormitories, and the remainder will find housing near campus.

  a. What is the ratio of commuting students to noncommuting students?

  b. What is the ratio of commuting students to the total? Interpret your answer.

Chapter 1

Name Date Class

TEST YOURSELF

Section

Evaluate each of the following:

1. $5 + 10 \times 3 - 16 \div 2$ 1.1

2. $625 \div 5 \div 5$ 1.1

3. $100\left(1 + 0.06 \times \dfrac{1}{2}\right)$ 1.1

4. $\dfrac{5}{4}Y - 3XY + 25$ if $Y = 2$ and $X = 3$ 1.3

Round to the nearest cent:

5. $26.051 6. $339.7974 1.2

7. Round to tenths, hundredths, and thousandths: 15.2749 1.2

8. In the algebraic expression $\dfrac{1}{2}T + P - 5$ 1.3

   the number of terms is _____;

   the coefficient of T is _____;

   the coefficient of P is _____;

   the constant is _____.

Simplify the following algebraic expressions by combining similar terms and removing parentheses.

9. $7Z - \dfrac{Z}{2} + Y - 6$ 10. $R(5 + 0.4) + (S - 0.1S)$ 1.3

In Problems 11–15, solve for the unknown and check.

11. $\dfrac{4H}{5} = 64$ 1.3

12. $8a = a + 14$ 1.3

	Section

13. $X + 5 = X\left(1 + 10 \times \dfrac{2}{5}\right)$ 1.3

14. $z + 0.5z = 60$ 1.3

15. $21 = \dfrac{a}{3} + 2a$ 1.3

Write the following ratios in fraction form reduced to lowest terms.

16. 8 to 32   17. $(19.5 + 0.5)$ to $(2 + 8)$ 1.4

Approximate by rounding to integers. State the answer as an integer.

18. $2.1\,(8.9) - 0.994\,(4.99)$ 1.5

19. $\dfrac{3.1(8.06)}{11.97}$ 1.5

In Problems 20 and 21, three choices are given for the answer. Without working out the problem, approximate the answer and indicate which choice is possible but not necessarily correct.

20. $3.04\,(12.01) - 1.99\,(7.11)$

   a. 2.24   b. 0.224   c. 22.4   1.5

21. $\dfrac{34.79}{6.9}$

   a. 5.1   b. 0.21   c. 49   1.5

22. Your checking account balance is $450.29 at the beginning of the semester. You issue checks for books ($211.95) and for college fees ($37). You also deposit two checks, one for $25 and the other for $185. About how much money would you have in your account after all these transactions have been completed?

   a. under $300   b. between $300 and $400   c. over $400   1.5

23. Metro Repair Service employs 42 repair persons. Twenty of these have special training in digital repairs.   1.4

   a. What is the ratio of employees with special training to those without this training?

   b. What is the ratio of the specially trained employees to all the repair employees?

2 ■ Percents and Their Applications

OBJECTIVES

Upon completion of this chapter, the student should be able to:

1. *Change a percent to its equivalent decimal or fraction.*
2. *Change a decimal or fraction to its equivalent percent.*
3. *Use the equation Rate × Base = Product to find rate, base, or product.*
4. *Solve word problems containing percents.*
5. *Find the percent change between two numbers.*

Pick up the newspaper any day and you will find statements like the following:

- Is tax-free income worth 20% to you?
- 28% off airfare.
- Retail sales were off 1.2%.
- Prime rate lowered from $6\frac{3}{4}$% to $6\frac{1}{2}$%.
- Only 21% in the survey favor candidate X.

Whether we are talking about politics, air fares, sales, or interest rates, the relationship between two quantities is frequently given in **percent** (%). The word percent comes from the Latin for "hundred" and represents a fraction whose denominator is 100. Thus, 35% could be written as $\frac{35}{100}$ or 0.35. Many business problems are stated, and answers given, in percent form. All work in business mathematics requires the ability to work with percents.

The most familiar use of percents is in our money system. A dollar, the basic unit of our currency, is divided into 100 equal parts, each of which is 1 percent of a dollar or 1 cent. All percents are basically the same as the cent and dollar. Any unit (population, money amount, cost of an item, etc.) may be thought of as divided into 100 equal parts. Each part is then 1% of the whole.

2.1 CHANGING PERCENTS TO DECIMALS AND FRACTIONS

Although the percent sign (%) is convenient and commonly used in writing, it is not used in calculation. It has a definite arithmetic value, and before any calculation is started, the quantity stated as a percent must be changed to an equivalent fraction or decimal. The arithmetic equivalent of % is $\frac{1}{100}$, or 0.01.

42 CHAPTER 2 PERCENTS AND THEIR APPLICATIONS

A percent may be changed to an equivalent decimal or fraction by substituting the value of the percent sign (%) for the sign itself—for example,

$$67\% = 67\,(0.01) = 0.67$$

or

$$67\% = 67\left(\frac{1}{100}\right) = \frac{67}{100}$$

We have changed the form but not the value of 67%. Mechanically, the change is accomplished in two steps.

Rule 2-1. To change a percent to a decimal, move the decimal point two places to the left and drop the percent sign (%).

Rule 2-2. To change a percent to a fraction, multiply by $\frac{1}{100}$ and drop the percent sign (%).

Note: The percent sign (%) may never be added or dropped without an additional step.

The following examples involve the arithmetic of fractions and decimals. Explanations are given for all the steps, but for a complete arithmetic review see Appendix A.

■ **Example 1**   Change 29% to a decimal.

$$29\% = 0.29\% = 0.29$$

Move the decimal point two places to the left and drop %.

Change 29% to a fraction.

$$29\% = 29 \times \frac{1}{100} = \frac{29}{100}$$

Multiply by $\frac{1}{100}$ and drop %.

Check: Convert $\frac{29}{100}$ to a decimal by dividing 29 by 100. The result is 0.29. We have changed the form but not the value of 29%. ■

■ **Example 2**   Change 4% to a decimal.

$$4\% = 0.04\% = 0.04$$

Add a zero to the left of 4. Then move the decimal point two places to the left and drop %.

Change 4% to a fraction.

$$4\% = 4 \times \frac{1}{100} = \frac{4}{100}$$

Multiply by $\frac{1}{100}$ and drop %.

$$4\% = \frac{4}{100} = \frac{1}{25}$$

Reduce the fraction by dividing the numerator and denominator by 4.

Check: Change $\frac{1}{25}$ to a decimal by dividing 1 by 25. The result is 0.04. ∎

CALCULATOR HINT

You can use your calculator to perform the checks that are indicated in Examples 1 and 2.

Example 1: Change $\frac{29}{100}$ to a decimal.

Example 2: Change $\frac{1}{25}$ to a decimal.

	ENTER	DISPLAY
Example 1	29 ÷ 100 =	0.29
Example 2	1 ÷ 25 =	0.04

You can use your calculator to change any fraction to a decimal. Just divide the numerator by the denominator.

■ **Example 3**

Change 318% to a decimal.

$$3.18\% = 3.18$$

Move the decimal point two places to the left and drop %.

Change 318% to a fraction.

$$318\% = 318 \times \frac{1}{100} = \frac{318}{100}$$

Multiply by $\frac{1}{100}$ and drop %.

$$318\% = \frac{318}{100} = 3\frac{18}{100} = 3\frac{9}{50}$$

Reduce the fraction by changing to a mixed number and then dividing the numerator and denominator by 2.

Check: Change $3\frac{9}{50}$ to a decimal. Divide 9 by 50, yielding 0.18. $3\frac{9}{50} = 3.18$. ∎

44 CHAPTER 2   PERCENTS AND THEIR APPLICATIONS

■ **Example 4**   Change 9.5% to a decimal.

$$9.5\% = 0.095\% = 0.095$$

Move the decimal point two places to the left and drop %.

Change 9.5% to a fraction.

$$9.5\% = 9.5 \times \frac{1}{100} = \frac{9.5}{100}$$

Multiply by $\frac{1}{100}$ and drop %.

$$= \frac{9.5}{100} = \frac{95}{1,000}$$

Remove the decimal point in the numerator by multiplying the numerator and denominator by 10.

$$9.5\% = \frac{95}{1,000} = \frac{19}{200}$$

Reduce the fraction by dividing the numerator and denominator by 5.

$$9.5\% = 0.095 = \frac{19}{200}$$

Check: Change $\frac{19}{200}$ to a decimal by dividing 19 by 200. The result is 0.095. ■

■ **Example 5**   Change 0.25% to a decimal.

$$0.25\% = 0.0025$$

Move the decimal point two places to the left, adding two zeros before the 2, and drop %.

Change 0.25% to a fraction.

$$0.25\% = 0.25 \times \frac{1}{100} = \frac{0.25}{100}$$

Multiply by $\frac{1}{100}$ and drop %.

$$= \frac{0.25}{100} = \frac{25}{10,000}$$

Remove the decimal point in the numerator by multiplying the numerator and denominator by 100.

$$0.25\% = \frac{25}{10,000} = \frac{1}{400}$$

Reduce the fraction by dividing the numerator and denominator by 25.

$$0.25\% = 0.0025 = \frac{1}{400}$$

Check: Change $\frac{1}{400}$ to a decimal. The result is 0.0025. ■

CHANGING PERCENTS TO DECIMALS AND FRACTIONS 45

■ **Example 6** Change $7\frac{1}{2}\%$ to a decimal.

$$7\frac{1}{2}\% = 7.5\%$$

Change the mixed number to a decimal. This does not affect the percent sign.

$$7\frac{1}{2}\% = 0.075$$

Move the decimal point two places to the left and drop %.

Change $7\frac{1}{2}\%$ to a fraction.

$$7\frac{1}{2}\% = \frac{15}{2}\%$$

Change the mixed number to an improper fraction. This does not affect the percent sign.

$$= \frac{15}{2} \times \frac{1}{100} = \frac{15}{200}$$

Multiply by $\frac{1}{100}$ and drop %.

$$7\frac{1}{2}\% = \frac{15}{200} = \frac{3}{40}$$

Reduce the fraction by dividing the numerator and denominator by 5.

$$7\frac{1}{2}\% = \frac{3}{40} = 0.075$$

Check: Convert $\frac{3}{40}$ to a decimal by dividing 3 by 40. The result is 0.075. ■

■ **Example 7** Change $\frac{5}{8}\%$ to a decimal.

$$\frac{5}{8}\% = 0.625\%$$

Divide 5 by 8 to change $\frac{5}{8}$ to a decimal. This does not affect the percent sign.

$$\frac{5}{8}\% = 0.00625$$

Move the decimal point two places to the left and drop %.

Change $\frac{5}{8}\%$ to a fraction.

$$\frac{5}{8}\% = \frac{5}{8} \times \frac{1}{100} = \frac{5}{800}$$

Multiply by $\frac{1}{100}$ and drop %.

$$\frac{5}{8}\% = \frac{1}{160}$$

Reduce the fraction by dividing the numerator and denominator by 5.

$$\frac{5}{8}\% = \frac{1}{160} = 0.00625$$

Check: Change $\frac{1}{160}$ to a decimal by dividing 1 by 160. The result is 0.00625. ■

In the preceding examples the percents were changed to fractions and decimals in two separate operations. The work can be done just as correctly in one continuous process. Thus, in Example 2,

$$4\% = 0.04 = \frac{4}{100} = \frac{1}{25}$$

or

$$4\% = \frac{4}{100} = \frac{1}{25} = 0.04$$

and in Example 6,

$$7\frac{1}{2}\% = 7.5\% = 0.075 = \frac{75}{1,000} = \frac{3}{40}$$

or

$$7\frac{1}{2}\% = \frac{15}{2}\% = \frac{15}{200} = \frac{3}{40} = 0.075$$

In doing business problems it will be necessary to change percents to either fraction or decimal form, and either may be used. However, in one case the fraction is more accurate: when there is no exact equivalent decimal for the percent. Consider, for example, the following illustration:

$$33\frac{1}{3}\% = \frac{100}{3}\%$$ *Change the mixed number to an improper fraction.*

$$= \frac{100}{3} \times \frac{1}{100}$$ *Multiply by $\frac{1}{100}$ and drop %.*

$$= \frac{1}{3}$$

$$33\frac{1}{3}\% = 0.333\overline{3}$$ *Changing $\frac{1}{3}$ to a decimal gives a repeating and unending decimal, which must be rounded for further calculations. This introduces an inaccuracy that can be avoided by using the fraction.*

Chapter 2

TRY THESE PROBLEMS (Set I)

Change the following percents to equivalent decimals and fractions in lowest terms.

1. 14%   2. 6%   3. 27%   4. 3%

5. 134%   6. 327%   7. 500%   8. 4.5%

9. 99.44%   10. 2.41%   11. 0.81%   12. 0.02%

13. 0.6%   14. $1\frac{3}{4}$%   15. $5\frac{1}{2}$%   16. $33\frac{1}{3}$%

17. $\frac{3}{8}$%   18. $\frac{2}{5}$%   19. $7\frac{2}{7}$%   20. $422\frac{1}{4}$%

2.2 CHANGING DECIMALS AND FRACTIONS TO PERCENTS

Although business problems are worked with decimals and/or fractions, answers are frequently converted to percents. The procedure for changing a decimal to a percent is the reverse of the method used to convert from a percent to a decimal. A fraction should be changed to a decimal first and then to a percent.

> **Rule 2-3.** To change a decimal to a percent, move the decimal point two places to the right and add the percent sign (%).

> **Rule 2-4.** To change a fraction to a percent, change it first to a decimal, then to a percent.

■ **Example 1**   Change 0.072 to a percent.

$$0.072 = 7.2\%$$

Move the decimal point two places to the right and add %. ■

■ **Example 2**   Change 4 to a percent.

$$4 = 4.00 = 4.00\%$$
$$4 = 400\%.$$

In a whole number where the decimal point is not shown, it is understood to be after the last digit. Add two zeros, move the decimal point two places to the right and add %. ■

■ **Example 3**   Change $\frac{3}{8}$ to a percent.

$$\frac{3}{8} = 0.375$$

Change the fraction to a decimal by dividing the numerator by the denominator.

$$\frac{3}{8} = 37.5\%$$

Move the decimal point two places to the right and add %. ■

■ **Example 4**   Change $0.47\frac{1}{2}$ to a percent.

$$0.47\frac{1}{2} = 0.475 = 47.5\%$$

First change the fraction to a decimal. Then move the decimal point two places to the right and add %. ■

■ **Example 5**   Change $1\frac{3}{5}$ to a percent.

$$1\frac{3}{5} = 1.6 = 160\%$$

First change the fraction to a decimal by dividing the numerator by the denominator. $\left(\frac{3}{5} = 0.6.\right)$
Then $1\frac{3}{5} = 1.6$. *Move the decimal point two places to the right and add %.*

or $\quad 1\frac{3}{5} = \frac{8}{5} = 1.6 = 160\%$

First change the mixed number to an improper fraction. Then divide the numerator (8) by the denominator (5). Move the decimal point two places to the right and add %. ∎

■ **Example 6**   Change $\frac{7}{12}$ to a percent.

$$\frac{7}{12} = 0.58\frac{1}{3}$$

Change the fraction to a decimal by dividing 7 by 12; $0.58\frac{1}{3}$ is an exact form.

$$\frac{7}{12} = 58\frac{1}{3}\%$$

Move the decimal point two places to the right and add %. The fraction $\frac{1}{3}$ is not the third decimal place. It is part of the second.

or

$$\frac{7}{12} = 0.583\overline{33}$$

$$\frac{7}{12} = 58.33\%$$

Depending on the requirements of the problem, the repeating decimal may be rounded to 58%, 58.3%, or 58.33% ∎

A few fractions and the equivalent decimals and percents are used frequently and should be learned. Many may already be familiar to you.

$\frac{1}{3} = 0.33\frac{1}{3} = 33\frac{1}{3}\%$ or 33.3% (rounded) $\frac{2}{3} = 0.66\frac{2}{3} = 66\frac{2}{3}\%$ or 66.7% (rounded)

$\frac{1}{2} = 0.5 = 50\%$ $\frac{1}{5} = 0.20 = 20\%$ $\frac{1}{8} = 0.125 = 12.5\%$

$\frac{1}{4} = 0.25 = 25\%$ $\frac{2}{5} = 0.40 = 40\%$ $\frac{3}{8} = 0.375 = 37.5\%$

$\frac{3}{4} = 0.75 = 75\%$ $\frac{3}{5} = 0.60 = 60\%$ $\frac{5}{8} = 0.625 = 62.5\%$

$\frac{4}{5} = 0.80 = 80\%$ $\frac{7}{8} = 0.875 = 87.5\%$

Chapter 2

TRY THESE PROBLEMS (Set II)

Change the following quantities to percents. Round to two decimal places where necessary.

1. 0.37
2. 0.03
3. 0.4
4. 1.45

5. 2.9
6. 0.634
7. 0.0058
8. 0.1

9. 0.01
10. 1
11. $0.42\frac{1}{2}$
12. $0.66\frac{2}{3}$

13. $\frac{5}{16}$
14. $\frac{3.7}{4.2}$
15. $3\frac{1}{5}$
16. $\frac{6}{20}$

2.3 THE BASIC EQUATION

Before considering the numerous applications of percents in business problems, it must be emphasized that a percent is a *relative* number; it has no meaning unless we also know the whole quantity of which it is a part. To say "Jane saved 5%" means nothing, but to say "Jane saved 5% of her $450 salary" is a complete and meaningful statement. If Jane's friend saved 5% of her $900 salary, she saved twice as much as Jane, although the percent was the same. Stated briefly,

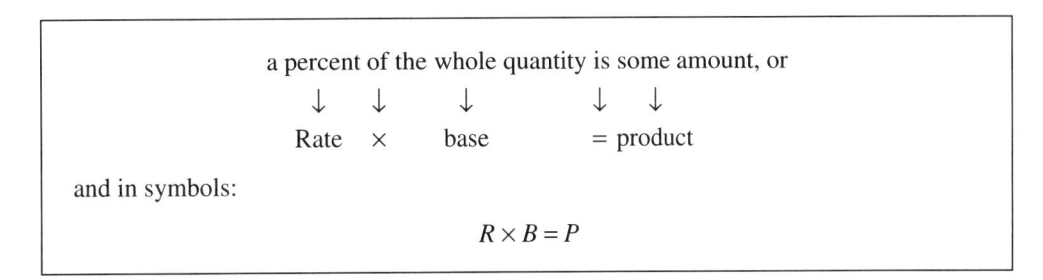

> a percent of the whole quantity is some amount, or
> $\downarrow$ $\quad$ $\downarrow$ $\quad$ $\downarrow$ $\quad\quad$ $\downarrow$ $\quad$ $\downarrow$
> Rate $\times$ base $\quad$ = product
>
> and in symbols:
>
> $$R \times B = P$$

The letter P in the formula is frequently read as percentage, but this causes unnecessary confusion with percent. We will call it *product* because it is the result of multiplying two numbers.

This equation ($R \times B = P$) can be solved for any one of the three parts if two are given. The essential step in solving such problems is to identify the parts. This process can be simplified by noting that:

- R (**rate**) is always given with a percent sign (%).

- B (**base**) is the whole quantity to which the rate is related and usually follows the word of (or on), as in 5% *of* her salary, 10% *of* the class, or 2% *of* sales.

- P (**product**) is the result of multiplying the other two terms.

To illustrate the use of the basic equation ($R \times B = P$), consider the following statements:
Jane saved 5% of her salary, which was $450 a week. In this way she was able to put $22.50 in the bank each week.

There are no unknowns in these statements, and the three variables (R, B, P) can be identified and given values:

- R, rate (the quantity with the %) is 5%.

- B, base (following *of*) is her salary, equal to $450 a week.

- P, product (the result of multiplying the rate and base) is the amount she puts in the bank each week, namely, $22.50.

We will now state and solve the problem with each of the variables unknown. *The quantities rate, base, and product remain the same whether or not their values are given.*

■ **Example 1**

Jane saved 5% of her salary, which was $450 a week. How much was she able to put in the bank each week?

R and B are given; P is the unknown. Before starting, change 5% to 0.05.

$$R \times B = P$$
$$0.05 \times 450 = P$$
$$\$22.50 = P \quad ■$$

54 CHAPTER 2   PERCENTS AND THEIR APPLICATIONS

■ **Example 2**   Jane saved 5% of her weekly salary. In this way she was able to put $22.50 in the bank each week. How much did she earn?

Note that "*of* her weekly salary" still indicates that the salary is the base, although there is no value given and the salary is the unknown to be found.

$$R \times B = P$$

$$0.05 \times B = 22.50$$

$$\frac{\cancel{0.05}}{\cancel{0.05}} B = \frac{22.50}{0.05}$$ *Divide both sides of the equation by the coefficient of B.*

$$B = \$450$$ ■

■ **Example 3**   Jane earned $450 a week and put $22.50 in the bank each week. What percent of her salary did she save?

The missing value is now the rate. (What percent?)

$$R \times B = P$$

$$R\,(450) = 22.50$$ *R (450) is the same as 450R.*

$$\frac{\cancel{450}}{\cancel{450}} R = \frac{22.50}{450}$$ *Divide both sides of the equation by the coefficient of the unknown.*

$$R = 0.05 = 5\%$$ ■

2.4 CHECKING PROBLEMS BY APPROXIMATION

Percent problems always involve moving decimal points, and answers should be checked by approximation. Several quickly calculated percents can indicate, by comparison, whether or not an answer is possible. We will use the number 500 as an example:

- 200% of a number is twice the number   (200% of 500 = 1,000)
- 100% of a number is the number itself   (100% of 500 = 500)
- 50% of a number is half the number   (50% of 500 = 250)
- 25% of a number is one-quarter the number   (25% of 500 = 125)
- 10% of a number is one-tenth the number   (10% of 500. = 50)

  (*move the decimal point one place to the left*)

- 1% of a number is one-hundredth the number   (1% of 500. = 5)

  (*move the decimal point two places to the left*)

Now let us locate different percents of 500 between the check points just established:

- 23% of 500 should be somewhat less than 25% of 500, or less than 125.
  23% of 500 = 0.23 × 500 = 115.

- 87% of 500 should be between 50% and 100% of 500, or between 250 and 500.
  87% of 500 = 0.87 × 500 = 435.

- 0.52% of 500 should be less than 1% of 500, or less than 5.
  0.52% of 500 = 0.0052 × 500 = 2.6.

- 132% of 500 should be between 100% and 200% of 500, or between 500 and 1,000.
  132% of 500 = 1.32 × 500 = 660.

Approximation does not check the calculation completely, but it does check the placing of the decimal point.

Check the answers to the following problems by approximation.

■ **Example 1**   Find 42% of $250.65.

$$R \times B = P$$
$$0.42 \times 250.65 = P$$
$$105.273 = P$$
$$P = \$105.27$$

Check: 42% is somewhat less than 50%.

$$42\% \text{ of } 250.65 \cong 50\% \text{ or } \frac{1}{2} \text{ of } 250 = \$125$$

The calculated answer, $105.27, is less than $125 but of the same magnitude. If the decimal point were misplaced (for example, $10.527 or $1,052.7) the answer would be obviously wrong. ■

■ **Example 2**   What percent of 20 is 32?

$$R \times B = P$$
$$R(20) = 32$$
$$R = \frac{32}{20} = 1.6 = 160\%$$

Check: The numerator of the fraction is larger than the denominator. Therefore, the answer must be more than 100%.

Chapter 2

TRY THESE PROBLEMS (Set III)

Find the following values without written arithmetic.

1. 100% of $672.81
2. 200% of 80
3. 50% of 60
4. 25% of 400

5. 10% of $982.26
6. 1% of $362

Find the following values and check by approximation.

7. 30% of 160
8. What percent of 800 is 360?

9. Sixty is what percent of 25?
10. 90% of $627.58

Find the missing quantity. Correct to two decimal places where necessary.

	RATE	BASE	PRODUCT		RATE	BASE	PRODUCT
11.	12%	430		16.		2.25	0.08
12.	6%	91		17.		891	500
13.	450%	20		18.	120%		420
14.		100	6	19.	$4\frac{1}{2}$%		800
15.		108	324	20.	0.5%		30

57

2.5 APPLICATIONS OF THE BASIC EQUATION

Direct uses of the basic relationship ($R \times B = P$) can be found in every phase of business operations. A few are illustrated in the following examples; others are included in the problems at the end of the chapter.

Commissions

Salespeople are frequently paid part or all of their compensation on the basis of how much they sell. This amount is called **commission.**

■ **Example 1**

A salesman is hired by the Xpand Co. The company agrees to pay him $500 a week and an additional 5% of his weekly sales as commission. During his first week of employment his sales were $1,920. How much did he earn for the week?

$$\text{Total earnings} = \$500 + \text{commission}$$

Since commission is equal to a percent of weekly sales, start with the basic equation:

Rate (R) = 5% = 0.05

Base (B) is weekly sales (5% *of* his weekly sales) = $1,920

Product (P) is commission, which is unknown.

$$R \times B = P$$
$$0.05 \times 1,920 = P$$
$$\$96 = P$$

Total earnings = $500 + 96 = $596   ■

Real Estate Taxes

Real property (residential and commercial) frequently has two different values. One is the market value, which is the amount for which the property could be sold. The second is the **assessed value,** which is used by the local government for taxing purposes. The assessed value is much smaller than the market value in many communities, and may be determined as a percent of market value.

The tax rate is usually given in dollars per hundred of assessed value, for example, $16.607 per $100. (The 7 in the third decimal place of $16.607 is in mills.) When stated in dollars per hundred, the tax rate is equivalent to a percent of assessed value. If a house in an area with a tax rate of $16.607 per hundred is assessed at $65,000, the taxes are 16.607% of $65,000, or $10,794.55.

$$R \times B = P$$
$$0.16607 \times 65,000 = P$$
$$\$10,794.55 = P \qquad ■$$

■ **Example 2**

1. Mr. and Mrs. Green own a home and have been paying a property tax of $11.20 per hundred of assessed value. (This is the same as saying the tax rate is 11.20% of assessed value.) If the property is assessed at $52,000, what is the annual property tax?

Since taxes are a percent of assessed value, the basic equation applies:

  a. Rate (R) is tax rate = 11.2% = 0.112.

  b. Base (B) is assessed value (*of* assessed value) = $52,000.

c. Product (P) is the amount of taxes and is unknown.

$$R \times B = P$$
$$0.112 (52{,}000) = P$$
$$\$5{,}824 = P$$

Annual taxes are $5,824.

2. Mr. and Mrs. Martinez decide to buy a house in a community where the tax rate is $10.90 per hundred of assessed value. They looked at a house priced at $97,000, for which taxes are $3,400 a year. Find the assessed value of the house:

a. Rate (R) is tax rate = 10.9% = 0.109.

b. Base (B) is assessed value and is unknown.

c. Product (P) is the amount of taxes = $3,400.

Note that the market value ($97,000), although given in the problem, is used only in checking the answer, not in the calculation.

$$R \times B = P$$
$$0.109 \times B = 3{,}400$$
$$\frac{\cancel{0.109}}{\cancel{0.109}} \times B = \frac{3{,}400}{0.109}$$

Divide both sides of the equation by the coefficient of the unknown.

$$B = \$31{,}193 \text{ (rounded to the nearest dollar)}$$

The assessed value is $31,193.

Check: The assessed value should be less than the market value. ∎

Space Allocation

■ **Example 3**

A specialty store selling women's clothing and accessories finds that the jewelry department has been selling more than expected and should be given more floor space. In the year just completed, jewelry sales amounted to 12% of the total store sales. It is decided to give the department the same proportion (percent) of the store's 20,000 square feet of space as it provides of total sales. How many square feet should the department be given?

- R is jewelry's percent of total sales = 12% = 0.12.
- B is total space (*of* the store's space) = 20,000 square feet.
- P is space for the department, which is unknown.

$$R \times B = P$$
$$0.12 \times 20{,}000 = P$$
$$2{,}400 = P$$

Floor space for the jewelry department should be 2,400 square feet. ∎

APPLICATIONS OF THE BASIC EQUATION

■ **Example 4**   The stockholders of the First State Bank expect a $2.5 million return on their $16 million investment. What rate of return is this?

- R is the percent of return, which is unknown.
- B is the investment = $16 million.
- P is the return on investment = $2.5 million.

$$R \times B = P$$
$$R \times 16 = 2.5 \text{ or}$$
$$16R = 2.5$$
$$\frac{\cancel{16}R}{\cancel{16}} = \frac{2.5}{16}$$

Divide both sides of the equation by the coefficient of the unknown.

$$R = 0.15625$$
$$R = 15.6\%$$

Move the decimal point two places to the right and add %. ■

CALCULATOR HINT

If your calculator has a % key, you can use it to find the product in any percent problem.

Look back to Example 1, where you were required to find 5% of 1,920.

	ENTER	DISPLAY
5% of 1,920	1,920 ⊠ 5 ⌘	96 (the % key also functions as an = key.)

Example 2 can be done as follows:

	ENTER	DISPLAY
11.2% of 52,000	52,000 ⊠ 11.2 ⌘	5,824

Remember, your calculator does not do your thinking for you. You must first determine whether the rate, base, or product is to be found and set up the appropriate equation.

Chapter 2

TRY THESE PROBLEMS (Set IV)

Solve the following problems. Round answers to two decimal places where necessary.

1. Last year tuition at Smalltown College increased by 5%. If the previous tuition was $8,520 per year,

    a. How much is the increase?

    b. What is the new tuition?

2. During the first quarter of the year Howell Company had sales totaling $2,095 million. Sales to the government amounted to $937 million. What percent of Howell's sales were made to the government?

3. A $115,000 house is located in a community that assesses property at 35% of its market value. What is the assessed value of the house?

4. Diversified Industries bought a majority interest in American Industries for $14 million. If American Industries was worth $27.9 million, what percent of the company did they buy?

5. A salesman is paid $200 a week plus a commission of 9% on all sales over $2,000. How much does he earn in a week when his sales are $5,540?

6. Mrs. Hong has been selling real estate in an area where her commission for selling a house is 3%. She hopes to earn $45,000 next year. How large must her sales be if she is to reach this goal?

7. In New City it was reported that 115 newly built houses have not been sold. If this represents $12\frac{1}{2}\%$ of the available housing, how many houses are there in New City?

8. Linda Johnson receives a paycheck of $494 after taxes every week. If she spends 25% of this for rent and 15% for food:

    a. How much will she spend for rent? for food?

    b. How much will she have for other expenses?

2.6 MEASURING CHANGE

The cost of attending a first-run movie in most areas has been increasing for many years. In Chicago, Illinois, for example, the price of a movie ticket at some theaters recently went from $7.00 to $8.00. At the same time a few theaters in Dayton, Ohio, raised their prices from $6.00 to $7.00.

Consider Chicago first. A person who sees two movies a week might measure the effect of the increase on his budget by thinking, "I'm going to have to pay $2.00 more each week, or if I take my friend, it is going to cost me $4.00 more." But if the theater owners wish to estimate the increase in revenues, they would probably use a different measure of change—the percent change.

The percent change (increase or decrease) is an application of the basic equation $(R \times B = P)$ in which:

- R is the percent change and is always the unknown.
- B is always the original value.
- P is the amount of change

Therefore $\quad R \times B = P$
$\quad\quad\quad\quad\quad \downarrow \quad\quad \downarrow \quad\quad \downarrow$

becomes $\quad$ % change × original value = amount of change

For movies in Chicago:

% change × original = amount of change

$R \times 7 = 1$ $\quad\quad\quad\quad$ $8.00 - 7.00 = 1.00$

$\dfrac{7}{7} R = \dfrac{1}{7}$ $\quad\quad\quad\quad$ *Divide both sides of the equation by the coefficient of the unknown.*

$R = 0.1428$

$R = 14.3\%$ increase $\quad\quad$ *Move the decimal point two places to the right, add %.*

The price of a movie ticket in Chicago was raised $1.00 or 14.3%.

Now consider the increase in movie ticket costs from $6.00 to $7.00 in Dayton. Did they go up more or less than those in Chicago? On an arithmetic basis, the increase for each was the same, $1.00. But the percent increase in the movie ticket price in Dayton is:

% change × original = amount of change

$R \times 6 = 1$

$R = \dfrac{1}{6}$ $\quad\quad\quad\quad$ *Divide both sides of the equation by the coefficient of the unknown.*

$R = 0.166\overline{6} = 16.7\%$ or $16\dfrac{2}{3}\%$ increase

Movie ticket prices in Dayton were raised 16.7%.

Measured in actual dollars, the cost of a movie ticket in Chicago and Dayton went up the same amount. Measured in *percent*, the cost in Dayton increased more because of the lower base or original price ($6.00 versus $7.00).

66 CHAPTER 2 PERCENTS AND THEIR APPLICATIONS

■ **Example 1**

The percent change is frequently a more useful measure than the arithmetic change. If an employee is given a $30 raise it may be satisfactory if he is earning $300 a week. This would be an increase of 10%.

$$\% \text{ change} \times \text{original} = \text{amount of change}$$

$$R \times 300 = 30$$

$$\frac{\cancel{300}}{\cancel{300}} R = \frac{30}{300} \qquad \text{\textit{Divide both sides of the equation by the coefficient of the unknown.}}$$

$$R = 0.10 = 10\% \text{ increase} \qquad \text{\textit{Move the decimal point two places to the right and add \%.}}$$

But the same $30 increase would probably be unacceptable if the employee was earning $2,000 a week. Then the $30 increase would mean only a 1.5% increase.

$$\% \text{ change} \times \text{original} = \text{amount of change}$$

$$R \times 2{,}000 = 30$$

$$\frac{\cancel{2{,}000}}{\cancel{2{,}000}} R = \frac{30}{2{,}000} \qquad \text{\textit{Divide both sides of the equation by the coefficient of the unknown.}}$$

$$R = 0.015$$

$$R = 1.5\% \text{ increase} \qquad \text{\textit{Change the decimal to a percent.}}$$

The method of finding the percent change when there is a decrease in the quantity being measured is the same as illustrated for an increase. In both cases the answer must be labeled in some way, such as "increase," "decrease," "went up," "declined," "+," or "−," to show the direction of the change. ■

■ **Example 2**

The U. Lose Co. was having difficulty selling its product and decided to cut the operating budget for the following year. The current year's budget was $250,000 and the proposed budget for the following year was $225,000. What percent decline was proposed?

$$\% \text{ change} \times \text{original} = \text{amount of change}$$

$$R \times 250{,}000 = 25{,}000 \qquad \text{\textit{Current budget is original. Amount of change is \$25,000. (\$250,000 − \$225,000.)}}$$

$$\frac{\cancel{250{,}000}}{\cancel{250{,}000}} R = \frac{25{,}000}{250{,}000} \qquad \text{\textit{Divide both sides of the equation by the coefficient of the unknown.}}$$

$$R = 0.10 = 10\% \text{ decrease} \qquad ■$$

Measuring Economic Performance

Percent changes are often used to measure economic performance whether we are considering a private business or the business of government. An illustration of each follows.

■ **Example 3**

One measure of inflation is the percent increase, from month to month, of the prices we pay for the goods and services we buy. These price changes are summarized in one figure, the Consumer Price Index,* compiled and published by the Bureau of Labor Statistics.

*For a description of the Consumer Price Index, see Chapter 12

In October, 1998, the index was 164.0, in October, 1997, it was 161.6. (These figures are compared with those from 1982–1984.) What was the percent change (to the nearest tenth)?

$$\% \text{ change} \times \text{original} = \text{amount of change}$$

$$R \times 161.6 = 2.4$$

The October, 1997, value is the original value from which the change occurred. The amount of change is 2.4 (164.0 − 161.6.)

$$\frac{\cancel{161.6}}{\cancel{161.6}} R = \frac{2.4}{161.6}$$

Divide both sides of the equation by the coefficient of the unknown.

$$R = 0.0148$$

Carry the division to four decimal places for a percent correct to the nearest tenth.

$$R = 1.5\% \text{ increase}$$

Move the decimal point two places to the right, add %, and round.

The index increased 2.8%.

Note: In percent change problems involving time, the earlier period is always the original value. ■

■ **Example 4**

The W. E. Win Co. management was pleased to find that its sales had increased from $1,000,000 in 1999 to $1,200,000 in 2000. But when they examined the profit data they were less satisfied, although there had been an increase from $50,000 to $55,000 in the same period. Why weren't they entirely satisfied?

The only way to make a meaningful comparison between the sales record and the profit record is to examine the percent change. The dollar increase in sales was $200,000 ($1,200,000 − $1,000,000), and no one would expect that much increase in a profit of $50,000. Therefore, we must use percent change to make the comparison.

For sales:

$$\% \text{ change} \times \text{original} = \text{amount of change}$$

$$R \times 1,000,000 = 200,000$$

The earlier period is the original value.

$$\frac{\cancel{1,000,000}}{\cancel{1,000,000}} R = \frac{200,000}{1,000,00}$$

$$R = 0.20 = 20\% \text{ increase}$$

For profits:

$$\% \text{ change} \times \text{original} = \text{amount of change}$$

$$R \times 50,000 = 5,000$$

The earlier period is the original value.

$$\frac{\cancel{50,000}}{\cancel{50,000}} R = \frac{5,000}{50,000}$$

$$R = 0.10 = 10\% \text{ increase}$$

The percent changes show that the increase in profit was only half the increase in sales. ■

CALCULATOR HINT

You can use the % key on your calculator to find the value of any number after a % increase or % decrease.

For example, suppose you want to find what the value of a $500 salary would be after a $6\frac{1}{2}$% increase.

ENTER	DISPLAY
500 [+] 6.5 [%]	532.50

The new salary is $532.50.

Find the cost of a $45 sweater after a 30% decrease in price.

ENTER	DISPLAY
45 [−] 30 [%]	31.5

The new price is $31.50.

Now go to Try These Problems (Set V) to practice measuring changes.

NEW RULES IN THIS CHAPTER

- **Rule 2-1.** To change a percent to a decimal, move the decimal point two places to the left and drop the percent sign (%).
- **Rule 2-2.** To change a percent to a fraction, multiply by $\frac{1}{100}$ and drop the percent sign (%).
- **Rule 2-3.** To change a decimal to a percent, move the decimal point two places to the right and add the percent sign (%).
- **Rule 2-4.** To change a fraction to a percent, change it first to a decimal, then to a percent.

NEW FORMULAS IN THIS CHAPTER

- $R \times B = P$ (rate × base = product)
- % change × original = amount of change

Chapter 2

Name	Date	Class

TRY THESE PROBLEMS (Set V)

Find the following percent changes. Round to the nearest tenth of a percent where necessary.

1. A salary is increased from $650 to $728.

2. There is a drop in the number of employees from 310 to 280.

3. A family accustomed to spending $90 a week for food finds it now takes $150 to buy the same products.

4. A jacket is marked down from $79 to $49.

5. The price of a movie ticket is increased from $6.00 to $8.00.

6. A $2,200 bank account is reduced by $700.

7. Average class size changes from 33 to 25 students.

8. Sales increase from $160,000 to $320,000.

9. A price changes from $3.29 a pound to $2.79 a pound.

10. A debt of $375 is reduced to $300.

Chapter 2

END-OF-CHAPTER PROBLEMS

Change the following percents to equivalent decimals and equivalent fractions in lowest terms.

1. 22%
2. 5%
3. 226%
4. 4%

5. 40%
6. 400%
7. 2.75%
8. 8.2%

9. 0.25%
10. 0.4%
11. 0.321%
12. 110%

13. $5\frac{1}{4}$%
14. $\frac{1}{2}$%
15. 50%
16. $\frac{5}{8}$%

17. $8\frac{1}{8}$%
18. $10\frac{3}{4}$%
19. $\frac{4}{5}$%
20. $133\frac{1}{3}$%

Change the following quantities to percents. Round to two decimal places where necessary.

21. 0.37
22. 0.01
23. 0.2
24. 0.532

25. 0.0618
26. 1
27. 2.4
28. 0.005

29. 3.00176
30. 0.7
31. $\frac{3}{10}$
32. $\frac{7}{8}$

33. $\frac{3}{11}$
34. $2\frac{1}{2}$
35. $\frac{23}{11}$
36. $\frac{4}{5}$

37. $\frac{4.2}{10.4}$
38. $1\frac{7}{17}$
39. $\frac{6}{10}$
40. $0.82\frac{1}{4}$

Find the missing quantity. Round to two decimal places where necessary.

	RATE	BASE	PRODUCT		RATE	BASE	PRODUCT
41.	7%	349		46.		360	420
42.	$3\frac{1}{2}\%$	520		47.		112	117.6
43.	105%	$410		48.	4%		3.36
44.		480	12	49.	$2\frac{1}{2}\%$		2.5
45.		257	35.98	50.	0.12%		0.86

Find the percent change to the nearest whole percent; indicate whether there is an increase or decrease.

	CHANGE IS FROM	TO		CHANGE IS FROM	TO
51.	30	45	57.	600	296
52.	15	45			
			58.	6.8	10.7
53.	50	20			
54.	$600	$675	59.	36.2	28
55.	12.8	14.2			
			60.	$10\frac{3}{4}$	$22\frac{1}{4}$
56.	296	600			

Chapter 2

END-OF-CHAPTER PROBLEMS

Solve the following problems. Round answers to two decimal places where necessary.

61. Find $42\frac{1}{2}\%$ of 400.

62. What is 3.6% of 940?

63. What percent of 30 is 40?

64. What percent of 40 is 30?

65. Fifteen percent of what number is 12.9?

66. 27 is 135% of what number?

67. A salesman earns $8\frac{1}{2}\%$ commission. If his sales last month were $42,500, how much were his earnings?

68. A homeowner pays $11.20 per hundred real estate taxes on his property, which is assessed at $38,000. How much is his tax payment?

69. Unemployment in the United States is measured as a percent of the labor force. In February, 1999, there were 139.3 million people in the labor force. Of these, 6.1 million were unemployed. Find the unemployment rate.

70. A company decided to move its factory to a different location. Twenty-four employees said they would not go with the company. If these employees were 15% of the number employed, how many people did the company employ?

71. A company executive receives 2% of the annual profits as a year-end bonus. If the bonus was $8,448 one year, what was the amount of the company's profits?

72. New Town collects a sales tax of 6%. What is the total cost of a portable compact disc player that costs $89.99 before tax?

73. A nearby town has a sales tax of 8%. For the week ending June 15 the H&H Dept. Store paid $6,560 as taxes on their sales. How much were their sales for the week?

74. An excise tax (a tax added before the item is sold to the consumer) of 10% is added to an item that is priced at $69.95 before taxes. What is the price, including tax?

75. At True-Vision Optical Company 39 employees are on vacation. If the firm employs 600 people, what percent of the staff is working?

76. See-It-All Tours advertises a "stupendous, all-expenses-included trip" at $779.* (The * is for a footnote that says, "Add 15% for service and taxes.")

   a. What is the charge for service and taxes?

   b. How much is the total cost of the tour?

77. A retailer sold a large lot of leftover blouses at 23% less than the original price. If the blouses had been sold originally at $27, what was the sale price?

78. Gloria Betan made a $66.00 deposit into her savings account. If this amount is $5\frac{1}{2}\%$ of her earnings, find her total earnings.

79. A company found that 4% of its employees applied for retirement. If 16 employees applied, how many people does the company employ?

80. An administrative assistant earning $410 a week received a $12\frac{1}{2}\%$ increase. What will her new salary be?

81. Last year Steve Lander paid taxes of $13,461.70. If his gross income was $38,642, what percent of his income did he pay in taxes?

82. State University has an entering class of 5,000 students. Of these, 1,200 are liberal arts majors, 1,500 are business majors, 700 have indicated other majors, and the remainder is undecided.

   a. What percent of the class wishes to major in liberal arts?

   b. What percent has chosen a major other than liberal arts or business?

   c. What percent is undecided?

83. At the end of the first year, there are 3,700 students at State University (Problem 82) who continue to the sophomore year. What percent is no longer in school?

84. When new machinery was introduced into a factory, a worker increased his production from 80 to 85 units. What was the percent increase?

85. A large supermarket ordered 60 cases of jellies. On verifying the order, it was found that only 56 were actually delivered. What percent of the order was missing?

86. A computer monitor is bought for $310 and sold for $595. What percent of the selling price is the cost?

87. Mr. and Mrs. Lucky sold their summer cottage for $140,000, although they had paid only for $84,000 for it a few years earlier. Find the percent change in the value of the home.

88. The Efficient Savings Bank decided to computerize its check verification procedures and found that they could reduce the staff of 410 employees to 25. What was the percent change in the number of employees?

Chapter 2

Name   Date   Class

END-OF-CHAPTER PROBLEMS

89. Academe University, forced to cut its budget by 18%, decided to cut the size of the entering freshman class by the same percent. There were 3,300 students in the previous class. How large is the new entering class?

90. Fine Fabrics Shop cut the price of a drapery fabric from $24 a yard to $13. What percent price cut could they advertise?

91. A production team in a factory produced 295 units one week, although their quota was only 255 units.

   a. By what percent was production higher than quota?

   b. If they are paid an additional $2.50 for every item produced above quota, how much additional compensation will they receive that week?

92. Bill Green made a down payment of $2,000 on a new car. After this payment, he owed 90% of the purchase price. What was the selling price of the car?

93. A long-distance communications company charges 15¢ for a 1-minute call within a 50-mile radius. However, after 5 P.M. the charge is only 10¢. What is the percent reduction in rate?

94. An automobile has a base price of $19,400. After the buyer added optional equipment, the total price was $24,600.

   a. By what percent has the addition of optional equipment increased the price?

   b. If the sale is in an area where the sales tax is 8%, what is the total cost of the car, including optional equipment and tax?

95. A group of 115 rats was fed large doses of a particular chemical in an experiment to test the chemical's safety. Five died of miscellaneous causes during the experiment, while twenty-one developed a certain type of cancer. What percent of the surviving rats developed the cancer?

96. A $160,000 house was bought with a down payment of $28,000 and a mortgage for the remainder. Closing costs of $8,000 were added to the purchase price. (Closing costs are not part of the mortgage.)

   a. What percent of the total purchase price was the mortgage?

   b. By what percent did the addition of closing costs increase the original price of the house?

97. A TV was priced at $445. Sue and Bill Jenkins purchased it with a down payment of 20% of the selling price. How much was their down payment? What percent of the price is the unpaid balance?

98. Hope Andreas bought 250 shares of stock at $12.25 a share. At the end of the year a dividend of 7.1% of the purchase price was paid to all stockholders. How much did Hope receive?

99. Leonard Rich was advised by an investment service to put 22% of his capital in government bonds, 51% in corporate stocks, and the remainder in real estate. If he invested $47,061 in real estate, what was the total value of his capital?

100. A credit agency managed to collect 80% of a $4,500 account that was long overdue. If they charged 12% for their services, how much did the agency receive?

101. The Norths used a real estate agent and sold their house for $75,800. The agent charged a 5% commission and $103 for expenses. How much did the Norths receive?

102. A family with a monthly income of $4,200 spends 95% of its earnings and saves the balance. What are the annual savings of this family?

103. Amy Pell owns 125 shares of Remco Stock which are valued at $38 per share. What was the total value of this stock after a new patent caused the value of the stock to increase by 24%?

104. The Brighter Light Bulb Company has a yearly sales goal of $3,260,000. By October, sales have reached $2,854,000. What percent of the sales goal remains?

SUPERPROBLEMS

Round answers to two decimal places where necessary.

1. A man earning $35,000 a year has a take-home pay of $22,750. In planning a budget he allows:

    35% of his monthly take-home pay for rent;

    $175 a month for utilities including telephone;

    $50 a week for food;

    10% of his take-home pay for clothing;

    20% of his take-home pay for entertainment, vacation, and miscellaneous expenses.

    a. What percent of his salary is deducted before he is paid?

    b. What percent of his take-home pay will he be able to save, or has he planned to spend more than he earns?

2. A retailer bought a portable digital cassette player for $45.50 and sold it for $80. When the cost to him was increased by 10%, he increased the selling price to $89.95. What was the percent increase in his markup (selling price − cost)?

Chapter 2

TEST YOURSELF

Section

1. Change the following percents to equivalent decimals and fractions in the lowest terms.   2.1

   a. 2%   **b.** 4.6%

   c. $19\frac{1}{4}\%$ **d.** 316%

   e. 0.5%   **f.** $\frac{2}{5}\%$

2. Change the following to percents. Round to two decimal places where necessary.   2.2

   a. 0.04   **b.** 0.6

   c. 1.3   **d.** 3

   e. $\frac{7}{20}$ **f.** $1\frac{1}{3}$

Find the missing quantity. Round to two decimal places where necessary.

	RATE	BASE	PRODUCT	
3.	6%	640		2.3
4.	$4\frac{1}{2}\%$		9.72	2.3

Find the percent change to the nearest whole percent; indicate whether there is an increase or decrease.

	CHANGE IS FROM	TO		
5.	800	936		2.6
6.	41.2	10.3		2.6

	Section
Solve the following problems. Round answers to two decimal places where necessary.	
7. Find $3\frac{1}{2}$ of 82?	2.3
8. What percent of 160 is 50?	2.3
9. 4.5 is what percent of 90?	2.3
10. 33 is 3% of what number?	2.3
11. What is $\frac{1}{4}$% of $2,500?	2.3
12. Blanes Department Store finds that about 1% of merchandise sold is returned. If $2,500 worth of goods was returned last week, what were the weekly sales?	2.5
13. The Alabaster Refining Company is employing only 80% of the usual number of workers. If they usually employ 465 people, how many are employed at present?	2.5
14. Bill Silver received $178 in dividends from his credit union. What was his annual rate of return if his balance was $3,560 for the entire year?	2.5
15. Last year Mark Mann paid $1,830 in income taxes. This year he owed $2,013. What is the percent change?	2.6

Chapter 2

COMPUTER SPREADSHEET EXERCISES

A spreadsheet is a computerized equivalent of an accounting ledger. Similar to a ledger, it consists of rows and columns that permit data to be organized in a useable and understandable format. Computers and spreadsheets allow us to compute changes to the data automatically, and faster and more accurately than could be accomplished by hand.

There is a wide variety of business applications for which spreadsheets are used, many of which will be shown throughout this text. When formulas and functions appear, they will be in Microsoft Excel format; however, students using Lotus 1-2-3 or other spreadsheet software will be fully able to follow the directions.

1. Lee needs help preparing a budget for her monthly expenses. Complete the spreadsheet below if:

    a. Her salary stays the same every month.
    b. $985 of her salary is spent on the mortgage payment.
    c. $470 a month is spent on food.
    d. 10% of her monthly salary is for clothing.
    e. Utility expenses total $311 a month.
    f. A one-time insurance expense of $495 occurs in September.
    g. $480 a month is spent on her car.
    h. 15% of her salary is for miscellaneous expenses.
    i. She saves 4% of her salary per month.

Enter the row labels so they are left-aligned and only the first letter is capitalized. Use formulas to calculate totals and balances. Make sure that the column headings appear as shown and that numbers are positioned correctly beneath the column headings. Use integer values (without dollar signs or commas) in your spreadsheet. Click on the Print Preview button to display your work as it will appear when printed. If necessary, adjust the cell widths or margin settings, or choose the Fit to Page scaling option so that the spreadsheet fits across the width of the page.

```
                    LEE'S MONTHLY BUDGET

                JUL    AUG    SEP    OCT    NOV    DEC    TOTALS

Salary          4000

Expenses:
Mortgage
Food
Clothing
Utilities
Insurance
Car Expenses
Miscellaneous
Savings

Total Expenses:

Balance:
```

2. Lee wishes to prepare another budget spreadsheet for January through June of the next year. Begin with the spreadsheet constructed in Problem 1. Make changes based on the following information:

    a. Change the labels in the month row to JAN, FEB, MAR, APR, MAY, JUN. Keep the labels underlined.

b. Lee wishes to begin a college fund for her young daughter. Insert a row below Miscellaneous and enter the label College Fund. Allocate $70 a month to the fund for the months April through June.

c. Her utility expenses will increase $15 per month beginning in February.

d. She pays a one-time insurance charge of $489 in March.

e. Lee decides to sell her car and use buses and taxis for transportation. Replace the label Car Expenses with the label Transportation. Change the monthly allotment to $360 per month beginning in May.

f. She will receive a 5% salary increase starting in March.

g. Lee's mortgage payment increases to $1,038 per month beginning in February.

h. Lee forgot to allow for expenses for leisure activities. She feels that $175 a month would be a proper allocation. Insert a row below Transportation, enter the label Leisure, and enter the value 175 for January through June.

i. Lee realizes she has a higher monthly balance than she needs to cover unforeseen expenses. She decides to increase her monthly saving. The current rate of savings is 4% of monthly salary. Change the percent used to calculate savings for each month except March, so that the balance for those months is as close as possible to, but no more than, $300. (Use only integer values for the savings rate; that is, 5%, 6%, 7%, and so on.)

3 ■ Pricing a Product

OBJECTIVES

Upon completion of this chapter, the student should be able to:

1. *Use C + M = S to find:*   *Use M = OH + NP to find:*

    cost;   *markup;*

    dollar markup;   *overhead;*

    selling price.   *net profit.*

2. *Calculate the percent markup on cost.*
3. *Calculate the percent markup on selling price.*
4. *Find the selling price that will give the required percent markup even though some of the merchandise is lost to spoilage or marked down.*
5. *Find the breakeven point, operating loss, and absolute loss.*

3.1 THE BASIC EQUATION

Businesses are established to make a profit by buying, or producing, and then selling products or services. Although business firms are very different in size, organization, and the type of products handled, they have many expenses in common: rent, taxes, utilities, equipment maintenance, and so on. To determine the selling price of a product, an amount sufficient to cover all these expenses and provide a profit is added to the cost. This added amount is called **markup**.*
The expenses are grouped together and called **overhead**. Any amount remaining after the overhead has been covered is the company's **net profit**.

In equation form these relationships are

and

$$\text{Markup} = \text{Overhead} + \text{Net Profit}$$
$$\downarrow \quad \quad \downarrow \quad \quad \downarrow$$
$$M = OH + NP$$

$$\text{Cost} + \text{Markup} = \text{Selling Price}$$
$$\downarrow \quad \quad \downarrow \quad \quad \downarrow$$
$$C + M = S$$

---

*Markup is sometimes called *margin* or *gross profit*.

82 CHAPTER 3   PRICING A PRODUCT

The formulas are illustrated below.

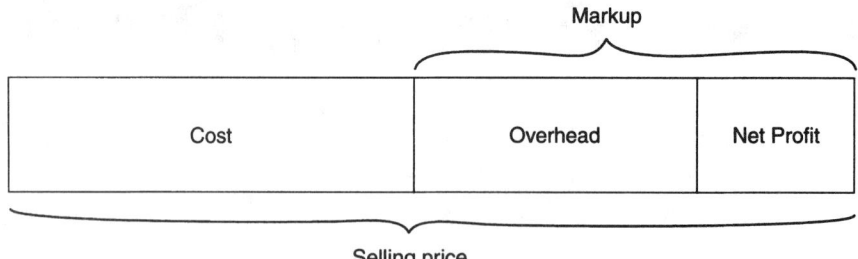

```
                              Markup
                         ⎧―――――――――――⎫
  ┌─────────────────────┬──────────────┬────────────┐
  │        Cost         │   Overhead   │ Net Profit │
  └─────────────────────┴──────────────┴────────────┘
  ⎩――――――――――――――――――――――――――――――――――――――――――――――――⎭
                        Selling price
```

■ **Example 1**

These relationships may be illustrated simply. If a producer manufactures an item that costs him $10, allows $3 for overhead expenses, and expects to make a net profit of $2, the selling price for the item may be calculated by:

$$M = OH + NP$$
$$= 3 + 2 \quad \text{\textit{Overhead is \$3 and net profit is \$2.}}$$
$$M = \$5$$
$$C + M = S$$
$$10 + 5 = S \quad \text{\textit{Cost is \$10 and markup, just calculated, is \$5.}}$$
$$\$15 = S$$

If, however, the item could not be sold at $15, but rather, was marked down and sold for $13, then:

$$C + M = S$$
$$10 + M = 13$$
$$M = \$3$$

Since we know that

$$M = OH + NP$$
$$3 = 3 + NP$$
$$\$0 = NP \text{ (No net profit)}$$

When prices are set, the markup is quoted as a percent rather than as a dollar and cents figure. The use of a percent does not change the basic relationships among cost, markup, and selling price. The equations $C + M = S$ and $M = OH + NP$ remain true and applicable whether we use dollars and cents and/or percents. ■

■ **Example 2**

The preceding problem could also have been stated as follows. A producer manufactures an item that costs him $10. If he adds 30% of his cost for overhead expenses and 20% of cost for net profit, what is the selling price of the product?

To translate this problem into algebraic form, note:

"That costs him $10" means $C = \$10$

"30% of cost for overhead expenses" means $OH = 0.30C$

"20% of cost for net profit" means $NP = 0.20C$

"What is the selling price?" means S is the unknown.

First, let us find the total markup as a percent of cost.

$$M = OH + NP$$
$$M = 0.30C + 0.20C$$
$$M = 0.50C$$

Now, substituting in the basic equation,

$$C + M = S$$
$$10 + 0.50(10) = S$$
$$10 + 5 = S$$
$$\$15 = S$$ ∎

3.2 MARKUP ON COST VERSUS MARKUP ON SELLING PRICE

When markup is stated as a percent it is always given as a percent of either cost or selling price. It is customary at the producer level to use markup as a percent of cost (as in the previous illustration). At the retail level, however, markup is generally quoted as a percent of selling price. Markup can be found by using one of the following equations:

$$M = (\%M \text{ on } C) \times C$$

or

$$M = (\%M \text{ on } S) \times S$$

The basic equation $C + M = S$ will serve to solve a variety of pricing problems, regardless of whether the markup is on cost or on selling price. As long as all the given facts can be stated in terms of a single unknown, the equation can be solved.

■ **Example 1**

A merchant uses a markup of $33\frac{1}{3}\%$ of cost. What selling price should be set for a briefcase that costs $48?

"Costs $48" means $C = \$48$

"Markup of $33\frac{1}{3}\%$ of cost" means $M = 33\frac{1}{3}\%$ of C or $\frac{1}{3}C$ or $0.\overline{3}C$

$$M = \frac{1}{3}(48) \text{ or } 0.\overline{3}(48)$$

"What price should be set?" means selling price (S) is unknown.

Substituting in the basic equation,

$$C + M = S$$
$$48 + 0.\overline{3}(48) = S$$
$$48 + 16 = S$$
$$\$64 = S$$

Check: S (selling price) is more than the cost and of the same magnitude. ∎

84 CHAPTER 3 PRICING A PRODUCT

■ **Example 2**

A manufacturer wishes to produce a coat to sell for $75. If he normally adds 55% of cost to cover all expenses and net profit, what is the most he can spend to produce the coat?

Note that "55% of cost to cover all expenses and net profit" means $M = 55\%$ of cost.

Substitute in the basic equation:

$$C + M = S$$

$$C + 0.55C = 75 \qquad \text{55\% of cost is } 0.55 \times C \text{ or } 0.55C.$$

$$1.55C = 75 \qquad \text{Similar terms are combined by adding the coefficients: } 1.00 + 0.55 = 1.55.$$

$$\frac{\cancel{1.55}}{\cancel{1.55}} C = \frac{75}{1.55} \qquad \text{Divide both sides of the equation by the coefficient of the unknown.}$$

$$C = 48.387$$

$$C = \$48.39 \qquad \text{Round the answer to the nearest cent.}$$

Check: C (cost) is lower than the selling price but of the same magnitude. ■

CALCULATOR HINT

Repeating decimals such as $0.\overline{3}$ can present a problem to calculator users. You can avoid any inaccuracies in calculations by using the equivalent fraction in your calculations instead of rounding the repeating decimal.

For example, suppose that you wish to use a calculator to compute $33\frac{1}{3}\%$ of 48, as occurred in Example 1. Since $33\frac{1}{3}\%$ equals the fraction $\frac{1}{3}$,

ENTER	DISPLAY*
1 ÷ 3 × 48 =	16

Inaccuracies that would have occurred had the rounded values 0.3 or 0.33 or 0.333, etc., been used in place of 1 ÷ 3 are avoided. For instance, had you entered 0.33 the answer would have been 15.84, which is a 0.16 or 16¢ difference from the correct result.

■ **Example 3**

A retailer wants to stock a videotape selling for $20 in order to meet the competition from a nearby store. He allows 30% of selling price for overhead expenses and 12% of selling price for net profit. How much can he afford to pay for the videotape if he wishes to obtain his normal markup?

"Selling for $20" means $S = \$20$

"30% of selling price for overhead expenses" means $OH = 0.30S$

"12% of selling price for net profit" means $NP = 0.12S$

"How much can he pay?" means cost (C) is unknown

---

*Some calculators cut off or truncate nonterminating decimals when the display is full. If this is the case, your calculator will be slightly less accurate and will display 15.999998 for this problem.

MARKUP ON COST VERSUS MARKUP ON SELLING PRICE   85

First, let us find the total markup as a percent of selling price.

$$M = OH + NP$$
$$M = 0.30S + 0.12S$$
$$M = 0.42S$$

Substituting in the basic equation,

$$C + M = S$$
$$C + 0.42(20) = 20$$
$$C + 8.40 = 20$$
$$C + 8.40 - 8.40 = 20 - 8.40$$ *Collect similar terms by subtracting 8.40 from both sides of the equation.*

$$C = \$11.60$$

Check: C (cost) is less than the expected selling price and about the same magnitude. ■

■ **Example 4**   A retailer bought a table for $50. She wishes to add a markup of 20% of the selling price to cover overhead and net profit. At what price should she sell the table?

$$C + M = S$$
$$50 + 0.20S = S$$ *$M = 20\%$ of selling price, or $0.20S$.*

$$50 + 0.20S - 0.20S = S - 0.20S$$ *Collect similar terms by subtracting $0.20S$ from both sides of the equation.*

$$50 = 0.8S$$ *Combine similar terms by subtracting the coefficients: $1.00 - 0.20 = 0.80$ or 0.8.*

$$\frac{50}{0.8} = \frac{0.8S}{0.8}$$ *Divide both sides of the equation by the coefficient of the unknown.*

$$\$62.50 = S$$

Check: S (selling price) is larger than the cost and about the same size. ■

Chapter 3

TRY THESE PROBLEMS (Set I)

Solve the following problems. Round dollar amounts to the nearest cent and rates to the nearest tenth of a percent where necessary.

1. If an item is bought for $6.49 and sold for $10.75, what is the dollar markup?

2. A bracelet bought at $32 is sold with a markup of $28.89. What is the selling price?

3. A portable radio is sold for $39.99. If the markup was $14.25, what was the cost?

4. A vase that costs $11.50 is sold for $19.98. If overhead expenses are $5.00:

    a. Find the markup;

    b. Find the net profit.

5. A table costs $75 to produce. If the manufacturer allows 35% of cost for overhead expenses and sells the table for $112.50:

    a. How much is the markup?

    b. How much is the net profit?

6. A pair of shoes costs $18.50 to manufacture. If 40% of the selling price is allowed for net profit and the merchandise is sold for $60:

    a. Find the markup;

    b. How much are overhead expenses?

7. A producer allows 13% of cost for overhead expenses and 20% of cost for net profit. What percent of cost is his markup?

8. A retailer allows 24% of selling price for overhead expenses and 12% of selling price for net profit. What percent of selling price is the markup?

9. A retailer adds a markup of 52% of selling price. If 36% of selling price is needed to cover overhead expenses, what percent of selling price is allowed for net profit?

10. A blouse cost $16 to manufacture. Find the selling price if the producer adds 61% of cost as a markup.

11. A manufacturer produces a scarf for $2.50 and adds 40% of cost as a markup. Find the selling price.

12. A retailer buys a handbag at $37.50 and adds a markup of 40% of selling price. Find the selling price.

13. Art-Works buys a picture frame for $12.60. Find their selling price if the markup is 65% of selling price.

14. Ace Music pays $8.50 for a compact disc. If they add 36% of selling price for overhead expenses and 10% of selling price for net profit, what is the selling price?

15. A computer monitor is sold for $200 when the markup is 60% of cost. What is the cost?

16. A retailer wishes to beat the competition by selling a calculator for $8.50. What is the most she can pay for the calculator if she uses a markup of 38% of selling price?

17. A stereo system sells for $1,200. If it is marked up 25% of cost, how much did it cost the store?

18. A new kitchen cabinet sells for $150 when a markup of 42% of selling price is used. What is the cost of the cabinet?

19. Atlas Tiles plans for 18% of cost to cover overhead expenses and 14% of cost for net profit. What was the cost of a wall unit that is sold for $750?

3.3 PERCENT MARKUP

In the previous problems the markup percent was given, and either the cost or selling price had to be found. In a different situation, it may be necessary to find the markup percent when cost and selling price are known.

For example, if the retailer who wishes to add a $20 videotape to his line (Section 3.2, Example 3) finds he can buy one for $12.20, will this cost give him his usual markup of 42% on selling price?

The basic relationship among cost, markup, and selling price can be used to solve this problem, but the solution is simplified by observing that the percent of markup on cost is the ratio of dollar markup to cost.*

$$\%M \text{ on } C = \frac{\$M}{C}$$

Similarly,

$$\%M \text{ on } S = \frac{\$M}{S}$$

To solve the problem, first find the dollar markup.

$$C + M = S$$
$$M = S - C$$

A more convenient form of the equation is found by subtracting C from both sides of the equation.

$$M = 20 - 12.20$$
$$M = \$7.80$$

Then

$$\%M \text{ on } S = \frac{\$M}{S}$$
$$= \frac{7.80}{20}$$
$$= 0.39$$
$$\%M \text{ on } S = 39\%$$

The answer is that the $12.20 videotape will give only a 39% markup on selling price, which is less than the usual 42% for this store. Of course, the retailer may decide that it is worth buying if he cannot find a suitable videotape for less. It may be important to him to complete his line and meet competition.

To find the percent markup on cost:

$$\%M \text{ on } C = \frac{\$M}{C}$$
$$= \frac{7.80}{12.20}$$
$$= 0.639$$

$\%M$ on $C = 63.9\%$, or 64% rounded to the nearest percent.

---

*Using $R \times B = P$, where R is the percent markup on cost, B is the cost, and P is the dollar markup:

$$\%M \text{ on } C \times C = \$M$$

$$\%M \text{ on } C = \frac{\$M}{C}$$

CHAPTER 3 PRICING A PRODUCT

Note that the same dollar markup, $7.80, gives very different percents on the basis of cost as compared to that of selling price. If the selling price is higher than the cost, the percent markup on selling price will be smaller than the percent markup on cost. The ratios have the same numerator, but the ratio to selling price will have the larger denominator and, therefore, the lower value.

■ Example 1

A software package cost a store $64 and was sold for $90.

1. What is the percent markup on cost, rounded to one decimal place?

$$M = S - C$$
$$= 90 - 64$$
$$M = \$26$$

$$\%M \text{ on } C = \frac{\$M}{C}$$
$$= \frac{26}{64}$$
$$= 0.4062$$

$\%M$ on $C = 40.6\%$ rounded to one decimal place

2. What is the percent markup on selling price, rounded to one decimal place?

$$\%M \text{ on } S = \frac{\$M}{S}$$
$$= \frac{26}{90}$$
$$= 0.2888$$

$\%M$ on $S = 28.9\%$ rounded to one decimal place

Check: The percent markup on cost (40.6%) is larger than the percent markup on selling price (28.9%). ■

■ Example 2

An item that costs $18 is sold at a price that allows a 40% markup on selling price.

1. Find the selling price.

$C + M = S$	
$18 + 0.4S = S$	$M = 40\%$ of S or $0.4S$.
$18 = S - 0.4S$	Collect like terms by subtracting $0.4S$ from both sides of the equation.
$18 = 0.6S$	Combine like terms by subtracting the coefficients: $1.0 - 0.4 = 0.6$.
$\dfrac{18}{0.6} = \dfrac{\cancel{0.6}S}{\cancel{0.6}}$	Divide both sides of the equation by the coefficient of the unknown.
$\$30 = S$	

Check: S (selling price) is more than cost and of the same magnitude.

2. What percent markup on cost was obtained?

$$M = S - C$$
$$= 30 - 18$$
$$M = \$12$$

$$\%M \text{ on } C = \frac{\$M}{C}$$
$$= \frac{12}{18}$$

$$\%M \text{ on } C = 0.666\overline{6} = 66.7\% \text{ or } 66\frac{2}{3}\%$$

Check: The percent markup on cost (66.7%) is larger than the percent markup on selling price (40%), which was given in the problem. ∎

■ **Example 3**   A wholesaler who uses a markup of 60% on cost wants to sell a set of glasses for $12.

1. What is the most he can pay for these?

$$C + M = S$$
$$C + 0.6C = 12 \qquad M = 60\% \text{ of } C, \text{ or } 0.6C.$$
$$1.6C = 12 \qquad \text{Combine like terms by adding the coefficients: } 1.0 + 0.6 = 1.6.$$
$$\frac{\cancel{1.6}C}{\cancel{1.6}} = \frac{12}{1.6} \qquad \text{Divide both sides of the equation by the coefficient of the unknown.}$$
$$C = \$7.50$$

Check: C (cost) is less than the selling price and of the same magnitude. ∎

2. What percent markup on selling price was obtained?

$$M = S - C$$
$$M = 12 - 7.50$$
$$M = \$4.50$$

$$\%M \text{ on } S = \frac{\$M}{S}$$
$$= \frac{4.50}{12}$$

$$\%M \text{ on } S = 0.375 = 37.5\%$$

Check: The percent markup on selling price (37.5%) is smaller than the percent markup on cost (60%), which was given in the problem. ∎

■ **Example 4**   A chair that cost $120 to manufacture was sold to a retailer for $195. The retailer then sold the chair for $325. Find the manufacturer's percent markup on cost and the retailer's percent markup on selling price.

92 CHAPTER 3 PRICING A PRODUCT

$$\text{Manufacturer's } \%M \text{ on } C = \frac{\$M}{C}$$

$$= \frac{195 - 120}{120}$$

$\$M = S - C$. *The manufacturer's cost is* $120; *his selling price is* $195.

$$= \frac{75}{120}$$

$$\%M \text{ on } C = 0.625 = 62.5\%$$

$$\text{Retailer's } \%M \text{ on } S = \frac{\$M}{S}$$

$$= \frac{325 - 195}{325}$$

$M = S - C$. *The retailer's cost is* $195; *his selling price is* $325.

$$= \frac{130}{325}$$

$$\%M \text{ on } S = 0.4 = 40\%$$

The manufacturer's markup on cost is 62.5%. The retailer's markup on selling price is 40%.

Note: In this case there is no expected relationship between the two markup percents because they result from different transactions. They have in common only the amount of $195, which is the manufacturer's selling price and the retailer's cost. ∎

■ **Example 5**

A retailer has an opportunity to buy a table from a manufacturer who is taking only a 45% markup instead of his usual 55% on his production cost of $48. The retailer wants to sell the table in a special sale for $110. If she must make 39% on selling price, should she buy the table? (Read the problem carefully; not all the numbers will be used in the solution.)

First find the manufacturer's selling price.

$$C + M = S$$
$$48 + 0.45(48) = S$$

45% markup on cost = 0.45×48.

$$48 + 21.60 = S$$
$$\$69.60 = S$$

The manufacturer's selling price ($69.60) is the retailer's cost. Is $69.60 low enough to give a 39% markup on a retail selling price of $110?

$$\%M \text{ on } S = \frac{\$M}{S}$$

$$= \frac{110 - 69.60}{110}$$

$\$M = S - C = 110 - 69.60$.

$$= \frac{40.4}{110} = 0.3672$$

$$\%M \text{ on } S = 36.7\%$$

The markup on selling price for the retailer would be only 36.7%. She would not buy the table if she needs a markup of 39% on selling price. ∎

Chapter 3

TRY THESE PROBLEMS (Set II)

Solve the following problems. Round dollar amounts to the nearest cent and rates to the nearest tenth of a percent where necessary.

1. A color printer that cost $460 to manufacture is sold for $550.

    a. What is the dollar markup?

    b. What percent of cost is this markup?

    c. What percent of selling price is this markup?

2. A flashlight that costs $18 is sold for $24. Find the percent markup on cost and on selling price.

3. At a garden sale, plants are priced at two for $5. If they cost $24 a dozen:

    a. What is the percent markup on cost?

    b. What is the percent markup on selling price?

4. A gas grill that cost a retailer $119 is sold for $199. Does the price give the retailer his usual markup of 48% of selling price?

5. A book costing $18 is sold with a markup of 25% of cost.

    a. What is the selling price?

    b. What is the percent markup on selling price?

6. A health food store carries a line of natural vitamins with a markup of 36% of selling price. If a bottle of vitamins is sold for $9:

   a. What is the cost?

   b. What is the percent markup on cost?

7. A package of batteries cost $4.50 and is sold with a markup of $33\frac{1}{3}$% of selling price.

   a. What is the selling price?

   b. What is the percent markup on cost?

8. A flea market dealer bought a large quantity of pens for 78¢ a dozen. If she uses a markup of 22% of selling price:

   a. What is her selling price per dozen?

   b. What is her percent markup on cost?

9. A clock is priced to sell at $27.30 after a markup of 30% on cost.

   a. What is the cost?

   b. What is the percent markup on selling price?

10. A laundry basket will sell for $10 after a markup of 50% of cost.

    a. What is the cost?

    b. What is the percent markup on selling price?

Chapter 3

Name _____ Date _____ Class _____

TRY THESE PROBLEMS (Set II)

11. A retailer bought towels for $84 a dozen for a special sale. Although her regular markup is 40% of selling price, for this special sale she will be using 30% of selling price. What percent markup on cost does the sale price represent?

12. A retailer needs to sell a new series of baseball cards at $2 each. He finds a supplier who will charge him $15 a dozen. Should he buy these if he must add 55% of selling price as his markup?

13. A travel kit is to retail for $25.30 after a markup of 15% on cost. Find the percent markup on selling price.

14. A wallet that cost $7.80 to produce is sold to a retailer for $11.00. The retailer sells it at $17.50. Find:

    a. The producer's percent markup on cost;

    b. The retailer's percent markup on selling price.

95

3.4 PRICING PERISHABLE PRODUCTS

A special problem in pricing occurs with perishable goods like fruits, vegetables, flowers, and pastry items, where part of the stock spoils and then cannot be sold at any price. How does the merchant price these products to provide the required markup for the entire shipment, although part is lost through spoilage?

■ **Example 1**

The owner of a small grocery store bought 120 pounds of potatoes at 30¢ a pound. He requires a markup of 60% of cost for the entire shipment, although 5% will probably spoil and be discarded. What price per pound would give the required markup?

1. Find the total cost and, from that, the total selling price required.

   Total cost = 120 × 0.30 = $36

   $C + M = S$

   $36 + 0.6(36) = S$ *M is 60% of cost = 0.6(36).*

   $36 + 21.60 = S$

   $\$57.60 = S$

2. The total selling price, $57.60, must be obtained from the sale of potatoes that do not spoil.

   Number to spoil: 5% of 120 pounds = 6 pounds

   Number to sell: 120 − 6 = 114 pounds

   $$\text{Price per pound} = \frac{\text{Total selling price}}{\text{Number to sell}}$$

   $$= \frac{57.60}{114} = 0.505 \text{ or } \$0.51 \text{ per pound}$$

   A price of 51¢ per pound for the potatoes would give the required markup and allow for 5% of the shipment to spoil and be discarded. ■

■ **Example 2**

A supermarket bought 120 pounds of bananas at 42¢ a pound. About 5% may spoil and be discarded. At what price per pound should they mark the bananas in order to make 30% of selling price on the entire purchase?

1. Find the total sales value the store must obtain. If all the bananas were sold at the original price, the selling price would be:

   $C + M = S$

   $0.42 + 0.30S = S$

   $0.42 = S - 0.30S$

   $0.42 = 0.7S$

   $\dfrac{0.42}{0.7} = \dfrac{\cancel{0.7}S}{\cancel{0.7}}$

   $\$0.60 = S$

   Total value of sales would be $0.60 × 120 = $72.

2. Find how much must be obtained from the bananas actually sold. Five percent of 120 pounds (or 6 pounds) may spoil and not be sold. Therefore, the entire $72 must be obtained from the remaining 114 pounds.

$$\text{Price per pound} = \frac{\text{Total selling price}}{\text{Number to sell}}$$

$$= \frac{72}{114}$$

$$\text{Price per pound} = \$0.631 = \$0.63$$

Check: A price of 63¢ is slightly higher than the 60¢ needed to give the required markup without any spoilage.

A common retail pricing practice should be noted. Many items are sold in established price lines. A price of 63¢ a pound would be very unusual for bananas, and the posted price would probably be 69¢. This would fit established price lines and take care of any additional spoilage beyond the expected 5%. ■

Metric Conversion. In countries where the metric system is used the bananas would be priced by the kilogram (kg), not the pound. Since a kilogram equals 2.2 pounds, a price of 63¢ per pound would translate to $1.39 per kilogram.

$$\$0.63 = 1 \text{ pound}$$

$$0.63\,(2.2) = 1\,(2.2)$$

$$\$1.39 = 2.2 \text{ pounds, or 1 kilogram}$$

A customer could buy half a kilogram, which is slightly more than 1 pound, for 70¢. (For a complete discussion of the metric system, see Appendix B.)

■ **Example 3**

The owner of a roadside fruit stand prices his peaches to obtain 52% of his cost as markup. If he expects to start with 50 boxes that cost him $1.60 each and to sell 90% before they become overripe and cannot be sold, at what price should he mark the peaches?

1. Find the total selling price.

$$\text{Total cost} = \$1.60 \times 50 = \$80$$

$$C + M = S$$

$$80 + 0.52(80) = S$$

$$80 + 41.60 = S$$

$121.60 = S$, the total selling price that must be obtained from the peaches actually sold.

2. Find how many boxes will be sold.

$$90\% \text{ of } 50 \text{ boxes} = 0.90 \times 50 = 45 \text{ boxes}$$

3. Find the price per box.

$$\text{Price per box} = \frac{\text{Total selling price}}{\text{Number to sell}} = \frac{121.60}{45} = 2.702, \text{ or } \$2.70 \text{ per box.}$$

Note: If the problem is worked using the cost of a single box, the selling price per box is $2.432 and should not be rounded.

$$C + M = S$$

$$1.60 + 0.52\,(1.60) = S$$

$$1.60 + 0.832 = S$$

$$2.432 = S, \text{ selling price per box}$$

$$\text{Total selling price} = 2.432 \times 50 = \$121.60 \qquad ■$$

Chapter 3

TRY THESE PROBLEMS (Set III)

Solve the following problems. Round dollar amounts to the nearest cent where necessary.

1. A grocer bought 100 pounds of fruit at 70¢ a pound. From experience it is known that 10% of this will spoil before it can be sold. If the usual markup is 48% of cost:

   a. Find the total cost.

   b. Find the total selling price.

   c. How many pounds will be available for sale?

   d. What price per pound should be charged for this fruit?

2. A bakery selling bread will take back any of its goods that are not sold by grocers within 4 days. The cost of production is 80¢ a loaf. The company distributes 500 loaves a week and about 8% are returned because they are in the grocers' stores too long. What price must each loaf be if the company's markup is 45% of cost?

3. The Greenhouse paid $5 each for 50 small rosebushes. They expect the sale of the entire lot to give them a markup of 36% of selling price. However, they guarantee the plants for 2 weeks and refund the purchase price if the bush dies. They expect about 6% to be returned under this guarantee.

   a. Find the total cost and total selling price.

   b. How many plants are expected to live and not be returned?

   c. What should the price of each plant be so that the company will get the required markup for the entire lot?

4. A delicatessen bought 45 pounds of cole slaw for 74¢ a pound. In warm weather 5 pounds usually spoils before it can be sold. What price per pound will give the desired markup of 40% of selling price?

5. The Value-Line Dairy purchased 400 quarts of milk at 60¢ each. Their markup is 65% of cost. If $7\frac{1}{2}$% of the milk will spoil before it can be sold, what regular price per quart will give the desired markup?

6. Quick Shop Fruit Store purchased 10 crates of berries at $28.80 a crate. Each crate contains 30 boxes of berries. The store has found that about 10% of the berries spoil and cannot be sold. What should be the price for each box of berries in order to obtain a markup of 55% of the selling price?

3.5 MARKDOWNS

Few businesses can count on selling all their products at a previously determined price, even if they are not perishable. In special promotions, with coupons and rebates, or in special sales like seasonal clearances, some portion of the merchandise will be sold below the original price. How does the merchant provide for this **markdown** and still obtain the markup originally set to cover expenses and net profit?

Let us start with a cost of $48 for a dress and a store markup of 40% on the selling price. The selling price is determined by:

$$C + M = S$$

$$48 + 0.4S = S$$

$$48 = S - 0.4S \quad \text{Subtract } 0.4S \text{ from both sides of the equation.}$$

$$48 = 0.6S \quad \text{Combine like terms: } 1.0 - 0.4 = 0.6.$$

$$\frac{48}{0.6} = \frac{0.6S}{0.6} \quad \text{Divide by the coefficient of the unknown.}$$

$$\$80 = S$$

This procedure is adequate if there is no markdown of merchandise. But if part of the stock is sold below $80, then the markup on the entire stock will be less than 40% on selling price.

For example, suppose the entire lot consists of 100 dresses. If they are priced at $80 each and all sold at that price, the store will take in $8,000 ($80 × 100). But this is not a realistic situation. Style merchandise will, to some extent, be sold below the original price. In this case, perhaps 20 dresses are finally sold at $60 instead of the original $80. These 20 bring in $1,200 (20 × $60). The total amount of money realized for the entire 100 dresses is:

$$80 \times 80 \text{ (the original price)} = \$6,400$$
$$\underline{20 \times 60} \text{ (the reduced price)} = \underline{1,200}$$
$$100 \qquad \text{Total} = \$7,600, \text{ compared with } \$8,000 \text{ originally expected.}$$

What markup percent on sales does this represent?

Percent markup can be found for a total number of items by the same ratio that was used for a single item. It is necessary, however, to use *total* sales, *total* costs, and *total* dollar markup.

$$M = S - C$$

$$= 7,600 - 4,800 \quad \text{Total cost} = 100 \times \$48.$$

$$M = \$2,800$$

$$\%M \text{ on } S = \frac{\text{Total } \$M}{\text{Total } S}$$

$$= \frac{2,800}{7,600}$$

$$\%M \text{ on } S = 0.368 \text{ or } 37\%$$

The markup of 37% on selling price is less than the 40% needed to cover costs and net profit. However, in setting the price originally, the merchant can allow for markdowns to be sure he realizes the 40% on selling price that he needs. He knows from experience approximately what proportion of the merchandise will be sold on sale and at about what price it will have to be sold. If he knew exactly how many and exactly which styles and sizes would not sell at the original price, he would not have the problem. But he cannot know this, and it is far better merchandising practice to have some excess than to lose a sale and probably a customer by not having enough or the right merchandise.

In the illustration, if the retailer estimates in advance that 20 dresses may be sold at $60, and he needs to realize $8,000 from the entire sale, then:

$$\$8,000 - \$1,200 \text{ (from the 20 dresses at } \$60) = \$6,800$$

This amount must be obtained from the sale of the other 80 dresses (100 − 20). Therefore, the original price must be:

$$\frac{\text{Value of sales at original price}}{\text{Number sold at original price}} = \frac{6,800}{80} = \$85$$

Using this original price of $85, the percent markup on sales is:

$$\%M \text{ on } S = \frac{\$M}{S}$$

$$= \frac{85 - 48}{85} \qquad \qquad \$M = S - C; \; S = \$85;\; C = \$48.$$

$$= \frac{37}{85} = 0.435$$

$$\%M \text{ on } S = 44\%$$

The store must use a markup of 44% on selling price to determine its original price in order to obtain 40% on total sales.

■ **Example 1**

ABC Department Store purchased 200 lamps at $55 each. The store expects to sell most at the original price but estimates that about 20% will have to be sold at $80. What must the original ticket price be if the store must have a 45% markup on selling price for the total sale?

1. Find the total sales value the store must obtain. If all the lamps could be sold at the original price, the selling price would be:

$$C + M = S$$

$$55 + 0.45S = S$$

$$55 = S - 0.45S \qquad \qquad \textit{Subtract 0.45S from both sides of the equation.}$$

$$55 = 0.55S \qquad \qquad \textit{Combine like terms: } 1.00 - 0.45 = 0.55.$$

$$\frac{55}{0.55} = \frac{0.55S}{0.55} \qquad \qquad \textit{Divide by the coefficient of the unknown.}$$

$$S = \$100$$

Total value of sales would be 200 × $100 = $20,000.

2. Find how much must be obtained from the lamps sold at the original price.

   a. Number to be sold at markdown price: 20% of 200 = 0.2 (200) = 40 lamps.

   b. Value of markdown sales: number × price = 40 × $80 = $3,200.

   c. Value of sales at original price, required total − total from markdown sales: $20,000 − $3,200 = $16,800.

3. Find the original price needed to give $16,800.

$$\text{Price per lamp} = \frac{\text{Value of sales at original price}}{\text{Number sold at original price}}$$

$$= \frac{16,800}{160}$$

200 bought, less 40 sold at the markdown price = 160.

Price per lamp = $105

Check: The price of $105 is somewhat higher than the $100 price needed if there were no markdowns. This is a reasonable answer. If more of the lamps were to be marked down, or if the markdown price was below $80, the answer would have to be more than $105. ∎

■ **Example 2**

The Kool Co. produced 250 small electric fans at a cost of $12 each. They hope to sell them all at a markup of 60% on cost. However, fans are a seasonal item, and they wish to allow for a markdown sale of 10% of the fans at $14 each. What should the original price be in order for the company to make 60% on cost for the total sale?

1. Find the total sales value the company must obtain. If all the fans were sold at the original price, the original price would be:

$$C + M = S$$
$$12 + 0.6\,(12) = S$$
$$12 + 7.20 = S$$
$$\$19.20 = S$$

Total value of sales without markdown would be 250 × $19.20 = $4,800.

2. Find how much must be obtained from the fans sold at regular price.

   a. Number sold at markdown price: 10% of 250 = 0.10 (250) = 25 fans.

   b. Value of markdown sales: 25 × $14 = $350.

   c. Value of sales at original price, required total − total from markdown sales: $4,800 − $350 = $4,450.

3. Find the original price needed to give $4,450.

$$\text{Price per fan} = \frac{\text{Value of sales at original price}}{\text{Number sold at original price}}$$

$$= \frac{4,450}{225}$$

250 produced, less 25 sold at markdown = 225.

Price per fan = $19.78

The repeating decimal, 19.777 . . . , is rounded to $19.78.

Check: A price of $19.78 is somewhat higher than the $19.20 that would have been high enough without markdowns. ∎

■ **Example 3**

A realistic situation might combine markdown with spoilage. A local farmer packs 300 boxes of strawberries for sale at his own roadside stand. He estimates the cost to him at $1.12 per box and wishes to clear 50% on cost for the entire lot. However, 5% are not top quality and will have to be sold at cost. An additional 5% will probably spoil by the end of the day and not be saleable. What price per box should he set?

1. Find the total sales amount the farmer must obtain. If all the strawberries were sold at one original price, the selling price would be:

$$C + M = S$$
$$1.12 + 0.5\,(1.12) = S$$
$$1.12 + 0.56 = S$$
$$\$1.68 = S$$

Total value of sales would be $\$1.68 \times 300 = \$504$.

2. Find how much must be realized from the boxes sold at the original price:

    a. 5% of 300 boxes will spoil; $0.05\,(300) = 15$ boxes will spoil and bring in no money at all.

    b. An additional 5%, or 15 boxes, will be sold at cost, which is $1.12; this sale will bring in $\$1.12 \times 15 = \$16.80$.

    c. 300 boxes less 15 that spoiled and 15 sold at cost (marked down) leaves 270 boxes to be sold at the regular price ($300 - 15 - 15 = 270$).

    Value of sales at original price, required total − total from markdown sales: $504 − $16.80 = $487.20.

3. Find the original price.

$$\text{Price per box} = \frac{\text{Value of sales at original price}}{\text{Number of boxes sold at original price}}$$

$$= \frac{487.20}{270}$$

Price per box = $1.804

The strawberries could be priced at $1.80 per box to give the required markup, but a more probable price is $1.89 or $1.99. This would give the farmer a little more insurance against loss and promote larger sales.

Check: The calculated price of $1.80 is higher than the $1.68 needed without allowance for markdowns and spoilage. ∎

3.6 BREAKEVEN POINT

In discussing markdowns, we have considered how a merchant allows for expected reductions in the original selling price and still realizes his required markup. However, not all reductions can be anticipated. Bad weather will affect the sales of many products, a new style may not sell as well as anticipated, or an unexpected competing product may cut into sales. The result can be that some sales produce no net profit and may even result in a loss.

Markup is planned to cover overhead (or operating expenses) and net profit. If the selling price is cut it may reduce the net profit, or if only overhead is covered, there is no net profit. In that case we say that the sale was made at the **breakeven point,** that is, there is no profit and no loss. If the selling price is cut further there is an **operating loss.** Any sale below actual cost results in an **absolute (or gross) loss.**

To illustrate these conditions, we will use a camera that cost a store $75 and is originally marked to sell for $130. The store estimates that the overhead for the camera is $25. The equation that we use to describe the profit and loss relationship for the camera is:

$$C + OH + NP = S$$

However, since we are interested in whether there is a profit or loss rather than in determining the selling price, we will use the equation in a more convenient form.

$$NP = S - C - OH$$

1. If the store were to sell the camera at the original selling price of $130 it realizes a net profit of $30.

$$NP = S - C - OH$$
$$NP = 130 - 75 - 25 = \$30$$

2. When the camera is marked down to $100, there is neither profit nor loss.

$$NP = 100 - 75 - 25 = \$0$$

One hundred dollars is therefore the *breakeven point.* The formula for breakeven is:

$$\text{Breakeven point} = \text{Cost} + \text{Overhead}$$

3. If the store reduces the price again and sells the cameras at $85 there will be an *operating loss* of $15.

$$NP = 85 - 75 - 25 = -\$15$$

The negative value for net profit indicates a loss. It is an operating loss because the original cost of $75 has been covered by the reduced selling price, but only part of the operating expenses have been recovered. A convenient equation to use is:

$$\text{Operating loss} = \text{Breakeven point} - \text{Markdown price}$$

4. Suppose that in a final effort to sell the camera, the store advertises a half-price sale at $65. Since this markdown price is less than the cost, there is not only an operating loss but also an *absolute loss*.

$$NP = 65 - 75 - 25 = -\$35$$

Again the negative value for net profit indicates an operating loss, in this case a loss of $35 per camera. Part of this loss is the absolute loss of $10, the difference between the cost ($75) and the markdown price ($65).

$$\text{Absolute loss} = \text{Cost} - \text{Markdown price}$$

This last situation is diagrammed as follows:

```
                                              Breakeven
                                              Point = $100
                            Absolute              ↓
                            Loss = $10
     |─────── Cost = $75 ───────⏜──| Overhead = $25 |
                                  ↑ ⏝⏝⏝⏝⏝⏝⏝⏝⏝
                                     Operating Loss = $35
                                Markdown
                                Price = $65
```

∎

106 CHAPTER 3 PRICING A PRODUCT

■ **Example 1**   The Exercise Shop bought 50 exercise bikes at $110 each. The overhead per bike was estimated at $40.

   1. What selling price would equal the breakeven point?

$$\text{Breakeven point} = \text{Cost} + \text{Overhead}$$
$$= 110 + 40$$
$$\text{Breakeven point} = \$150$$

   2. If the selling price is reduced to $125, what is the operating loss?

$$\text{Operating loss} = \text{Breakeven point} - \text{Markdown price}$$
$$= 150 - 125$$
$$\text{Operating loss} = \$25$$

   3. If the bike is sold at $99, how large is the absolute loss?

$$\text{Absolute loss} = \text{Cost} - \text{Markdown price}$$
$$= 110 - 99$$
$$= \$11$$

Now go to Try These Problems (Set IV).

NEW FORMULAS IN THIS CHAPTER

- $M = OH + NP$ (Markup = Overhead + Net profit)
- $C + M = S$ (Cost + Markup = Selling price)
- $\%M \text{ on } C = \dfrac{\$M}{C}$ $\left(\% \text{ Markup on cost} = \dfrac{\text{Dollar markup}}{\text{Cost}}\right)$
- $\%M \text{ on } S = \dfrac{\$M}{S}$ $\left(\% \text{ Markup on selling price} = \dfrac{\text{Dollar markup}}{\text{Selling price}}\right)$
- $NP = S - C - OH$ (Net profit = Selling price − Cost − Overhead)
- Breakeven point = Cost + Overhead
- Operating loss = Breakeven point − Markdown price
- Absolute loss = Cost − Markdown price

Chapter 3

TRY THESE PROBLEMS (Set IV)

Solve the following problems. Round dollar amounts to the nearest cent where necessary.

1. A T-shirt manufacturer is pricing the season's new styles. A shirt costs $15 to produce, and the normal markup is 55% of cost. This year 500 garments are produced, and at the end of the season any leftover merchandise will be sold for $18 each. If 11% are sold at the reduced price:

    a. Find the total sales value the manufacturer must obtain.

    b. How many will be sold at the markdown price?

    c. What is the value of markdown sales?

    d. How much money must be obtained from the merchandise that is sold at the original price?

    e. How many are sold at the original price?

    f. What price should be marked on each shirt that is sold at the original price?

2. The Pottery Shed always runs a summer sale in which they mark 15% of their merchandise at cost. If they purchased 100 bowls at $2.40 each and they use a markup of 40% of selling price:

    a. How much money must they obtain from the sale of all the bowls?

    b. What is the total value of the marked down goods?

    c. How much must be obtained from the bowls that are sold at regular price?

    d. How many will be sold at the regular price?

    e. What is the regular selling price for each bowl?

3. A supermarket bought 120 pounds of strawberries at 75¢ a pound. From past experience they know that 5% of these will become overripe and have to be sold at 30¢ a pound. At what price per pound would the store make 35% on cost?

4. A retailer sells sweaters with a markup of 45% of selling price. Each sweater costs $22, and about 10% of them will be sold at $5 above cost after the Christmas season. If 10 dozen sweaters are purchased, what original price on each one will give the required markup?

5. The Fish Market bought 200 pounds of fish at $3.60 a pound and will price the fish to make 36% of selling price as a markup. However, there is usually some fish left at the end of the day. Since this is a very perishable product, they sell about 10% of the fish to the diner next door and discard 4% of the fish. They charge the diner $3.80 a pound.

   a. How much money must they obtain from the sale of all the fish?

   b. How many pounds will be sold to the diner?

   c. How much money will they obtain from the sale to the diner?

   d. How many pounds will be discarded?

   e. How many pounds will be available to be sold at full price?

   f. At what price per pound should they sell these fish to get their desired markup on the entire lot?

6. The Bagel Factory baked 200 dozen bagels at a cost of $3.60 a dozen. Five dozen bagels become stale and cannot be sold, and 12 dozen are sold at a reduced price of $1.80 a dozen. Find the selling price per dozen bagels that will give the bakery their required markup of $66\frac{2}{3}\%$ of cost.

7. A discount food store purchased 10 cases of canned soup at $8.40 a case. Each case contained 24 cans, and when the cases were unpacked it was found that two cases contained dented cans that could not be sold at the regular price. The store manager decided to mark the dented goods at 35¢ a can. If the store uses a markup of 40% of selling price, at what price should the good cans be sold so that they can maintain their markup on the entire shipment?

Chapter 3

TRY THESE PROBLEMS (Set IV)

8. A computer store, enthusiastic about a new software program, stocked 50 copies. The cost to the store was $145 per package and the marked selling price was $295. The overhead per package is $70.

   a. At what selling price would the store break even?

   b. If the selling price is reduced to $199, does the store have a net profit or an operating loss? How large?

   c. Below what selling price would there be an absolute loss?

9. The Game Store purchased 200 Super-Video game sets at $170 each and planned to sell them at a 58% markup on selling price. One hundred forty games were sold at the original price but the remaining ones were sold at $169. The overhead per game is $50.

   a. What was the original selling price?

   b. Was the sale price at, above, or below the breakeven point?

   c. Did the sale price give an operating loss or an absolute loss? How much?

   d. Compute the net profit or loss for the entire shipment.

Chapter 3

	Name	Date	Class

END-OF-CHAPTER PROBLEMS

Complete the following problems by filling in the blank spaces. Round answers to the nearest cent where necessary.

	COST	OVERHEAD	NET PROFIT	MARKUP	SELLING PRICE
1.	$35.00	$9.40	$7.50		
2.		$10.50	$8.00		$58.50
3.	$12.00	$6.18			$22.50
4.	$39.50		$13.60	$25.00	
5.		$15.25		$22.00	$42.00
6.			$9.35	$17.50	$48.00
7.	$60.00	20% of C	12% of C		
8.		32% of S	18% of S		$125.00
9.	$200.00	26% of C			$375.00
10.	$45.00	28% of S	27% of S		
11.	$85.00	25% of S	$12\frac{1}{2}$% of S		
12.		30% of C	10% of C		$84.00
13.			13% of C	30% of C	$65.00

111

	COST	OVERHEAD	NET PROFIT	MARKUP	SELLING PRICE
14.			$16.95	42% of S	$75.00
15.	$42.50		10% of S		$80.00

Complete the following problems by filling in the blank spaces. Round dollar amounts to the nearest cent and rates to the nearest tenth of a percent where necessary.

	COST	MARKUP	SELLING PRICE	%M on C	%M on S
16.	$70	$14			
17.	$56		$65		
18.		$34	$92		
19.	$45	$30			
20.	$57			$33\frac{1}{3}\%$	
21.			$72		48%
22.			$120		45%
23.	$37			60%	
24.	$70		$145		
25.			$40		23%
26.	$180				64%

Chapter 3

END-OF-CHAPTER PROBLEMS

	COST	MARKUP	SELLING PRICE	%M on C	%M on S
27.			$240	25%	
28.			$32.50	30%	
29.	$24				40%
30.		$12		20%	
31.		$4.20			35%
32.		$20		50%	
33.		$8.40			70%

Complete Problems 34 through 36 by filling in the blank spaces. Round answers to the nearest cent where necessary.

	QUANTITY BOUGHT	COST PER UNIT	TOTAL COST	MARKUP	TOTAL SALES	AMT. TO SPOIL	NO. TO SELL	SELLING PRICE PER UNIT
34.	50	$3 ea		40% of S		4%		
35.	250 lb	$0.45 lb		45% of C		10%		
36.	90 dz	$0.85 dz		32% of C		3 dz		

In Problems 37 through 39 find:

    a. Total cost.

    b. Total sales.

113

c. Value of reduced sales.

d. Amount at regular price.

e. Regular selling price.

Round dollar amounts to the nearest cent where necessary.

	QUANTITY BOUGHT	COST PER UNIT	MARKUP	AMOUNT REDUCED IN PRICE	REDUCED PRICE
37.	300 lb	$0.70 lb	30% of C	5%	$0.50 lb
38.	75	$2.50 ea	70% of S	5%	$2.00
39.	40 dz	$8.00 dz	52% of C	10%	$7.00 dz

Solve the following problems. Round dollar amounts to the nearest cent and rates to the nearest tenth of a percent where necessary.

40. A hardware store owner bought a shovel at $10.80. He expects to make a net profit of $4 and sell the shovel for $17.98. How much has he allowed for overhead expenses?

41. A music store sold a used electric guitar for $100. If the guitar cost $85 and expenses are 20% of sales, how much profit was made on the sale?

42. An electric can opener cost $17.80 to produce and is sold with a 70% markup on cost. What is the selling price?

43. A retailer expects to sell a sofa for $550 and to obtain a markup of 60% on selling price. What is the most she can pay for the sofa?

44. A manufacturer of canned goods uses a markup of 28% of cost to price his merchandise. If a can of fruit cost $1.69 to produce, at what price should it be sold?

Chapter 3

END-OF-CHAPTER PROBLEMS

45. Peoples Department Store sells a set of knives for $29.95. They take a markup of 40% on selling price in this department. What is the most they can afford to pay for the set?

46. A refrigerator that cost $218 was sold for $429.99. What is the percent markup on: (a) cost, and (b) selling price?

47. A kitchen chair that cost $29.32 was sold for $52.95. Find the percent markup on: (a) cost, and (b) selling price.

48. A calculator cost a store $7 and was originally marked to sell for $14. However, there were competing calculators available retailing at $12 and the store had to cut its price to meet competition.

  a. What percent reduction in price had to be made?

  b. What percent markup on selling price did the store obtain at the new price?

49. A wholesaler sells a set of glasses at $18. If this includes a markup of $33\frac{1}{3}\%$ on cost, how much did the glasses cost?

50. Edison Manufacturing sold a wall unit for $918. If they allow 20% of cost for overhead expenses and 15% of cost for net profit, find the cost.

51. A manufacturer wants to produce a floor lamp to sell for $180. He normally operates with a markup of 25% of cost. How much can he afford to pay for producing the lamp if he wishes to get his usual markup?

52. The manufacturer in Problem 51 sold several lamps to Supreme Lighting Company, which added a markup of $120. What percent markup does this represent on: (a) cost, and (b) selling price?

53. A man's jacket is to be sold for $50. The producer allows 40% of cost as overhead and 22% of cost as net profit. What is the cost?

54. Eddie's Electronics uses a markup of 25% of selling price. A desktop computer cost them $660. Find the selling price.

55. An auto supply store has a markup of 38% of selling price. If the cost of a wheel rim is $36.50, find: (a) the selling price, and (b) the percent markup on cost.

56. Shutter Shop bought a camera for $89.32 to be sold at a net profit of 12% on selling price. If they allow 30% of selling price for overhead, what should the selling price of the camera be?

57. A furniture store bought a picture for $33.80. If the store needs 23% of the selling price to cover overhead expenses and 12% of the selling price for net profit, find: (a) the selling price, and (b) the percent markup on cost.

58. The Ski Barn knows that their customers will not pay more than $28 for a ski hat. If their markup is 40% of cost, what is the most it can cost them to purchase these hats?

59. Plates are sold for $9 when the markup is 70% of cost. What is the cost of each plate? What is the percent markup on selling price?

60. A retailer who uses a markup of 40% of selling price bought a gold bracelet for $108. At what retail price should he sell the bracelet? What percent markup on cost does this represent?

61. An iron was marked up 50% of cost. If the markup was $11, find: (a) the cost, (b) the selling price, and (c) the percent markup on selling price.

62. A home supply company uses a markup of 18.5% of selling price. A wading pool was marked up $2.22. Find: (a) the selling price, (b) the cost, and (c) the percent markup on cost.

63. A hosiery manufacturer adds a markup of 58% of cost to determine selling price. One item in her line costs $6 a dozen to produce. The retailer who buys the hosiery adds a markup of 38% of selling price. At what price per pair should the hosiery be sold at retail to give the required markup?

Chapter 3

Name Date Class

END-OF-CHAPTER PROBLEMS

64. Have-A-Hobby Shop sells a woodworking vise for $26.00. For a special Father's Day sale they reduced the price to $20. If the normal markup was $37\frac{1}{2}\%$ of selling price: (a) how much did the vise cost the store, and (b) what percent markup on selling price will the sale price give them?

65. Window Wonders, Inc. will feature drapes in a special sale at $50. Since they expect a large volume at that price, their supplier agreed to cut the price to the store from $40 to $35.

    a. What percent markup on selling price will they get?

    b. What percent reduction in price does the cut from $40 to $35 represent?

66. The Greenhouse bought 200 small plants at $1.08 each. If they allow for 15 plants to die before being sold, at what price must they sell the healthy plants if they require a markup of 46% of cost?

67. A farmer operating a roadside stand has 80 boxes of tomatoes for sale. He hopes to sell the entire lot at a markup of 55% on his cost of 60¢ a box. However, he usually has some produce left at the end of the day, and he allows for 5 boxes to be discarded. At what price should he sell the tomatoes to get his required markup?

68. Plants-N-Things purchases 40 flowering azalea plants at $5 each for Mother's Day. Last year about 5% of their purchases lost their bloom and could not be sold. What must they charge per plant in order to make 75% on selling price?

69. Jules Specialties buys 500 costumes at $20 each. They find that 8% of the costumes are damaged and cannot be sold. If they use a markup of 36% of selling price, what ticket price must they put on the costumes that are saleable?

70. A gift shop bought 75 picture frames at $4 each for a special Christmas promotion. The owner allowed for 15 to be left after Christmas and to be sold marked down at $4.20. At what price would he have to sell the frames before Christmas if he expects to make 40% of selling price as a markup?

71. Snappy Clothes, Inc. bought 200 dresses in assorted styles at $64 each. The manager believes that if the weather is normal the store will sell 85% of the dresses at the original marked price. The other 15% will be sold at $50. What should the original marked price be if the store is to get its normal markup of 36% on selling price for the entire shipment?

72. Ace Tire recaps tires at a cost of $20 per tire. Past experience shows that 12% of the recapped tires must be sold as seconds for $25. If they recap 400 tires and a markup of 100% of cost is desired, find the regular selling price per tire.

73. An athletic shoe manufacturer makes a running shoe at a cost of $38 per pair. Based on past experience, 9% of the shoes will be defective and must be sold as irregulars for $48 per pair. If the manufacturer produces 500 pairs of shoes and uses a markup of 80% of cost, what is the selling price per pair?

74. Fancy Foods, Inc. decided to add fresh bean sprouts to their line of vegetables. They started with a shipment of 50 packages for which they paid 78¢ a package. Since they had no experience with demand for this item, they allowed 20% for spoilage. What price should they charge to get a markup of 53% on cost?

75. A & B Supermarket tried the same bean sprouts as Fancy Foods, Inc. (Problem 74), but they allowed for 10% to be sold at a reduced price of $1.17 and only 10% to be discarded. At what price should they sell the packages to realize the same markup of 53% of cost for the 50 packages they had for sale?

76. A supermarket paid $30 per box for 5 boxes of lettuce with 50 heads in each box. When the boxes are opened, it is found that about 10% of the heads are small. What original price would give the store a markup of 65% on cost if the small heads are sold at 75¢?

77. A bakery produced six dozen muffins at a cost of 45¢ each. If 10 muffins will be sold "day old" at 50¢ each, what must the regular price be if the bakery is to make 42% on cost as markup for all the production?

78. A store sells yogurt marked with a last date of sale. The owners pay 31¢ a container and use a markup of 38% of selling price. On the last day of sale they reduce the price to 33¢, and continue to sell them for an additional two days. After that any remaining containers are discarded. In a shipment of 100 containers they expect to sell 5% at the reduced price and to discard about 3%. What original price would give the store the required markup for the entire shipment?

79. Howell Meats buys 200 pounds of ground beef each week at $1.85 a pound. On weekends 10% of the meat is placed on sale at $2.25 a pound. By the end of the weekend an additional 5% of the meat is thrown out due to spoilage. If Howell Meats maintains a 30% markup on cost, find the regular price of a pound of ground beef.

80. A delicatessen made 40 brownies at a cost of 25¢ each. About 15% of these will be sold "day old" at 10¢ each and another 5% will be discarded. Find the regular selling price for the delicatessen to make 35% on cost.

Chapter 3

END-OF-CHAPTER PROBLEMS

81. The Eat-Well Supermarket received a shipment of 10 cases of canned pears. Each case contained 24 cans and cost $18.40. During the unloading some cases dropped and 15% of the cans were dented. The pears usually sell for 98¢ a can, and the manager decided to sell the dented stock at 78¢. What percent markup on selling price will the store realize for the entire shipment?

82. A bookstore sold a group of leftover books on a special counter at $5 each. There are 100 books in the lot. Of these, 30 had cost the store $2.50 each, 20 had cost $4 each, and the remainder had cost $5.

   a. How much net profit or loss would result from the sale if they allowed 20% of cost for overhead?

   b. Find the percent markup on selling price.

83. A store bought 100 coats at $158.15 each and planned to sell them at a 53% markup on selling price. Eighty-five coats were sold at the original selling price but the remaining ones were sold at $185. Operating costs are estimated at 20% of cost.

   a. What was the original selling price?

   b. Was the sale price at, above, or below the breakeven point?

   c. Did the sale price give an absolute loss or an operating loss? How many dollars was this?

   d. Compute the net profit or loss for the entire lot of 100 coats.

84. A remote-controlled toy truck is sold for $49.95 during the Christmas season. The store adds $12 for operating expenses to the cost of $25.

   a. What percent of selling price is the net profit?

   b. On January 1, the truck is put on sale priced at the breakeven point. What is the selling price?

   c. If the last of the trucks must be reduced to $23.95, how large are the operating loss and the absolute loss?

SUPERPROBLEM

Solve the following problem. Round dollar amounts to the nearest cent and rates to the nearest tenth of a percent where necessary.

The manufacturer's selling price for a dress is $104, which includes a markup of 60% on cost. A retailer bought 50 of the dresses and planned to sell them at $179.95. However, they did not sell well and 20% were sold in a special promotion at $150.

  a. Find the manufacturer's cost.

  b. What markup on selling price did the retailer get for the entire lot of 50 dresses?

  c. What original price would have given the retailer the 42% markup on selling price that he usually gets?

Chapter 3

TEST YOURSELF

Complete the following problems by filling in the blank spaces. Round answers to the nearest cent where necessary.

	Cost	Overhead	Net Profit	Markup	Selling Price	Section
1.	$32	$12	$15.20	_____	_____	3.1
2.	_____	_____	$176	$480	$960	3.1

Complete the following problems by filling in the blank spaces. Round dollar amounts to the nearest cent and rates to the nearest tenth of a percent where necessary.

	Cost	Markup	Selling Price	%M on C	%M on S	Section
3.	$40	$25	_____	_____	_____	3.2
4.	$89.50	_____	_____	44%	_____	3.2

Solve the following problems. Round dollar amounts to the nearest cent and rates to the nearest tenth of a percent where necessary.

5. A leather briefcase that costs $75 to manufacture is sold with a markup of 68% of cost. What is the selling price? *3.2*

6. A set of fine china costs $500 to produce. If the manufacturer allows 40% of cost for overhead expenses and sells the china for $775: *3.1*

   a. How much is the markup?

   b. How much is the net profit?

7. A lamp currently selling for $75 has a markup of 39% of the selling price. What is the cost of the lamp? *3.2*

8. The Jeans Shop purchased a new line of pants for $36 a pair. At what price should these pants be sold if the store wishes to obtain a markup of 40% of the selling price? What percent markup on cost does this represent? *3.3*

121

	Section

9. A cassette player is sold for $56. What is the cost of this player if the markup is 40% of cost? 3.2

10. A ten-speed bicycle that cost $400 is sold for $500. 3.3

    a. What is the percent markup on cost?

    b. What is the percent markup on selling price?

11. The Green Market bought 200 pounds of vegetables at $1.20 a pound. The owners expect about 10% to spoil before they can be sold. What price per pound will give the required markup of 45% of cost? 3.4

12. The Cookie Corner bought 50 boxes of chocolate chip cookies at $1.10 a box. If business is slow, they will have to mark down at least 8 boxes to $2.00 a box. What price should they mark all the boxes so that they are sure to get their usual markup of 75% of selling price? 3.5

13. Finest Foods bought 120 pounds of fruit at $1.30 a pound. Based on past experience, they expect to sell about 10% at a reduced price of $1.59 a pound, and it is likely that 5% will spoil and have to be thrown out. At what price should they mark the fruit so that they will obtain their usual markup of 60% of cost? 3.5

14. An outdoor swing set costs a store $620 and the overhead per set is estimated at $75.

    a. What selling price would equal the breakeven point? 3.6

    b. What percent markup on cost would this selling price give the store? 3.3

    c. If a set is sold at half of the breakeven price at the end of the season, how large is the absolute loss? 3.6

Chapter 3

Name _____ Date _____ Class _____

COMPUTER SPREADSHEET EXERCISES

Dan's Health Food Store has 10 items that it presently keeps in stock. The item code and the cost of each item are given in the following spreadsheet.

Item	Cost	Markup	Selling Price
A6413	$3.50		
B2104	$1.95		
G3200	$6.50		
C9110	$12.17		
J2511	$8.95		
B5645	$21.15		
K7899	$42.33		
F1011	$2.81		
A6207	$1.06		
R5048	$7.25		

When completing each of the following three spreadsheet exercises, use $ signs with all numbers that represent dollars and cents, and round all dollar amounts to the nearest cent. Enter formulas to find the missing numbers. Make sure that cell entries are located beneath the correct column headings and that decimal points in columns are aligned.

1. If Dan requires a 60% markup on cost to cover his overhead expenses and net profit, complete the above spreadsheet.

2. Given the same Item and Cost columns, construct another spreadsheet for Dan with identical column headings if he now needs a markup of 40% on selling price for each of the 10 items.

3. Dan sells the following amounts of each item:

Item	Number Sold
A6413	3
B2104	5
G3200	5
C9110	0
J2511	2
B5645	2
K7899	3
F1011	6
A6207	6
R5048	3

Prepare a new spreadsheet by adding three columns to the spreadsheet obtained in Exercise 2. Enter the number sold (use integers) in the first new column, the total cost for each item sold (cost × number sold) in the second, and the total $ sales per item (selling price × number sold) in the third. Give totals for each of the new columns. When completed, your spreadsheet will contain seven columns, with totals shown for the last three.

4 ■ Commercial Discounts

OBJECTIVES

Upon completion of this chapter, the student should be able to:

1. *Find the list price or net price using a single trade discount or a series of discounts.*
2. *Find the percent of trade discount when the list price and net price are given.*
3. *Find the single equivalent discount and the net cost rate factor.*
4. *Find the additional percent discount that will make two prices equal.*
5. *Use sales terms given on an invoice to determine:*
   whether a cash discount can be taken;
   the amount of the cash discount;
   the net payment;
   the amount credited; and
   the amount still due.
6. *Find the amount credited and the amount still due when a partial payment is made.*

Manufactured products usually go through several hands before they reach us. The **producer** (manufacturer) may sell directly to a retail business or store, which then sells to the consumer. Sometimes a producer may sell a product to a **wholesaler** (a business that buys merchandise to sell to other businesses, and not to the general public). The wholesaler then resells the merchandise to various **retailers.** This is one single marketing path out of many variations. The path is diagrammed as follows:

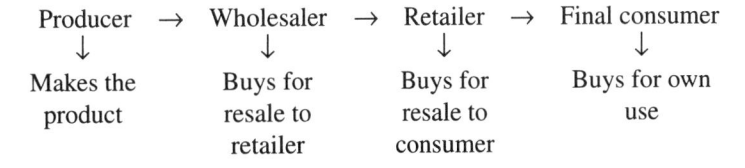

Producer	→	Wholesaler	→	Retailer	→	Final consumer
↓		↓		↓		↓
Makes the product		Buys for resale to retailer		Buys for resale to consumer		Buys for own use

4.1 TRADE DISCOUNTS

In Chapter 3 we considered some of the factors that determine product prices. For instance, in Section 3.2, Example 2, a coat manufacturer decided to sell a given type of coat at $75. A company may, however, sell the same article at different prices to different customers under special conditions, for example:

- If a producer sells to retailers as well as to wholesalers, the price to the wholesaler is usually lower than to the retailer.
- A customer buying an unusually large quantity will expect a price reduction.
- A producer may wish to carry out a joint advertising program with a particular retailer. The retailer may be given a price allowance as the producer's contribution to the joint program.

Thus, the stated price ($75) may be paid by some customers, but less will be paid by others. The amount deducted is called a **trade discount.** Transactions in which trade discounts are given may involve producers, wholesalers, and retailers, but not consumers.

Trade discounts are also used in industries where the products are so numerous that they are shown in catalogs to which customers refer when ordering merchandise. The hardware trade is a good illustration. The list prices in the catalog may be suggested retail prices and are high enough to allow for trade discounts to be given. Changing the price becomes a simple matter of changing a discount.

The term **list price,** narrowly used, means the suggested retail price set by the manufacturer, sometimes printed on the package. It is the manufacturer's effort to prevent price cutting for certain products. The price paid by the retailer or wholesaler is then determined by taking a trade discount off this list price. We will use the term list price in a broader sense to include any quoted price.

A *trade discount is always a percent of the stated or list price.* It is deducted from the list price to determine what the customer actually pays. The amount paid is the **net price** (to the seller) or the **net cost** (to the buyer).

4.2 COMPUTING DISCOUNT AND NET PRICE

We may use the basic percent equation to determine the amount of a trade discount.

$$R \times B = P$$
$$\downarrow \quad \downarrow \quad \downarrow$$

% Discount × List price = Amount of discount
$$\downarrow \quad \quad \downarrow \quad \quad \downarrow$$

or

$$\%D \times L = \$D$$

Note that when the *discount percent* is used as the rate, the product is the *amount of discount.*

The net price is now found by subtracting the discount from the list price. In other words,

---

List price − Discount = Net price

or

$$L - \$D = N$$

■ Example 1

A wholesaler bought a file cabinet from a manufacturer. It is listed in the catalog at $59, less a trade discount of 23%. What is the net price?

First let us find the amount of discount.

$$\%D \times L = \$D$$
$$0.23 \times 59 = \$D$$
$$\$13.57 = \$D$$

Using the net price equation, we obtain:

$$\text{List} - \text{Discount} = \text{Net price}$$
$$59.00 - 13.57 = \text{Net price}$$
$$\$45.43 = \text{Net price}$$

Rewriting this last equation in vertical form, we have:

	IN DOLLARS	IN PERCENT
List price	$59.00	100%
− Discount	− 13.57	− 23%
Net price	$45.43 (amount paid)	77% (percent paid)

The list price is always the base, or 100%. When a trade discount is given, the list price is divided into two parts—the discount and the part that is actually paid (net price). These two parts must add up to the list price, whether they are given in dollars or in percents. When given in percents, the two parts are said to be *complements* because they add up to 100%.

We now can find the net price directly by using the percent paid as the rate.

%Paid × List price = Amount paid (net price)

or

%Pd × L = N

Substitution of the previous data into the formula gives

$$0.77 \times 59 = N$$

77% is the complement of the 23% discount.
(100% − 23% = 77%)

$$\$45.43 = N$$

When the *percent paid* is used as the rate, the product is the *amount paid.*
The amount of discount can be found by subtracting the net price from the list price.

List price − Net price = Discount

This is the net price equation solved for the discount.

$$59.00 − 45.43 = \$D$$

$$\$13.57 = \$D$$ ∎

■ **Example 2**   A box of dusting powder has a suggested retail price of $2.50. A wholesaler who supplies many small drugstores can buy the product from the manufacturer at 55% off list price. What is the cost to the wholesaler?

%Pd × L = N

This equation gives the net price directly.

$$0.45 \times 2.50 = N$$

The percent paid is the complement of the discount.
(100% − 55% = 45%, or 0.45)

$$\$1.13 = N$$

$1.125 is rounded to $1.13.

Check: The net price is smaller than the list price. ∎

128 CHAPTER 4 COMMERCIAL DISCOUNTS

■ **Example 3** A hardware store has a choice between two metal shelves from different suppliers. One is listed at $14.25 with a trade discount of 35%. The other is listed at $13.75 with a trade discount of $33\frac{1}{3}\%$. Which will give the store the larger amount of discount? Which will give the lower net price?

There are two ways to solve this problem.

1. *First shelf:* $\%D \times L = \$D$ *This equation gives the discount.*

$$0.35 \times 14.25 = \$4.99$$

List price − Discount = Net price

$$14.25 - 4.99 = \$9.26$$

Second shelf: $\%D \times L = \$D$

$$0.\overline{3} \times 13.75 = \$4.58$$ $33\frac{1}{3}\% = \frac{1}{3} = 0.\overline{3}$

List price − Discount = Net price

$$13.75 - 4.58 = \$9.17$$

or

2. *First shelf:* $\%Pd \times L = N$ *This equation gives the net price.*

$$0.65 \times 14.25 = \$9.26$$ $100\% - 35\% = 65\% = 0.65$

List price − Net price = Discount

$$14.25 - 9.26 = \$4.99$$

Second shelf: $\%Pd \times L = N$

$$0.\overline{6} \times 13.75 = \$9.17$$ $100\% - 33\frac{1}{3}\% =$
$$66\frac{2}{3}\% = \frac{2}{3} = 0.\overline{6}$$

List price − Net price = Discount

$$13.75 - 9.17 = \$4.58$$

More discount is given on the first shelf. However, the second shelf has the lower net price.

Check: The two discounts and net prices should be very close because both the list prices and the discount percents are very close. ■

■ **Example 4** A retailer paid $7.50 for a book listed at $12. What percent was the trade discount?

$$\%D \times L = \$D$$ *Use the discount equation and solve for % discount.*

$$\%D \times 12 = 4.50$$ *List price − net price = discount.* $(12.00 - 7.50 = 4.50)$

$$\frac{\cancel{12}\,(\%D)}{\cancel{12}} = \frac{4.50}{12}$$ *Divide both sides of the equation by the coefficient of the unknown.*

$$\%D = 0.375$$

$$\%D = 37.5\%$$

Move the decimal point two places to the right and add %.

The trade discount was 37.5%.

Check: The net price is more than half the list price. Therefore, the discount must be less than 50%. ∎

4.3 COMPUTING LIST PRICE

■ **Example 1**

An office furniture manufacturer wishes to sell a desk chair at $91 and to list it in a catalog with a trade discount of 30%. What should the catalog price be?

1. To translate the problem into an equation, note that:

    a. "Sell a chair at $91" means net price = $91.

    b. "Catalog price" is the same as list price and is unknown.

2. Use the equation giving the relationship between net and list prices.

$$\%Pd \times L = N$$

$$0.70 \times L = 91$$

% paid is the complement of 30% discount.
(100% − 30% = 70%, or 0.70)

$$\frac{\cancel{0.70}L}{\cancel{0.70}} = \frac{91}{0.70}$$

Divide both sides of the equation by the coefficient of the unknown.

$$L = \$130$$

The catalog price should be $130.

Check: The list price is higher than the net price but of about the same magnitude. ∎

Chapter 4

TRY THESE PROBLEMS (Set I)

Fill in the blank spaces. Round dollar amounts to the nearest cent and rates to the nearest tenth of a percent where necessary.

	LIST	% TRADE DISCOUNT	AMOUNT OF DISCOUNT	NET PRICE
1.	$15.80	25%		
2.	$75.00	40%		
3.	$330	$33\frac{1}{3}\%$		
4.	$375			$300.00
5.			$12.00	$28
6.	$48.00			$16
7.	$66.00		$23.10	
8.		10%		$90
9.		27%		$423.40
10.	$72.00		$18.00	
11.			$2.40	$13.60

131

Solve the following problems. Round dollar amounts to the nearest cent and rates to the nearest tenth of a percent where necessary.

12. A washer that lists for $350 is sold with a 30% trade discount.

    a. How much discount is being given?

    b. What is the net price after the discount is allowed?

13. An office supply manufacturer allows a 15% discount to all customers who buy in large quantities. What net price does the ABC Company pay for 10 reams of paper that list for $15 each?

14. A pair of binoculars is listed at $84 and sold after trade discounts for $52.08.

    a. What amount has been allowed as a trade discount?

    b. What percent trade discount has been given?

15. A dictionary is listed at $18 and sold with a trade discount for $11.25. What percent trade discount has been allowed?

16. Find the net price of a lamp listed at $24 and sold to a wholesaler with a trade discount of $42\frac{1}{2}\%$.

17. What percent was allowed as a trade discount when an item that lists for $81 was sold for $74.52?

18. The manufacturer of a lighted makeup mirror wants to list it in his catalog with a 5% trade discount. If he sets a net price of $23.75, what should the catalog (list) price be?

19. At what price should a tape deck be listed in a wholesaler's catalog if he wishes to allow a trade discount of 28% and to sell the item for $72?

SERIES DISCOUNTS   133

4.4 SERIES DISCOUNTS

Trade discounts are frequently quoted in a **series** of two or more discounts rather than as a single percent. For example, a manufacturer may sell to both wholesalers and retailers. Instead of stating a single discount for each type of customer, he may specify that the retailer will be allowed 30% and that the wholesaler will get an additional 15%. A second wholesaler, buying in very large quantities, might be given an additional 5%. Accordingly,

- The retailer pays list price less 30%.
- The first wholesaler pays list price less 30%, less 15%.
- The second wholesaler pays list price less 30%, less 15%, less 5%. (The series may also be written 30%, 15%, 5%, or 30/15/5.)

For the second wholesaler this means that the net price would be determined by first deducting 30% of the list price, then deducting 15% of *what remains,* then deducting an additional 5% of *what remains after the second deduction.*

The following calculation of net price illustrates this procedure for a chair listed at $80 with trade discounts of 30%, 15%, and 5%. This problem can be solved using either $\%D \times L = \$D$ or $\%Pd \times L = N$. With the first formula we have:

$\%D \times L = \$D$ List $80
                Less 30%   −24   (30% of 80 = 24)
                                 56
                Less 15%   −8.40   (15% of 56 = 8.40)
                               47.60
                Less 5%   −2.38   (5% of 47.60 = 2.38)
                Net price = $45.22

Notice that the single discounts in the series are *never* added together.

Using $\%Pd \times L = N$, however, computes the net price more quickly.

For the first discount:

$\%Pd \times L = N$

$0.70 \times 80 = $ 1st N *%Pd is the complement of the first discount.*
                                       (100% − 30% = 70% = 0.70)

$\$56 = $ 1st N

For the second discount:

$\%Pd \times L = N$

$0.85 \times 56 = $ 2nd N *%Pd is the complement of the second discount.*
                                       (100% − 15% = 85% = 0.85)

$\$47.60 = $ 2nd N

For the third discount:

$\%Pd \times L = N$

$0.95 \times 47.60 = $ 3rd N *%Pd is the complement of the third discount.*
                                       (100% − 5% = 95% = 0.95)

$\$45.22 = $ 3rd N

134 CHAPTER 4 COMMERCIAL DISCOUNTS

Note that each of the series discounts is based on a different amount. Therefore, *the discounts may never be added and used as one discount.*

A much more convenient form of the calculation makes it possible to do different kinds of problems with one equation. The three-step solution can be rewritten as one multiplication instead of three.

$$\%Pd_1 \times \%Pd_2 \times \%Pd_3 \times L = N$$

Where $\%Pd_1$ is the first percent paid, $\%Pd_2$ is the second percent paid, and $\%Pd_3$ is the third percent paid.

This is the form that we will use to solve series discount problems. It is an expanded form of the equation used before with a single discount. When the three percents paid are multiplied together, the result is the final percent paid after taking three successive discounts.

■ **Example 1**

A cordless telephone listed at $120 is sold with discounts of 20/15/10. What is the net price?

$$\%Pd_1 \times \%Pd_2 \times \%Pd_3 \times L = N$$
$$0.80 \times 0.85 \times 0.90 \times 120 = N$$
$$0.612 \times 120 = N$$
$$\$73.44 = N$$

The net price is $73.44.

This illustration is based on a series of three discounts. The method is exactly the same whether there are two, three, or four discounts in the series. There would then be two, three, or four discount complements, respectively, to be multiplied together to give the final percent paid.

The number that is the product of all the "percents paid" is called the **net cost rate factor.** In Example 1 the net cost rate factor is 0.612. This number is multiplied by the list price to determine the net price.

All items with the same discounts have the same net cost rate factor. Thus, a buyer finds the appropriate net cost rate factor and instead of reworking the entire problem for each item, merely multiplies it by each list price to find the net price. ■

■ **Example 2**

Find the net price of an $800 stove and a $650 television that each have discounts of 30/15/10.

The net cost rate factor for these discounts would be:

$$0.70 \times 0.85 \times 0.95 = 0.56525$$

Therefore, the net price of the stove is:

$$0.56525 \times 800 = N$$
$$\$452.20 = N$$

And the TV has a net price of:

$$0.56525 \times 650 = N$$
$$\$367.41 = N$$ ■

Example 3

A furniture store bought a couch at $700 less 30/12. What was the net cost?

$$\%Pd_1 \times \%Pd_2 \times L = N$$

There are two discounts and, therefore, two complements.

$$(0.7)(0.88)(700) = N$$

The complement of 30% is 70% = 0.7. The complement of 12% is 88% = 0.88.

$$0.616 \times 700 = N$$

Multiply the complements before multiplying by the list price and do not round the decimal. This product is the net cost rate factor.

$$\$431.20 = N$$

The net cost of the couch was $431.20.

Check: The net cost of $431.20 is less than the list price of $700 but of the same magnitude. ■

Example 4

A cosmetics manufacturer wishes to sell a bottle of nail polish at a net price of $2.55 after discounts of 30/10/10. At what price should the nail polish be listed?

$$\%Pd_1 \times \%Pd_2 \times \%Pd_3 \times L = N$$

There are three discounts and therefore three complements.

$$0.7 \times 0.9 \times 0.9 \times L = 2.55$$

The complements are:
$(100\% - 30\% = 70\% = 0.7)$
$(100\% - 10\% = 90\% = 0.9)$
$(100\% - 10\% = 90\% = 0.9)$

$$0.567 \times L = 2.55$$

$$\frac{\cancel{0.567}}{\cancel{0.567}} L = \frac{2.55}{0.567}$$

Divide both sides of the equation by the coefficient of the unknown.

$$L = \$4.50$$

Division gives 4.497, which is rounded to 4.50.

The list price is $4.50.

Check: This list price is higher than the net price and therefore possible for this item. ■

CALCULATOR HINT

If your calculator has Memory keys ($\boxed{M+}$ and $\boxed{MR}$ for this example), you can use it to solve Example 4.

	ENTER	DISPLAY
0.7 $\boxed{\times}$ 0.9 $\boxed{\times}$ 0.9 $\boxed{\times}$ L $\boxed{=}$ 2.55	0.7 $\boxed{\times}$ 0.9 $\boxed{\times}$ 0.9 $\boxed{M+}$	0.567
	2.55 $\boxed{\div}$ $\boxed{MR}$ $\boxed{=}$	4.497

4.497 rounds to $4.50. This is the list price.

Remember to Clear Memory before you do another example.

136 CHAPTER 4 COMMERCIAL DISCOUNTS

■ **Example 5**   Hardware, Inc., has been selling a cordless drill to retailers at $50 less 35%. One of their customers notifies them that he can get the same drill from another supplier at $50 less 30% and 10%. What additional discount would bring the net price down to the same level as the competitor's?

There are several ways of doing this problem; a simple, direct method is shown here.

1. Determine Hardware, Inc.'s price and the competitor's price.

HARDWARE, INC.	COMPETITOR
$\%Pd \times L = N$	$\%Pd_1 \times \%Pd_2 \times L = N$
$0.65 (50) = N$	$0.7 \times 0.9 \times 50 = N$
$\$32.50 = N$	$0.63 \times 50 = N$
	$\$31.50 = N$

2. Find the discount that will reduce $32.50 to the required $31.50.

$$\%D \times L = \$D$$ *Use the discount equation and solve for % discount.*

$$\%D \times 32.50 = 1.00$$ *The price that must be reduced is the list price for this calculation. The difference between the two prices $(32.50 - 31.50 = \$1.00)$ is the amount of discount.*

$$\frac{\cancel{32.50}\,(\%D)}{\cancel{32.50}} = \frac{1.00}{32.50}$$ *Divide both sides by the coefficient of the unknown.*

$$\%D = 0.0307 = 3.1\%$$

Hardware, Inc., must offer discounts of 35% and 3.1% to meet the competitor's price.

Check: The additional discount should be small since the list prices were the same and the competitor gave a smaller first discount. ■

■ **Example 6**   Which net price is lower, $130 less 20/10, or $150 less 30/20/2? How much discount is allowed in each case?

First net price:

$$\%Pd_1 \times \%Pd_2 \times L = N$$ *There are two discounts and therefore two complements:*
$100\% - 20\% = 80\% = 0.8$
$100\% - 10\% = 90\% = 0.9$

$$0.8 \times 0.9 \times 130 = N$$

$$0.72 \times 130 = N$$

$$\$93.60 = N$$

First amount of discount:

$$\$D = L - N$$

$$= 130 - 93.60$$ *The list price is given. The net price was found in the first part of the problem.*

$$\$D = \$36.40$$

Second net price:

$$\%Pd_1 \times \%Pd_2 \times \%Pd_3 \times L = N$$

$$0.7 \times 0.8 \times 0.98 \times 150 = N$$

$$0.5488 \times 150 = N$$

$$\$82.32 = N$$

There are three discounts and three complements:

$100\% - 30\% = 70\% = 0.7$
$100\% - 20\% = 80\% = 0.8$
$100\% - 2\% = 98\% = 0.98$

Second amount of discount:

$$\$D = L - N$$
$$= 150 - 82.32$$
$$\$D = \$67.68$$

The second net price is lower. The second list price is much higher than the first, but to the buyer the important comparison is that between the two net prices.

Check: There is no check on which price should be lower, but the two prices should not be far apart, and both should be lower than their respective list prices. ■

■ **Example 7**

Lights N Such paid a manufacturer $850 less 20/10 for track lighting. If they want to maintain their usual markup of $33\frac{1}{3}\%$ of selling price, what ticket price should they use?

The solution to this problem has two parts.

1. Find the net cost of the lighting.

$$\%Pd \times L = N$$
$$0.8 \times 0.9 \times 850 = N$$
$$0.72 \times 850 = N$$
$$\$612 = N$$

The net cost is $612.

2. Find the selling price by using the net cost and the given markup.

$$C + M = S$$
$$612 + 0.\overline{3}S = S$$
$$612 = 0.\overline{6}S$$

Subtract $0.\overline{3}S$ from both sides of the equation.

$$\frac{612}{0.\overline{6}} = \frac{0.\overline{6}S}{0.\overline{6}}$$

Divide both sides of the equation by $0.\overline{6}$.

$$\$918 = S$$

The selling price is $918.

Check: The net cost is less than the list price but the same magnitude. The selling price is more than the cost but also of the same magnitude. ■

4.5 SINGLE EQUIVALENT DISCOUNT

It is often useful, particularly when making price comparisons, to know one discount that will give the same net price as a series of two, three, or more discounts. This one discount is called the **single equivalent discount**. No new calculation or formula is required, only a different

138 CHAPTER 4   COMMERCIAL DISCOUNTS

application of a fact already used, namely, that the discount percent and the percent paid (or net cost rate factor) are complements. (They add up to 100%.)

In the formula

$$\%Pd_1 \times \%Pd_2 \times \%Pd_3 \times L = N$$

the product of the three complements ($\%Pd_1 \times \%Pd_2 \times \%Pd_3$) is the final percent paid. Therefore, the complement of this product is the single equivalent discount.

Written in symbols, this statement is:

$$\text{Single equivalent discount} = \text{Complement of } (\%Pd_1 \times \%Pd_2 \times \%Pd_3)$$

$$SED = 100\% - (\%Pd_1 \times \%Pd_2 \times \%Pd_3)$$

or

$$SED = 1.00 - \text{Net cost rate factor}$$

In this case, just understanding the relationship will be as useful as using the formula.

■ **Example 1**   A retail store bought a chair for $70 less 20/10. What was the net price? What single discount would give the same net price?

$\%Pd_1 \times \%Pd_2 \times L = N$	*There are two discounts and therefore two complements.*
$0.8 \times 0.9 \times 70 = N$	*The complement of 20% is 80%. The complement of 10% is 90%.*
$0.72 \times 70 = N$	*Multiply the complements first.*

Observe that 0.72, or 72%, is the net cost rate factor or final percent paid and that the complement, 28% (100% − 72%), is the single equivalent discount.

Completing the multiplication, we have:

$$0.72 \times 70 = N$$
$$\$50.40 = N$$

Check: Use the single equivalent discount, 28%, to check the net price.

$\%Pd \times L = N$	
$0.72 \times 70 = N$	*The complement of 28% is 72%, or 0.72.*
$\$50.40 = N$	

or

$$\%D \times L = \$D$$
$$0.28 \times 70 = 19.60$$
$$70 - 19.60 = \$50.40$$

An additional check is the fact that the single equivalent discount is always *less than* the sum of the series.

If the problem requires only the single equivalent discount, the list price is not needed and does not enter into the calculation. ■

■ **Example 2** Two suppliers offer identical sleds at the same list price. However, the SlipSlide Co. gives trade discounts of 20/5/5, while Jolly Toys Co. offers 15/10/5. Which is the better discount for the buyer?

SlipSlide Co. (with formula):

$$SED = 100\% - (\%Pd_1 \times \%Pd_2 \times \%Pd_3)$$
$$= 1.00 - (0.80 \times 0.95 \times 0.95)$$
$$= 1.00 - 0.722 = 0.278$$
$$SED = 27.8\%$$

Jolly Toys Co. (without formula)

$$\text{Product of complements} = 0.85 \times 0.90 \times 0.95$$
$$= 0.72675 = 72.675\%$$
$$\text{Complement of } 72.675\% = 100\% - 72.675\%$$
$$= 27.325\%$$
$$SED = 27.3\%, \text{ rounded to the nearest tenth}$$

SlipSlide Co. is offering a slightly better discount for the buyer, 27.8% compared with 27.3% from Jolly Toys.

Check: The two equivalent discounts should be very close because the two series discounts are very similar. As noted before, the single equivalent discount is always less than the sum of the series. In this case the two series have the same sum: 20% + 5% + 5% = 30% and 15% + 10% + 5% = 30%. SlipSlide Co.'s single equivalent discount is a little bit larger because their first discount of 20% is larger than Jolly Toy's first discount of 15%. ■

Chapter 4

Name _____ Date _____ Class _____

TRY THESE PROBLEMS (Set II)

Fill in the blank spaces. Round dollar amounts to the nearest cent and rates to the nearest tenth of a percent where necessary.

	LIST PRICE	TRADE DISCOUNTS	NET COST RATE FACTOR (% Paid)	NET PRICE	SINGLE EQUIVALENT DISCOUNT
1.	$650	12% and 10%			
2.	$175	20/10/10			
3.	$93	$33\frac{1}{3}\%$, 4%			
4.		15/10		$114.75	
5.		20/10/5		$2,736	
6.	$140			$98	—
7.		9/5/3		$268.34	
8.	$6,750			$5,771.25	—
9.		10/5		$239	

Solve the following problems. Round dollar amounts to the nearest cent and rates to the nearest tenth of a percent where necessary.

10. Hairdryers are listed in a beauty supply catalog for $15 less 10%. The wholesaler found himself overstocked and sent a notice to his customers that an additional 4% discount could be taken. What is the new net price for a hairdryer?

11. Which series of discounts is better for the buyer, 20/15/5 or 35/5? (Can you do this problem without calculations?)

12. Find the price at which an electric pencil sharpener should be listed in a catalog if the desired net price is $23.80 and the trade discounts are 30/15.

13. A furniture manufacturer wishes to price a chair so she can give her best customers discounts of 20/15/10 and still have a net price of $91.80. What price should she list in her catalog?

14. A denim blouse is listed at $24 less 35%. What additional percent discount must be offered to bring the net price down to $12, the price being charged by a competitor?

15. Acme Wholesalers currently sells radios for $80 less 25%. Their competitors are now selling the same product for $48.

   a. What additional discount must Acme offer in order to meet their competitor's price?

   b. What single discount would give the same net price?

16. Able Company sells a set of outdoor furniture for $300 less 20%. Baker Company advertises the same items for $360 less 25% and 20%.

   a. Find the current net price of each wholesaler.

   b. What further discount must be offered by the higher company in order to meet the competitor's price?

17. A manufacturer of outdoor barbeque grills sells one model for $90 less discounts of 10% and 8%. A competitor sells a similar model for $90 less 20%. What additional percent discount must the manufacturer offer to meet the competitor's price?

18. Find the net cost rate factor and the single equivalent discount for the following series. (Do not round your answers.)

   a. 10/10/10

   b. 15/10/10

   c. $12\frac{1}{2}$/10/5

19. An emerald bracelet cost a jeweler $1,200 less 25/20. If he wishes to obtain his usual markup of 20% of selling price, what price should he mark the bracelet?

20. A manufacturer has a cost of $150 for each piece of sculpture that he produces. If his usual markup is 65% of cost, what list price should he set so that he can offer his customers discounts of 15/10?

4.6 CASH DISCOUNTS

We have now considered two stages in the pricing of merchandise: establishing a quoted or list price and using trade discounts that reduce this price. Is the net price the final cost to the customer?

The answer to this question depends on the credit terms given when the sale is made. Frequently, the customer is not required to pay the bill immediately but is given a specified time before payment is due. This is a familiar situation to anyone who has had a charge account for a department store, for gasoline, or the like.

In commercial credit arrangements it is frequently possible to anticipate the bill—that is, to pay it before it is due and get a discount on the cost of the merchandise. This discount, granted for the payment of a bill before it is due, is called a **cash discount.** *Cash discounts are given only on the value of merchandise, never on freight, insurance, storage, or any other charge for a service supplied by a party other than the seller.*

Figure 4-1 is a sample **invoice** (bill for merchandise). A fabric wholesaler bought fabric from a mill for resale to its customers. The total of the charges for merchandise was $1,519.19 list. (See area labeled 1 in Figure 4-1.) A trade discount of 5% (area 2 in Figure 4-1) was allowed and this reduced the net price to $1,443.23 (area 3). At the top of the invoice is the statement of the terms: 2/10 net 60 (area 4). This means that if the bill is paid within 10 days of the invoice date the customer may deduct 2% from the cost of the merchandise. If paid after the

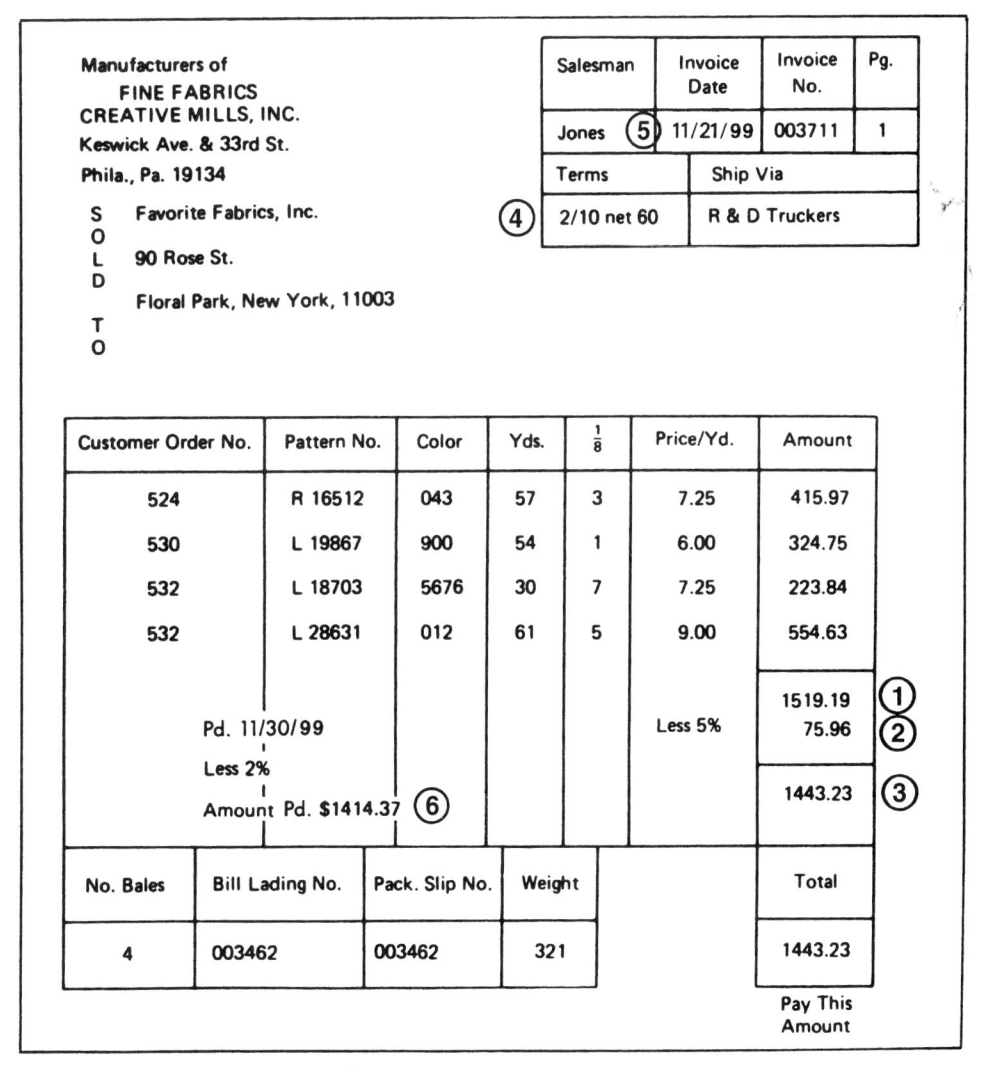

FIGURE 4-1 Invoice from manufacturer.

tenth day, no discount is allowed and payment is due within 60 days. After that there may be a charge for late payment, as in the case of a charge account.

To calculate the amount of the cash discount, first check the date of the bill and determine the number of days between the invoice date and the payment date. In this case the bill is dated 11/21/99 (area 5). If it is paid on 11/30/99, then, since there are 9 days between the two dates, the customer is entitled to a 2% discount on the cost of merchandise.

Since the discount is a percent of the cost of the merchandise, the basic equation ($R \times B = P$) may be used. For a cash discount problem it becomes:

> % Discount × Cost of merchandise = Amount of discount
>
> or
>
> $\%D \times C = \$D$

For the invoice in Figure 4-1,

$$0.02 \times 1{,}443.23 = \$D$$

$$\$28.86 = \$D$$

The customer may deduct $28.86, and the net payment is

$$1{,}443.23 - 28.86 = \$1{,}414.37$$

The important point to note is that the customer had paid less than the full amount of the bill but will receive credit for paying the entire bill. He owes nothing more. The amount paid is $1,414.37 (area 6). The amount the customer gets credit for is $1,443.23. The difference is the discount.

As in trade discount problems, the amount paid can be found directly without finding the discount first. *The base (100%) is always the amount credited.* In this problem it is the full amount of the invoice because the entire bill is settled and the amount owed is for merchandise only.

	IN DOLLARS	IN PERCENT
Amount credited	$1,443.23	100%
Amount of discount	−28.86	−2%
Amount paid	$1,414.37	98% (percent paid)

The equation may be written

> % Paid × Amount credited = Amount paid
> ↓   ↓   ↓
> %Pd   ×   Cr   = $Pd

For the previous problem,

$$0.98 \times 1{,}443.23 = \$1{,}414.37$$

■ **Example 1**   The Handy Merchandise Center received an invoice for $656.50 with terms of 2/10, n/30. If the invoice was received on March 12 and paid on March 20: (a) what amount must be paid, and (b) how much cash discount is allowed?

There are 8 days between March 12 and March 20, so the 2% cash discount may be taken.

ORDINARY DATING AND OTHER CREDIT TERMS   145

This problem may be solved using either formula.

$$\%D \times Cr = \$D \qquad \%Pd \times Cr = \$Pd$$
$$0.02 \times 656.50 = \$D \qquad 0.98 \times 656.50 = \$643.37$$
$$\$13.13 = \$D$$

The amount paid is:

$$\$656.50 - \$13.13 = \$643.37$$

The discount can be found by subtracting the amount paid from the amount credited.
$656.50 − $643.37 = $13.13

In cash discount problems the time periods are usually short, and it is not difficult to determine the number of days between the start of the credit period and the date of payment.

- If the two dates are in the same month, subtract the beginning of the credit period (date of bill, receipt of goods, or end of month) from the payment date.

```
  October 19     date paid
− October  9     date of invoice
   10 days
```

- If the date of payment is the month after the start of the credit period, find the number of days between the start of the credit period and the end of the month; then add the number of days in the next month up to the payment date. For example, if a bill dated November 24 is paid on December 3, the number of days between the two dates is

November:	6 days remaining (November 24 to November 30)
December:	3 days
Total	9 days

Note that the date of the invoice is not counted in calculating the days in the discount period.

The number of days between two dates can also be found by referring to the Calendar Table on page 596. The table will be used frequently in later chapters and may be helpful here as well. ■

4.7 ORDINARY DATING AND OTHER CREDIT TERMS

The cash discount terms described in the previous invoice indicated that the discount period started with the date of the invoice. Such terms are referred to as **ordinary dating** terms.

Terms may be more detailed than in this invoice, with different discounts allowed for different periods of time. An example would be credit terms of 3/10, 2/20, *n*/30. This means that if a payment is made within 10 days of the billing date a discount of 3% is allowed. If payment is made between the eleventh and twentieth day, a discount of 2% is allowed. After that no discount is allowed but payment is due within 30 days.

Credit periods do not always start with the date of the invoice. There are additional ways in which the time may be stated.

E.O.M. or PROX.

The terms may be 4/10, 3/15, *n*/30 E.O.M. Terms of 4/10, 3/15, *n*/30 PROX. are treated the same way. The letters E.O.M. stand for "end of month" while PROX. is from the Latin word *proximo* meaning "next month after present." Both indicate that the 10- and 15-day credit periods begin on the first day of the month after the date of the bill. Thus, if a bill is dated March 18 and the terms are 4/10, 3/15, *n*/30 E.O.M., the 10- and the 15-day credit periods start with the first day of April. The customer could take a 4% discount if he paid anytime through April 10 and 3% if he paid on April 11 through April 15.

146 CHAPTER 4   COMMERCIAL DISCOUNTS

With the E.O.M. and PROX. methods of finding cash discounts, it is a common business practice to add an extra month to the cash discount period when the date of an invoice is after the twenty-fifth of the month. For example, if an invoice is dated August 26, and has credit terms of 2/10, n/30 PROX., the cash discount period would be extended to October 10. This allows merchants enough time to receive and pay the invoice. Note that the practice does not apply to any of the other cash discount methods discussed in this chapter.

R.O.G.

The terms may be stated 3/15, 2/30, n/60 R.O.G. The letters R.O.G. stand for "receipt of goods" and indicate that the credit period starts when the merchandise is received by the buyer. If the invoice is mailed at the same time the goods are shipped, it may arrive much in advance of the goods if the shipment is by surface transportation over a long distance.

Extra

The terms may be 2/10–90X. The 90X means that a cash discount may be taken if payment is made within 100 days (10 days plus 90 extra days) after the date on the invoice. Such dating is used to induce customers to buy out-of-season merchandise such as air conditioners in the winter or Christmas decorations in the summer. With delayed payments, retailers are willing to place orders sooner, and in many cases these retailers sell the merchandise before the cash discount period expires.

■ **Example 1**

An invoice is dated September 23 and the merchandise is received on October 30. Find the last date of the cash discount if the terms are:

    a. 2/10, n/30   **b.** 2/10 E.O.M.   **c.** 2/10 R.O.G.
    d. 2/10–30X   **e.** 2/10 PROX.

a. The last date of the discount period falls 10 days after September 23, that is, October 3.

September:	7 days remaining (Sept. 23 to Sept. 30)
October:	3 days
	10 days

b. The term E.O.M. means that the discount period starts with the first day of the next month. Thus, a discount can be taken until October 10.

c. Since the goods were received on October 30, the customer has 10 days after that date to obtain the discount. The last day to obtain the discount will be November 9.

October:	1 day remaining
November:	9 days
	10 days

d. The customer has 40 days from September 23 to obtain the discount. November 2 is 40 days after September 23.

e. This is the same as question b. The discount period starts with the first day of the next month. A discount can be taken until October 10. ■

■ **Example 2**

An invoice for merchandise is dated April 8 and has credit terms 3/15, 2/20, n/30 E.O.M.

1. If it is paid on May 15, what percent discount may be taken?

Since the terms state E.O.M., the credit period begins on May 1. If the bill is paid on or before May 15, a 3% cash discount may be taken.

2. If it is paid on May 17, what percent discount may be taken?

If the bill is paid on May 17, a 2% discount may be taken. May 17 is more than 15 days, but less than 20 days, from the end of April.

ORDINARY DATING AND OTHER CREDIT TERMS 147

3. What is the last date on which any discount may be taken?

The last date on which any discount may be taken is May 20, which is 20 days from the end of April. ■

■ **Example 3** On June 1 the ABC Co. paid two bills owed to the XYZ Co. Both bills had credit terms 2/10, n/30. The first bill was dated May 15 and was for $315.75. The second bill was dated May 23 and was for $405.80. What amount will settle both bills?

First bill: There are 17 days between May 15 and June 1 (16 days left in May and 1 in June). Therefore, no discount may be taken, and the bill must be paid in full.

Amount paid on first bill = $315.75

Second bill: There are 9 days between May 23 and June 1 (8 days in May and 1 in June). Therefore, a 2% discount may be taken and the percent paid is 98%.

$$\%Pd \times Cr = \$Pd$$

$$0.98 \times 405.80 = \$Pd$$

$$\$397.68 = \$Pd$$

Amount paid on second bill = $397.68

Total paid for both bills: 315.75 + 397.68 = $713.43

Note: In Example 3 the 2% discount for the second bill can be found first, in the following way.

$$\%D \times Cr = \$D$$

$$0.02 \times 405.80 = \$D$$

$$\$8.12 = \$D$$

Payment = 405.80 − 8.12 = $397.68

However, this method can be used only in a limited number of problems. As will be illustrated in the next section, the first method, using percent paid rather than percent discount, is more generally useful. ■

■ **Example 4** A retailer received an invoice for $682.25, which showed $667 for merchandise and $15.25 for special handling charges. The bill was dated October 1 and had credit terms 4/10, 2/20, n/30 R.O.G. The merchandise arrived on October 8. The retailer sent a check for $668.91 on October 27 in full payment of the invoice.

1. Is the check amount correct?

In this problem the date of the invoice has no effect on the credit period since the terms include R.O.G. The credit period begins on October 8, when the merchandise arrived. There are 19 days between October 8 and October 27, when the bill is paid, and the buyer is entitled to a 2% discount on the cost of merchandise. The buyer pays 98% of this cost.

$$\%Pd \times Cr = \$Pd$$

$$0.98 \times 667 = \$Pd \qquad \textit{Cash discounts are allowed only on merchandise costs.}$$

$$\$653.66 = \$Pd$$

The charge for special handling also must be paid, but no discount may be taken.

$$\begin{aligned}\text{Total payment: Merchandise} \quad & \$653.66 \\ \text{Special handling} \quad & +\ 15.25 \\ \hline \text{Total} \quad & \$668.91\end{aligned}$$

The amount of the check was correct.

2. How much was saved by paying the bill before the due date?

$$\begin{aligned}\text{Cost of merchandise} \quad & \$667.00 \\ \text{Payment for merchandise} \quad & -653.66 \\ \hline \text{Savings (discount)} \quad & \$\ 13.34\end{aligned}$$

3. How much is still owed after the payment?

Nothing is owed. The buyer is credited with paying the entire invoice amount of $682.25 ($667 + $15.25).

Check: When a discount is allowed, the amount paid must be smaller than the amount credited, but since cash discounts are generally less than 5%, the difference should not be large. ■

■ **Example 5**

An invoice for smoke detectors showed a list price of $550 with trade discounts of 20/10 and terms of 2/10, *n*/30. If the invoice is paid in time to take both the trade and cash discounts, what is the net price?

First, the trade discounts are taken.

$$\%Pd \times L = N$$
$$0.8 \times 0.9 \times 550 = N$$
$$\$396 = N$$

Next, the cash discount is taken on the net price of $396.

$$\%Pd \times Cr = \$Pd$$
$$0.98 \times 396 = \$Pd$$
$$\$388.08 = \$Pd$$

Note: Since %*Pd* is used in both formulas, the calculations may all be done at the same time.

$$0.8 \times 0.9 \times 0.98 \times 550 = \$388.08$$

A check for $388.08 will pay the $550 invoice in full.

Remember, freight and insurance charges do not qualify for either cash or trade discount. They must be repaid in full. ■

Chapter 4

TRY THESE PROBLEMS (Set III)

In Problems 1 through 3 determine: (a) the last date on which the larger discount may be taken, and (b) the last date on which the smaller discount may be taken. Assume no leap years.

	DATE OF BILL	CREDIT TERMS
1.	September 23	4/10, 2/30, n/60
2.	June 5	3/10, 2/20, n/30 E.O.M.
3.	February 14	2/15, 1/30, n/45

In Problems 4 through 6 determine the last date on which the discount may be taken.

	DATE OF BILL	CREDIT TERMS
4.	March 8	2/10–40X
5.	July 19	3/15 PROX.
6.	November 16	4/10, n/30 E.O.M.

Solve the following problems. Round answers to the nearest cent where necessary.

7. A bill dated July 22 is paid on August 6. The amount of the invoice is $248 and the credit terms are 3/15, n/60. What amount must be paid?

8. Jules Auto Shop received a shipment of mufflers and tailpipes on April 14. The terms were 2/10, n/30 E.O.M. and the invoice totaled $2,050. What amount must be paid on May 10? Is anything still owed?

9. A bill for some appliances was dated July 2 and the goods were received on July 12. The bill is for $538.28 and the terms are 2/10, n/30 R.O.G. What amount is due if payment is made on: (a) July 22, (b) August 10?

149

10. Bargain Books paid two invoices on September 13. The first was for $205 dated August 22 with terms 2/10, 1/20, *n*/30. The second was for $246 dated August 30 and had the same credit terms. What was the total amount paid? How much discount was taken?

11. Empire Lighting Company received an invoice for $5,250. This amount included a freight charge of $250. The terms of the bill were 4/10, 2/30, *n*/60. If the invoice was dated May 13 and paid on May 23, what amount must be paid?

12. An invoice dated December 15 shows $620 for merchandise and $22.50 for prepaid freight and insurance. The terms are 3/15, *n*/30 and the invoice is paid on December 22. Determine the amount that will pay the bill in full.

13. A pair of stereo speakers is listed at $275.50 with trade discount of 20/10 and credit terms of 3/10, *n*/30 R.O.G. If a retailer pays the invoice soon enough to qualify for the cash discount, what will the net price be?

14. The list price of a programmable calculator is $80. If the manufacturer offers trade discounts of 30/10 and terms of 2/15, *n*/30, find the net price assuming that all discounts are taken.

15. The owner of a small store has $800 available for paying bills at the end of July. He has two bills outstanding and would like to settle both. One is for $395, dated 6/15, with credit terms 2/15, *n*/45. The second is for $426, dated 6/17, with credit terms 3/10, 2/30, *n*/60 E.O.M. Can he settle both bills on July 30 with the $800? If not, how much more does he need?

16. Plumb Line Co. owes $320 for merchandise. The bill is dated November 22 and has credit terms 4/20, 2/30, *n*/60. If half the bill is paid on December 12 and the remainder on December 20, how much discount will the company take in both payments?

Chapter 4

Name _____ Date _____ Class _____

TRY THESE PROBLEMS (Set III)

17. The bookkeeper of Jolly Toy Co. received a check for $285.70 in payment of an invoice for $296.60, which included $3.20 for freight. The customer paid early enough to take a 3% cash discount. Was the amount of the check correct?

18. For the invoice shown in Figure 4-2:

   a. What is the list price?

   b. What is the net price?

   c. What is the total amount of trade discounts?

   d. If the invoice is paid on June 9, what is the cash discount?

   e. If the invoice is paid on June 9, what amount must be paid?

	Eastside Plumbing Supply	
Sold to: Smith Plumbing 109 Oak Street Columbus, OH 43232		Invoice Date: May 20 Invoice No.: 876152 Ship Via: UPS Terms: 2/10, n/30 PROX

Item No.	Quantity	Description	Unit Price	Amount
427	36	Copper tubing	$13.95	
689	125	Copper fittings	$5.79	
214	200	Plastic elbows	$0.98	
471	50	Misc. washers	$0.12	
			List Price	
			Less 15%	
			Net Price	
			Cash Discount	
			Amount Due	

FIGURE 4-2 Sample invoice.

4.8 PARTIAL PAYMENTS

If the buyer wants to pay the invoice in advance of the due date in order to get the discount but does not have sufficient funds to pay the entire bill, the buyer may pay part of the bill and take the discount on the part paid. In this case the amount credited depends on how much the payment is, and it is less than the amount of the invoice. The buyer still owes an amount equal to the invoice less the amount credited.

■ **Example 1**

The I.O.U. Store would like to settle part of a $3,370 invoice for merchandise. The invoice is dated May 5 and the credit terms are 3/15, n/30. After making all other payments due at the time, the owner finds she can send a check for $2,500 as a partial payment. The check is sent on May 20.

1. How much does the store still owe after the partial payment?

    First, determine whether a discount may be taken. Since May 20 is within 15 days of May 5, the customer is entitled to a 3% discount. The payment represents 97% of the amount for which she will get credit:

    $\%Pd \times Cr = \$Pd$

    $0.97 \times Cr = 2{,}500$ *The unknown is the amount credited (Cr).*

    $\dfrac{0.97\ Cr}{0.97} = \dfrac{2{,}500}{0.97}$ *Divide both sides of the equation by the coefficient of the unknown.*

    $Cr = \$2{,}577.32$

    Amount owed − Amount credited = Amount still due

    $\$3{,}370 \quad - \quad \$2{,}577.32 \quad = \quad \$792.68$

    After $2,577.32 is credited to the store's account, $792.68 is still owed.

2. How much was saved by paying part of the bill early?

    The amount saved (cash discount) is the difference between the payment and the amount credited.

    $Cr - \$Pd = \$D$

    $2{,}577.32 - 2{,}500 = \$77.32$

    The store saved $77.32 by paying part of the bill early.

    Check: The amount credited should be less than the invoice amount because this is a partial payment. The payment should be less than the credit to the store's account because a discount was allowed. ■

■ **Example 2**

The Button-Up Shop usually anticipates bills in order to take advantage of cash discounts offered. However, the manager has found that it is not worth the time and work involved if the discount is less than $50. An invoice for buttons amounting to $2,500 has credit terms 3/10, 2/20, n/30 and is dated December 2. Would a partial payment of $1,000 on December 12 be worth making?

A payment on December 12 is within 10 days of the invoice date, and a 3% discount is allowed. Then 97% of the amount credited will be paid.

$\%Pd \times Cr = \$Pd$

$0.97 \times Cr = \$1{,}000$

154 CHAPTER 4 COMMERCIAL DISCOUNTS

$$\frac{\cancel{0.97}\, Cr}{\cancel{0.97}} = \frac{1,000}{0.97}$$

Divide both sides of the equation by the coefficient of the unknown.

$$Cr = \$1,030.93$$

Partial payment would not be made because only $30.93 ($1,030.93 − $1,000) would be saved. ∎

Now go to Try These Problems (Set IV).

NEW FORMULAS IN THIS CHAPTER

- $\%D \times L = \$D$ (Percent discount × List price = Amount of discount)
- $L - \$D = N$ (List price − Amount of discount = Net price)
- $\%Pd \times L = N$ (Percent paid × List price = Net price)

Trade Discount

- $\%Pd_1 \times \%Pd_2 \times \%Pd_3 \times L = N$ (for series discounts; First percent paid × Second percent paid × Third percent paid × List price = Net price)
- NET COST RATE FACTOR = $(\%Pd_1 \times \%Pd_2 \times \%Pd_3)$
- SED = $100\% - (\%Pd_1 \times \%Pd_2 \times \%Pd_3)$ (single equivalent discount is the complement of the product of the percents paid)

Cash Discount

- $\%Pd \times Cr = \$Pd$ (Percent paid × Amount credited = Amount paid)
- $\%D \times Cr = \$D$ (Discount percent × Amount credited = Amount of discount)

Chapter 4

TRY THESE PROBLEMS (Set IV)

Solve the following problems. Round answers to the nearest cent where necessary.

1. An invoice for $970 was reduced by a payment of $455.90. If a 3% cash discount was allowed on the partial payment, how much was credited to the account? How much is still owed?

2. R.T. Interiors owed $3,050 to a fabric dealer. The bill had credit terms 2/10, n/30, E.O.M. and was dated April 20. On May 9 the store sent a check for $1,500. How much credit will they receive for the payment? How much do they still owe?

3. New City Plumbing Co. received an $8,700 invoice dated March 28 and containing sales terms 3/15, n/30 E.O.M. On April 15, New City decided to pay enough to reduce their bill by $4,000. What amount must be paid? How much is still owed?

4. An invoice of $5,700, dated March 9, has terms 2/10 PROX. A partial payment of $3,600 is made on April 5. Find: (a) the amount credited, (b) the balance due on the invoice, and (c) the cash discount earned.

5. Leon's Supply received an invoice for $785 dated January 8 with sales terms of 2/10–40X. On February 15 Leon decided to pay an amount that would reduce the balance owed by $500. What payment should be made?

6. How much will a retailer save by making a partial payment of $408 on an invoice for $595 if a discount of 4% is allowed?

7. A retailer received an invoice for $800 less trade discounts of 10/10. There was an additional freight charge of $30. The invoice was dated October 8 and the credit terms were 3/15, n/30. If a partial payment of $400 was sent on October 23, how much is still owed?

8. Imperial Designs received an invoice for $1,500 less 20/10. The terms were 3/15, 1/20, n/45 and the date was July 3. On July 12 Imperial sent a check for $897.25. How much do they still owe?

155

Chapter 4

Name Date Class

END-OF-CHAPTER PROBLEMS

Complete the following problems by filling in the blank spaces. Round dollar amounts to the nearest cent and rates to the nearest hundredth of a percent where necessary.

	LIST PRICE	TRADE DISCOUNT	NET PRICE	AMOUNT OF DISCOUNT	NET COST RATE FACTOR	SINGLE EQUIVALENT DISCOUNT
1.	$600	25%				—
2.	$112		$78.40			—
3.	$620	20%, 10%				
4.		25/20/10	$81			
5.	$1,500	15/10/2				
6.	$200		$170			—
7.		30/5	$13.30			
8.	$738		$492			—

For Problems 9–16, find the net payment where indicated, and the credit toward account and the amount still due. Round answers to the nearest cent where necessary.

	INVOICE AMOUNT	DATE OF INVOICE	DATE OF PAYMENT	CREDIT TERMS	NET PAYMENT
9.	$770	June 6	June 15	2/10, n/30	
10.	$538.42	April 23	May 14	3/10, 1/15, n/30 E.O.M.	

157

INVOICE AMOUNT	DATE OF INVOICE	DATE OF PAYMENT	CREDIT TERMS	NET PAYMENT
11. $4,000	December 6	January 24	2/10–50X	
12. $765	February 24	March 10	2/10, *n*/30 PROX.	$392
13. $711.79	May 27	June 9	1/10, *n*/30	$582
14. $435	September 21	October 10	2/10, *n*/30 E.O.M.	
15. $480.48	September 20	October 15	3/15, 1/30, *n*/60	
16. $3,787.20	January 4	January 25	4/10, 2/20, *n*/45	$2,000

Solve the following problems. Round dollar amounts to the nearest cent and rates to the nearest tenth of a percent where necessary.

17. An air conditioner is listed at $590 with a trade discount of 25%. Find the amount of trade discount and the net price.

18. The list price of a television set is $450. It is sold with trade discounts of 10% and 20%. What is the net price?

19. A tennis racket is priced at $30 by the producer and sold to a wholesaler with a trade discount of 15%. If the discount is changed to 12% in order to raise the price, how much more will the wholesaler pay?

20. A stereo system listed at $960 is sold with discounts of $33\frac{1}{3}$%, 10%, and 5%. What is the net price?

21. Fine Furniture bought a sofa for a net price of $360. The manufacturer's list price was $600. What percent trade discount did the store receive?

Chapter 4

Name Date Class

END-OF-CHAPTER PROBLEMS

22. A set of adjustable wrenches is priced at $16.80 less 10/5.

   a. What is the net price?

   b. What single discount would give the same net price?

23. A manufacturer of luggage offers trade discounts of 20/10/10.

   a. Find the net price if the list price is $59.

   b. What single discount would give the same net price?

24. What percent discount has been allowed if a jacket that lists for $63 is sold for $42?

25. A jewelry box is listed at $48 with discounts of 30%, 20%, and 5%.

   a. What is the net price?

   b. What single discount would give the same net price?

   c. What is the net cost rate factor?

26. What percent discount has been offered if a lawn mower that lists for $260 is sold for $182?

27. A manufacturer wishes to list a tool kit in his catalog with a discount of 12%. At what price should he list it if the net price he wants is $74.80?

28. An office furniture manufacturer wishes to sell a file cabinet for $68 after allowing trade discounts of 10/10. At what price should the file cabinet be listed?

29. After discounts of 30/10 the net price of an item was $37.80. Find the list price.

30. Find the list price of a ceiling light fixture if, after discounts of 25% and 16%, the net price is $151.20.

31. Mollie's Bed and Bath Shoppe is selling a mirrored table for $140 less 20%.

   a. What additional percent discount must they offer to meet the price of $95.20 that is being charged by their competitors?

   b. What single discount would give the same net price?

32. Fast Camera, Inc., has an auto-focus camera for $80 less 25/10. Price-Rite has the same model for $75 less 20/20/10.

   a. What net price is each firm charging?

   b. What further percent discount must be given by the higher company in order to meet their competitor's price?

33. ABC Distributors has been selling a lawn mower for $216 less 24%. Their competitors are now selling a similar mower for $144. What additional percent discount will ABC have to offer to meet their competitor's price?

34. A paint manufacturer sells paint at $14 a gallon less 10% to hardware stores. A chain store offers to buy a large quantity at $12 a gallon. What additional discount will the manufacturer be giving the chain if the sale is made?

35. Hardware, Inc., sells a set of screwdrivers to distributors at $15 less 20%. A mail-order retailer offers to buy the set in a large quantity, but only if the price is reduced to $10.20. What additional percent discount would bring the price down to this level?

36. A retailer can buy a coffeemaker from two distributors. One price is $44 less 20% and 10%. The second price is $46 less 20% and 15%. Which gives the buyer the lower net cost?

37. The list price of a video game is $85. It is sold by two different manufacturers. One gives discounts of 15/10/10, while the other offers 20/15.

   a. Which series of discounts gives the lower net price?

   b. Find the difference in prices.

Chapter 4

Name Date Class

END-OF-CHAPTER PROBLEMS

38. A new running shoe is listed at $89.95. Wholesalers receive discounts of 20/10/10, while the discount to retailers is 20/10.

   a. What is the wholesaler's price?

   b. What is the retailer's price?

   c. What is the difference between the two prices?

39. Harry Hopeful has just opened an auto supply shop and is deciding from which supplier to order his stock. The prices of two suppliers are very close, but one offers trade discounts of 10/10/5 and the other allows 20/4. Which offers the better single equivalent discount?

40. Two distributors offer the same item at the same list price. One offers a trade discount of 20%. The second offers discounts of 10/8/2. Which is the better offer for the buyer? (Can you answer this question without calculation?)

41. A retailer buys a set of glasses at $50 less 20% and 10%. She then adds a markup of 45% of selling price. Find the selling price.

42. Dante Furniture buys an oak end table for $550 less discounts of 25/20. Find their selling price if they use a markup of 45% of cost.

43. A retailer usually buys sheets at $20 less 10/10. For a special sale the supplier offers the same product at $15.

   a. What additional percent discount is being offered?

   b. If the retailer takes a markup of 40% of selling price, at what price can he sell the sheets?

44. A manufacturer produces a crib at a cost of $120 and adds a markup of 65% of cost to determine the net price at which the crib will be sold. If he wishes to allow a 20% trade discount for his best customers, what list price should he set for the crib?

45. All-For-Autos Co. makes a car vacuum that costs $30 to produce. They add a 60% markup on cost to get the lowest price at which they will sell the product. At what price should they list the vacuum if they allow their best customers trade discounts of 10%?

46. A retailer wants to add a line of bike carriers to his sports department and plans to sell one model at $39.99. If he can buy a good product for $32.50 less 20%, will he be able to get his usual markup of 42% on selling price and sell the carrier at the price he set?

47. An invoice for computer hardware totals $852.40. It was dated July 7 and the credit terms were 3/15, *n*/30 R.O.G. If the shipment was received on August 28 and payment was made on September 11, find the amount due.

48. Push N Pull Toys offers discounts of 2/10, 1/15, *n*/30 on all purchases. If an invoice dated May 17 for $863.50 is paid on June 1, what amount would represent full payment?

49. The Beautie Shoppe received an invoice for supplies dated January 7 with credit terms 3/10, 2/20, *n*/30. The amount of the invoice was $202.60.

a. What amount should be paid if the bill is paid on January 25?

b. What is the last date for payment according to the credit terms?

50. An invoice for kitchen appliances is dated December 23 and is paid on January 14 of the following year. Credit terms are 2/15, *n*/45 E.O.M. The total amount, $497.48, includes a charge for freight and insurance of $62.34.

a. What amount should be paid?

b. How much is credited to the buyer's account?

c. Is any money still owed after the payment?

Chapter 4

Name   Date   Class

END-OF-CHAPTER PROBLEMS

51. An invoice totaling $629.60 includes prepaid freight and insurance of $29.60. It is dated November 7 and has terms 2/15 PROX.

    a. What is the last day on which the cash discount may be taken?

    b. Find the amount due if the invoice is paid on November 26.

52. Jones Surgical Company received a shipment of home health care products on May 29. The invoice was dated May 25, had sales terms of 2/10, n/30 R.O.G., and showed $850 for merchandise with an additional $45 for freight charges. On June 7, Jones decided to pay enough to reduce their balance by $600.

    a. What payment must they make?

    b. How much do they still owe?

53. The Miller Company wants to pay three invoices, which all have sales terms of 2/10, 1/20, n/30. The first, for $630, is dated May 5. The second, for $350, is dated May 17 and the last one, dated May 22, is for $440. If the invoices are all paid on June 1, how much is due?

54. An invoice dated July 25 lists $800 for merchandise and $30 for freight. Trade discounts of 10/10 are deducted and the credit terms are 4/15, 2/30, n/45 E.O.M. If the bill is paid on August 14:

    a. What is the amount due?

    b. How much is saved by paying the bill early?

55. Valley Stores receives a shipment of ski equipment with a list price of $2,540. Trade discounts of 25/10/10 are allowed and the credit terms are 3/15, n/45. If payment is made on October 28 and the invoice is dated October 13, what amount must be paid? How much will be credited to Valley Stores' account?

56. Trade discounts of 20/10 may be taken on an invoice that totals $938.50. This amount includes prepaid freight of $26.50. Sales terms are 2/10, n/30. Find the amount due if the invoice is paid within the discount period.

57. An invoice amounting to $960 is dated July 23 and has credit terms 2/10, 1/20, n/45. If a partial payment of $479 is made on August 4, find: (a) the credit given for the payment, and (b) the amount still due.

58. A partial payment of $300 is made on an invoice for merchandise. The invoice amount is $722 and the payment is made early enough for the customer to take a 2% cash discount.

  a. How much is credited to the account for this payment?

  b. How much is still owed on this invoice?

59. NYX Industries sent a $2,500 bill with terms 3/15, 1/30, n/45 to Lindy Company. The invoice was dated April 3. NYX received a check for $1,455 dated April 12.

  a. How much credit should Lindy receive for this partial payment?

  b. What is the balance still due on Lindy's account?

60. An invoice for $617.29 is dated September 30 and has credit terms 4/10, 2/20, n/30. The following two payments are made: (1) on October 10 a check is sent for $144, and (2) on October 20 a check is sent for $100.

  a. What is the last date on which the invoice may be paid without being overdue?

  b. How much is still due on the invoice after both payments are made?

61. Ann Apple owns a prosperous dress store and usually takes advantage of all cash discounts. She has two bills to pay. The first one is for $1,200 with a discount of 4% if paid by June 10. The other is for $1,400 with a discount of 3% if paid by the same date. Because of unusual expenses, she has only $2,000 available on June 10.

  a. Which bill should she pay? Why?

  b. After paying one bill (part a), Ann sends the remainder of her available cash as a partial payment on the remaining bill. How much does she still owe?

Chapter 4

SUPERPROBLEM

Solve the following problem. Round answers to the nearest cent where necessary.

A manufacturer produces a calculator at a cost of $19 and adds a markup of 55% of cost to determine his selling price after all trade discounts have been allowed. He lists the calculator with trade discounts of 10/5/2. A distributor buying a large quantity receives the maximum discount. He sells 20 calculators at a 45% markup on cost to a retailer with terms 3/20, n/30 E.O.M. The bill is dated July 22 and has an additional shipping charge of $3.10.

a. At what price does the manufacturer list the calculator?

b. At what price does the dealer sell to the retailer?

c. What is the total amount the retailer pays if he sends a check for half the debt on August 19 and pays the remainder on August 29?

Chapter 4

Name　　　　　　　　　Date　　　　　　　　　Class

TEST YOURSELF

Complete the following problems by filling in the blank spaces. Round dollar amounts to the nearest cent and percents to the nearest tenth where necessary.

Section

	List Price	% Trade Discount	Amount of Discount	Net Price	
1.	_____	_____	$431.20	$1,108.80	4.2
2.	_____	15%	_____	$13.60	4.3

Determine the last date on which the discount may be taken.

	Invoice Date	Credit Terms	Date Goods Received	Last Date to Take Discount	
3.	August 5	3/10, n/60 R.O.G.	August 27	_____	4.7
4.	February 8	2/10, n/30 E.O.M.	February 17	_____	4.7
5.	October 13	4/15, n/30	November 2	_____	4.6

Find the net payment. Round dollar amounts to the nearest cent where necessary.

	Invoice Amount	Date of Invoice	Date of Payment	Credit Terms	Net Payment	
6.	$14,550	March 9	May 18	3/10 – 60X	_____	4.7
7.	$742.38	June 20	July 19	2/10, 1/20, n/30 PROX	_____	4.7

Solve the following problems. Round dollar amounts to the nearest cent and rates to the nearest tenth of a percent where necessary.

8. A leather briefcase that lists for $210 is sold with a 40% trade discount.　　　4.2

   a. What is the net price?

   b. How much discount has been given?

9. A camera that lists for $128 is sold for $99.84. What percent trade discount has been given?　　　4.2

167

	Section
10. At what price should a tape deck be listed in a wholesaler's catalog if he wishes to allow a discount of 28% and sell it for $144?	4.3
11. A dining room table has a list price of $500 with discounts of 12/10. Find:	4.4
a. The net price.	4.5
b. The amount of discount.	
c. The net cost rate factor (% paid).	
d. The single equivalent discount.	
12. Find the price at which a radio should be listed if the desired net price is $68.40 and the discounts are 20/5.	4.3 4.4
13. A lawn chair is listed at $35 less 20%. What additional percent discount must be given to bring the price down to $25, the price being charged by a competitor with similar merchandise?	4.4
14. An invoice for $2,225 is dated September 22. The terms are 2/10, n/30 E.O.M. What amount must be paid on October 9?	4.6 4.7
15. Valley Stores receives an invoice for $846, including $32 for freight and insurance. The terms are 4/10, 1/20, n/45. If the invoice is dated October 2 and paid in full on October 17, find:	4.7
a. The amount paid.	
b. The amount credited to Valley Stores' account.	
c. The amount still owed.	

Chapter 4

TEST YOURSELF

	Section
16. For the invoice shown in Figure 4-3.	4.6
a. What is the list price?	4.7

    b. What is the net price?

    c. What is the total amount of trade discounts?

    d. If the goods are received on July 25 and the invoice is paid on August 22, what amount must be paid?

Computer Supply Co.				
Sold to: The Electronics Store 1442 W. Fifth Avenue San Antonio, TX 78278			Invoice Date: July 12 Invoice No.: 514152 Ship Via: Roadway Express Terms: 3/15, 2/30, n/60 R.O.G.	
Item No.	Quantity	Description	Unit Price	Amount
584 563 755 478	12 45 20 10	Sound cards - 16 bit 56K V.90 int. fax/modem 27.2 GB Hard drives 5 GB tape backup units	$62.30 $197.25 $189.95 $214.99 List Price Less 10%, 5% Net Price Cash Discount Amount Due	

FIGURE 4-3 Sample invoice.

17. Finest Fabrics owed $219 to a supplier. The bill had credit terms 2/10, *n*/30 and was dated January 22. On February 1 the store sent a check for $109.76.   4.8

    a. How much was credited to their account?

    b. How much was still owed?

Chapter 4

Name   Date   Class

COMPUTER SPREADSHEET EXERCISES

On each of the following spreadsheets, place all numbers beneath the appropriate column headings. Use $ signs with dollar amounts and round to the nearest cent. Align decimal points in all columns where they occur.

1. Reproduce the following spreadsheet with the missing values filled in the last three columns. Use formulas to obtain these values. Display decimal values to the hundredths position in the Net Cost Rate Factor column.

List Price	Trade Discount	Net Cost Rate Factor	Amount of Discount	Net Price
$64.00	20%			
$1,693.00	15%			
$42.50	19%			
$96.00	25%			
$23.50	10%			
$9.17	25%			
$41.75	10%			
$2,160.00	12%			
$1,750.00	15%			
$19.43	22%			

2. Reproduce the following spreadsheet with correct values in the last two columns. Assume that all cash discounts begin at the invoice date and that net is paid if the discount is not taken. To calculate the number of days in the Days from Invoice Date to Date Paid column, subtract the invoice date from the date paid. For example,

=E7−D7

calculates the exact number of days between cells E7 and D7. If cell E7 was 01-Feb-99 and cell D7 was 01-Jan-99, the result would be 31. The dates must be entered in any of the acceptable Microsoft Excel date formats, one of which is shown here.* Be sure that the cell containing the calculation is in the General or Number format. Use the IF function to calculate the values in the Amount Due column. This function,

=IF(condition,x,y)

produces the value x if the condition is true or the value y if the condition is false. For instance,

=IF(DAYS FR INV DATE TO DATE PD<=DAYS IN DISC PER,INV AMT*(1-DISC RATE),INV AMT)

will display the correct amount due when the appropriate cell addresses are substituted for the column headings written in the formula.

Invoice Amount	Days in Discount Period	Discount Rate	Invoice Date	Date Paid	Days from Invoice Date to Date Paid	Amount Due
$2,678.12	10	4.0%	08-Aug-99	15-Aug-99		
$901.03	30	1.0%	28-Dec-99	25-Jan-00		
$5,242.91	15	3.0%	11-Mar-00	31-Mar-00		
$1,655.43	15	2.0%	17-Jun-00	01-Jul-00		
$2,981.88	10	2.0%	23-Jul-99	02-Aug-99		
$307.17	20	1.5%	10-Feb-00	02-Mar-00		
$8,946.23	30	3.0%	29-Aug-98	28-Sep-98		
$1,276.98	8	4.0%	17-Dec-99	26-Dec-99		
$4,522.54	10	2.5%	02-Jan-01	11-Jan-01		
$653.32	15	3.0%	22-Nov-00	01-Dec-00		

*To display a date in any of the Excel date formats, use the Format Cells command and select Date.

5 ■ Accounting Topics: Depreciation and Valuation of Inventory

OBJECTIVES

Upon completion of this chapter, the student should be able to:

1. *Compute depreciation, cost recovery, and appropriate schedules using the following methods:*

   straight line;

   sum-of-years' digits;

   declining balance; and

   Modified Accelerated Cost Recovery (MACRS).

2. *Find the value of ending inventory using:*

   FIFO;

   LIFO; and

   average cost.

5.1 DEPRECIATION

Individuals and businesses buy and own assets (things of value) for different reasons. An individual buys a car to get from one place to another, and perhaps for status, if the car is new and expensive. A dishwasher has obvious value as a labor-saving device. As these personal assets perform their functions and become less useful, they are eventually discarded or traded.

Business **assets**—buildings, production equipment, land, record-processing machines, office furniture and similar properties—are all purchased because they will aid in producing revenue in the present as well as in the future. A salesman's car will get him from one place to another, but the aim of the trip is to sell something and thereby create revenue. The dishwasher in a restaurant is a labor-saving device, but it will help produce income because it reduces labor expenses. As business assets are used to generate revenue, their cost is periodically matched against the revenues produced. The cost of the asset is thus "used up" until the asset is sold or scrapped. This process is called **depreciation.**

This traditional view of depreciation was altered in two revisions of the tax law. In the Economic Recovery Tax Act of 1981, depreciation was replaced by the concept of **cost recovery,** where the entire cost of the asset is "recovered" or allowed as a deduction, over a period of years, regardless of any actual value the asset may have at the end of the period. The Tax Reform Act of 1986 modified the 1981 method, generally lengthening the time required for a business to recover its investment cost.

There are several methods of determining each year's depreciation for income tax and accounting purposes. These result in a predetermined, regular schedule of depreciation or cost recovery. A company does not have to use the same method of depreciation for all of its various assets. Furthermore, it can use one method to prepare its financial statements, and another for income tax returns. The method chosen for internal accounting purposes will be based on the financial objectives of management, but Internal Revenue Service (IRS) guidelines must be followed when preparing federal tax returns. Once the appropriate federal income tax method of depreciation has been selected, it must be used throughout the useful life of the asset.*

In recent times, several methods of computing depreciation have been used including **straight-line, sum-of-years'-digits,** and **declining-balance.** These methods can be used to compute depreciation for accounting purposes within the business regardless of an asset's date of acquisition, but they normally are used for federal tax purposes only on assets that were placed in service before 1981.

The most recent methods of depreciation are the **Accelerated Cost Recovery System (ACRS),** used to depreciate assets purchased from 1981 through 1986, and the **Modified Accelerated Cost Recovery System (MACRS),** used on assets placed into service after 1986. Both methods are used for federal income tax purposes; businesses often use another method of depreciation for internal accounting.

Four of the methods of depreciation used for tax calculations and/or financial accounting purposes will be discussed here. The following terms are common to each of the methods.

- **ESTIMATED USEFUL LIFE FOR THE ASSET:** No one can be sure how long a particular piece of equipment will last, but a reasonable estimate must be made. The IRS either has such estimates available or sets the number of years according to the type of asset.

- **SALVAGE OR TRADE-IN VALUE:** What will the equipment be worth at the end of its useful life? Will it be sold for continued use, or will it have value as scrap? Again, this cannot be accurately determined in advance, but a reasonable estimate is made. The salvage or scrap value may be zero.

- **ACCUMULATED DEPRECIATION:** Annual depreciations are added to give total depreciation from the time of purchase to date. Thus, if an asset depreciates $200 each year, the accumulated depreciation for three years will be $600 ($200 + $200 + $200).

- **TOTAL DEPRECIATION:** Original cost less salvage or resale value.

- **BOOK VALUE:** The original cost less the accumulated depreciation gives the value that the equipment still has in the company's accounting records. Book value does not have any relationship to market value.

Straight-Line Method

The **straight-line method** of depreciation is the simplest to calculate and results in the amount of depreciation each year the equipment is owned. For example, a piece of equipment that will depreciate $1,000 during 5 years will depreciate $200 ($\frac{\$1,000}{5}$) each year. In general:

$$\text{Annual depreciation} = \frac{\text{Original cost} - \text{Salvage value}}{\text{Life in years}}$$

■ **Example 1**

A machine that cost $5,000 has an estimated useful life of 5 years. At the end of that time it is expected that the machine will have a salvage value of $500. How much will it depreciate during each of the 5 years if the straight-line method is used?

---

*Depreciation is used only with assets having a useful life of more than 1 year. Land cannot be depreciated because it can be used indefinitely.

$$\text{Annual depreciation} = \frac{\text{Original cost} - \text{Salvage cost}}{\text{Life in years}}$$

$$= \frac{5{,}000 - 500}{5}$$

$$\text{Annual depreciation} = \frac{4{,}500}{5} = \$900$$

Depreciation each year is $900. The annual rate of depreciation is $\frac{1}{5}$, or 20%, of total depreciation.

The depreciation of assets during their useful lives is shown in a **depreciation schedule,** a table that summarizes depreciation and book value. The depreciation schedule for the machine in Example 1 is shown in Table 5-1. For simplicity it is assumed that the machine is owned for 5 whole, taxable years.

TABLE 5-1 DEPRECIATION SCHEDULE (STRAIGHT-LINE METHOD)

(1) END OF YEAR	(2) ANNUAL DEPRECIATION	(3) ACCUMULATED DEPRECIATION (END OF YEAR)	(4) BOOK VALUE (END OF YEAR)
—	—	—	$5,000 (cost)
1	$900	$ 900	4,100 (5,000 − 900)
2	900	1,800 (900 + 900)	3,200 (5,000 − 1,800)
3	900	2,700	2,300
4	900	3,600	1,400
5	900	4,500	500

The starting book value in column 4 is the cost of the machine before depreciation has begun. For each year of use:

1. Enter annual depreciation (column 2).

2. Add annual depreciation to the last total under Accumulated Depreciation (column 3).

3. Subtract accumulated depreciation from cost, the first figure under Book Value (column 4).

Check: The depreciation schedule should always be checked; an error at any point will make subsequent figures incorrect.

Note: The final book value must equal the salvage value in this method. The last figure under Accumulated Depreciation is the sum of all annual depreciations and must equal total depreciation (original cost − salvage value). For this problem,

$$4{,}500 = 5{,}000 - 500$$

$$\$4{,}500 = \$4{,}500$$

■

Example 2 A $44,000 piece of equipment is expected to last 6 years and to have a scrap value of $8,000. Find the annual depreciation by the straight-line method and set up the depreciation schedule.

$$\text{Annual depreciation} = \frac{\text{Original cost} - \text{Salvage value}}{\text{Life in years}}$$

$$= \frac{44,000 - 8,000}{6}$$

$$\text{Annual depreciation} = \frac{36,000}{6} = \$6,000$$

Check: In Table 5-2 the final book value equals the scrap value ($8,000). The final figure under Accumulated Depreciation ($36,000) equals total depreciation ($44,000 − $8,000 = $36,000). ■

TABLE 5-2 DEPRECIATION SCHEDULE (STRAIGHT-LINE METHOD)

(1) END OF YEAR	(2) ANNUAL DEPRECIATION	(3) ACCUMULATED DEPRECIATION (END OF YEAR)	(4) BOOK VALUE (END OF YEAR)
—	—	—	$40,000 (cost)
1	$6,000	$ 6,000	38,000
2	6,000	12,000	32,000
3	6,000	18,000	26,000
4	6,000	24,000	20,000
5	6,000	30,000	14,000
6	6,000	36,000	8,000

Chapter 5

TRY THESE PROBLEMS (Set I)

Use the straight-line method of depreciation in these problems. In Problems 1 through 3, set up the depreciation schedules for the equipment.

	COST	SALVAGE OR RESALE VALUE	USEFUL LIFE IN YEARS
1.	$12,000	$1,000	5
2.	2,500	400	4
3.	3,000	750	3

4. A delivery truck cost $36,000 and will be used for 4 years. If the truck has an estimated salvage value of $8,000, find:

   a. Annual depreciation.

   b. Accumulated depreciation after 2 years.

5. A machine cost $15,000 and has an estimated life of 8 years. It will then have a salvage value of $1,400. What is the book value at the end of 3 years?

6. A personal computer cost $1,920. The company expects to use it for 6 years, but at the end of that time it will be worthless. Find the accumulated depreciation and the book value after 4 years.

Sum-of-Years'-Digits Method

The **sum-of-years'-digits** method may be used if the company finds it advantageous to take a large part of the depreciation at the beginning of the period. The *digits* referred to in the name of the method are the first year, second year, third year, and so on, of the life of the asset. If the equipment is expected to last 5 years, for example, the digits that are added are $1 + 2 + 3 + 4 + 5$. The sum of the digits is 15.

The sum can also be found by a formula that is convenient when the number of years (n) is large:

$$\text{Sum of the years' digits} = \frac{n(n+1)}{2}$$

For 15 years this formula would give: $\frac{15(16)}{2} = 120$.

Each year's depreciation rate is a fraction whose numerator is one of the year digits, taken in reverse order (5, 4, 3, 2, 1 for a 5-year life) and whose denominator is the sum of the years' digits (15). For equipment with 5 years of useful life the depreciation fractions would be:

$\frac{5}{15}$ for the first year

$\frac{4}{15}$ for the second year

$\frac{3}{15}$ for the third year

$\frac{2}{15}$ for the fourth year

$\frac{1}{15}$ for the fifth year

The annual depreciation for each year is found by multiplying the appropriate fraction by the total depreciation. Thus, the largest single annual depreciation is assigned to the first year, with a decreasing amount for each subsequent year. The formula to calculate annual depreciation for the sum-of-years'-digits method is:

$$\text{Annual depreciation} = \text{Depreciation fraction} \times (\text{Original cost} - \text{Salvage value})$$

■ **Example 3**

An asset that cost $44,000 is expected to last 6 years and have a salvage value of about $2,000 at the end of that time. Find the depreciation for each of the 6 years using the sum-of-years'-digits method and set up the depreciation schedule.

$$\text{Total depreciation} = \text{Original cost} - \text{Salvage value}$$

$$\text{Total depreciation} = \$44,000 - \$2,000 = \$42,000$$

$$\text{Sum of the years' digits} = 1 + 2 + 3 + 4 + 5 + 6 = 21$$

or

$$\text{Sum of the years' digits} = \frac{n(n+1)}{2} = \frac{6 \times 7}{2} = 21$$

180 CHAPTER 5 ACCOUNTING TOPICS: DEPRECIATION AND VALUATION OF INVENTORY

To find annual depreciation, set up the fractions and multiply by total depreciation, $42,000. Use the digits in order but start with the largest. The depreciation schedule is worked out as it was in the straight-line method and is given in Table 5-3.

TABLE 5-3 DEPRECIATION SCHEDULE (SUM-OF-YEARS'-DIGITS METHOD)

(1) YEAR	DEPRECIATION FRACTION	×	TOTAL DEPRECIATION	=	(2) ANNUAL DEPRECIATION	(3) ACCUMULATED DEPRECIATION (END OF YEAR)	(4) BOOK VALUE (END OF YEAR)
—					—	—	$44,000 (cost)
1	$\frac{6}{21}$	×	42,000	=	$12,000	$12,000	32,000 (44,000 − 12,000)
2	$\frac{5}{21}$	×	42,000	=	10,000	22,000 (12,000 + 10,000)	22,000 (44,000 − 22,000)
3	$\frac{4}{21}$	×	42,000	=	8,000	30,000	14,000
4	$\frac{3}{21}$	×	42,000	=	6,000	36,000	8,000
5	$\frac{2}{21}$	×	42,000	=	4,000	40,000	4,000
6	$\frac{1}{21}$	×	42,000	=	2,000	42,000	2,000

Check: The last entry under Book Value ($2,000) should be equal to trade-in or scrap value ($2,000). The last entry under Accumulated Depreciation ($42,000) must equal Total Depreciation ($42,000). ∎

CALCULATOR HINT

Many calculators can be used as shown below to solve Example 3. Try your calculator to see if you get the annual depreciation for each year.

	ENTER	DISPLAY
	42,000 ÷ 21 × 6 =	12,000
	5 =	10,000
	4 =	8,000

Continue with 3 =, 2 =, and 1 =.

Chapter 5

TRY THESE PROBLEMS (Set II)

Use the sum-of-years'-digits method of depreciation in these problems. In Problems 1, 2, and 3, set up depreciation schedules for the equipment.

	COST	SALVAGE OR RESALE VALUE	USEFUL LIFE IN YEARS
1.	$1,800	$120	6
2.	4,200	700	4
3.	3,000	600	3

4. A machine cost $9,800 and has an estimated useful life of 7 years. At the end of that time it will have no salvage value.

   a. What is the annual depreciation during the third year?

   b. What is the accumulated depreciation after 3 years?

5. Some office furniture was purchased at a cost of $2,600. It has a life expectancy of 6 years, after which time it should have a value of $500. What is the annual depreciation during the third year?

   a. What is the annual depreciation for the second year?

   b. What is the accumulated depreciation after 2 years?

   c. What is the book value after 2 years?

181

6. A company car purchased for $25,200 has an estimated life of 3 years, at which time it should have a scrap value of $6,000.

   a. Find the annual depreciation for each of the 3 years.

   b. What is the book value at the end of 3 years?

7. Fast Foods Restaurant purchased a deep fryer for $1,500. If the estimated life is 5 years and the scrap value is $100, find the book value at the end of 3 years.

DEPRECIATION

Declining-Balance Method

The **declining-balance method** may also be used when depreciation should be highest in the early years. In this method each year's depreciation is a *constant percent of the previous year's book value*. The percent used may vary but may not exceed

$$2\left(\frac{1}{\text{Life in years}}\right)$$

Although salvage value is not used in the same way as in the other two methods, it should be determined before depreciation is started. The salvage value limits the decrease in book value: *book value may not go below salvage value.*

We will use the maximum permitted percent (often referred to as **double-declining-balance**) to illustrate the declining-balance method for the machine in Example 1, where depreciation was calculated by the straight-line method. The annual depreciation amounts are found using the following formula:

$$\text{Annual depreciation} = \text{Rate} \times \text{Previous year's book value}$$

■ **Example 4**

A machine that costs $5,000 has an estimated useful life of 5 years. At the end of that time it is expected that the machine will have a salvage value of $500. Find the depreciation during each of the 5 years and set up the depreciation schedule if the double-declining-balance method is used.

$$\text{Double-declining-balance rate} = 2 \times \frac{1}{\text{Life in years}}$$

$$\text{Double-declining-balance rate} = 2 \times \frac{1}{5} = \frac{2}{5} = 40\%$$

The rate remains the same each year, with the rate being multiplied by the previous year's book value to determine the annual depreciation amount. (*Do not* subtract the salvage value from the cost before calculating depreciation.) The depreciation schedule is given in Table 5-4.

TABLE 5-4 DEPRECIATION SCHEDULE (DOUBLE-DECLINING-BALANCE METHOD)

(1) END OF YEAR	(2) ANNUAL DEPRECIATION (RATE × PREVIOUS YEAR'S BOOK VALUE)	(3) ACCUMULATED DEPRECIATION (END OF YEAR)	(4) BOOK VALUE (END OF YEAR)
—	—	—	$5,000 (cost)
1	0.4 × 5,000 = 2,000	2,000	3,000 (5,000 − 2,000)
2	0.4 × 3,000 = 1,200	3,200 (2,000 + 1,200)	1,800 (5,000 − 3,200)
3	0.4 × 1,800 = 720	3,920	1,080
4	0.4 × 1,080 = 432	4,352	648
5	0.4 × 648 = ~~259.20~~ 148	4,500	500

184 CHAPTER 5 ACCOUNTING TOPICS: DEPRECIATION AND VALUATION OF INVENTORY

Check: The checks used in the other two methods do not apply here. The last book value will come out equal to the salvage value only if the rate is determined in advance by a more complex formula. If the highest permitted percent is used, the last book value may be somewhat higher than scrap value, but not lower.

Note particularly the fifth-year calculation. If depreciation were found in the same way as for previous years, it would equal $259.20 and reduce the book value to $388.80 ($648.00 − $259.20). This is less than the $500 salvage value. Since this is not permitted, depreciation for the last year is limited to the amount that will reduce the book value to the salvage value of $500. Recall that this adjustment is never needed in the straight-line or sum-of-years'-digits methods. ■

Chapter 5

Name Date Class

TRY THESE PROBLEMS (Set III)

Use the double-declining-balance method in these problems. In Problems 1, 2, and 3 set up the depreciation schedules for the equipment.

	COST	SALVAGE OR RESALE VALUE	USEFUL LIFE IN YEARS
1.	$1,400	$100	5
2.	5,000	400	4
3.	2,000	200	5

4. An asset that cost $9,100 is to be depreciated over 7 years. The scrap value is estimated at $850. Find the book value at the end of 2 years.

5. A medical laboratory purchased new equipment for blood analysis. The cost was $50,000 and the estimated useful life is 6 years. It will then have a trade-in value of about $3,000. What is the accumulated depreciation at the end of the first 3 years of use?

6. Office furniture valued at $15,500 has a useful life of 10 years, but at the end of that time it will have no salvage value.

   a. What percent of the book value can the company take as depreciation each year?

   b. How much will the furniture depreciate during the third year?

7. Dix Realty paid $12,000 for a new air conditioner. Constant use will wear the machine out in about 6 years, at which time it will have no salvage value. Find the book value after 3 years.

Modified Accelerated Cost Recovery System (MACRS)

The depreciation rules already discussed, straight-line, sum-of-years'-digits, and declining-balance, are often used by businesses for financial accounting purposes. They also are used for income tax purposes, but only for property acquired before 1981. The Economic Tax Recovery Act of 1981 created another set of tax rules, which apply to assets acquired between January 1, 1981, and December 31, 1986, called the **Accelerated Cost Recovery System (ACRS).** For property acquired after December 31, 1986, a third set of rules called the **Modified Accelerated Cost Recovery System (MACRS),** is applicable.*

The ACRS rules enacted in 1981 provided an economic stimulus because the cost-recovery amounts were based on accelerated methods and artificially shortened the useful life of an asset. Businesses were able to recover 100% of the cost of assets over a period of time that was substantially shorter than the property's useful economic life. (For recovery rates under this method see Appendix Table C-10.) The Tax Reform Act of 1986 modified the ACRS system. The new method, the Modified Accelerated Cost Recovery System (MACRS), retained many of the original features while returning to a recovery period that more nearly approaches a true life. Depreciable property placed in service after December 31, 1986, is categorized in one of nine cost-recovery classes: 3, 5, 7, 10, 15, 20, 27.5, 31.5, or 39 years. A half-year of depreciation is assumed in the year of acquisition. Zero salvage value is also assumed.

Since 100% of the purchase price is recovered during the class time period, the MACRS method will permit an asset to have a book value below its actual value. Other methods of depreciation consider the trade-in value or the scrap value of an asset before determining depreciation.

The following MACRS classifications apply:

3-year	Tractor units for over-the-road use, special tools, race horses over 2 years old, any other horse over 12 years old.
5-year	Automobiles, taxis, buses, office machinery, light-duty trucks, computers, research and experimental equipment, and semiconductor manufacturing equipment.
7-year	Office furniture and fixtures, most types of machinery and manufacturing equipment, appliances, carpet, and furniture used in residential rental properties. This class covers assets not specifically assigned elsewhere.
10-year	Barges, vessels, petroleum, and food equipment.
15-year	Billboards, service stations, and land improvements.
20-year	Certain farm buildings, utilities and sewers.
27.5-year	Residential rental real estate such as rental houses and apartments.
31.5-year	Nonresidential rental real estate such as office buildings, stores, and warehouses placed in service before May 13, 1993.
39-year	Nonresidential rental real estate such as office buildings, stores, and warehouses placed in service after May 12, 1993.

Under MACRS, the depreciation deduction can be calculated in two different ways. The first method uses a double-declining-balance method to depreciate the 3-, 5-, 7-, and 10-year classes with a switch to straight-line about halfway through the recovery period. The 15- and 20-year classes use a one and one-half declining-balance method, also switching later to straight-line, while the 27.5-, 31.5-, and 39-year classes exclusively use the straight-line method. Because the calculations require additional knowledge and are repetitive, an asset's cost recovery is often computed using tables. Table 5-5 shows the annual rate allowed for cost

---

*MACRS is also known as the General Depreciation System (GDS).

188 CHAPTER 5 ACCOUNTING TOPICS: DEPRECIATION AND VALUATION OF INVENTORY

TABLE 5-5 MACRS DEPRECIATION

YEAR	3-YEAR CLASS	5-YEAR CLASS	7-YEAR CLASS	10-YEAR CLASS
1	0.3333	0.2000	0.1429	0.1000
2	0.4445	0.3200	0.2449	0.1800
3	0.1481	0.1920	0.1749	0.1440
4	0.0741	0.1152	0.1249	0.1152
5	—	0.1152	0.0893	0.0922
6	—	0.0576	0.0892	0.0737
7	—	—	0.0893	0.0655
8	—	—	0.0446	0.0655
9	—	—	—	0.0655
10	—	—	—	0.0655
11	—	—	—	0.0328

recovery in the 3-, 5-, 7-, and 10-year classes. Each proportion is multiplied by the original cost to determine the cost recovery for any given year. In other words,

$$\text{Annual depreciation} = \text{Depreciation rate} \times \text{Original cost}$$

MACRS depreciation calculations are based on a **half-year convention.*** This means that it is assumed that an asset is acquired halfway through the year, regardless of when the asset is placed in service. Thus, the first-year depreciation rate is less than the second-year depreciation rate.

■ **Example 5**

A computer (5-year class) that cost $20,000 was placed into service in March 2000. Find the amount of depreciation that can be claimed each year using the MACRS method and set up the depreciation schedule.

Use the rates in the 5-year class of the MACRS table. Multiply the decimal for each year by the cost of the computer.

YEAR	COST × RATE	ANNUAL DEPRECIATION
1	$20,000 × 0.2	$ 4,000
2	20,000 × 0.32	6,400
3	20,000 × 0.192	3,840
4	20,000 × 0.1152	2,304
5	20,000 × 0.1152	2,304
6	20,000 × 0.0576	1,152
	Total	$20,000

Notice that the entire cost of $20,000 is depreciated.

*A 1989 update of MACRS requires that the company use a different method that depreciates assets according to the quarter they were acquired if more than 40% of all assets are placed in service during the last 3 months of the calendar year. Residential rental property (27.5 years) is depreciated according to the month it was acquired.

TABLE 5-6 DEPRECIATION SCHEDULE
MODIFIED ACCELERATED COST-RECOVERY SYSTEM (MACRS)

(1) END OF YEAR	(2) ANNUAL DEPRECIATION	(3) ACCUMULATED DEPRECIATION	(4) BOOK VALUE (END OF YEAR)
—	—	—	$20,000 (cost)
1	$4,000	$4,000	16,000 (20,000 − 4,000)
2	6,400	10,400	9,600
3	3,840	14,240	5,760
4	2,304	16,544	3,456
5	2,304	18,848	1,152
6	1,152	20,000	0

The depreciation schedule for the computer is shown in Table 5-6. As was the case with previous methods, the beginning book value (column 4) is the cost of the machine before depreciation has begun. The steps of constructing the schedule remain the same as before:

1. Enter the annual depreciation (column 2).
2. Add the annual depreciation, year by year, to obtain the accumulated depreciation (column 3).
3. Subtract accumulated depreciation from cost to obtain the book value (column 4).

Check: The final book value should equal 0, and the accumulated cost recovery should equal the original cost. ■

CALCULATOR HINT

Many calculators work in a way that will help you solve Example 5 most efficiently. If you enter $20,000 and the multiplication symbol ⨯ first, you do not have to press these keys again. This certainly saves time and removes a source of human error. Press the keys shown below:

ENTER	DISPLAY
20,000 ⨯ 0.2 =	4,000
0.32 =	6,400
0.192 =	3,840
0.1152 =	2,304
0.0576 =	1,152

Chapter 5

TRY THESE PROBLEMS (Set IV)

Use the MACRS method in these problems. In Problems 1, 2, and 3, prepare depreciation schedules for the equipment.

ITEM	COST	CLASS YEARS
1. Copying machine	$22,000	5
2. Special tools	8,000	3
3. Lighting system	5,000	7

4. United Transport purchased a barge for $250,000. Using the 10-year recovery class, find:

   a. The annual depreciation in the fourth year.

   b. The accumulated depreciation after 4 years.

   c. The book value after 4 years.

5. In 1999, Northwest Pipeline, Ltd., purchased some heavy equipment in the 7-year class. Find the amount of depreciation allowed on their income tax returns for 1999, 2000, and 2001 if the total cost was $85,000.

191

5.2 VALUATION OF INVENTORY

Inventory, the merchandise on hand at a specific time, is an important business asset, and good management of inventory levels helps make a business profitable. If too much inventory is on hand, the business has money invested that might be needed for another purpose. But if the inventory is too small or of the wrong kind, the firm may lose sales and even customers.

The value placed on inventory is quite important since it is used in the calculation of net income and taxes, and it is to the company's advantage to place a value on its inventory that will keep taxes down. For this reason inventories are valued at cost rather than at selling price. Some retailers find it more convenient to keep inventory records in terms of selling price because markup, sales, returns, and other records are in terms of selling price rather than cost. When recording the store's net worth and income, however, the inventory value is reduced to the cost level by using the $C + M = S$ relationship with an average markup percent for the store.*

Two methods can be used to determine the amount of merchandise that remains in stock. Some firms use a **perpetual inventory system,** which provides an accurate count of in-stock merchandise at any time. When this system is used, larger businesses normally rely on electronic cash registers that are connected to a computer to maintain inventory and sales records. Smaller firms might record information on inventory cards, which are manually updated when the amount of inventory changes. However, it is often necessary for a business to count its physical inventory at regular intervals. This method is called a **periodic inventory system.** It may seem that this should not be necessary if an initial count is made and accurate records are kept after that, but as a practical matter this does not work. There may be losses due to damage or pilferage, and there may be mistakes in record keeping.

The question we are then faced with is, Once the inventory is counted and listed, what value should be placed on it? In **specific identification,** the actual cost of an item is matched with the item itself. The value placed on ending inventory is the sum of these costs. Companies that sell high-cost items such as furniture, automobiles, jewelry, and so on, usually use this method, since this type of item can be easily cost-coded and identified. However, if inventories contain many items purchased at different times for different amounts, it may be impractical to code each item and identify the cost of the ones that remain in stock. (This is why many businesses prefer taking inventory at retail, since the retail value of all identical merchandise is the same). This section discusses three other methods for a periodic inventory evaluation.

FIFO (First In-First Out)

The assumption in the FIFO method of inventory valuation is that the merchandise is sold in the same order as it is bought: The first ones bought are also the first ones sold. If prices are rising, this leaves the high-cost items in stock and results in higher income and taxes.

■ **Example 1**

The Buy-2-Sell Co. (Table 5-7) had 55 tire pumps remaining in stock at the end of the year. Evaluate the inventory using the FIFO method.

TABLE 5-7 BUY-2-SELL CO.: PURCHASES OF TIRE PUMPS

DATE	NUMBER BOUGHT	PRICE PER ITEM
March 10	30	$3.00
June 5	150	2.80
September 30	40	3.50
December 4	50	3.70

*For a review of markup, see Chapter 3.

194 CHAPTER 5 ACCOUNTING TOPICS: DEPRECIATION AND VALUATION OF INVENTORY

By the FIFO method, the 55 items remaining in inventory consist of the last 50 tire pumps and 5 from the next-to-last purchase. To find the inventory value, multiply the number of items times the price per item and add.

	NUMBER		PRICE PER ITEM		
	50	×	$3.70	= $185.00	value of last 50
	5	×	3.50	= 17.50	value of previous 5
Total	55			$202.50	

The value of the tire pump inventory using FIFO is $202.50. ■

LIFO (Last In-First Out)

The assumption in the LIFO method of inventory valuation is that the last items bought are the first ones sold. Therefore, the items in stock are those purchased first. This may not appear to be a reasonable assumption, but it is permitted, and in a rising market gives the business lower income and taxes.

■ Example 2

For the 55 tire pumps remaining in the Buy-2-Sell Co. inventory, use the LIFO method to place a value on inventory.

By the LIFO method, the 55 items remaining in inventory consist of the first 30 tire pumps and 25 from the second purchase.

	NUMBER		PRICE PER ITEM		
	30	×	$3.00	= $ 90	value of first 30
	25	×	2.80	= 70	value of next 25
Total	55			$160	

Using LIFO, the tire pump inventory is valued at $160. ■

Average Cost

Closest to actual cost as a basis for inventory valuation is average cost, a weighted average* that can be determined from company purchasing records. According to the average-cost method, the total cost for a given set of items is divided by the number of items bought, and this result is multiplied by the number of items still in stock, yielding the total value of the items.

■ Example 3

Table 5-8 again shows the record of tire pumps purchased by the Buy-2-Sell Co., this time with accumulated totals. Find the value of the ending inventory of 55 tire pumps using the average-cost method.

TABLE 5-8 BUY-2-SELL CO.: PURCHASE OF TIRE PUMPS

DATE	NUMBER BOUGHT	PRICE PER ITEM	TOTAL COST
March 10	30	$3.00	$ 90
June 5	150	2.80	420
September 30	40	3.50	140
December 4	50	3.70	185
Total	270		$835

*See Chapter 12.

VALUATION OF INVENTORY

$$\text{Average cost per item} = \frac{\text{Total cost}}{\text{Number bought}} = \frac{835}{270} = \$3.09$$

If the inventory includes 55 tire pumps, each is priced at the average of $3.09. The total value of the 55 tire pumps can be found using the following formula:

$$\text{Number in stock} \times \frac{\text{Average cost}}{\text{per item}} = \frac{\text{Inventory value}}{\text{using average cost}}$$

For our example,

$$55 \times 3.09 = \$169.95$$

is the value of the tire pump inventory using the average cost method. ∎

■ **Example 4**

A stationery store, after checking inventory at the end of the year, found that there were 65 Kwik Calculators still in stock. According to the store's records, the calculators purchased during the year were:

DATE	NUMBER	PRICE
January 5	25	$14
March 2	30	13
June 12	20	13
September 29	50	10
November 1	40	8

What is the dollar value of the 65 calculators according to:

- The average-cost method
- The FIFO method
- The LIFO method

CALCULATOR HINT

You can use the memory keys on your calculator to help you solve Example 3 most efficiently.

ENTER	DISPLAY
30 ⊠ 3 M+	90 and the letter M (The M + key also functions as an = key.)
150 ⊠ 2.80 M+	420
40 ⊠ 3.50 M+	140
50 ⊠ 3.70 M+	185
MR ÷ 270 =	3.0925925

This is the average price. You must round it to $3.09 before you multiply it by 55.

| 55 ⊠ 3.09 = | 169.95 |

The inventory of 55 pumps is valued at $169.95 by the average-cost method.

FIFO method. The late purchases remain in inventory. The stock of 65 calculators is made up of 40 from the last purchase and 25 from the September 29 purchase.

	NUMBER		PRICE PER ITEM	
	40	×	$8	= $320
	25	×	10	= 250
Total	65			$570

The value of inventory by the FIFO method is $570.

LIFO method. The early purchases remain in inventory. The stock of 65 calculators is made up of 25 from the first purchase, 30 from the second, and 10 from the third.

	NUMBER		PRICE PER ITEM	
	25	×	$14	= $350
	30	×	13	= 390
	10	×	13	= 130
Total	65			$870

The value of inventory by the LIFO method is $870.

Average-cost method. Complete the record of purchases by determining the total cost for each purchase (column 4) and totaling columns 2 and 4. Then find the average cost of the calculators.

(1) DATE	(2) NUMBER	(3) PRICE PER ITEM	(4) TOTAL COST
January 5	25	$14	$ 350
March 2	30	13	390
June 12	20	13	260
September 29	50	10	500
November 1	40	8	320
Total	165		$1,820

$$\text{Average cost per item} = \frac{\text{Total cost}}{\text{Number bought}} = \frac{1{,}820}{165} = \$11.03$$

To find the value of the 65 calculators in stock use:

$$\text{Number in stock} \times \frac{\text{Average cost}}{\text{per item}} = \frac{\text{Inventory value}}{\text{using average cost}}$$

or

$$65 \times 11.03 = \$716.95.$$

The value of inventory by the average-cost method is $716.95.
In summary, the inventory of 65 calculators is valued at:

$570.00   by the FIFO method

$870.00   by the LIFO method

$716.95   by the average-cost method

Check: The three methods are expected to give different answers, but the three results should be of the same magnitude.

Note the effect of the decline in prices. The FIFO method now gives the lowest inventory value. In the previous illustrations, when prices were rising, LIFO gave the smallest value. ∎

Now go to Try These Problems (Set V).

NEW FORMULAS IN THIS CHAPTER

- Total depreciation = Original cost − Resale or salvage value
- Book value = Original cost − Accumulated depreciation
- Annual depreciation (straight-line) = $\dfrac{\text{Original cost} - \text{Resale or salvage value}}{\text{Life in years}}$
- Sum-of-years' digits = $\dfrac{n(n+1)}{2}$, where n is useful life in years
- Annual depreciation (sum-of-years' digits) = Depreciation fraction × (Original cost − Salvage value)
- Maximum percent (declining-balance-method) = $2\left(\dfrac{1}{\text{Life in years}}\right)$
- Annual depreciation (declining-balance) = Rate × Previous year's book value
- Annual depreciation (MACRS) = Depreciation rate × Original cost

Chapter 5

TRY THESE PROBLEMS (Set V)

Find the value of these inventories by the FIFO, LIFO, and average-cost methods.

INVENTORY	PURCHASES MADE DURING YEAR
1. Twenty blouses	50 at $10.20 in April 75 at $11.00 in May
2. Eighty ballpoint pens	100 at 92¢ in January 100 at 70¢ in May 75 at 78¢ in August
3. Twenty-six sleeping bags	20 at $50 on October 30 20 at $55 on November 2 25 at $62 on November 25
4. Fourteen lighting fixtures	12 at $24 in March 20 at $25 in May 20 at $30 in June
5. Thirteen paint sprayers	5 at $69.60 on February 14 8 at $72.00 on April 12 10 at $77.40 on May 21

6. The J & G Toy Shoppe bought new dolls at different times during the Christmas season. Their orders showed 20 at $25.50, 18 at $30, and 25 at $32. January's inventory showed 20 dolls still in stock. Evaluate this inventory using each of the following methods:

   a. FIFO.

   b. LIFO.

   c. Average cost.

Chapter 5

	Name	Date	Class

END-OF-CHAPTER PROBLEMS

Set up the depreciation schedules for the following equipment:

	COST	SALVAGE VALUE	ESTIMATED USEFUL LIFE	METHOD
1.	$ 8,000	$ 300	4 years	Double-declining-balance
2.	37,500	None	5	Sum-of-years'-digits
3.	1,842	75	6	Straight-line
4.	3,440	150	4	Sum-of-years'-digits
5.	40,000	600	10	Double-declining-balance for first 3 years
6.	1,000	200	5	MACRS
7.	2,500	300	4	Straight-line
8.	12,000	1,500	10	MACRS for first 4 years

9. A manufacturer uses the double-declining-balance method of depreciation for his equipment. A machine bought at $80,000 has an estimated life of 10 years and a salvage value of $1,500. By what amount will the equipment depreciate during the second year?

10. A delivery truck that was bought for $24,000 has an expected trade-in value of $3,000 at the end of its 6 years of useful life.

   a. What is the book value at the end of the 6 years if depreciation is by the sum-of-years'-digits method? (Do not calculate).

   b. Using the straight-line method, find annual depreciation and set up the depreciation schedule.

11. A salesman's car cost $24,735 and will have a trade-in value at the end of 4 years of $2,235. Using the sum-of-years'-digits depreciation, set up a schedule for the 4 years.

12. All-Around Town, Inc., purchased $7,500 of office furniture in 1999. If this asset falls in the 7-year class, use MACRS to prepare a depreciation schedule for the first 3 years.

13. Artie's Auto Shop purchased some office equipment for $2,500. It is estimated that the salvage value will be $400 at the end of a 5-year life. Prepare depreciation schedules using each of the following methods:

   a. MACRS.

   b. Straight-line.

   c. Double-declining-balance.

   d. Sum-of-years'-digits.

14. The air-conditioning unit in a restaurant lasts about 8 years and then has no salvage value. The machine cost $3,800 including installation. After 3 years, the restaurant decided to modernize and replace the unit. What was the book value at that time if the straight-line method of depreciation was used?

15. Ron Riter bought a personal computer to be used in his work as an author. The cost was $2,400 and it has an estimated useful life of 4 years. At that time the salvage value should be about $250.

   a. What is the book value at the end of 3 years using the double-declining-balance method?

   b. How much will the machine depreciate in the fourth year?

16. A delivery service buys new trucks after keeping them for their estimated useful life of 5 years. Each truck costs $31,000 with an estimated trade-in value of $4,000. Using the sum-of-years'-digits method, find the depreciation for the second year.

17. During 1999, Thompson Research, Inc., purchased various assets in the 3-year class that totaled $18,000. Use the MACRS method to prepare a depreciation schedule.

Chapter 5

END-OF-CHAPTER PROBLEMS

18. It is estimated that machinery costing $36,895 will have a salvage value of $4,500 at the end of 11 years of use. Find the annual depreciation and the book value at the end of 6 years, using the straight-line method.

19. A machine that cost $22,470 is estimated to have a useful life of 6 years. At that time it will have no salvage value. Using the sum-of-years'-digits method, find:

   a. Annual depreciation during the fourth year.

   b. Accumulated depreciation for the first 4 years.

   c. Book value at the end of 4 years.

20. Daily Data Co. purchased a computer for $80,000. The expected useful life of the equipment is 5 years and the trade-in value is expected to be about $5,000. Set up the depreciation schedule by the double-declining-balance method.

21. A used truck purchased at the beginning of January 2000 cost a store $12,000. It is to be depreciated over a 5-year period at the end of which it will be sold for about $1,600.

   a. Using the straight-line method, what fraction of total depreciation can be taken each year?

   b. Using the sum-of-years'-digits method, find the depreciation during 2002.

   c. Using the MACRS method, find the depreciation during 2002.

   d. What is the rate that can be used in the double-declining-balance method?

22. A store made two purchases of dining room chairs during the year and paid $30 a chair for each purchase. At the end of the year there were 10 chairs left in stock. Will it make any difference which method of inventory valuation the store uses? Why?

23. A retailer of bicycles ends the year with 60 bicycles of a particular model in stock. Costs have been rising during the year. Would LIFO or FIFO give the store a higher inventory valuation? Why?

24. The P & A Supermarket had 6 cases of canned green beans in inventory at the end of the year. They had bought 4 cases in February at $8.40, 10 in September at $8.80, and 6 in December at $9.00.

   a. Using the average-cost method, what is the value of the inventory?

   b. What is the value of the 6 cases using FIFO?

   c. Evaluate the inventory using LIFO.

   d. Which method would give the best tax advantage?

25. While making a spot check of the inventory count (Problem 24), the supervisor discovered that among the cases of green beans there had been included 2 cases of beets. What values should be given to the remaining cases of beans using:

   a. Average cost?

   b. LIFO?

   c. FIFO?

26. The QED Company bought the following units:

April 1	30 units at $64
September 30	16 units at $68
October 19	24 units at $70
November 26	5 units at $74

At the last inventory, 29 units remained. Evaluate this inventory using each of the following methods:

   a. Average cost.

   b. FIFO.

   c. LIFO.

Chapter 5

END-OF-CHAPTER PROBLEMS

27. The Linen Store had 30 cotton sheets in inventory at the end of the year. Purchases of this item were:

    1 dozen on May 20 at $66.00 per dozen.

    3 dozen on June 3 at $68.40 per dozen.

    2 dozen on September 2 at $72.00 per dozen.

    Evaluate this inventory using:

    a. Average cost.

    b. FIFO.

    c. LIFO.

28. The Bath Boutique's records show the following purchases of bathroom towels:

    | February 11 | 10 at $5.00 each |
    | June 14 | 10 at $6.00 each |
    | November 4 | 10 at $6.50 each |

    At the end of the year 12 towels were still unsold. What value should they be given if the method of inventory valuation is:

    a. Average cost?

    b. FIFO?

    c. LIFO?

Chapter 5

SUPERPROBLEMS

1. A machine that has an initial value of $120,000 has a useful life of 10 years. It will then be worth $22,000 as a trade-in for a newer model. Without setting up depreciation schedules or calculating depreciation each year, find the book value of the machine at the end of 3 years by the straight-line and the sum-of-years'-digits methods. Which method should give the larger total depreciation for the first 3 years? Why?

2. The owner of the Fun and Games Store decided to add a new computer game to his stock. The first purchase of 20 games was a trial to see if it would sell. The games cost $27.50 each. He paid the bill early and was able to take a 2% discount. After selling a few of the games quickly, he found a different distributor who priced the item at $28 but offered a trade discount of 8% with an additional 2% for large orders. The retailer bought 50 games and was allowed both discounts. A shipping charge of $12 was added. The bill was dated October 15 with terms 3/10, *n*/30, R.O.G. The merchandise arrived on October 21 and he paid the bill on October 31. Christmas business was disappointing and 12 of the games remained in inventory. What is the value of the games, calculated by the average-cost method?

Chapter 5

TEST YOURSELF

1. A machine that was bought for $23,000 has an expected trade-in value of $5,000 at the end of its 4 years of useful life. Set up a depreciation schedule for this asset using the straight-line method.

End of Year	Annual Depreciation	Accumulated Depreciation	Book Value (End of Year)
—			_____
1	_____	_____	_____
2	_____	_____	_____
3	_____	_____	_____
4	_____	_____	_____

Section 5.1

2. Joe's Catering Service purchased a delivery truck for $31,300. It is estimated that the salvage value will be $8,500 at the end of its 3 years of useful life. Set up a depreciation schedule using the sum-of-years'-digits method.

End of Year	Annual Depreciation	Accumulated Depreciation	Book Value (End of Year)
—			_____
1	_____	_____	_____
2	_____	_____	_____
3	_____	_____	_____

5.1

3. A manufacturer uses the double-declining-balance method of depreciation for his equipment. A machine bought at $80,000 has an estimated life of 5 years and a scrap value of $8,000. Prepare a depreciation schedule for the machine.

End of Year	Annual Depreciation	Accumulated Depreciation	Book Value (End of Year)
—			_____
1	_____	_____	_____
2	_____	_____	_____
3	_____	_____	_____
4	_____	_____	_____
5	_____	_____	_____

5.1

Section 5.1

4. During the year, Bighorn Steakhouse purchased some assets in the 7-year class that totaled $16,000. Use the MACRS method to prepare a depreciation schedule for the first 3 years. (The first three annual rates in the 7-year class of the MACRS table are 0.1429, 0.2449, and 0.1749.)

End of Year	Annual Depreciation	Accumulated Depreciation	Book Value (End of Year)
—			_____
1	_____	_____	_____
2	_____	_____	_____
3	_____	_____	_____

5. An art supply store's records show the following purchases of stencilling brushes:

> 8 at $2.50 in January
> 3 at $3.00 in February
> 5 at $4.00 in March
> 6 at $4.50 in July
> 10 at $5.00 in November

At the end of the year 9 brushes remained in stock. Find the dollar value of these brushes using each of the following methods:

a. Average cost.   5.2

b. FIFO.   5.2

c. LIFO.   5.2

Chapter 5

COMPUTER SPREADSHEET EXERCISES

Use formulas where appropriate to find the missing values on the following spreadsheets.

All columns and column headings should appear on your spreadsheet exactly as they are shown here. Be sure to use $ signs for dollar amounts and round to the nearest cent where necessary. Make sure your spreadsheet fits across the width of the page.

Use absolute references to match the cell entries entered for the cost, salvage value, and useful life in Exercises 2, 3, and 4. An absolute reference keeps the same cell in a formula throughout the copy operation and is specified with a $ in front of the column and row designation. For example, assume that the cost of the asset is in cell C7 at the top of the spreadsheet. If you wish to use that value in a formula, and you do not want the cell address to change when you copy the formula, enter C7 for the cell address. Preparing depreciation schedules with absolute references enables you to recompute an entire schedule for a different asset automatically simply by changing the referenced cells.

1. Use the MACRS method to prepare a depreciation schedule for the following asset. MACRS depreciation rates can be found in Table 5-5. Your preliminary schedule should appear as follows:

    Asset: Office Equipment
    Class: 5-year
    Cost: $10,000

End of Year	Depreciation Rate	Annual Depreciation	Accumulated Depreciation	Book Value
----	----	----	----	
1				
2				
3				
4				
5				
6				

2. Use the straight-line method of depreciation to prepare a depreciation schedule for the following asset:

    Asset: Office Equipment
    Useful Life (in years): 10
    Cost: $14,700
    Salvage Value: $2,100

    Type this information somewhere above the depreciation schedule you are about to construct. Place the titles Asset, Useful Life, Cost, and Salvage Value in a separate column from Office Equipment, 10, $14,700 and $2,100 so that the numbers can be used in formulas. Use the straight-line depreciation function (with absolute cell references for the cost, salvage, and life entries required in parenthesis) to compute the annual depreciation amounts in your spreadsheet. In Microsoft Excel the function

    =SLN(cost,salvage,life)

    calculates the straight-line annual depreciation allowance of an asset with initial value of cost, an expected useful life, and a final value of salvage. For example,

    =SLN(1000,200,5)

calculates the annual depreciation of an asset that costs $1,000, has a salvage value of $200, and a useful life of 5 years. Use the following titles for your spreadsheet columns:

```
     End          Annual       Accumulated      Book
   of Year     Depreciation    Depreciation     Value
```

3. Prepare a depreciation schedule for the asset in Exercise 2 using the sum-of-years'-digits method. Use the same column titles and the function

$$=SYD(cost,salvage,life,period)$$

to calculate the sum-of-years'-digits annual depreciation for a specified year (period). For instance,

$$=SYD(8000,500,4,3)$$

calculates the third-year depreciation of an asset that costs $8,000, has a salvage value of $500, and a useful life of 4 years.

4. Prepare a depreciation schedule for the asset in Exercise 2 using the double-declining-balance method. Use the same column titles and =DDB(cost,salvage,life,period) to calculate annual depreciation amounts.

5. Carly's Office Warehouse needs help in completing the following spreadsheet showing purchase information and the number of items remaining in inventory for a certain model calculator during an inventory period. Use $ signs for dollar amounts and round to the nearest cent. Give totals for the Number and Total Cost columns. All headings, underlinings, and numbers should appear as shown below. Be sure to use boldface type to display the Carly's Office Warehouse heading and use formulas where appropriate. (Try writing IF statements for the FIFO and LIFO ending inventory values.)

Carly's Office Warehouse

Number of items in ending inventory: 35

Date	Number	Price per Item	Total Cost
January 10	20	$11.35	
March 1	25	$12.00	
May 20	30	$12.50	
June 30	40	$11.70	
August 15	30	$10.45	_____
Total			

	FIFO	LIFO	Ave. Cost
End. Inv. Value			

6 ■ Accounting Topics: Partnership Profits and Payroll

OBJECTIVES

Upon completion of this chapter the student should be able to:

1. Divide the profits (or losses) of a business among the members of a partnership according to:

    equal shares;

    fixed percents;

    fixed ratios;

    original investment;

    average investment;

    interest on investment; and

    combination of methods.

2. Calculate gross earnings based on:

    salary;

    hourly rates;

    commission;

    production quotas; and

    combination of methods.

3. Compute employees' net earnings using the following payroll deductions:

    FICA;

    federal income tax; and

    pension, health insurance, and other deductions.

6.1 DIVISION OF PROFITS IN A PARTNERSHIP

When two or more people become co-owners of a business, they have formed a **partnership.** Each makes a contribution to the business, usually money and/or some special skill, and each expects to share in the net income and net losses of the business. The way in which these profits and losses are shared is agreed to by all the partners and should be included in the *articles of copartnership,* the legal document stating the conditions according to which the business will be run.

There are many methods for dividing profits and losses to which the partners may agree. To illustrate some of these, consider the partnership of Al, Ben, and Cal. The procedures would be the same for two, four, or more partners. Six possible partnership agreements follow.

Agreement I:
Equal Shares

If all the partners feel that their contributions are about equally valuable, they may agree to divide the profits and losses equally. By this method a $12,000 profit in a business with three partners would give each $4,000 $\left(\dfrac{\$12,000}{3}\right)$.

Agreement II:
Fixed Percents

The three partners have agreed to give each a fixed percent of profits. Al will get 20%, Ben will get 50%, and Cal will get 30%. (The percents must add to 100%.) During the first year of operations the business has a profit of $8,000. How much should each partner receive as his share?

$$R \times B = P$$

Al gets 20% of $8,000	$0.20 \times 8,000 = \$1,600$
Ben gets 50% of $8,000	$0.50 \times 8,000 = \ \ 4,000$
Cal gets 30% of $8,000	$0.30 \times 8,000 = \ \ 2,400$
	Total = $8,000

This total checks the arithmetic.

CALCULATOR HINT

Many calculators work as illustrated below. If your calculator operates in this way it is very simple and efficient to use it for Agreement II.

Agreement II   Once you press 8,000 and ⊠, you do not need to touch these keys again.

ENTER	DISPLAY
8,000 ⊠ 0.2 ⊟	1,600
0.5 ⊟	4,000
0.3 ⊟	2,400

The profits are $1,600, $4,000, and $2,400.

You can use your Memory keys for Agreement II.

ENTER	DISPLAY
8,000 ⊠ 0.2 ⊟ M+	1,600
0.5 ⊟ M+	4,000
0.3 ⊟ M+	2,400
MR	8,000

The individual profits are $1,600, $4,000, and $2,400. Total profits are $8,000.

Agreement III:
Fixed Ratios

The three partners have agreed to divide the profits according to a fixed ratio of 3 : 5 : 4. (3 : 5 : 4 **is** the short way of stating that the ratio of Al's share to Ben's is $\frac{3}{5}$, Ben's to Cal's is $\frac{5}{4}$, and Al's to Cal's is $\frac{3}{4}$.)

To divide $8,000 profit according to this ratio, think of the $8,000 as divided into 12 (3 + 5 + 4) equal parts. Al would get 3 of the 12 parts, Ben would get 5 of the 12 parts, and Cal would get 4 of the 12 parts.

Al receives:	3/12 × 8,000 =	$2,000.00
Ben receives:	5/12 × 8,000 =	3,333.33
Cal receives:	4/12 × 8,000 =	2,666.67
	Total =	$8,000.00

This total checks the arithmetic.

CALCULATOR HINT

You can use the Memory keys on your calculator to solve Agreement III.

ENTER	DISPLAY	
3 ⊠ 8,000 ⊟ 12 M+	2,000	
5 ⊠ 8,000 ⊟ 12 M+	3,333.3333	(Round to 3,333.33 only for your notes.)
4 ⊠ 8,000 ⊟ 12 M+	2,666.6666	(Round to 2,666.67 only for your notes.)
MR	7,999.999	(This is total profits. Round to 8,000.)

The individual profits are $2,000, $3,333.33, and $2,666.67. The total profits are $8,000.

Agreement IV: Original Investment

When the business was started, Al invested $12,000, Ben invested $15,000, and Cal invested $13,000. They agreed to share profits according to the ratios of each partner's investment to the total investment. The total investment is $40,000 ($12,000 + $15,000 + $13,000).

$$\text{Al receives:} \quad \frac{12,000}{40,000} \times 8,000 = \$2,400$$

$$\text{Ben receives:} \quad \frac{15,000}{40,000} \times 8,000 = 3,000$$

$$\text{Cal receives:} \quad \frac{13,000}{40,000} \times 8,000 = \underline{2,600}$$

$$\text{Total} = \underline{\underline{\$8,000}}$$

This total checks the arithmetic.

Chapter 6

	Name	Date	Class

TRY THESE PROBLEMS (Set I)

1. Joanne, Maryann, and Kathryn have started a small business. Joanne invested $9,000, Maryann invested $15,000, and Kathryn invested $6,000. During the first year the business showed a profit of $8,400. Divide this profit among the three partners according to each of the following agreements.

   a. In the ratio of 2 : 4 : 2.

   b. By fixed percents: 25%, 60%, 15%.

   c. In the ratios of the original investments.

   d. $500 to each and the remainder in the ratio of 3 : 4 : 1.

   e. $800 to each and the remainder by fixed percents: 35%, 50%, 15%.

2. If the partners of Problem 1 agree to divide the profits equally, how much will each partner receive?

3. Three partners agreed to share profits or losses from their partnership in the ratio 7 : 3 : 4.

   a. How much would each partner receive if there was a net profit of $21,000?

   b. What is each one's share of an $18,000 loss?

DIVISION OF PROFITS IN A PARTNERSHIP

**Agreement V:
Average
Investment
Per Month**

When the partners' investments change during the life of the business, the original investments (Agreement IV) may not be a fair basis for the division of profits. Instead, the agreement could state that the average investment per month should be used.

Suppose, for example, that although Al made the smallest initial investment, he finds it possible to add more money twice during the second year of operations. Suppose also that Ben makes no additional investment, Cal adds only once to his investment during the year, and the second-year profit is $11,000. Table 6-1 shows the month-by-month record of each partner's investment.

TABLE 6-1 RECORD OF TOTAL AMOUNT INVESTED

	Al	Ben	Cal
January	$ 12,000	$ 15,000	$ 13,000
February	12,000	15,000	13,000
March	12,000	15,000	13,000
April	12,000	15,000	13,000
May	14,000	15,000	16,000
June	14,000	15,000	16,000
July	14,000	15,000	16,000
August	14,000	15,000	16,000
September	16,000	15,000	16,000
October	16,000	15,000	16,000
November	16,000	15,000	16,000
December	16,000	15,000	16,000
Total	$168,000	$180,000	$180,000
Average per month (divide the total by 12)	$ 14,000	$ 15,000	$ 15,000

The average investments per month for the year are now used as the basis for the division of profits in the same way as the original investments were used in Agreement IV. Each partner receives his share according to the ratio of his average investment to the new total. The new total investment is the sum of the average investments.

Al: $14,000

Ben: $15,000

Cal: $15,000

Total: $44,000

Al receives: $\dfrac{14,000}{44,000} \times 11,000 = \$ \ 3,500$

Ben receives: $\dfrac{15,000}{44,000} \times 11,000 = \ \ \ 3,750$

Cal receives: $\dfrac{15,000}{44,000} \times 11,000 = \ \ \ 3,750$

Total profits = $11,000   *This total checks the arithmetic.*

220 CHAPTER 6 ACCOUNTING TOPICS: PARTNERSHIP PROFITS AND PAYROLL

Agreement VI: Interest on Investment

The agreement for dividing profits may be that each partner receives interest at a specified percent on his investment, and the remainder of the profits is divided according to another agreed-on basis. (Interest will be discussed in Chapter 7, but because the time is one year the calculation is simply a percent of the investment.)

For example, suppose that as an original investment, Dave invested $12,000, Ed invested $15,000, and Frank invested $13,000. This time, however, the agreement specifies that each partner receive 8% simple interest on his original investment and that the remainder of the profits be divided equally. Assuming a profit for the first year of $8,000, first year division of profits can be computed as follows:

Dave receives interest on $12,000 at 8% for 1 year:

$$(\text{Interest}) \; I = (0.08)12,000$$
$$I = \$960 \qquad \$\;960$$

Ed receives interest on $15,000 at 8% for 1 year:

$$I = (0.08)15,000$$
$$I = \$1,200 \qquad 1,200$$

Frank receives interest on $13,000 at 8% for 1 year:

$$I = (0.08)13,000$$
$$I = \$1,040 \qquad \underline{1,040}$$

Profits distributed as interest $3,200

This leaves $4,800 ($8,000 − $3,200) to be divided equally among the three partners. Each receives $\dfrac{\$4,800}{3}$ = $1,600 in addition to his interest payment.

The totals received are:

	INTEREST	+	EQUAL SHARE	= TOTAL PAYMENT
Dave:	960	+	1,600	= $2,560
Ed:	1,200	+	1,600	= $2,800
Frank:	1,040	+	1,600	= $2,640
			Total profit	= $8,000

Although the investments used in this illustration were the original amounts, a partnership agreement could have specified that the average investments per month be used instead.

Combination of Methods

Many other combinations of the previously discussed methods for dividing profits in a partnership are possible. Two are illustrated in Examples 1 and 2.

■ **Example 1**

Isabelle and Jerry, as partners in a new pet shop, invested $28,000 and $20,000, respectively. They agreed to share equally in any profit or loss up to $6,000. Any amount over that would be divided 55% for Isabelle and 45% for Jerry.

1. During their first year they lost $3,200. For how much is each partner responsible?
    The agreement specifies equal shares up to $6,000. Therefore, each partner would be liable for $1,600 of the loss.

2. During the second year they showed a profit of $11,000. How much does each partner get?

DIVISION OF PROFITS IN A PARTNERSHIP 221

The first $6,000 is divided equally; each receives $3,000. The remaining $5,000 ($11,000 − $6,000) is divided according to the percents in the agreement.

Isabelle receives: 55% of $5,000 = 0.55 (5,000) = $2,750
+ 3,000
Total $5,750 $5,750

Jerry receives: 45% of $5,000 = 0.45 (5,000) = $2,250
+ 3,000
Total $5,250 $5,250

Total profits = $11,000 ■

■ **Example 2**   Keith, Lola, and Moe started a local business with a total capital of $225,000. Keith invested $100,000, Lola invested $75,000, and Moe invested $50,000. They did not change their investments during the first year they were in business, but at the beginning of the second year Keith added $25,000 to his investment, while Lola withdrew $15,000 of hers. Moe did nothing further. Their agreement stated that second-year profits would be divided by giving each 10% on their average investment per month with the remainder divided in the ratio of 5 : 3 : 2. If the profit is $40,000, how much does each partner receive?

Part of the division of profits is based on average investment per month and this must be determined for each partner.

Keith: Added $25,000 at the beginning of the year and left it for the entire year, making his average investment equal to $125,000 per month.

Lola: Withdrew $15,000 at the beginning of the year, leaving $60,000 invested for the entire year. The average investment is $60,000 per month.

Moe: The original investment of $50,000 remained during the entire year. The average investment is $50,000 per month.

Interest on average investment:

Keith: 10% of $125,000 = (0.1) 125,000 = $12,500
Lola: 10% of $ 60,000 = (0.1) 60,000 =   6,000
Moe: 10% of $ 50,000 = (0.1) 50,000 =   5,000

Total paid in interest = $23,500
Profit left for distribution = $40,000 − $23,500 = $16,500

Distribution by ratio of 5 : 3 : 2 (The total of $1,650 is divided into 10 equal shares [5 + 3 + 2]):

Keith gets 5 of the 10 = 5/10 × 16,500 = $ 8,250
Lola gets 3 of the 10 = 3/10 × 16,500 =   4,950
Moe gets 2 of the 10   = 2/10 × 16,500 =   3,300

Total distributed by ratio = $16,500

The partners receive:

	KEITH	LOLA	MOE
Interest	$ 12,500	$ 6,000	$ 5,000
By ratio	8,250	4,950	3,300
Total	$ 20,750	$ 10,950	$ 8,300

Check: $20,750 + $10,950 + $8,300 = $40,000. ■

Chapter 6

	Name	Date	Class

TRY THESE PROBLEMS (Set II)

Table 6-2 shows the investments of Nat and Oscar in their partnership. Distribute the following profits.

TABLE 6-2 MONTHLY PARTNERSHIP INVESTMENTS

Beginning of:	Nat	Oscar
January	$ 9,000	$ 6,000
February	9,000	7,000
March	8,000	7,000
April	8,000	7,000
May	8,000	7,000
June	10,000	8,000
July	10,000	8,000
August	10,000	9,000
September	12,000	9,000
October	12,000	9,000
November	12,000	9,000
December	12,000	10,000

1. $8,640 according to the partners' average investment per month.

2. $8,640 by first giving each 10% interest on average investments and then dividing the remainder equally.

3. $8,640 by first giving each 12% on average investments and then giving Nat 60% and Oscar 40% of the remainder.

4. $16,300 by giving each partner 9% interest on average investments and then dividing the remainder in the ratio 7 : 3.

5. $20,000 by dividing $8,100 according to average investments and the remainder by giving Nat 65% and Oscar 35%.

6. Harry and Larry became partners on January 1 and agreed to distribute profits by paying 8% interest on their average investments per month and dividing the remaining profit equally. Harry invested $54,000 on January 1 and did nothing further. Larry invested $62,000 to start and another $8,000 on May 1. On September 1 he withdrew $10,000.

   a. Determine each partner's average investment.

   b. What is each partner's share of a $162,000 profit?

6.2 PAYROLL

The preparation of the payroll is now a computerized function in large U.S. businesses. In smaller firms the payroll may be prepared by a bookkeeper or payroll clerk. Regardless of the size of the firm, however, the payroll is generally a major company expense, and understanding the steps by which total or gross pay is determined and then reduced to "take-home" or net pay is important to both the employer and employee.

The **Fair Labor Standards Act,** which has now been extended to cover the majority of full-time employees in the nation, was passed by Congress in 1938. It sets a minimum hourly wage rate and specifies the number of hours per week after which overtime rates must be paid.* There are several exemptions from these wage and hour provisions, the most important being executive, administrative, and professional employees and outside salespeople. The overtime regulation requires at least $1\frac{1}{2}$ times the employee's regular rate of pay for all hours worked in excess of 40 in the work week. The minimum wage has been increased over the years, and as of 1999 was $5.15 per hour.

Salaries, wages, and commissions are the most important means of compensating employees and will be considered here in determining **gross earnings** (earnings before deductions). The terms salaries and wages, although sometimes used interchangeably, are properly applied to different kinds of work.

Salaries

A **salary** is a fixed amount of money per employee pay period, regardless of the actual number of hours worked. Most managerial and professional positions are salaried, and many of these workers are not covered by the Fair Labor Standards Act. Although frequently quoted on an annual basis, salaries are commonly paid weekly, biweekly, semimonthly, or monthly. The gross amount for each pay period will depend on how many pay periods there are during the year.

COMMON PAY PERIODS	NUMBER OF ANNUAL PAYCHECKS
Weekly (once a week)	52
Biweekly (every 2 weeks)	26
Semimonthly (twice a month)	24
Monthly (once a month)	12

To determine gross earnings for a salaried employee, use the following formula:

$$\text{Gross earnings} = \frac{\text{Annual salary}}{\text{Number of paychecks per year}}$$

■ **Example 1**

Tonya Turner is an accountant earning a $42,000 annual salary.

1. Determine her gross earnings each pay period if she is paid monthly.
   There are 12 monthly pay periods. Tonya's gross earnings will be

   $$\frac{42,000}{12} = \$3,500$$

2. Determine her gross earnings each pay period if she is paid biweekly.
   There are 26 biweekly pay periods. Tonya's gross earnings will be

   $$\frac{42,000}{26} = \$1,615.38 \quad ■$$

*This is a very general statement of the provisions of the Fair Labor Standards Act. There are additional provisions for specific industries. Many states also have labor laws.

226 CHAPTER 6   ACCOUNTING TOPICS: PARTNERSHIP PROFITS AND PAYROLL

Hourly Wages

The majority of working Americans are paid an **hourly wage.** Under this method, the employee's gross earnings are determined by the number of hours worked during the pay period and the hourly rate paid by the employer. Employees receiving an hourly rate are generally covered by the Fair Labor Standards Act. For the first 40 hours of work, gross earnings for an hourly worker are calculated as follows:

$$\text{Gross earnings} = \text{Hours worked} \times \text{Rate per hour}$$

The usual **overtime** rate is $1\frac{1}{2}$ times the hourly rate. In most instances a worker is required to have 40 hours of work before receiving overtime; however, some workers are paid overtime when an employee works more than 8 hours on a given day. There are some situations where **double time** might be paid for work on Sundays or holidays, but it is not required under federal law. To compute gross earnings with **time and a half** for overtime, the formula is

$$\text{Gross earnings} = \text{Regular hours} \times \text{Regular rate} + \text{Overtime hours} \times \text{Regular rate} \times 1.5$$

Of course, it is possible for a worker to have overtime hours at several different overtime rates.

■ **Example 2**

Determine the gross earnings for Nathan Joseph, whose base rate is $8 an hour with time and a half for work in excess of 40 hours. He worked the following number of hours during one week:

EMPLOYEE	M	T	W	TH	F	S
Joseph, N.	$8\frac{1}{2}$	7	10	$7\frac{1}{2}$	9	5

First find the hours worked for the week. This gives a total of 47 hours. Break the 47 hours into 40 regular and 7 overtime at time and a half.

$$\text{Time-and-a-half rate} = \$8 \times 1.5 = \$12 \text{ an hour}$$

Gross wages: 40 hours at $8 per hour (40 × 8) = $320
7 hours at $12 per hour (7 × 12) = 84
47 hours $404 gross wages ■

■ **Example 3**

Determine the gross pay for Maria Alvarez. Her regular rate of pay is $10.50 per hour and her company pays time and a half for all time worked over 8 hours in one day, no matter how many hours are worked in a week. She worked the following hours during the last pay period:

EMPLOYEE	M	T	W	TH	F	S
Alvarez, M.	12	7	10	—	5	—

In this case there are 6 hours of overtime, with 12 − 8 = 4 on Monday and 10 − 8 = 2 on Wednesday. There are 28 regular hours consisting of 8 on Monday, 7 on Tuesday, 8 on Wednesday, and 5 on Friday. The gross pay calculation would be:

Time-and-a-half rate = 10.50×1.5 = $15.75 an hour

Gross wages: 28 hours at $10.50 per hour = $294.00

6 hours at $15.75 per hour = 94.50

34 hours $388.50 gross wages ■

CALCULATOR HINT

You may find it convenient to use the memory keys on your calculator for solving Example 3.

ENTER	DISPLAY	
28 $\boxed{\times}$ 10.50 $\boxed{M+}$	294	(Remember, the M+ key also functions as an = key)
6 $\boxed{\times}$ 15.75 $\boxed{M+}$	94.5	
$\boxed{MR}$	388.5	

The gross earnings are $388.50.

If a worker is paid a weekly amount for a specified number of hours a week, the overtime rate is determined by first converting the weekly earnings to an equivalent hourly rate. For example, a clerk paid $360 a week for 40 hours earns $360 ÷ 40, or $9.00 an hour.

Time and a half would be 9.00×1.5 = $13.50

Double time would be 9.00×2 = $18.00

Piecework Wages

When wages are determined by a worker's actual production, they are called **piecework wages** and are also subject to the hours and wage provisions of the Fair Labor Standards Act. The basic **straight piecework** plan provides that the worker is paid a specified amount for each unit he produces that passes inspection. In this case gross earnings are found using the following formula:

> Gross earnings = Number of units produced × Rate per unit

■ **Example 4**

A worker operating a lathe and producing table legs is paid $12 per piece. During one week he produced 72 legs, of which 1 was rejected after inspection. What were his gross earnings?

Since he will not be paid for the rejected piece, his wages will be based on 71 pieces at $12 each. Therefore,

Gross earnings = 71×12 = $852

Many variations of straight piecework have been introduced to encourage greater productivity. For example, a **quota** that is considered reasonable for the average worker may be set. An employee who produces more than the quota may then be paid a premium for the extra production or a higher rate for the entire production. Another commonly used method is the **differential piece-rate plan,** where the rate paid per item depends on the number of items produced. ■

Example 5

Suppose that a quota has been set at 65 pieces per week for the worker in Example 4. Determine the gross earnings if:

1. All production is paid for at $12.50 each, if total production exceeds the quota.
   Since 71 accepted pieces were produced, the gross amount would be:

   $$71 \times 12.50 = \$887.50$$

2. A bonus is paid for production above the quota. Additional production is worth $13.00 instead of $12.00.
   The gross amount for the 71 pieces would be calculated as follows:

   65 pieces at $12 (65 × 12) = $780

   6 pieces at $13 (6 × 13) = 78

   71                $858 total gross earnings ∎

Commission

Commission, or a combination of salary and commission, is the usual method of paying salespeople. As in piecework rates, commissions tend to encourage greater productivity, and there are many variations in the arrangements made. Commissions are generally calculated as a percent of net sales (sales less returns and discounts), but may be based on total sales or sales above a specified amount, or may be on a sliding scale. We will consider the main types of commission arrangements.

The simplest case is **straight commission,** where the salesperson is paid a fixed percent of net sales. Under this plan, gross earnings are found with the following formula.*

$$\boxed{\text{Gross earnings} = \text{Rate of commission} \times \text{Net sales}}$$

Example 6

A salesman for a chemical company works on a straight commission of 14% of his net sales. During one week he had net sales of $6,000. Find his gross earnings.

$$\text{Gross earnings} = 0.14 \times 6,000 = \$720$$

A widely used combination is **salary plus commission.** Commonly used by large retail stores, the salesperson is paid a fixed amount per pay period, plus a commission on all sales. The formula for gross earnings with salary plus commission is

$$\boxed{\text{Gross earnings} = \text{Salary} + (\text{Rate of commission} \times \text{Net sales})} \quad \blacksquare$$

Example 7

Smith Manufacturing pays all sales personnel a salary of $450 per week plus a 5% commission on net sales. During the week Cindy Kane had $7,000 of net sales. What were her gross earnings?

$$\text{Gross earnings} = 450 + (0.05 \times 7,000)$$
$$= 450 + 350$$
$$= \$800$$

---

*This formula is based on the percentage formula, $R \times B = P$, which is discussed in Section 2.3.

Some commissioned personnel, especially those who receive a salary, are required to sell a specified amount (or **quota**) before the commission begins. This guarantees a certain performance level for the salary paid. Under these circumstances, the formula becomes

$$\text{Gross earnings} = \text{Salary} + (\text{Rate of commission} \times \text{Net sales over quota}) \quad \blacksquare$$

■ **Example 8**

Sam Sells is paid a salary of $500 a week plus 4% of all net sales over $4,000. Find his gross earnings for the week in which he had net sales of $8,800.

$$\text{Gross earnings} = 500 + [0.04 \times (8{,}800 - 4{,}000)]$$
$$= 500 + 192$$
$$= \$692$$

Many companies are willing to pay a higher rate of commission once a salesperson has made sufficient sales to cover job-related expenses, or to encourage greater productivity. As net sales increase, the rate of commission increases. This plan is known as **sliding-scale commission.** The gross earnings will be the sum of the commission earned at each rate. ■

■ **Example 9**

Willard's Restaurant Supply uses a sliding scale to determine the salespeople's commissions. On a weekly basis the rate is 10% on the first $3,000, 14% on the next $2,000, and 18% on any amount over $5,000. During the week one salesman sold $10,800 worth of supplies, but an order for $900 was returned. Find his gross earnings.

Subtract the return from sales to find net sales:

$$\text{Net sales} = 10{,}800 - 900 = \$9{,}900$$

Gross earnings would be:

$$10\% \text{ of } 3{,}000 = \$\ \ 300$$
$$14\% \text{ of } 2{,}000 = \ \ \ \ 280$$
$$18\% \text{ of } 4{,}900 = \ \ \ \underline{882}$$
$$\$9{,}900 \quad \$1{,}462 \text{ gross earnings} \quad \blacksquare$$

Because commission earnings are normally based on monthly sales, many companies allow their salespersons to receive a **draw** (an advance against commissions) during the pay period. The draw is deducted from the employee's commission at the time the employee's earnings are determined. If the commission does not equal the draw, the employee owes the company the difference, which is normally charged against future commissions.

■ **Example 10**

This past month, Melody Cunningham recorded net sales of $39,000. Determine her gross earnings if she is paid a commission of 7% of net sales and received a draw of $500 during the period.

Subtract the draw from the commission to find gross earnings.

$$\text{Gross earnings} = \text{Commission} - \text{Draw}$$
$$= (0.07 \times 39{,}000) - 500$$
$$= 2{,}730 - 500$$
$$= \$2{,}230 \quad \blacksquare$$

CHAPTER 6 ACCOUNTING TOPICS: PARTNERSHIP PROFITS AND PAYROLL

■ **Example 11** Fisch and Herring, Inc., are importers of seafood. The office staff consists of an office manager, an administrative assistant, and a clerk who addresses envelopes for direct mail advertising. Calculate the gross earnings of each if all are paid weekly and the basic work week is 40 hours. Overtime is paid at time and a half for all work in excess of 40 hours in one week.

The manager's salary is $40,000 a year, but she receives no overtime payments. Her weekly gross earnings are:

$$\frac{40,000}{52} = \$769.23$$

The administrative assistant earns $412 a week. She worked 40 hours during the week and came in on Saturday for 4 hours to do a special report. She earns $412 plus 4 hours overtime at $1\frac{1}{2}$ times her hourly rate.

$$\text{Hourly rate} = \frac{\text{Weekly salary}}{\text{Number of hours per week}} = \frac{412}{40} = \$10.30$$

$$\text{Overtime rate} = 1.5 \times 10.30 = \$15.45$$

Her gross earnings are:

$$\text{Regular earnings} = \$412.00$$
$$\text{Overtime earnings } (4 \times \$15.45) = \underline{61.80}$$
$$\$473.80 \text{ gross earnings}$$

The clerk is paid on a piecework basis, 15¢ each for the first 1,500 envelopes she addresses and 21¢ each for additional ones. No check is made on accuracy, but the manager checks returns to make sure they are not excessive. The clerk typed 2,000 envelopes during the week. Her gross pay would be:

First 1,500 at $0.15 each (1,500 × 0.15) = $225
Additional 500 at $0.21 each (500 × 0.21) = 105
             2,000 $330 gross earnings

The salesman is paid $475 a week and is expected to sell about $8,500 worth of seafood. If he sells more than his quota, he receives a commission of 9% on the additional sales. He sold $9,900 worth of merchandise during the week. His gross earnings are:

$$\text{Gross earnings} = \text{Salary} + \text{Commission on sales over quota}$$
$$= 475 + [0.09 \times (9,900 - 8,500)]$$
$$= 475 + 126$$
$$= \$601 \text{ gross earnings} \quad ■$$

Chapter 6

| Name | Date | Class |

TRY THESE PROBLEMS (Set III)

Determine the gross periodic earnings for each of the following employees.

1. A college professor earning $55,500 is paid:

   a. Monthly.

   b. Semimonthly.

2. An office manager earning $49,000 is paid:

   a. Biweekly.

   b. Weekly.

3. A worker assembling machine parts earns $11.80 an hour for a 40-hour week. He worked 46 hours during one week and was paid double time for the additional hours because they were worked on Sunday.

4. Green Lawns, Inc., pays its employees $60 for an 8-hour day. Work in excess of 8 hours is considered overtime and is paid at $1\frac{1}{2}$ times the regular rate. Fred Flower did not work on Monday, worked 8 hours on Tuesday and Wednesday, and took 10 hours on Thursday and Friday to complete his work.

5. A woman pressing shirts in a dry cleaning store receives $280 a week and an additional 40¢ per shirt if she exceeds her quota. She completes 98 shirts above the quota during the week.

6. A machine operator works under a piecework system that pays $1.50 each for the first 500 units and $1.75 each for production over this quota. During the week he produced 580 accepted units.

7. A factory has the following differential piece-rate plan. For the first 200 units produced the worker is paid $390. If production is more than 200 units, the next 20 are worth $2.25 each, and for any after that payment goes up to $2.50 a piece. Sue Shnell produced 291 units during the week.

8. A real estate salesman operating on a straight 3% commission on sales sold two houses during the month. One house was sold for $165,000 and the other for $94,900.

9. Gamble's Department Store pays its salespeople $120 a week plus 6% on net sales (sales less returns) over specified amounts in each department. In the furniture department the quota is $2,000 a week. One salesman sold $8,500 worth of furniture, but an order for a $310 chair was canceled before shipment.

10. A large plumbing supply company uses a sliding scale to determine the salespeople's commissions. On a weekly basis the rate is $2\frac{1}{2}$% for the first $1,000, 5% for the next $3,000, and 7% for sales above $4,000. During the week one salesman sold $9,250 worth of supplies, but one order for $216 was returned.

11. Maria Hernandez is a manager of a woman's clothing store. She receives a salary of $150 a week plus a 5% commission on her net sales and a 1% commission on the net sales of the sales associates working under her. Her sales last week were $6,300 with no returns, while the other employees had sales of $28,950 with returns of $740. She also received a draw of $300 during the week. Find her gross earnings.

12. Complete the following payroll record for C. J. Wholesalers. All employees are required to work 40 regular hours after which they receive overtime at time and a half. (Use the table below.)

| EMPLOYEE | \multicolumn{6}{c}{HOURS WORKED} | \multicolumn{2}{c}{TOTAL HRS.} | REG. RATE | \multicolumn{3}{c}{GROSS EARNINGS} |
	M	T	W	TH	F	S	REG.	O.T.		REG.	O.T.	TOTAL
Able, M.	$5\frac{1}{2}$	10	9	10	8	—			$6.80			
Green, S.	8	8	$7\frac{1}{2}$	9	8	$4\frac{1}{2}$			$9.40			
Lee, C.	9	9	8	10	—	—			$8.50			

Deductions

Deductions must be calculated after gross earnings have been determined. Some deductions are required by law and some are voluntary. Federal law requires the deduction of a Federal Insurance Contribution Act tax, better known as social security and medicare taxes, and an income tax. In addition, many states and some cities impose income taxes, and these, as well as the federal income tax, are withheld; that is, they are deducted before salaries and wages are paid. Other deductions include union dues, health insurance, pension payments, and charitable contributions.

The only required deductions that we will consider here are those for the federal government. These are collected, or withheld, by the employer and are paid directly to the government. State and city taxes may be determined from tables or as a fixed percent of income. Voluntary contributions are generally a fixed amount or percent for each pay period.

FICA taxes. The **Federal Insurance Contributions Act,** enacted in 1937, requires most employers to withhold a tax at a specified rate based on a taxable amount of gross earnings accumulated during the calendar year. Employers send the tax to the Internal Revenue Service (IRS) to finance the Social Security Fund. The fund provides payments for retirement, disability, survivor benefits, and health insurance for the elderly (**medicare**). The original plan, now referred to as **social security,** was designed to provide a minimum income for retired workers and their survivors. As benefits have been expanded, and as the number of retirees and their life expectancies have grown, so has the rate and the taxable base used to compute social security and medicare taxes. From the original 1% of the first $3,000 in 1937, both the rate and taxable base have been steadily increased by Congress until an employee in 1999 paid 7.65% (6.2% social security and 1.45% medicare) on the first $72,600 of earnings.*

For many years there was one tax for both social security and medicare; however, since 1991 these tax rates have been expressed individually. Table 6-3 shows the tax rates and maximum earnings on which an employee pays social security and medicare taxes. All social security and medicare deductions are matched by an equal deduction from the employer. Self-employed people pay taxes equal to the combined rate of the employee and employer.

Social security and medicare deductions can be determined using the following equations, which are a practical application of the percentage formula ($R \times B = P$) presented in Chapter 2:

Social security tax = Social security tax rate × Gross earnings

Medicare tax = Medicare tax rate × Gross earnings

TABLE 6-3   SOCIAL SECURITY AND MEDICARE DEDUCTIONS

	SOCIAL SECURITY TAX		MEDICARE TAX	
YEAR	EMPLOYER AND EMPLOYEE RATE	WAGE CEILING	EMPLOYER AND EMPLOYEE RATE	WAGE CEILING
1993	6.2%	$57,600	1.45%	$135,000
1994	6.2%	$60,600	1.45%	none
1995	6.2%	$61,200	1.45%	none
1996	6.2%	$62,700	1.45%	none
1997	6.2%	$65,400	1.45%	none
1998	6.2%	$68,400	1.45%	none
1999	6.2%	$72,600	1.45%	none

*For 1999, social security taxes stop at $72,600; however, there is no wage ceiling for medicare taxes.

234 CHAPTER 6   ACCOUNTING TOPICS: PARTNERSHIP PROFITS AND PAYROLL

Accumulated earnings must be checked so that social security deductions do not continue past the legal limit. For example, if one employee earned $72,600 in 1999 and another earned $200,000, both would pay the same social security tax. However, the formula can always be used to compute medicare taxes since they are deducted regardless of accumulated earnings.

Note: All the calculations in the following examples, and the answers to problems in the text, are based on a 6.2% social security tax rate up to $72,600, and a 1.45% medicare tax rate with no wage ceiling.

■ **Example 12**

Sally Super is an account executive in an advertising agency. Her salary is $74,880 a year, and she is paid twice a month. Find her social security and medicare deductions for the pay periods ending on May 15 and December 31.

Before calculating the deduction for any date, determine the gross earnings per pay period. There are 24 pay periods and the gross salary for each is:

$$\frac{74{,}880}{24} = \$3{,}120$$

May 15 is the ninth pay period, and accumulated earnings through that date are

$$3{,}120 \times 9 = \$28{,}080$$

This total is less than the legal limit of $72,600, and therefore 6.2% of the May 15 salary must be deducted for social security.

$$\text{Social security tax} = 0.062 \times 3{,}120 = \$193.44$$

Medicare taxes have no ceiling and are deducted on all earnings at 1.45%.

$$\text{Medicare tax} = 0.0145 \times 3{,}120 = \$45.24$$

December 31 is the twenty-fourth pay period. Accumulated earnings up to that date are

$$3{,}120 \times 23 = \$71{,}760$$

Only $840 (72,600 − 71,760) is subject to social security taxes. However, all gross earnings are subject to medicare taxes.

$$\text{Social security tax} = 0.062 \times 840 = \$52.08$$

$$\text{Medicare tax} = 0.0145 \times 3{,}120 = \$45.24 \quad ■$$

■ **Example 13**

An executive secretary is paid $910 a week. What proportion of her salary is subject to social security and medicare taxes? How much is deducted each week for each tax?

To determine the proportion of her salary that is subject to the tax deduction, find her annual salary. Since she is paid for 52 weeks, her annual earnings are

$$910 \times 52 = \$47{,}320$$

This total is less than $72,600 and therefore 100% of her salary is subject to the social security deduction. The medicare deduction is not subject to a maximum.

Therefore, the weekly deductions are

$$\text{Social security tax} = 0.062 \times 910 = \$56.42$$

$$\text{Medicare tax} = 0.0145 \times 910 = \$13.20 \quad ■$$

■ **Example 14**

A factory machinist is paid \$990 a week for a 40-hour week. By the first pay period in December he has worked enough overtime to give him accumulated earnings of \$72,281. If he has worked only the regular 40 hours during the current week, how much will be deducted for social security and medicare taxes?

Only the difference between the legal limit of \$72,600 and the accumulated earnings of \$72,281 is subject to the social security tax deduction. The fact that he has worked only the regular 40 hours during the week has no bearing on the solution of the problem because his base pay brings accumulated earnings over the legal maximum.

$$\text{Earnings subject to social security tax} = 72{,}600 - 72{,}281 = \$319$$

Using the formula:

$$\text{Social security tax} = 0.062 \times 319 = \$19.78$$

However, the entire weekly gross is subject to medicare tax, since there is no wage ceiling for medicare.

$$\text{Medicare tax} = 0.0145 \times 990 = \$14.36 \quad ■$$

Federal income tax withholding. Employers are required to make deductions from each employee's earnings for **federal income tax** (also called federal withholding tax), which is the largest single source of revenue for the federal government. The tax is collected on a pay-as-you-earn basis. Some amount is withheld each pay period from the employee's earnings.

To determine income tax withholding the employer must take into account:

- Gross earnings during the payroll period.
- The marital status of the employee.
- The number of claimed exemptions.

An **exemption,** also called **a withholding allowance,** is an amount of gross earnings that is not subject to tax. An employee is permitted one personal exemption, one for a spouse if the spouse does not work, and one for each dependent (such as a child or relative who is supported

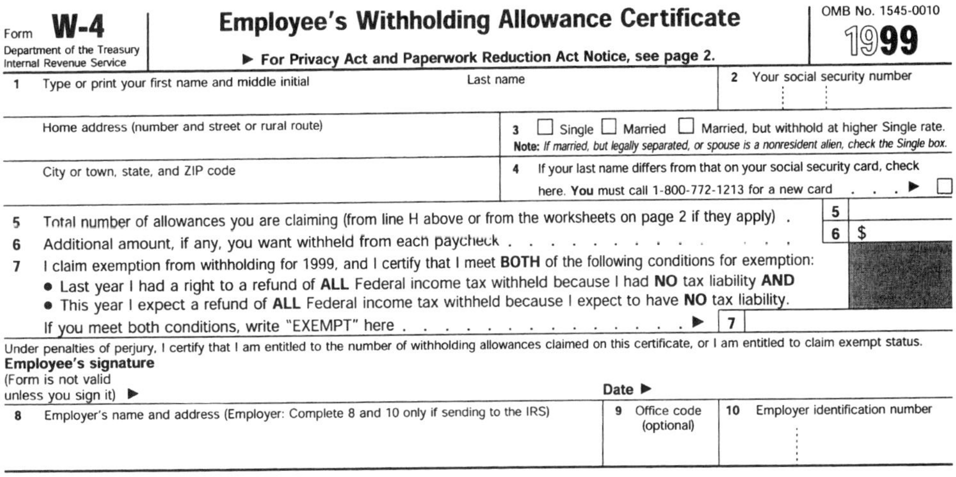
Form W-4
Department of the Treasury Internal Revenue Service
Employee's Withholding Allowance Certificate
► For Privacy Act and Paperwork Reduction Act Notice, see page 2.
OMB No. 1545-0010
1999
1 Type or print your first name and middle initial
Last name
2 Your social security number
Home address (number and street or rural route)
3 ☐ Single ☐ Married ☐ Married, but withhold at higher Single rate.
Note: If married, but legally separated, or spouse is a nonresident alien, check the Single box.
City or town, state, and ZIP code
4 If your last name differs from that on your social security card, check here. You must call 1-800-772-1213 for a new card . . . ► ☐
5 Total number of allowances you are claiming (from line H above or from the worksheets on page 2 if they apply) . 5
6 Additional amount, if any, you want withheld from each paycheck . . . . 6 \$
7 I claim exemption from withholding for 1999, and I certify that I meet BOTH of the following conditions for exemption:
• Last year I had a right to a refund of ALL Federal income tax withheld because I had NO tax liability AND
• This year I expect a refund of ALL Federal income tax withheld because I expect to have NO tax liability.
If you meet both conditions, write "EXEMPT" here . . . . ► 7
Under penalties of perjury, I certify that I am entitled to the number of withholding allowances claimed on this certificate, or I am entitled to claim exempt status.
Employee's signature
(Form is not valid unless you sign it) ►
Date ►
8 Employer's name and address (Employer: Complete 8 and 10 only if sending to the IRS)
9 Office code (optional)
10 Employer identification number

FIGURE 6-1 Employee W-4 Form.

236 CHAPTER 6 ACCOUNTING TOPICS: PARTNERSHIP PROFITS AND PAYROLL

by the taxpayer). The number of exemptions may be lowered if the employee wants to be sure that no additional tax will be due on April 15, or the number may be raised if he has been receiving a refund of income taxes. A husband and wife, both working, may split the allowed exemptions between them. The number of claimed exemptions is recorded on a W-4 form that the employee will fill out when hired. (See Figure 6-1.)

The tax deduction can be found in **wage bracket tables** supplied by the federal government. There are separate tables for single and married employees and for weekly, biweekly, semimonthly, monthly, and miscellaneous payroll periods. Only the tables for a weekly payroll period are included here (Table C-1, pages 590–593). The others are used in the same way.

A second method of calculating the income tax deduction, the **percentage method,** is used by companies with computerized payroll. This eliminates the need to store the several pages of tables needed with the wage bracket method. For this method Table C-2, pages 594–595 will be used.

■ **Example 15** Use the wage bracket tables to determine the federal income tax deduction for an employee who earned $542.70 during the week if:

    a. He is single and claims one exemption.

    b. He is married and claims one exemption.

    c. He is married and claims three exemptions.

To make the solutions easy to follow, the parts of the wage bracket tables that are used in these problems are reproduced in Tables 6-4 and 6-5.

    a. In Table 6-4, for single persons, locate the gross earnings of $542.70 between $540 and $550. In the column headed "1" for one allowance (column 4) find the deduction of $66.00.

    b. In Table 6-5, for married persons, locate the gross earnings of $542.70 between $540 and $550. In the column for one allowance find the deduction of $55.00. Note that this deduction is $11.00 less than for the single worker earning the same amount with the same number of exemptions.

    c. In Table 6-5, locate $542.70 and move across to the column headed "3." The tax is $39.00. ■

Other Deductions. Additional deductions are usually necessary to cover such items as state and city income taxes, health insurance, union dues, pension plans, and so on. They are typically either a fixed amount per pay period, like union dues, or a fixed percent, like contributions to a pension plan. In practice, the simplest procedure is to add all the deductions and then subtract the total from gross earnings.

■ **Example 16** Use the percentage method to determine the federal income tax deduction for a married employee who earned $1,184.29 during the week if he claims two withholding allowances.

Table 6-6 shows two tables that are issued by the federal government for calculating income tax using the percentage method. The table at the top shows how much to subtract from gross earnings for each withholding allowance claimed. The table at the bottom shows the amount of tax to be withheld.

    a. Locate the weekly withholding for one allowance of $52.88 in the right column of the withholding allowance table. Multiply this number by 2.

$$52.88 \times 2 = \$105.76$$

    b. Subtract this amount from gross earnings.

$$1,184.29 - 105.76 = \$1,078.53$$

    c. Go to the "married person's weekly" section of the other table. Since $1,078.53 falls between $913 and $1,894 the tax is:

TABLE 6-4 WEEKLY WAGE BRACKET WITHHOLDING TABLE FOR SINGLE PERSONS

(For Wages Paid in 1999)

If the wages are—		And the number of withholding allowances claimed is—										
At least	But less than	0	1	2	3	4	5	6	7	8	9	10
		The amount of income tax to be withheld is—										
$0	$55	0	0	0	0	0	0	0	0	0	0	0
55	60	1	0	0	0	0	0	0	0	0	0	0
60	65	2	0	0	0	0	0	0	0	0	0	0
65	70	2	0	0	0	0	0	0	0	0	0	0
70	75	3	0	0	0	0	0	0	0	0	0	0
75	80	4	0	0	0	0	0	0	0	0	0	0
80	85	5	0	0	0	0	0	0	0	0	0	0
85	90	5	0	0	0	0	0	0	0	0	0	0
90	95	6	0	0	0	0	0	0	0	0	0	0
95	100	7	0	0	0	0	0	0	0	0	0	0
100	105	8	0	0	0	0	0	0	0	0	0	0
105	110	8	1	0	0	0	0	0	0	0	0	0
110	115	9	1	0	0	0	0	0	0	0	0	0
115	120	10	2	0	0	0	0	0	0	0	0	0
120	125	11	3	0	0	0	0	0	0	0	0	0
125	130	11	4	0	0	0	0	0	0	0	0	0
130	135	12	4	0	0	0	0	0	0	0	0	0
135	140	13	5	0	0	0	0	0	0	0	0	0
140	145	14	6	0	0	0	0	0	0	0	0	0
145	150	14	7	0	0	0	0	0	0	0	0	0
150	155	15	7	0	0	0	0	0	0	0	0	0
155	160	16	8	0	0	0	0	0	0	0	0	0
160	165	17	9	1	0	0	0	0	0	0	0	0
165	170	17	10	2	0	0	0	0	0	0	0	0
170	175	18	10	2	0	0	0	0	0	0	0	0
175	180	19	11	3	0	0	0	0	0	0	0	0
180	185	20	12	4	0	0	0	0	0	0	0	0
185	190	20	13	5	0	0	0	0	0	0	0	0
190	195	21	13	5	0	0	0	0	0	0	0	0
195	200	22	14	6	0	0	0	0	0	0	0	0
200	210	23	15	7	0	0	0	0	0	0	0	0
210	220	25	17	9	1	0	0	0	0	0	0	0
220	230	26	18	10	2	0	0	0	0	0	0	0
230	240	28	20	12	4	0	0	0	0	0	0	0
240	250	29	21	13	5	0	0	0	0	0	0	0
250	260	31	23	15	7	0	0	0	0	0	0	0
260	270	32	24	16	8	0	0	0	0	0	0	0
270	280	34	26	18	10	2	0	0	0	0	0	0
280	290	35	27	19	11	3	0	0	0	0	0	0
290	300	37	29	21	13	5	0	0	0	0	0	0
300	310	38	30	22	14	6	0	0	0	0	0	0
310	320	40	32	24	16	8	0	0	0	0	0	0
320	330	41	33	25	17	9	1	0	0	0	0	0
330	340	43	35	27	19	11	3	0	0	0	0	0
340	350	44	36	28	20	12	4	0	0	0	0	0
350	360	46	38	30	22	14	6	0	0	0	0	0
360	370	47	39	31	23	15	7	0	0	0	0	0
370	380	49	41	33	25	17	9	1	0	0	0	0
380	390	50	42	34	26	18	10	3	0	0	0	0
390	400	52	44	36	28	20	12	4	0	0	0	0
400	410	53	45	37	29	21	13	6	0	0	0	0
410	420	55	47	39	31	23	15	7	0	0	0	0
420	430	56	48	40	32	24	16	9	1	0	0	0
430	440	58	50	42	34	26	18	10	2	0	0	0
440	450	59	51	43	35	27	19	12	4	0	0	0
450	460	61	53	45	37	29	21	13	5	0	0	0
460	470	62	54	46	38	30	22	15	7	0	0	0
470	480	64	56	48	40	32	24	16	8	0	0	0
480	490	65	57	49	41	33	25	18	10	2	0	0
490	500	67	59	51	43	35	27	19	11	3	0	0
500	510	68	60	52	44	36	28	21	13	5	0	0
510	520	70	62	54	46	38	30	22	14	6	0	0
520	530	71	63	55	47	39	31	24	16	8	0	0
530	540	74	65	57	49	41	33	25	17	9	1	0
540	550	77	66	58	50	42	34	27	19	11	3	0
550	560	80	68	60	52	44	36	28	20	12	4	0
560	570	82	69	61	53	45	37	30	22	14	6	0
570	580	85	71	63	55	47	39	31	23	15	7	0
580	590	88	73	64	56	48	40	33	25	17	9	1
590	600	91	76	66	58	50	42	34	26	18	10	2

TABLE 6-5 WEEKLY WAGE BRACKET WITHHOLDING TABLE FOR MARRIED PERSONS

(For Wages Paid in 1999)

If the wages are—		And the number of withholding allowances claimed is—										
At least	But less than	0	1	2	3	4	5	6	7	8	9	10
		The amount of income tax to be withheld is—										
$0	$125	0	0	0	0	0	0	0	0	0	0	0
125	130	1	0	0	0	0	0	0	0	0	0	0
130	135	1	0	0	0	0	0	0	0	0	0	0
135	140	2	0	0	0	0	0	0	0	0	0	0
140	145	3	0	0	0	0	0	0	0	0	0	0
145	150	4	0	0	0	0	0	0	0	0	0	0
150	155	4	0	0	0	0	0	0	0	0	0	0
155	160	5	0	0	0	0	0	0	0	0	0	0
160	165	6	0	0	0	0	0	0	0	0	0	0
165	170	7	0	0	0	0	0	0	0	0	0	0
170	175	7	0	0	0	0	0	0	0	0	0	0
175	180	8	0	0	0	0	0	0	0	0	0	0
180	185	9	1	0	0	0	0	0	0	0	0	0
185	190	10	2	0	0	0	0	0	0	0	0	0
190	195	10	2	0	0	0	0	0	0	0	0	0
195	200	11	3	0	0	0	0	0	0	0	0	0
200	210	12	4	0	0	0	0	0	0	0	0	0
210	220	14	6	0	0	0	0	0	0	0	0	0
220	230	15	7	0	0	0	0	0	0	0	0	0
230	240	17	9	1	0	0	0	0	0	0	0	0
240	250	18	10	2	0	0	0	0	0	0	0	0
250	260	20	12	4	0	0	0	0	0	0	0	0
260	270	21	13	5	0	0	0	0	0	0	0	0
270	280	23	15	7	0	0	0	0	0	0	0	0
280	290	24	16	8	0	0	0	0	0	0	0	0
290	300	26	18	10	2	0	0	0	0	0	0	0
300	310	27	19	11	3	0	0	0	0	0	0	0
310	320	29	21	13	5	0	0	0	0	0	0	0
320	330	30	22	14	6	0	0	0	0	0	0	0
330	340	32	24	16	8	0	0	0	0	0	0	0
340	350	33	25	17	9	1	0	0	0	0	0	0
350	360	35	27	19	11	3	0	0	0	0	0	0
360	370	36	28	20	12	4	0	0	0	0	0	0
370	380	38	30	22	14	6	0	0	0	0	0	0
380	390	39	31	23	15	7	0	0	0	0	0	0
390	400	41	33	25	17	9	1	0	0	0	0	0
400	410	42	34	26	18	10	2	0	0	0	0	0
410	420	44	36	28	20	12	4	0	0	0	0	0
420	430	45	37	29	21	13	5	0	0	0	0	0
430	440	47	39	31	23	15	7	0	0	0	0	0
440	450	48	40	32	24	16	8	1	0	0	0	0
450	460	50	42	34	26	18	10	2	0	0	0	0
460	470	51	43	35	27	19	11	4	0	0	0	0
470	480	53	45	37	29	21	13	5	0	0	0	0
480	490	54	46	38	30	22	14	7	0	0	0	0
490	500	56	48	40	32	24	16	8	0	0	0	0
500	510	57	49	41	33	25	17	10	2	0	0	0
510	520	59	51	43	35	27	19	11	3	0	0	0
520	530	60	52	44	36	28	20	13	5	0	0	0
530	540	62	54	46	38	30	22	14	6	0	0	0
540	550	63	55	47	39	31	23	16	8	0	0	0
550	560	65	57	49	41	33	25	17	9	1	0	0
560	570	66	58	50	42	34	26	19	11	3	0	0
570	580	68	60	52	44	36	28	20	12	4	0	0
580	590	69	61	53	45	37	29	22	14	6	0	0
590	600	71	63	55	47	39	31	23	15	7	0	0
600	610	72	64	56	48	40	32	25	17	9	1	0
610	620	74	66	58	50	42	34	26	18	10	2	0
620	630	75	67	59	51	43	35	28	20	12	4	0
630	640	77	69	61	53	45	37	29	21	13	5	0
640	650	78	70	62	54	46	38	31	23	15	7	0
650	660	80	72	64	56	48	40	32	24	16	8	0
660	670	81	73	65	57	49	41	34	26	18	10	2
670	680	83	75	67	59	51	43	35	27	19	11	3
680	690	84	76	68	60	52	44	37	29	21	13	5
690	700	86	78	70	62	54	46	38	30	22	14	6
700	710	87	79	71	63	55	47	40	32	24	16	8
710	720	89	81	73	65	57	49	41	33	25	17	9
720	730	90	82	74	66	58	50	43	35	27	19	11
730	740	92	84	76	68	60	52	44	36	28	20	12

TABLE 6-6 PERCENTAGE METHOD TABLES

Payroll Period	One Withholding Allowance
Weekly	$ 52.88
Biweekly	105.77
Semimonthly	114.58
Monthly	229.17
Quarterly	687.50
Semiannually	1,375.00
Annually	2,750.00
Daily or miscellaneous (each day of the payroll period)	10.58

(For Wages Paid in 1999)

TABLE 1—WEEKLY Payroll Period

(a) SINGLE person (including head of household)—

If the amount of wages (after subtracting withholding allowances) is: The amount of income tax to withhold is:

Not over $51 . . . . . $0

Over—	But not over—		of excess over—
$51	—$525	15%	—$51
$525	—$1,125	$71.10 plus 28%	—$525
$1,125	—$2,535	$239.10 plus 31%	—$1,125
$2,535	—$5,475	$676.20 plus 36%	—$2,535
$5,475	. . . . .	$1,734.60 plus 39.6%	—$5,475

(b) MARRIED person—

If the amount of wages (after subtracting withholding allowances) is: The amount of income tax to withhold is:

Not over $124 . . . . $0

Over—	But not over—		of excess over—
$124	—$913	15%	—$124
$913	—$1,894	$118.35 plus 28%	—$913
$1,894	—$3,135	$393.03 plus 31%	—$1,894
$3,135	—$5,531	$777.74 plus 36%	—$3,135
$5,531	. . . . .	$1,640.30 plus 39.6%	—$5,531

TABLE 2—BIWEEKLY Payroll Period

(a) SINGLE person (including head of household)—

If the amount of wages (after subtracting withholding allowances) is: The amount of income tax to withhold is:

Not over $102 . . . . $0

Over—	But not over—		of excess over—
$102	—$1,050	15%	—$102
$1,050	—$2,250	$142.20 plus 28%	—$1,050
$2,250	—$5,069	$478.20 plus 31%	—$2,250
$5,069	—$10,950	$1,352.09 plus 36%	—$5,069
$10,950	. . . . .	$3,469.25 plus 39.6%	—$10,950

(b) MARRIED person—

If the amount of wages (after subtracting withholding allowances) is: The amount of income tax to withhold is:

Not over $248 . . . . $0

Over—	But not over—		of excess over—
$248	—$1,827	15%	—$248
$1,827	—$3,788	$236.85 plus 28%	—$1,827
$3,788	—$6,269	$785.93 plus 31%	—$3,788
$6,269	—$11,062	$1,555.04 plus 36%	—$6,269
$11,062	. . . . .	$3,280.52 plus 39.6%	—$11,062

TABLE 3—SEMIMONTHLY Payroll Period

(a) SINGLE person (including head of household)—

If the amount of wages (after subtracting withholding allowances) is: The amount of income tax to withhold is:

Not over $110 . . . . $0

Over—	But not over—		of excess over—
$110	—$1,138	15%	—$110
$1,138	—$2,438	$154.20 plus 28%	—$1,138
$2,438	—$5,492	$518.20 plus 31%	—$2,438
$5,492	—$11,863	$1,464.94 plus 36%	—$5,492
$11,863	. . . . .	$3,758.50 plus 39.6%	—$11,863

(b) MARRIED person—

If the amount of wages (after subtracting withholding allowances) is: The amount of income tax to withhold is:

Not over $269 . . . . $0

Over—	But not over—		of excess over—
$269	—$1,979	15%	—$269
$1,979	—$4,104	$256.50 plus 28%	—$1,979
$4,104	—$6,792	$851.50 plus 31%	—$4,104
$6,792	—$11,983	$1,684.78 plus 36%	—$6,792
$11,983	. . . . .	$3,553.54 plus 39.6%	—$11,983

TABLE 4—MONTHLY Payroll Period

(a) SINGLE person (including head of household)—

If the amount of wages (after subtracting withholding allowances) is: The amount of income tax to withhold is:

Not over $221 . . . . $0

Over—	But not over—		of excess over—
$221	—$2,275	15%	—$221
$2,275	—$4,875	$308.10 plus 28%	—$2,275
$4,875	—$10,983	$1,036.10 plus 31%	—$4,875
$10,983	—$23,725	$2,929.58 plus 36%	—$10,983
$23,725	. . . . .	$7,516.70 plus 39.6%	—$23,725

(b) MARRIED person—

If the amount of wages (after subtracting withholding allowances) is: The amount of income tax to withhold is:

Not over $538 . . . . $0

Over—	But not over—		of excess over—
$538	—$3,958	15%	—$538
$3,958	—$8,208	$513.00 plus 28%	—$3,958
$8,208	—$13,583	$1,703.00 plus 31%	—$8,208
$13,583	—$23,967	$3,369.25 plus 36%	—$13,583
$23,967	. . . . .	$7,107.49 plus 39.6%	—$23,967

$$\$118.35 \text{ plus } 28\% \text{ of the excess over } \$913$$
$$\text{Excess} = 1{,}078.53 - 913 = \$165.53$$
$$\text{Tax} = 118.35 + 0.28\,(165.53)$$
$$= 118.35 + 46.35$$
$$= \$164.70$$

The federal income tax withholding amount would be $164.70 using the percentage method. This result may differ slightly from the result that would be obtained from the wage bracket tables, but the amounts will be close enough to be acceptable. ■

■ **Example 17**

A married employee who claims three withholding allowances earned $595.30 during the week. In addition to the required social security, medicare, and federal income tax deductions, she contributes 4% of her earnings to a pension plan, 1% toward the purchase of savings bonds, and $7.88 for union dues. What are her net earnings after these deductions have been made? (Use the wage bracket tables to determine the federal income tax and assume that social security taxes must still be deducted.)

$$\text{Social security tax} = 0.062 \times 595.30 \quad = \$ 36.91$$
$$\text{Medicare tax} = 0.0145 \times 595.30 \quad = 8.63$$

Income tax: $595.30 is between $590 and $600.
In Table 6-5, in the column headed "3" the tax is $47.   47.00

$$\text{Pension plan contribution}$$
$$(4\% \text{ of earnings}) = 0.04 \times 595.30 \quad = 23.81$$
$$\text{Bond purchase } (1\% \text{ of earnings}) = 0.01 \times 595.30 = 5.95$$
$$\text{Union dues} \quad = \underline{7.88}$$
$$\text{Total deductions} \quad \$130.18$$

Net earnings = Gross earnings − deductions
= 595.30 − 130.18

Net earnings = $465.12 ■

Chapter 6

Name	Date	Class

TRY THESE PROBLEMS (Set IV)

Solve the following problems. Use a 6.2% rate up to a maximum of $72,600 for social security tax and a 1.45% rate with no wage ceiling for medicare tax.

For Problems 1–4, find the social security tax and the medicare tax for each employee. Assume that social security taxes must still be deducted.

EMPLOYEE	GROSS WEEKLY EARNINGS
1. Selby, S.	$734.19
2. Hughes, K.	$492.53
3. Golden, H.	$876.16
4. Rivera, J.	$1,227.44

For Problems 5–8, find the social security and medicare tax for each employee for the current pay period. Year-to-date earnings do not include earnings for the current period.

EMPLOYEE	YEAR-TO-DATE GROSS EARNINGS	GROSS EARNINGS FOR CURRENT PAY PERIOD
5. Allen, D.	$72,500	$591.62
6. Hammond, J.	$70,152	$2,800
7. Jones, M.	$72,747	$3,641.97
8. Stutz, C.	$52,204	$417.44

For Problems 9–12, use the wage bracket tables to determine the federal income tax deduction for each employee.

EMPLOYEE	GROSS WEEKLY EARNINGS	MARITAL STATUS	EXEMPTIONS
9. Avens, C.	$643.29	M	3
10. Brooks, J.	$988.17	S	4

241

EMPLOYEE	GROSS WEEKLY EARNINGS	MARITAL STATUS	EXEMPTIONS
11. Esposito, T.	$349.00	S	0
12. Lo, Y.	$774.61	S	1

For Problems 13–16, fill in the missing quantities. Use the percentage method to find federal income tax. Assume that social security taxes must still be deducted. The number of exemptions and the marital status are listed after each employee's name.

EMPLOYEE	GROSS EARNINGS	PAY PERIOD	FEDERAL INCOME TAX	SOCIAL SECURITY	MEDICARE TAX	TOTAL DEDUCTIONS	NET PAY
13. Bell, 5, M	$6,093.26	monthly					
14. Chang, 4, M	$462.91	weekly					
15. Presley, 0, S	$8,958.23	semimonthly					
16. Zorn, 1, S	$1,231.56	biweekly					

17. Sean O'Keefe, a rate analyst for an insurance company, is paid a biweekly salary of $1,470. He is single and claims four exemptions. In addition to social security, medicare, and federal income tax deductions, he contributes 5% of gross earnings to a pension plan, 1% toward the purchase of savings bonds, $79 for health insurance, and $20 to the United Way. Find his net pay for the pay period ending December 31. (Use the percentage method to find federal income tax.)

18. Jungle Excursions pays its manager Julie Kressup a monthly salary of $3,600 plus a commission of 3% on total monthly sales over $40,000. In the month of January, Jungle Excursions has total sales of $91,880. Kressup is married and claims two exemptions. Her deductions include social security, medicare, and federal income tax, 10% of gross earnings to a retirement plan, a credit union payment of $400, dental insurance of $18, group health insurance of $92, and charitable contributions of $25. Find her net pay for the month. (Use the percentage method to find federal income tax.)

Chapter 6

END-OF-CHAPTER PROBLEMS

In Problems 1 through 8, divide the partnership profits as indicated.

For Problems 1 through 5:

PARTNERS	INVESTMENT
Ruth	$16,000
Sue	11,000
Ted	15,000

Net profit for the year = $6,000

1. Ratio of investments.

2. Ratio of 4 : 1 : 3.

3. $500 to each and the remainder in the ratio of 2 : 3 : 4.

4. Twelve percent interest on investments and the remainder equally divided.

5. First $2,000 to be divided equally; the remainder in the ratio of investments.

For Problems 6 through 8:

PARTNERS	INVESTMENT
Joe	$ 8,000
Roy	12,000
Claire	7,500

Net profit for the year = $16,500

6. Ratio of investments.

7. Eight percent interest on investments, and the remainder divided 40%, 45%, 15%.

8. Ten percent interest on investments, and the remainder divided in the ratio of 3 : 2 : 2.

9. Dan and Doreen are partners operating a small delicatessen. Dan spends full time in the store and Doreen works half-time. Therefore, they agreed to divide the profits in the ratio of 2 : 1. How much does each receive in a year when the profit is $9,900?

10. Penny and Benny opened a flower shop and agreed to divide their profits and losses in the ratio of 3 : 5.

   a. How much debt is each responsible for in a year when they lost $24,000?

   b. How much did each receive in a year when their profit was $44,000?

243

11. The four Finch sisters agreed to divide the profits from their hair salon in the ratio of 2 : 5 : 6 : 7. Last year the profits were $120,000. How much will each receive?

12. Ben, Boris, and Bertha have agreed to distribute their partnership profits and losses by giving Ben 45%, Boris 20%, and Bertha 35%.

    a. How much will each receive in a year when the profit is $12,600?

    b. How much will each be responsible for if the loss is $1,800?

13. Ursula, Vera, and Will were partners in a restaurant. They had each made the same initial investment, but because Vera was the most experienced and Will could not give full time to the business they decided to divide their profits by fixed percents. Ursula was to get 30%, Vera 50%, and Will 20%.

    a. Divide their $8,700 profit according to this agreement.

    b. If they had agreed to divide the profit in the same ratio as their investments, how would the $8,700 be divided?

14. Linda, Larry, and Louis decided to distribute profits or losses by giving 10% to Linda, 30% to Larry, and the remainder to Louis. What was each partner's share of a $36,000 net profit?

15. Karen and Keith started a new travel agency. Since Karen runs the agency she will get a salary of $25,000. Any profits will be distributed in the ratio 1 : 4. If the profit is $14,000 how much will each receive?

16. Carol and Carl have invested $5,000 and $4,000, respectively, in a partnership. Their agreement states that they will share profits and losses in the ratio of their investments. During the first year of business they make a small profit of $720. How much does each receive in total?

17. Ellie and Shelley formed a partnership. At the beginning of the year, Ellie invested $10,000 and Shelley invested $5,000. On July 15 Shelley added $9,000. They agreed to divide profits or losses in the ratio of their original investments. How much would each receive if $16,000 was earned?

Chapter 6

END-OF-CHAPTER PROBLEMS

18. Abigail and Aaron are partners who have agreed to share net income and net losses in their business in the ratio of 3 : 4. Distribute the following profits and losses.

    a. $6,300 loss.

    b. $14,560 profit.

19. Eve, Enis, and Edna run a women's accessories store. Their copartnership agreement states that each will receive a share of the profits determined by the ratio of their average investments per month for the year. At the beginning of the year their investments were $13,000, $10,000, and $20,000, respectively. On April 1, Eve added $2,000 and on July 1 she added another $1,000. Enis made no change in her investment, but Edna withdrew $2,000 on July 1.

    a. The profit for the year was $8,800. How would it be divided?

    b. If no changes had been made in investments during the year, how would a $9,030 profit have been divided?

20. Merry Music Shop is run as a partnership by Frank and Frances. They each made an initial investment of $15,000 and agreed to add to it when possible in order to build the business. On July 1 Frank added $5,000 to his investment and Frances added $4,000. The partnership agreement states that each will receive 12% on the average investment per month and the remainder will be divided in the ratio of 5 : 4. Divide the profit according to the agreement if it is:

    a. $17,300.

    b. $5,000.

21. Elaine Diamond invested $150,000 in a small electronics company. Her partner, Bill Williams, is going to run the business. They agree that she will receive a 15% return on her investment, and any additional profits will be divided in the ratio 3 : 4. How will a profit of $80,000 be divided?

22. Jon and Jim have made the following agreement for division of their partnership profits and losses. Each will get 10% on his average investment per month for the year. If profits are not large enough to cover this amount, the profit (or loss) will be shared in the ratio of the average investment. Any profits remaining after the interest payment is made will be divided in fixed percents, 65% to Jon and 35% to Jim. Jon's investment at the beginning of the year is $13,000 and Jim's is $15,000. No changes are made during the year. Divide the following profits:

    a. $5,600.

    b. $2,000.

23. Xavier, Yolanda, and Zena were partners in a business in which they had invested $22,000, $24,000, and $28,000, respectively. Their copartnership agreement contained the following terms:

> If there was sufficient profit, each would get a year-end bonus of $4,000.
>
> From the remaining profit each would receive 10% on the original investment.
>
> Any profit remaining after that would be divided in the ratio of 2 : 2 : 4.
>
> The net profit in one year was $20,000. How much money did each partner receive in total? (The terms of the agreement are carried out in the order stated.)

24. Jennifer and Jeffrey decided that profits from their business would be distributed according to their average investment per month. Jennifer's original investment of $10,000 remained unchanged during the year. Jeffrey invested $5,000 on January 1 and $2,400 more on April 1. On October 1, he deposited an additional $600. He made no other changes during the year. How would Jennifer and Jeffrey divide a net profit of $47,460?

25. Amber Hill has an annual salary of $47,000. Determine her gross earnings each pay period if she is paid:

    a. Biweekly.

    b. Semimonthly.

26. Compute the gross wages for Paul Pyros. His rate is $9 an hour with time and a half for work in excess of 40 hours. He worked the following number of hours during one week:

EMPLOYEE	M	T	W	TH	F	S
Pyros, P.	$5\frac{1}{2}$	11	9	$7\frac{1}{2}$	8	5

27. Determine the gross pay for Marilyn Masterson. Her regular rate of pay is $8.50 per hour and her company pays time and a half for all time worked over 8 hours in 1 day, no matter how many hours are worked in a week. She worked the following hours during the last pay period:

EMPLOYEE	M	T	W	TH	F	S
Masterson, M.	12	$8\frac{1}{2}$	10	$5\frac{1}{2}$	—	—

28. A salesman for a computer supply company works on a straight commission of 5% of net sales. During one week he sold $8,900 worth of supplies, but one order for $412 was returned. Find his gross earnings.

29. Reginald Burns receives a salary of $650 a week plus a 4% commission on net sales. Find his gross earnings for the week in which he had net sales of $8,300.

Chapter 6

END-OF-CHAPTER PROBLEMS

30. Lauren Adams is a sales manager for a pharmaceutical company. She receives a salary of $350 a week plus a 6% commission on her net sales and a 2% commission on the net sales of her sales associates working under her. Her sales last week were $5,300 with returns of $270 while the other employees had sales of $22,880 with returns of $400. She also received a draw of $200 during the week. Find her gross earnings.

31. A steelworker producing automobile body parts operates under a piecework system that pays $13 for each piece completed. During one week he completed 67 pieces, of which 3 were rejected after inspection. What were his gross earnings?

32. Suppose that a quota has been set at 60 pieces per week for the steelworker (from Problem 31). Determine the gross earnings if:

    a. All production is paid for at $13.50 each, if total production exceeds the quota.

    b. A bonus is paid for production above the quota. Additional production is worth $15.00 instead of $13.00.

33. Rick Scanlon, a salesman, is paid a salary and commission. If he sells only the basic quota of $10,000 or less during the month, he is paid $2,400. On sales over that amount he gets a commission of 8%. How much does he earn during a month when his sales reach $18,550?

34. Ann Laury works for a marketing research firm obtaining answers to questionnaires. She is paid $2.70 each for the first 75 questionnaires completed during the week and $3.20 for any additional ones. During the week she obtained 110 questionnaires. What were her gross earnings?

35. Complete the following payroll record for Mountain Boot Company. All employees are required to work 40 regular hours, after which they receive overtime at time and a half.

EMPLOYEE	M	T	W	TH	F	S	REG.	O.T.	REG. RATE	REG.	O.T.	TOTAL
Duff, Z.	$5\frac{1}{2}$	10	9	$9\frac{1}{2}$	8	—			$8.50			
Hollis, K.	8	8	8	$8\frac{1}{2}$	8	$4\frac{1}{2}$			$9.90			
Smith, R.	9	9	7	10	—	—			$7.70			

HOURS WORKED / TOTAL HRS. / GROSS EARNINGS

36. Evelyn Green is a sales clerk at Harold's Department Store. She is paid $8.50 an hour plus $1\frac{1}{2}$% commission on all sales. Last week Evelyn worked 26 hours and had sales of $3,000. What was her gross pay?

37. Langston Corporation pays its employees on a graduated commission scale. The rate is 1% on sales up to $40,000, 2% on sales from $40,000 to $80,000, and $3\frac{1}{2}$% on sales greater than $80,000. If Doris Ditty had sales of $184,000 last month, what commission did she earn?

38. Jack Horner works for $12 an hour with time and a half for overtime after 40 hours. In a typical week he works 42 hours. If he is given the option of being paid $6.20 per piece but working no more than 40 hours, how many pieces would he have to produce to earn the same amount?

For Problems 39–59, use a 6.2% rate up to a maximum of $72,600 for social security tax and a 1.45% rate with no wage ceiling for medicare tax.

For Problems 39–42, find social security tax and medicare tax for each employee. Assume that social security tax must still be deducted.

EMPLOYEE	GROSS WEEKLY EARNINGS	EMPLOYEE	GROSS WEEKLY EARNINGS
39. Boyer, B.	$621.14	40. Frazier, G.	$318.85
41. Lightyear, B.	$780.56	42. Yabroff, J.	$834.94

For Problems 43–46, find the social security tax for each employee for the current pay period. Year-to-date earnings do not include earnings for the current period.

EMPLOYEE	YEAR-TO-DATE GROSS EARNINGS	GROSS EARNINGS FOR CURRENT PAY PERIOD
43. Congrove, W.	$29,530	$693.04
44. Keller, M.	$72,012	$2,388.92
45. Russo, G.	$81,745	$3,436.13
46. Stevens, T.	$66,378	$1,084.19

Chapter 6

Name   Date   Class

END-OF-CHAPTER PROBLEMS

For Problems 47–50, use the wage bracket tables to determine the federal income tax deduction for each employee.

	EMPLOYEE	GROSS WEEKLY EARNINGS	MARITAL STATUS	EXEMPTIONS
47.	Drumm, E.	$487.21	M	3
48.	Hickey, S.	$591.04	S	4
49.	McCurry, N.	$745.69	M	0
50.	Stamos, M.	$1,230.37	M	1

For Problems 51–54, fill in the missing quantities. Use the percentage method to find federal income tax. Assume that no employee has exceeded $72,600 so far this year. The number of exemptions and the marital status are listed after each employee's name.

	EMPLOYEE	GROSS EARNINGS	PAY PERIOD	FEDERAL INCOME TAX	SOCIAL SECURITY	MEDICARE TAX	TOTAL DEDUCTIONS	NET PAY
51.	Avon, 5, M	$6,212.55	biweekly					
52.	Juarez, 1, S	$1,987.93	monthly					
53.	Kasten, 1, M	$1,361.17	semimonthly					
54.	West, 3, S	$1,147.29	weekly					

For Problems 55–57 use the wage bracket tables to find federal income tax.

55. Scott Thomas works under a differential piece-rate plan that pays 80¢ apiece for the first 100 items produced, 90¢ each for the next 100, and $1 for production over 200 pieces. He is married, has two children, and claims four income tax allowances. His deductions include social security, medicare, and federal income tax, but there is no state or city income tax in his area. He also contributes 5% of his earnings to his pension plan, $5.50 toward union dues, and 1% for savings bonds. Find his net earnings during a week when he produces 419 acceptable pieces.

56. Yvette Marbrey is an administrative assistant at The Three Bears Co., producers of stuffed animals. Her salary is $390 per week, but she frequently works overtime and is paid time and a half after 40 hours. She is single but supports her mother and claims two exemptions for tax calculations. Her deductions include social security, medicare, and federal income tax, a state income tax of 3.5% of gross earnings, and group health insurance of $30. Find her net earnings for a week in which she worked 47 hours.

57. Jack is a bachelor earning $420 a week and claiming one exemption. Jill is also single, earning $440 a week, and claiming two exemptions. By how much would they increase their combined take-home pay if they got married and maintained the same number of exemptions each?

For Problems 58 and 59 use the percentage method to find federal income tax.

58. Robert Burns is a management employee earning $81,600 a year. He is married and can claim three exemptions. He is paid on a semimonthly basis but receives no overtime regardless of the number of hours he works. In addition to social security, medicare, and federal income tax deductions, his state requires a withholding tax of 2% of gross earnings and he has authorized a deduction of $\frac{1}{2}$% for the United Fund, 6% for pension, and 1% to repay a loan from his pension fund. Calculate his net salary for the pay period ending December 31.

59. Ms. Muffet earns $73,400 a year and is paid on a weekly basis. Find her social security, medicare, and federal income tax deductions for the first week in February and the last week in December if she is single and claims two exemptions.

SUPERPROBLEMS

1. Ella, Frank, and George started a small business at the beginning of the year. Ella invested $12,000, Frank invested $15,000, and George invested $18,000. At the beginning of July Ella increased her investment to $18,000 and at the beginning of September Frank put in an additional $3,000. George made no changes to his initial investment. At the end of the year there was a profit of $35,000. How much money did each partner receive if their partnership agreement contained the following terms:

   The first $12,000 would be divided evenly.

   Any amount over $12,000 would be used first to give each partner 8% on their original investment. If the remaining amount was insufficient to cover this 8% distribution, the amounts paid would be determined by the ratio of the amount remaining to the amount needed, multiplied by the amount due each partner.

   If any profit remained after the 8% distribution, it would be divided according to the partners' average investment per month for the year.

2. Belle Bryte is a computer programmer earning $47,600 a year. She is married, has two children and claims herself and the children as exemptions. In addition to social security, medicare, and federal income tax she has the following deductions:

   2% for state tax;

   1% for city tax;

   5% pension contribution;

   $9 a week for union dues; and

   $20 each paycheck for savings bonds.

   Her employer uses the percentage method to find federal income tax.

   a. How much is her take-home pay each week?

   b. If she did not claim the two children as exemptions, how much less would her take-home pay be?

   c. Find the percent increase in net weekly salary that will result from a 5% increase in gross salary (assuming three exemptions).

Chapter 6

TEST YOURSELF

Section

1. Sam Small and Tom Tall decided to open a Health Food Shop. Sam invested $32,000 and Tom added $40,000 to this. At the end of the first year the shop had a profit of $24,000. Divide this profit between the two partners according to the following agreements:

   a. Ratio of 3 : 5. 6.1

   b. Fixed percents of 40% and 60%. 6.1

   c. Ratio of the original investments. 6.1

   d. $2,000 to each and the remainder in the ratio of 3 : 7. 6.1

2. Jenny invested $11,000 when she and Penny became partners on January 1. Penny invested $9,400 at that time. Jenny made no changes to her initial investment. Penny added $8,600 at the beginning of March but withdrew $3,000 at the beginning of July. How much of their $10,000 profit will go to each partner if:

   a. Profits are divided according to the partners' average investment per month for the year? 6.1

   b. Each partner gets 12% interest on average investment and the remainder is divided equally? 6.1

3. Joyce Moss is a department supervisor earning an annual salary of $37,350. Determine her gross earnings each pay period if she is paid: 6.2

   a. Biweekly.

   b. Semimonthly.

   c. Monthly.

4. David Drake earns $8.40 an hour for a 40-hour week. Overtime is paid at time and a half for weekdays and Saturdays and double time for Sundays. If David worked the following schedule, what would his gross pay be? 6.2

EMPLOYEE	M	T	W	TH	F	S	SU
Drake, D.	8	$8\frac{1}{2}$	9	$8\frac{1}{2}$	8	5	$6\frac{1}{2}$

253

5. Cindy Hundley operates a drill press and is paid $12.75 an hour for a 40-hour week or $0.45 a unit, whichever is higher. If she produced 230 units on Monday, 285 units on Tuesday, 240 units on Wednesday, 220 units on Thursday, and 205 units on Friday, determine her gross earnings for the week. 6.2

6. Frank Griffin is paid $11.00 an hour and is expected to produce 160 units during a normal 40-hour workweek. A bonus of $2.50 a unit is paid for each acceptable unit above this quota. Find Frank's gross earnings when his output is 183 units. 6.2

7. Roberta Roberts earns $320 a week plus 4% commission on sales in excess of $5,000. If Roberta sells $14,000, how much are her gross earnings? 6.2

8. Upshaw Building Supply uses a sliding scale to determine the salespeople's commissions. On a weekly basis the rate is 5% on the first $3,000, 8% on the next $4,000, and 11% on any amount over $7,000. During the week Kerri Miller sold $12,300 worth of supplies, had an order for $1,400 that was returned, and had a draw of $300. Find her gross earnings. 6.2

For Problems 9–18, use a 6.2% rate up to a maximum of $72,600 for social security tax and a 1.45% rate with no wage ceiling for medicare tax.

For Problems 9 and 10, find the social security tax and the medicare tax for each employee. Assume that social security taxes must still be deducted.

EMPLOYEE	GROSS WEEKLY EARNINGS	
9. Dilts, G.	$891.95	6.2
10. Leasure, M.	$443.17	6.2

For Problems 11 and 12, find the social security and medicare tax for each employee for the current pay period. Year-to-date earnings do not include earnings for the current period.

EMPLOYEE	YEAR-TO-DATE GROSS EARNINGS	GROSS EARNINGS FOR CURRENT PAY PERIOD	
11. Hayes, C.	$71,100	$1,881.62	6.2
12. Stake, R.	$79,332	$3,450.81	6.2

Chapter 6

TEST YOURSELF

For Problems 13 and 14, use the wage bracket tables to determine the federal income tax deduction.

EMPLOYEE	GROSS WEEKLY EARNINGS	MARITAL STATUS	EXEMPTIONS	
13. Glenny, J.	$1,090.59	S	3	6.2
14. Shanley, M.	$788.17	M	1	6.2

For Problems 15 and 16, use the percentage method to find federal income tax.

EMPLOYEE	GROSS EARNINGS	PAYROLL PERIOD	MARITAL STATUS	EXEMPTIONS	
15. Berry, B.	$6,093.26	biweekly	M	5	6.2
16. Hsu, J.	$1,462.91	semimonthly	S	0	6.2

17. Lee Joseph, a project manager at Citizen's Power, is paid a salary of $1,125 per week. He is married and claims three exemptions. Find his weekly net earnings after deductions for social security, medicare, and federal income tax. (Use the wage bracket tables to find federal income tax.) 6.2

18. Malcolm Evans earns $74,360 a year. He is single and claims two exemptions. He is paid biweekly and in addition to social security, medicare, and federal income tax deductions he has 3% for state tax, $5\frac{1}{4}$% for pension, and $81 for health insurance withheld from his paycheck. Find his net earnings for the last two weeks in October. (Use the percentage method to find federal income tax.) 6.2

Chapter 6

Name	Date	Class

COMPUTER SPREADSHEET EXERCISES

Find the missing values in the following spreadsheets. Assume a 6.2% rate on gross earnings for social security tax and a 1.45% rate on gross for medicare tax. Use formulas where possible (see specific instructions below) and $ signs for dollar amounts. Align decimal points on dollar values and round to the nearest cent where necessary. All columns and numbers in your spreadsheet should appear as shown in each problem. If your spreadsheet is too wide to fit across the width of the page, choose File Page Setup and use the Landscape orientation and the Fit Columns to Page scaling option to print across the length of the paper. Customizing your column widths to fit the data may also improve the appearance of your printout. Use the following percentage method withholding tables to compute federal income tax for Exercises 1 and 2. (See Exercise 1 for specific instructions.)

Weekly Allowance per Exemption =	$52.88

Single Employee - Weekly Payroll

Taxable Earnings

Over	But not over	Amount to be withheld
$51	$525	15% of excess over $51
$525		$71.10 plus 28% of excess over $525

Married Employee - Weekly Payroll

Taxable Earnings

Over	But not over	Amount to be withheld
$124	$913	15% of excess over $124
$913		$118.35 plus 28% of excess over $913

Biweekly Allowance per Exemption =	$105.77

Single Employee - Biweekly Payroll

Taxable Earnings

Over	But not over	Amount to be withheld
$102	$1,050	15% of excess over $102
$1,050		$142.20 plus 28% of excess over $1,050

Married Employee - Biweekly Payroll

Taxable Earnings

Over	But not over	Amount to be withheld
$248	$1,827	15% of excess over $248
$1,827		$236.85 plus 28% of excess over $1,827

1. The following biweekly payroll record for Pollard Electronics is incomplete. Fill in the missing numbers. Assume that social security taxes must still be deducted. Give totals for all columns except Employee, Marital Status, and Exemptions. The calculation to find taxable earnings (column 3) is

$$\text{Taxable earnings} = \text{Gross earnings} - \left(\text{Number of exemptions} \times \text{Allowance per exemption}\right)$$

(Putting the allowance per exemption in a separate cell will enable you to use it in a formula.) The taxable earnings amount can then be used in an IF function to compute federal income tax. For example, a Microsoft Excel IF function to compute biweekly federal income tax for a single employee would be:

=IF(MS="S",IF(TE<=1050,0.15*(TE–102),142.2+0.28*(TE–1050)),IF(TE<=1827,0.15*(TE–248),236.85+0.28*(TE–1827)))

where the cell addresses of marital status and taxable earnings are substituted for MS and TE in the formula.

257

Pollard Electronics

Employee	Marital Status	Exemptions	Gross Earnings	Taxable Earnings	Federal Income Tax	Social Security	Medicare	Total Deductions	Net Earnings
Arthur, E.	S	5	$2,067.93						
Meyer, J.	M	0	$911.35						
Moss, J.	M	1	$5,348.59						
Price, V.	M	3	$1,133.76						
Riley, S.	S	2	$1,953.05						
Stamos, M.	S	4	$1,114.66						
Thacker, R.	M	3	$861.00						
Ward, W.	S	2	$1,745.15						
Totals	XXXXX	XXXXX	$15,135.49						

2. The following weekly payroll record shows the hours worked per employee for the current payroll period at Duska Enterprises. Fill in the missing numbers. Overtime is paid at time and a half for all work in excess of 40 hours. Assume a pension fund contribution of 4% of gross earnings and a medical insurance deduction of $19 for single employees or $34 for married employees, and exclude these deductions as well as the appropriate exemption deduction from federal income tax. Use formulas for all calculations with IF statements for medical insurance, federal income tax, and social security tax calculations. Year-to-date earnings are prior to the current pay period. Use a wage ceiling of $72,600 for social security tax with no wage ceiling for medicare tax.

Duska Enterprises

Employee	Marital Status	Exemp- tions	Year-to-Date Earnings	M	T	W	Th	F	S	Total Hrs.	Rate per Hour	Gross Earnings	Taxable Earnings	Federal Inc. Tax	Social Security	Medicare Tax	Medical Insurance	Pension Fund	Total Deductions	Net Earnings
Adams, R.	M	2	$43,361.88	4.5	8	10	9	12	0		$15.71									
Butler, G.	M	1	$74,689.33	9	7	5.5	8	8	6		$25.34									
Davis, J.	S	3	$32,109.64	11	9.5	10	8	8	0		$13.86									
Imes, E.	S	0	$29,227.09	12	9	9	7	8	4		$12.69									
Jarvis, M.	M	6	$37,997.26	8	0	8	4	6	9		$19.31									
Koonce, A.	S	1	$40,632.19	9	9	9.5	8	4	5		$17.59									
Nelson, R.	M	4	$46,797.67	7.5	7	9	10	8	0		$14.92									
Quarles, T.	M	3	$71,681.48	8	12	8	8	7.5	8		$28.87									
Watson, L.	M	7	$41,169.70	7	11	8.5	10.5	5	0		$18.62									
Zale, V.	S	2	$35,783.03	8	8	8	8	8	0		$14.42									

3. During the week the amounts of gross sales and returns recorded by salespeople at Murray Pharmaceuticals were as follows:

Murray Pharmaceuticals

Employee	Region	Gross Sales	Returns	Net Sales	Commission	Bonus	Gross Earnings
Barker, B.	1	$6,681.65	$374.27				
Costello, E.	2	$19,075.68	$428.22				
Elkin, C.	2	$23,085.01	$93.48				
Hodge, R.	3	$4,953.08	$950.56				
Lopez, G.	1	$30,527.38	$891.87				
Miller, K.	3	$20,650.49	$398.61				
Nock, E.	1	$12,816.09	$407.00				
Peacock, R.	3	$11,504.56	$0.00				
Price, J.	2	$16,773.39	$732.52				
Short, R.	1	$8,743.04	$1,002.88				
Totals		$154,810.35	$5,279.41				

Murray pays its salespeople on a sliding commission scale:

5% on the first $5,000 of net sales

7% on amounts over $5,000

Complete the spreadsheet for Murray by using formulas to compute net sales and gross earnings using the IF function to compute commission and bonus amounts.

For instance,

$$=IF(Net\ Sales>5000, 0.07*(Net\ Sales-5000)+0.05*5000, 0.05*Net\ Sales)$$

would compute a 5% commission on the first $5,000 of net sales plus a 7% commission on the amount over $5,000. Some regions have exceeded their sales quota resulting in bonuses for some of the salespeople. The bonus amounts are as follows:

Region 1: $50
Region 2: $100
Region 3: $0

Enter a nested IF formula to determine the correct amounts in the Bonus column. A nested IF formula contains an IF inside of another IF. For example,

$$=IF(Region=1, x, IF(Region=2, y, z))$$

produces the value x if the Region number is 1, the value y if the Region is 2, and the value z if the Region is 3.

Give totals for all columns except Employee and Region.

7 ■ Borrowing Money: Interest

OBJECTIVES

Upon completion of this chapter, the student should be able to:

1. Use the simple-interest formula $I = Prt$ to find:

   interest;

   principal;

   rate; and

   time.

2. Use the formula $S = P + I$ to find:

   maturity value;

   principal; and

   interest.

3. Determine the exact time between any two dates.
4. Determine the maturity date of a loan.
5. Calculate ordinary and exact interest.
6. Use the formula $P = \dfrac{S}{1 + rt}$ to find the present value:

   on the day the note is drawn; and

   at any time before maturity.

Some of the earliest recorded documents show evidence of the borrowing and lending of money and the calculation of interest charges. Today the business of borrowing and lending is a large part of the U.S. economy. "Buy now and pay later" is a common way of purchasing items—for individuals, businesses, and even governments. An individual may wish to borrow to purchase a house or car or to go on to college, while a business owner may have to borrow money to expand the business or make improvements. We borrow through charge accounts, mortgage companies, banks, businesses, and from individuals.

Unless we borrow from a friend, we expect to pay for the use of other people's property. The cost of borrowing money depends on the rate charged, the amount borrowed, and the length of time it is kept. This cost is called **interest** and can be simply defined as money paid for the use of money. The principles in this chapter apply whether paying interest or receiving interest.

Two types of interest are commonly used today: simple interest and compound interest. In this chapter we will discuss **simple interest,** which is interest calculated only on the amount

262 CHAPTER 7   BORROWING MONEY: INTEREST

borrowed. Simple interest is used mainly for short-term debts of one year or less. *Compound interest,* which is computed on the original amount and previous interest, will be covered in Chapter 10.

When a lender charges interest, it is collected using either of two methods:

- by adding the interest to the amount borrowed and receiving this total as repayment at the end of the period; or

- by deducting the interest from the amount to be repaid and giving the borrower the difference.

The first method, sometimes called **add-on interest,** will be covered in this chapter. The second method, *discount,* will be covered in Chapter 8.

7.1 SIMPLE INTEREST

The formula for calculating simple interest includes four quantities:

- **Interest**—the money paid for the use of money.
- **Principal**—the amount of money borrowed.
- **Rate**—a percent of the principal for a stated unit of time.
- **Time**—the period for which the money is borrowed.

The relationship among these four quantities is:

$$\text{Interest} = \text{Principal} \times \text{Rate} \times \text{Time}$$

or

$$I = Prt$$

Given any three of the quantities in the equation, the fourth can be found.

When the loan is repaid, the interest is added to the principal to give the total amount repaid, or the **maturity value.**

$$\text{Maturity value} = \text{Principal} + \text{Interest}$$

or

$$S = P + I$$

It is important to note that r (rate) and t (time) must use the same unit of time. Both may be either days, weeks, months, or years, but they must be the same. If the problem states the rate and time in different units, one must be converted to conform to the other. For example, if the rate is 6% per year and the time is 3 months, the time is converted to $\frac{3}{12}$ or $\frac{1}{4}$ of a year. Then both r and t are in years.

In commercial transactions the rate is usually quoted per annum (for a year), and the time should also be used in the formula as part of a year. If the rate is given only as a percent with no time stated, it is assumed to be rate per year.

SIMPLE INTEREST

Find the Interest

■ **Example 1**   A man borrowed $1,200 for 5 months and was charged 9% simple interest.

1. How much interest did he pay?

   I (interest) is unknown.

   P (principal) is the amount borrowed = $1,200.

   r (rate) is 9% per year. Change to 0.09 before substituting.

   t (time) is 5 months. Change to $\frac{5}{12}$ before substituting.

   $$I = Prt$$
   $$= 1,200 \times 0.09 \times \frac{5}{12}$$
   $$= \frac{540}{12} = \$45$$

   The interest charge is $45.

Find the Maturity Value

2. How much did he repay at the end of the 5 months?

   $$S = P + I$$
   $$= 1,200 + 45$$
   $$S = \$1,245$$

   The amount repaid is $1,245.

Check: The amount of interest for a short period should be small compared with the principal. The amount repaid must be larger than the principal. ■

CALCULATOR HINT

If you use your calculator to solve Example 1, follow the procedure shown below.

ENTER	DISPLAY
1,200 ⊠ 0.09 ⊠ 5 ⊡ 12 ⊟	45 (This is $45.00)

It is important that you NOT change the time, $\frac{5}{12}$, to a decimal first. Just do the example from left to right, doing all the multiplications first and the division by 12 last. Do NOT round any numbers until the final step.

If your calculator has a percent key, you may solve Example 1 as shown below.

ENTER	DISPLAY
1,200 ⊠ 9 % ⊠ 5 ⊡ 12 ⊟	45

The interest is $45.00.

264 CHAPTER 7 BORROWING MONEY: INTEREST

Find the Rate

■ **Example 2**   A debt of $2,600 was repaid at the end of 3 months with an additional $52 as interest. What was the interest rate?

I is the amount of interest = $52.

P is the amount borrowed = $2,600.

r is unknown.

$t = 3$ months, or $\frac{3}{12}$ or 0.25 of a year.

$$I = Prt$$

$$52 = 2{,}600 \times r \times \frac{3}{12}$$

$$52 = 650r \qquad \textit{Multiply 2,600 by } \frac{3}{12} \textit{ to simplify the coefficient of r.}$$

$$\frac{52}{650} = \frac{650r}{650} \qquad \textit{Divide both sides of the equation by the coefficient of r.}$$

$$r = 0.08 = 8\%$$

The rate is 8% a year. Since the time was stated as part of a year, the rate is also based on a year.

Check: Simple interest rates vary but should be checked if they are less than 5% or more than 15%, except in charge accounts and installment transactions. ■

CALCULATOR HINT

If your calculator has memory (M+, M−, MR) keys, you can use them to help you solve Example 2.

ENTER	DISPLAY
2,600 ⊠ 3 ⊞ 12 M+	650 and the letter M
52 ⊞ MR ⊟	0.08

0.08 = 8%. The rate is 8%.

If you would like to use your % key so the final answer will be a percent, follow the procedure shown below.

ENTER	DISPLAY
2,600 ⊠ 3 ⊞ 12 M+	650 and the letter M
52 ⊞ MR %	8

The rate is 8%.

The % key also includes "=" and it gives the final answer in percent form.

REMEMBER: Clear the memory by pressing the AC key or the MR key, by turning the calculator off, or by keying 0 into memory *before* starting each new example.

SIMPLE INTEREST

Find the Time

■ **Example 3**

A debt of $480 was repaid with a check for $498. If the interest rate was $7\frac{1}{2}\%$, for how long was the money borrowed?

$P = \$480$

$r = 7\frac{1}{2}\% = 0.075$

t is the unknown

$S = \$498$

$I = 498 - 480 = \$18$ *To find the interest, subtract the principal from the maturity value.*

TO FIND THE TIME IN YEARS USE "t"	TO FIND THE TIME IN MONTHS USE $\frac{"t"}{12}$
$I = Prt$	$I = Prt$
$18 = 480 \times 0.075 \times t$	$18 = 480 \times 0.075 \times \frac{t}{12}$
$18 = 36t$	$18 = 3t$
$\frac{18}{36} = \frac{36t}{36}$	$\frac{18}{3} = \frac{3t}{3}$
$0.5 = t$ or $\frac{1}{2} = t$	$6 = t$

The time is $\frac{1}{2}$ year or 6 months.

Check: The time is less than a year. This will be true for most simple interest problems. ■

CALCULATOR HINT

If you are using a calculator with memory keys, solve Example 3 as shown below.
   If you are solving $I = Prt$:

ENTER	DISPLAY
480 ⊠ 0.075 M+	36 and the letter M
18 ÷ MR =	0.5

0.5 year is the time.

If you are solving $I = P \times r \times \frac{t}{12}$:

ENTER	DISPLAY
AC	
480 ⊠ 0.075 ÷ 12 M+	3 and the letter M
18 ÷ MR =	6

6 months is the time.

Find the Principal

■ **Example 4** How much was borrowed if the interest is $27, the rate is 9%, and the time is 2 months?

$$I = \$27$$

P is unknown

$$r = 9\% = 0.09$$

$t = 2$ months, or $\frac{2}{12}$ of a year. The fraction $\frac{2}{12}$ will be used instead of the equivalent repeating decimal, namely, $0.1666666\overline{6}$.

$$I = Prt$$

$$27 = P \times 0.09 \times \frac{2}{12}$$

$$27 = P(0.015) \qquad\qquad \textit{Multiply } 0.09 \textit{ by } 2 \textit{ and divide the result by } 12.$$

$$\frac{27}{0.015} = \frac{P(\cancel{0.015})}{\cancel{0.015}} \qquad \textit{Divide both sides by the coefficient of P.}$$

$$\$1{,}800 = P$$

The principal is $1,800.

Check:
$$I = Prt$$
$$= 1{,}800 \times 0.09 \times \frac{2}{12}$$
$$I = \$27 \qquad\qquad\qquad\qquad ■$$

Chapter 7

TRY THESE PROBLEMS (Set I)

Find the missing quantities. Express the time in months.

	PRINCIPAL	RATE	TIME	INTEREST	MATURITY VALUE
1.	$3,900	12%	8 months		
2.	$4,500	10%	11 months		
3.	$12,000		6 months	$570	
4.	$240	5%		$5	
5.	$10,000		3 months		$10,150
6.	$2,000	$6\frac{1}{4}$%			$2,031.25
7.		9%	$1\frac{1}{2}$ years	$675	
8.		$7\frac{1}{2}$%	8 months	$1,048.75	

9. Sally Burns borrowed $6,300 for 10 months at $11\frac{1}{4}$% simple interest.

   a. How much interest will be owed on this principal?

   b. How much must Sally repay after 10 months?

10. If interest of $85 is charged on a loan of $3,000 for 4 months, what is the interest rate?

11. An investment pays $17 interest every 6 months. If the rate is 8%, how large is the principal?

12. A debt of $5,000 with interest at 12% was repaid with a check for $5,300. What was the term of the loan (in months)?

7.2 EXACT TIME BETWEEN DATES

The time during which a loan is *outstanding* (the period between borrowing and repayment) may be indicated by dates. For example, "A loan was made on January 5 and repaid on March 3." To calculate interest it is then necessary to find the exact number of days between the two dates. In this problem no year was mentioned, and it must be assumed that it is not a leap year, when February has 29 days. Between century years, leap years occur every four years. They coincide with presidential election years such as 1992 and 1996. All leap years are evenly divisible by 4.

The number of days can be found in either of two ways.

- Without the use of a table, either the beginning or final date is counted, never both. We will count the final date. Between January 5 and March 3 there are:

In January:	26 days	*There are* 31 *days in January;* 31 − 5 = 26.
In February:	28 days	
In March:	3 days	
Time:	57 days	

TABLE 7-1 NUMBER OF EACH DAY OF THE YEAR

DAY OF MONTH	JAN.	FEB.	MAR.	APR.	MAY	JUNE	JULY	AUG.	SEPT.	OCT.	NOV.	DEC.	DAY OF MONTH
1	1	32	60	91	121	152	182	213	244	274	305	335	1
2	2	33	61	92	122	153	183	214	245	275	306	336	2
3	3	34	62	93	123	154	184	215	246	276	307	337	3
4	4	March 3		94	124	155	185	216	247	277	308	338	4
5	5	36	64	95	125	156	186	217	248	278	309	339	5
6	6	37	65	96	126	157	187	218	249	279	310	340	6
7	7	38	66	97	127	158	188	219	250	280	311	341	7
8	8	39	67	98	128	159	189	220	251	281	312	342	8
9	9	40	68	99	129	160	190	221	252	282	313	343	9
10	10	41	69	100	130	161	191	222	253	283	314	344	10
11	11	42	70	101	131	162	192	223	254	284	315	345	11
12	12	43	71	102	132	163	193	224	255	285	316	346	12
13	13	44	72	103	133	164	194	225	256	286	317	347	13
14	14	45	73	104	134	165	195	226	257	287	318	348	14
15	15	46	74	May 14		166	196	227	258	288	319	349	15
16	16	47	75	106	136	167	197	228	259	289	320	350	16
17	17	48	76	107	137	168	198	229	260	290	321	351	17
18	18	49	77	108	138	169	199	230	November 17			352	18
19	19	50	78	109	139	170	200	231				353	19
20	20	51	79	110	140	171	201	232	263	293	324	354	20
21	21	52	80	111	141	172	202	233	264	294	325	355	21
22	22	53	81	112	142	173	203	234	265	295	326	356	22
23	23	54	82	113	143	174	204	235	266	296	327	357	23
24	24	55	83	114	144	175	205	236	267	297	328	358	24
25	25	56	84	115	145	176	206	237	268	298	329	359	25
26	26	57	85	116	146	177	207	238	269	299	330	360	26
27	27	58	86	117	147	178	208	239	270	300	331	361	27
28	28	59	87	118	148	179	209	240	271	301	332	362	28
29	29		88	119	149	180	210	241	272	302	333	363	29
30	30		89	120	150	181	211	242	273	303	334	364	30
31	31		90		151		212	243		304		365	31

Note: In leap years, after February 28, add 1 to the tabular number.

- The calendar table on page 596 gives us a much more convenient method, which will be used in all the examples. The calendar table is also shown in Table 7-1, and the method of locating the number associated with the date is illustrated for three dates used in the problems that follow. When the two dates are in the same year, locate both in the table and subtract the numbers that correspond to them.

January 5: Find January in the heading at the top of the page and 5 under "Day of Month." The number in both the row and column is 5, indicating that January 5 is the fifth day of the year.

March 3: Locate March 3 in the same way. The corresponding number in the table is 62, which means March 3 is the 62nd day of the year. The number of days between the two dates is:

$$\begin{array}{rr} \text{March 3:} & 62 \\ \text{January 5:} & -5 \\ \hline \text{Time:} & 57 \text{ days} \end{array}$$

When the dates are in different years, do the problem in two parts. Find the number of days left in the first year and add the number of days needed in the second year.

■ **Example 1**

Find the exact number of days between November 17, 1998, and February 12, 1999.

$$\begin{array}{rr} \text{From the table, December 31:} & 365 \\ \text{November 17:} & -321 \\ \hline \text{Days left in 1998:} & 44 \\ \text{Days in 1999 up to February 12:} & +43 \\ \hline \text{Time:} & 87 \text{ days} \end{array}$$

Whenever a particular year is given in the problem, check whether it is a leap year. If it is, add 1 (for the extra day in February) to the table number for any date *after* February 28. ■

■ **Example 2**

Find the number of days between January 15, 1996, and May 14, 1996. 1996 was a leap year. Add 1 to the table number for May 14.

$$\begin{array}{rr} \text{May 14:} & 135 \; (134 + 1) \\ \text{January 15:} & -15 \\ \hline \text{Time:} & 120 \text{ days} \end{array}$$

■

Maturity Date

When the time is given in months, the maturity date (the date payment is due) is on the same day the required number of months later. For example, the maturity date of a 3-month loan dated May 2 is August 2. However, the due date of a 1-month loan dated May 31 is June 30 because June 31 does not exist. When a situation like this occurs, the maturity date is the last day of the month.

When the time is in days, the calendar table is used to find the maturity date.

■ **Example 3**

A sum of money is borrowed on July 11 and must be repaid in 60 days. On what date is payment due?

$$\begin{array}{rl} \text{July 11 is:} & \text{192nd day of year} \\ \text{60 days later is:} & +60 \\ \hline \text{Due date is:} & \text{252nd day of year} \end{array}$$

The 252nd day of the year is September 9; the maturity date is September 9.

Check: 60 days is approximately 2 months, and the maturity date must be near September 11. ■

■ **Example 4** A sum of money is borrowed on November 11, 2000, for 90 days. What is the maturity date? The time period in this problem involves 2 years.

1. Find the number of days left in 2000.

December 31:	365
November 11:	−315
Days remaining in 2000:	50

2000 was a leap year, but both dates are after February 28, and adding 1 to both table numbers would not change the result.

2. Find the number of additional days needed.

$$90 - 50 = 40$$

3. Locate the 40th day in the table. It is February 9. The maturity date is February 9, 2001.

Check: 90 days is approximately 3 months, and the maturity date should be approximately 3 months from November 11. ■

Chapter 7

TRY THESE PROBLEMS (Set II)

Find the number of days between:

1. March 15 and July 29.

2. October 27, 2001, and March 18, 2002.

3. February 13, 2000, and July 7, 2000.

4. October 19, 2000, and February 2, 2001.

5. January 15, 2001, and June 25, 2001.

Find the maturity dates for the following loans dated:

6. July 12 for 8 months.

7. March 3 for 60 days.

8. January 31 for 3 months.

9. September 30, 2000, for 150 days.

10. December 15, 1998, for 150 days.

273

7.3 ORDINARY AND EXACT INTEREST: TIME IN DAYS

When the period of a loan is in days, it would seem reasonable to use 365 days as the denominator of the time fraction, just as we used 12 (for 12 months) when the time was in months. This, however, is not the universal practice. The year is frequently converted to 360 days, because the number 360 greatly simplified calculations before calculators became common. 360 is evenly divisible by 2, 3, 4, 5, 6, 9, 10, 12, 15, etc.

When the year is used with 360 days (30 days for each month), the period is in **ordinary time.** When the year is used with 365 days, the period is in **exact time.** From these we get the terms for two methods of calculating interest:

- **Ordinary interest,** in which the time fraction has 360 in the denominator.
- **Exact interest,** in which the time fraction has 365 days in the denominator.

The time fractions are:

$$\text{For ordinary interest: } t = \frac{\text{number of days}}{360} \text{ (banker's rule)}$$

$$\text{For exact interest: } t = \frac{\text{number of days}}{365}$$

The use of exact time and ordinary interest is called the **banker's rule,** since banks and other financial institutions commonly compute their interest in this manner. The obvious reason for its historical use is that banks can obtain the most interest when lending money by using 360 instead of 365 in the denominator. Some lenders have justified the practice by stating that it simplifies calculations. However, many consumers have questioned its continued use, since widespread computerization makes the simplified calculation no longer necessary, and have pressured lawmakers to pass laws requiring the use of exact interest. In at least one case, a state court has held that the fact that banks have customarily calculated interest in this way is not a legitimate defense for overcharging on interest payments. Some banks have changed their policy and now calculate interest using 365 days. Government agencies and the Federal Reserve Bank also use exact interest. Although the use of ordinary interest over exact interest may appear to be a trivial amount for each borrower, the difference is substantial for a long-term loan such as a mortgage. For example, for a $150,000 loan at 9% for 25 years, the overpayment in interest would approximate $2,000.

Note: For simplicity, if the type of interest is not specified in a problem, it is assumed to be ordinary interest. In problems where both types of interest are calculated, exact interest will always be less than ordinary for the same principal, rate, and time.

■ **Example 1**

Find the exact and ordinary interest on a loan of $4,000 at 12% for 60 days.

ORDINARY

$I = Prt$

$= 4{,}000 \times 0.12 \times \dfrac{60}{360}$

$= \dfrac{28{,}800}{360}$

$I = \$80$

EXACT

$I = Prt$

$= 4{,}000 \times 0.12 \times \dfrac{60}{365}$

$= \dfrac{28{,}800}{365}$

$I = \$78.90$

Ordinary interest is $80; exact interest is $78.90.

276 CHAPTER 7   BORROWING MONEY: INTEREST

Check: For a given principal, interest rate, and time, the exact interest is always less than the ordinary interest. However, the difference should be small. ■

Find the Interest

■ **Example 2**

Find the interest on a loan of $3,500 made on April 4 and due on May 19 if the interest rate is 8%.

$$\begin{aligned} \text{May 19:} &\quad 139 \\ \text{April 4:} &\quad -94 \\ \text{Exact time:} &\quad 45 \text{ days} \end{aligned}$$

$$I = Prt$$
$$= 3{,}500 \times 0.08 \times \frac{45}{360}$$
$$I = \$35$$

Banker's rule is used if exact interest is not specified.

Check: The amount of interest should be small compared with the principal because the time (45 days) is short. ■

Find the Maturity Value

■ **Example 3**

Mr. Scrooge lent $2,000 to an employee on February 18, 1996, with interest at $7\frac{1}{2}\%$.

The debt was to be repaid on December 24, 1996. How much money will Mr. Scrooge receive on the maturity date?

Days between February 18 and December 24, 1996:

$$\begin{aligned} \text{December 24:} &\quad 359\ (358 + 1) \\ \text{February 18:} &\quad -49 \\ \text{Exact time:} &\quad 310 \text{ days} \end{aligned}$$

1996 was a leap year; 1 day must be added to table numbers after February 28. February 18 is before the leap year day; do not add 1.

$$I = Prt$$
$$= 2{,}000 \times 0.075 \times \frac{310}{360}$$
$$I = \$129.17$$
$$S = P + I$$
$$= 2{,}000 + 129.17$$
$$= \$2{,}129.17$$

$7\frac{1}{2}\% = 7.5\% = 0.075$

Mr. Scrooge will receive $2,129.17 on the maturity date.

Check: The interest is small compared with the principal, and the amount repaid is larger than the principal. ■

Find the Time

■ **Example 4**

A debt of $2,000 bearing 8% interest was repaid with a check for $2,040. For how long was the money owed?

$$I = 2{,}040 - 2{,}000 = \$40$$

$$I = Prt$$

$$40 = 2{,}000 \times 0.08 \times t$$

$$40 = 160t$$

$$t = \frac{40}{160} = 0.25 \text{ years}$$

Note: If t is used to represent time, the answer will be time in years or a part of a year. If $\frac{t}{12}$ is used, the answer will be in months, and if $\frac{t}{360}$ is used, the answer will be in days. The arithmetic is usually simplest when t is used with no denominator. The answer may be converted to months or days. For this example,

$$0.25 \text{ years} = 0.25 \times 12 \text{ months} = 3 \text{ months}$$

$$0.25 \text{ years} = 0.25 \times 360 \text{ days} = 90 \text{ days}$$

Using t with a denominator:

$I = Prt$	$I = Prt$
$40 = 2{,}000 \times 0.08 \times \dfrac{t}{12}$	$40 = 2{,}000 \times 0.08 \times \dfrac{t}{360}$
$40 = 13.333333t$	$40 = 0.444444t$
$t = \dfrac{40}{13.333333}$	$t = \dfrac{40}{0.444444}$
$t = 3$ months	$t = 90$ days

■

Chapter 7

	Name	Date	Class

TRY THESE PROBLEMS (Set III)

Find the ordinary and exact interest.

1. $2,000 is borrowed for 93 days at 8% interest.

2. $1,500 is borrowed at 10% interest for 150 days.

3. $900 is borrowed on April 10 and repaid on August 30. The rate is 11%.

Find the missing quantities. Time can be months, days, or years.

	PRINCIPAL	RATE	TIME	INTEREST	MATURITY VALUE
4.	$3,500	11%			$4,270
5.	$4,620	8%	10 months		
6.	$6,000	$9\frac{1}{2}\%$	March 10 to June 8		
7.	$846	7%	50 days		
8.	$2,200	8%			$2,420
9.			30 days	$14.74	$1,622.74
10.		10%	9 months	$189.00	
11.	$3,276		March 4 to May 3		$3,311.49
12.	$4,200		6 months	$225.75	
13.	$8,169	8%		$571.83	

279

14. Rock City met its end-of-month payroll with a loan of $2 million at $6\frac{3}{4}$% interest. If the loan is repaid in 60 days, how much will the city pay in interest charges?

15. On May 2 Tony Jones borrowed money at $8\frac{1}{2}$% interest for 45 days.

   a. When must the debt be repaid?

   b. If $12.75 interest was charged, how much did he borrow?

16. An $8,000 debt is discharged with a payment of $8,110. If the money was held for 45 days, what was the interest rate?

17. A loan of $530 was made on November 5, 2000. If the interest rate was 9% and the interest charged was $15.90:

   a. What was the term of the loan (in months)?

   b. What was the maturity date of the loan?

7.4 PRINCIPAL: PRESENT VALUE

In the last section we found the future or maturity value for a given principal. There are also instances where businesses and individuals are faced with the opposite problem; that is, computing the principal when the future or maturity value is known. The principal being found is often referred to as the **present value.** Present value can also be thought of as the amount which would have to be invested *today* (at the present) in order to have a given maturity value in the future.

The present value at simple interest can be found using the formula:

$$S = P(1 + rt)*$$

which can also be written as

$$P = \frac{S}{1 + rt}$$

where

P = Present value or principal

S = Maturity value

r = Interest rate

t = Time

Although either version may be used to find present value when S, r, and t are known, the last version is more convenient, since it is in solution form.

■ **Example 1**

A 40-day loan with interest at 9% was repaid with a check for $909. How large was the loan?

This problem asks for the principal that must be invested *today* at 9% interest to have $909 in 40 days.

$$P = \frac{S}{1 + rt}$$ — *Use the present value formula.*

$$= \frac{909}{1 + 0.09 \times \frac{40}{360}}$$

$$= \frac{909}{1 + 0.01}$$ — *Complete arithmetic in denominator first. Multiplication is done before addition.*

$$1.00 + .01 = 1.01$$

$$= \frac{909}{1.01}$$

$$P = \$900$$

---

*$I = Prt$

$S = P + I$

$ = P + Prt$ *Substitute Prt for I.*

$S = P(1 + rt)$ *Factor.*

The loan was $900. The entire problem is diagrammed below.

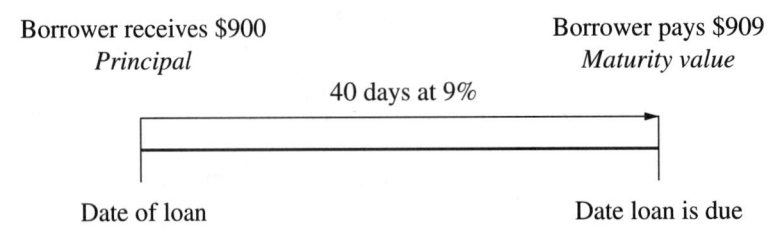

```
   Borrower receives $900                    Borrower pays $909
          Principal                             Maturity value
                        40 days at 9%
       ┌─────────────────────────────────────────────┐
       │                                             │
       └─────────────────────────────────────────────┘
       Date of loan                              Date loan is due
```

Check: The principal is less than the maturity value, but the difference is small. ■

CALCULATOR HINT

If you are using a calculator with memory keys, follow the procedure shown below to solve Example 1.

ENTER	DISPLAY
0.09 ⊠ 40 ÷ 360 + 1 M+	1.01 and the letter M

(Do the work in the denominator first. Multiplication and division precede addition.)

| 909 ÷ MR = | 900 |

The principal is $900.

■ **Example 2**

What is the present value of $1,456 due in 6 months if 8% interest is paid?

Another way of asking this question is: What principal will amount to $1,456 in 6 months at 8% interest?

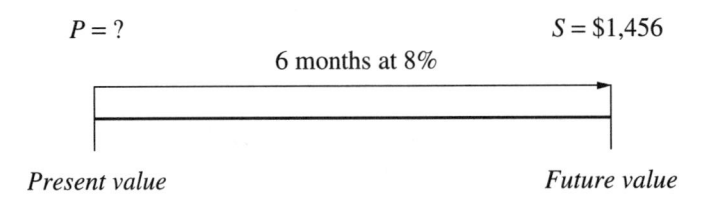

```
      P = ?                                    S = $1,456
                    6 months at 8%
      ┌─────────────────────────────────────────────┐
      │                                             │
      └─────────────────────────────────────────────┘
      Present value                            Future value
```

$P = \dfrac{S}{1 + rt}$ *S, r, and t are given, but not I. This formula must be used.*

$= \dfrac{1{,}456}{1 + 0.08 \times \dfrac{6}{12}}$

$= \dfrac{1{,}456}{1 + 0.04}$ *Complete arithmetic in denominator first.*

$= \dfrac{1{,}456}{1.04}$ *1.00 + 0.04 = 1.04*

$P = 1{,}400$

The present value is $1,400.

Check: The present value is less than the future value. ■

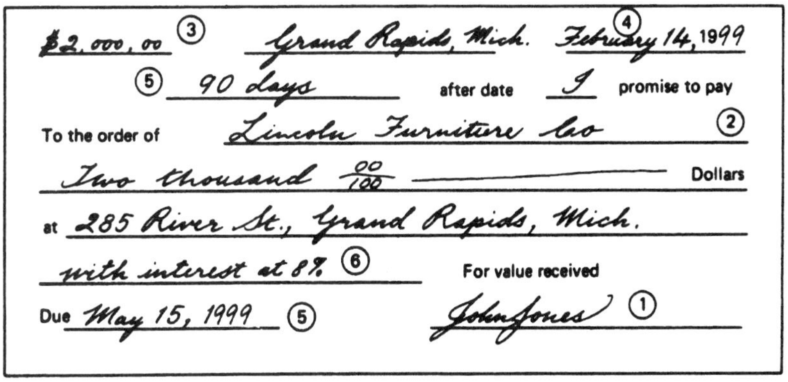

FIGURE 7-1 Interest-bearing note.

7.5 PROMISSORY NOTES

It is customary in business loans to write a short document showing all details of the agreement. This written promise to pay is called a **promissory note.** It is legally binding and negotiable (can be sold). An illustration of an interest-bearing promissory note is shown in Figure 7-1.

This note (in Figure 7-1) includes the following information:

①	**Maker:** the borrower who signs the note and promises to pay	John Jones
②	**Payee:** person or firm to whom the money is owed	Lincoln Furniture Co.
③	**Face value:** the amount of the debt	$2,000.00
④	**Date of loan**	February 14, 1999
⑤	**Term** and/or **maturity date**	Term is 90 days. Maturity date is May 15, 1999.
⑥	**Interest rate:** if there is an interest charge	8%

If the note is non-interest-bearing, the **maturity value** (amount repaid) is the same as the **face value** (amount borrowed). If the note bears interest, the maturity value is the face value plus the interest.

$$\text{Maturity value} = \text{Face value} + \text{Interest}$$
$$S = P + I$$

Because the interest for this note is found by using the simple interest formula, it is also referred to as a *simple interest note.*

Maturity Value of a Promissory Note

■ **Example 1**

A 4-month note dated July 7 with a $960 face value bears interest at $11\frac{1}{4}\%$. Find the maturity value.

First find the interest.

$$11\frac{1}{4}\% = 11.25\% = 0.1125$$

$$I = Prt$$
$$= 960 \times 0.1125 \times \frac{4}{12}$$
$$= \frac{432}{12}$$
$$I = \$36$$

Then find the maturity value.

$$S = P + I$$
$$= 960 + 36$$
$$S = \$996$$

The maturity value (*S*) of the note is $996.

Check: The maturity value must be larger than the face value of an interest-bearing note. However, the difference (interest) should be small compared with either one. ■

Face Value of a Promissory Note

■ **Example 2**

Mr. Rich expects $3,584 on July 5 in payment of a note he owns. The note was drawn on April 24 and bears interest at 12%. What was the face value?

April 24 to July 5 (114 to 186) is 72 days

The problem situation can be illustrated as follows:

$P = ?$ $S = \$3,584$

72 days at 12%

April 24   July 5

We are finding the face value (or principal, or present value) when the maturity value, rate, and time are given.

$$P = \frac{S}{1 + rt}$$ *Use this formula because S, r, and t are given.*

$$= \frac{3{,}584}{1 + 0.12 \times \frac{72}{360}}$$

$$= \frac{3{,}584}{1 + 0.024}$$

$$= \frac{3{,}584}{1.024}$$

$$= \$3{,}500$$

The face value of the note is $3,500.

Check: The face value is less than the maturity value and about the same size. ■

Chapter 7

Name _____ Date _____ Class _____

TRY THESE PROBLEMS (Set IV)

In the note in Figure 7-2:

1. The maker is _____.

2. The payee is _____.

3. The face value is _____.

4. The term of the note is _____.

5. The maturity date is _____.

6. The interest is _____.

7. The maturity value is _____.

8. Fran Friendly loaned her sister $350 to repair the car she uses to get to school. She asked her sister to sign a note that required that the debt be repaid in 2 months with 4% interest. How much must her sister repay?

9. In Problem 8, identify the maker and the payee of the note.

10. A note dated February 5 is due in 80 days. The face value is $1,244 and the interest rate is $9\frac{1}{2}\%$.

    a. What is the maturity date?

    b. What is the maturity value?

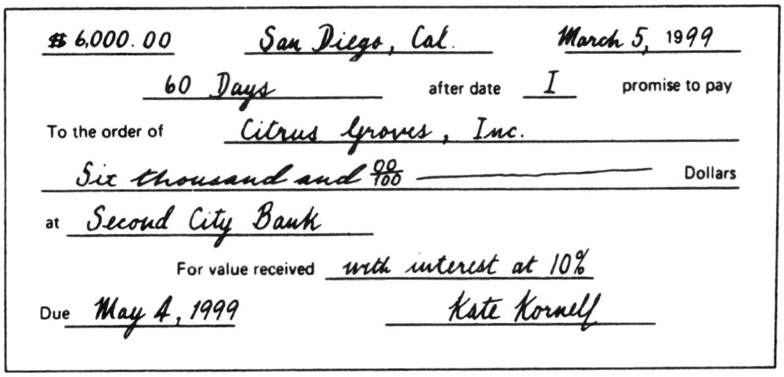

$6,000.00 San Diego, Cal. March 5, 1999

60 Days after date I promise to pay

To the order of Citrus Groves, Inc.

Six thousand and 00/100 —————— Dollars

at Second City Bank

For value received with interest at 10%

Due May 4, 1999   Kate Kornell

FIGURE 7-2 Sample note.

285

11. Mr. Brown borrowed $2,700 from Usury, Inc., on July 27. He signed a note agreeing to pay 6% interest and to repay the loan on October 30.

   a. Who is the maker?

   b. Who is the payee?

   c. How much interest will be paid?

   d. What is the maturity value?

12. What principal will amount to $1,701 in 50 days if the interest rate is 9%?

13. A 90-day loan with interest at 8% was repaid with a check for $2,040. How large was the loan?

14. What is the present value of a debt that has a maturity value of $2,767.50 if the term is 6 months and the interest rate is 5%?

15. On May 1, Mr. Brown received $972 in payment of a note dated March 12. He had charged interest at 9%. What was the face value of the note?

Present Value of a Promissory Note

An interest-bearing note increases in value the longer it is held, until the maturity date. It can therefore be regarded as having one value at maturity and other values at different times before maturity. The value at *any time* before maturity is also called **present value.** Present value was used before only for the value on the initial date of the debt and is extended here to mean the value on any date before maturity, including the initial date.

If the note is sold, it may be worth more or less than its face value (or face value plus accumulated interest if sold after the date of the note) depending on the average rate of interest being paid by banks and other financial institutions, referred to as the **money market rate.** For example, if the money market rate is 5%, then a 7% note is desirable to a buyer, and the value of the note would be greater than the face value plus accumulated interest. But if interest rates are generally above 7%, then a 7% note is less desirable, and the value would decline.

■ **Example 3**

A note for $1,000 for 6 months bears interest at 9%. What was the present value of the note on the day it was drawn if the market rate for money is 10%?

Solving this problem involves two steps:

Find the maturity value of the note *and*

Use the market rate to find the present value

1. The maker received $1,000 on the day it was drawn and 6 months later she repays $1,045 ($1,000 + $45 interest).

$$I = Prt$$

$$= 1,000 \, (0.09) \left(\frac{6}{12}\right)$$

$$I = \$45$$

$$S = P + I$$

$$= 1,000 + 45$$

$$S = \$1,045$$

Regardless of interest rates in the money market, the term and maturity value of the note remain the same. These are part of the original contract and are not changed if the note is sold.

2. The present value, however, is based on the market value of money, not the interest rate stated on the note.

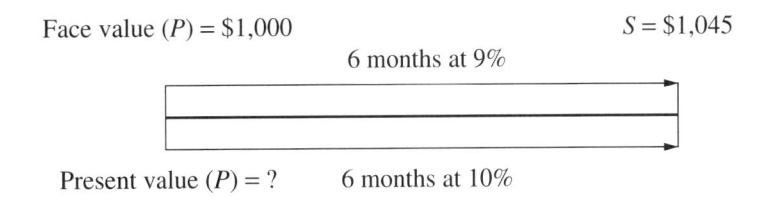

Face value (P) = $1,000   S = $1,045

6 months at 9%

Present value (P) = ?   6 months at 10%

288   CHAPTER 7   BORROWING MONEY: INTEREST

$$P = \frac{S}{1+rt}$$

S, r, and t are given.

$$= \frac{1{,}045}{1 + 0.10 \times \frac{6}{12}}$$

r = 10% because money was worth 10%.

$$= \frac{1{,}045}{1 + 0.05}$$

Complete arithmetic in denominator.

$$= \frac{1{,}045}{1.05}$$

1.00 + 0.05 = 1.05

$$P = \$995.24$$

The present value of the note on the day it was drawn was $995.24 at 10% interest.

This means that $995.24 at 10% has the same maturity value as $1,000 at 9% after 6 months. It also means that if the payee had to sell the note on the day it was drawn she would expect to get $995.24 not $1,000.

Check: The present value when the note is drawn must be less than face value if the market rate for money is higher than the interest rate charged. ■

■ **Example 4**

Mr. Goodheart loaned a friend $620 for 4 months and accepted a non-interest-bearing note in exchange. If Mr. Goodheart could have obtained 9% interest in the money market, what was the present value of the note on the day the loan was made?

The present value is based on maturity value. In this problem there is no interest on the debt and the maturity value equals the face value.

$$P = \frac{S}{1+rt}$$

$$= \frac{620}{1 + 0.09 \times \frac{4}{12}}$$

r = money market rate = 9%.

$$= \frac{620}{1 + 0.03}$$

$$= \frac{620}{1.03}$$

$$P = \$601.94$$

The present value of the note was $601.94.

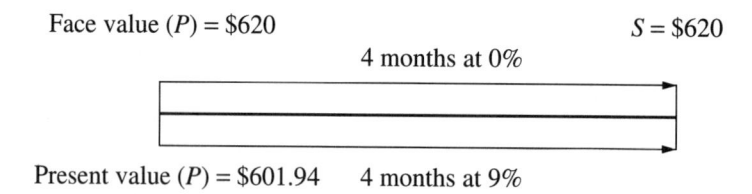

Face value (P) = $620   S = $620

4 months at 0%

Present value (P) = $601.94   4 months at 9%

This means that Mr. Goodheart could have loaned $601.94 at the market rate (9%) and received the same maturity value ($620).

Check: Present value is less than face value and maturity value for a non-interest-bearing note. ■

■ **Example 5**

The Old Stuff Co. sells made-to-order antiques. One of its customers went bankrupt and they had to resell merchandise priced at $1,200. As a concession to the new customer the company accepted a 90-day note with interest at 5% instead of cash when the payment was due. What was the present value of the note on the day it was drawn if money was worth 8%?

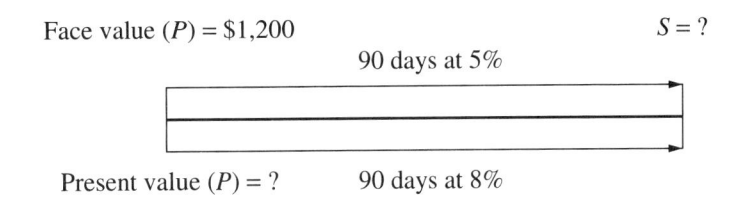

Face value (P) = $1,200   S = ?
                 90 days at 5%

Present value (P) = ?   90 days at 8%

Present value is based on maturity value, and maturity value must be found.

First, find the maturity value using the rate on the note:

$$I = Prt$$
$$= 1,200 \times 0.05 \times \frac{90}{360}$$

r is interest on note = 5%; t is term of note = 90 days.

$$I = \$15$$
$$S = 1,200 + 15 = \$1,215$$

Then find the present value using the market rate for money:

$$P = \frac{S}{1 + rt}$$
$$= \frac{1,215}{1 + 0.08 \times \frac{90}{360}}$$
$$= \frac{1,215}{1 + 0.02}$$
$$= \frac{1,215}{1.02}$$
$$P = \$1,191.18$$

r is the money market rate = 8%; t is the term of the note = 90 days.

The present value was $1,191.18.

This means that $1,191.18 invested at the market rate of 8% would give the same maturity value as $1,200 at 5% interest after 90 days. It also means that the merchandise was really sold for $1,191.18.

Check: The present value is less than the face value because the money market rate is higher than the interest rate on the note. ■

Present value was previously defined as any value before maturity. Thus far we have found the present value only on the date the note is drawn. The same method can be extended to find the present value on any other date before maturity. One additional fact must be considered, however: When the present value is found on the date the note is drawn (as in Examples 3, 4, and 5), the time is the same as the term of the note. For any other date, *the time is the period from the present-value date to the date of maturity* and is shorter than the term of the note.

■ **Example 6**

A note with interest at 6% is drawn on April 15 and matures on August 25. The face value is $500. If money is worth 8%, find the present value on July 26.

A time diagram is essential for understanding the problem situation.

290 CHAPTER 7   BORROWING MONEY: INTEREST

Face value (P) = $500   132 days at 6%   S = ?

30 days at 8%

April 15—Table No. (105)   July 26 (207)   August 25 (237)
Present value (P) = ?

First find the *maturity value* using the rate on the note.

$$I = Prt$$

r is the interest rate on the note; t is the term of the note (from the diagram).

$$= 500 \times 0.06 \times \frac{132}{360}$$

$$I = \$11$$

$$S = 500 + 11 = \$511$$

Then find the *present value* using the money market rate.

$$P = \frac{S}{1 + rt}$$

r is the money market rate of 8%; t is the time from the present-value date to maturity = 30 days (from the diagram).

$$= \frac{511}{1 + 0.08 \times \frac{30}{360}}$$

$$= \frac{511}{1 + 0.006666}$$

$$= \frac{511}{1.006666}$$

Do not round the repeating decimal.

$$P = \$507.62$$

The present value on July 26 is $507.62 if money is worth 8%.

Check: The present value is less than the maturity value of $511. ∎

Now go to Try These Problems (Set V).

NEW FORMULAS IN THIS CHAPTER

- $I = Prt$ (Interest = Principal × Rate × Time)

- $S = P + I$ (Amount or maturity value = Principal + Interest)

- $P = \dfrac{S}{1 + rt}$ or $S = P(1 + rt)$

- For ordinary interest: $t = \dfrac{\text{Number of days}}{360}$ (banker's rule)

- For exact interest: $t = \dfrac{\text{Number of days}}{365}$

Chapter 7

TRY THESE PROBLEMS (Set V)

Determine whether the present value will be less than, equal to, or more than the face value of the notes in Problems 1 through 3.

	INTEREST RATE FOR THE NOTE	MARKET RATE FOR MONEY
1.	$8\frac{1}{2}\%$	$7\frac{1}{2}\%$
2.	7	7
3.	$9\frac{1}{4}$	$10\frac{1}{2}$

4. Find the present value of a 150-day note with a maturity value of $8,200 if money is worth 6%.

5. What amount should a lender accept today if the maker wishes to settle a 6-month loan with a maturity value of $4,800 if the market rate for money is $9\frac{1}{2}\%$?

6. An 80-day note with interest at 9% has a maturity value of $3,570.

   a. Find the face value of the note.

   b. Find the present value if money is worth 10%.

7. A $450 note with interest at 12% has a term of 9 months. What is its present value on the day it is drawn if money is worth 10%?

8. A 90-day note for $2,200 is to be repaid with interest at 8%. What is its present value halfway through the term of the note if money is worth 10%?

9. A note with a face value of $4,000 dated September 9 is due on November 20 with interest at 11%. After the note is signed the market rate rises to 12%. What is the present value of the note on October 21 at the market rate?

10. A loan of $5,120 was made on May 13 at 11%. It was due on December 24. Find the present value of the loan on July 27 if the market rate for money is $10\frac{1}{2}\%$.

11. The Lion Toy Company borrowed $9,000 at 7% for 5 months. After three months they found they had excess cash and wished to pay the debt. How much cash did they need if the market rate for money was then 6%?

12. The Barnes Bookshop owed a publisher $727 for a shipment of books. The owner signed a note for the amount owed with $8\frac{1}{2}\%$ interest to be paid in 6 months. Two months before the note was due, the owner offered to pay the note with accumulated interest.

   a. How much was he offering?

   b. The publisher said they would accept the present value at the market rate of 7%. What payment would they accept?

Chapter 7

	Name	Date	Class

END-OF-CHAPTER PROBLEMS

Find the missing quantity.

	FROM	TO	NUMBER OF DAYS
1.	February 13, 2000	June 26, 2000	
2.	October 19, 2000	April 18, 2001	
3.	July 7		65
4.	November 4, 2001		82
5.	August 30	November 10	
6.	May 9		120

Find the exact and ordinary interest on:

7. $1,095 at 10% for 80 days.

8. $720 at 9% from August 10 to September 19.

9. $1,250 at 12% from December 2 to March 2.

Find the missing quantities. Time can be in months or days.

	PRINCIPAL	RATE	TIME	INTEREST	MATURITY VALUE
10.	$2,000	9%	50 days		
11.	$4,000	14%	9 months		

	PRINCIPAL	RATE	TIME	INTEREST	MATURITY VALUE
12.		9%	4 months	$4.80	
13.		12%	120 days	$18.00	
14.	$720		40 days		$727.60
15.	$500		3 months	$13.75	
16.	$960	8%			$979.20
17.	$1,600	$7\frac{1}{2}$%		$20.00	
18.		10%	20 days		$905.00
19.		14%	9 months		$4,420.00

Find the present value of the following notes.

	MATURITY VALUE	DAYS BEFORE MATURITY	MONEY MARKET RATE
20.	$1,454	50	7%
21.	569	40	9

22. The Fast Funds Finance Company charges customers 15% interest on loans.

   a. How much interest would they collect on a loan of $2,500 for 9 months?

   b. How much would the customer repay on the due date?

Chapter 7

END-OF-CHAPTER PROBLEMS

23. Aaron Morrison borrowed $600 on June 1 and promised to pay back the money in 120 days with interest at $9\frac{1}{2}\%$.

    a. What amount must be repaid?

    b. What is the due date of the loan?

24. Find the maturity value of a note dated April 20 if it bears interest at 9% and matures on June 25. The face value is $2,240.

25. A note dated February 23 was due on August 22. Find the interest and amount due if the face value was $419 and the interest rate was 8%.

26. The owner of Good Food Store is just starting business. He needs fixtures and obtains credit from the manufacturer for an indefinite period. He buys $5,000 worth of equipment and gives the supplier a non-interest-bearing note for 90 days. However, they agree that if credit must be extended past 90 days there will be an interest charge of 6% for the next 60 days and 8% after that.

    a. What is the maturity value of the debt after 90 days?

    b. What is the maturity value of the debt after 120 days?

    c. What is the maturity value of the debt after 180 days?

27. Robin Stone borrows $1,460 on January 16, 2001. She agrees to pay interest at $8\frac{1}{2}\%$ and to repay the loan 100 days later.

    a. On what date must the loan be repaid?

    b. If she pays exact interest, how much will she repay?

    c. If she pays ordinary interest, how much will she repay?

295

28. On February 2, 2000 (a leap year), Karl Kruger borrowed $4,280 at 9%. The loan was repaid on May 16 of the same year. Find the maturity value if:

   a. Exact interest was paid.

   b. Ordinary interest was paid.

29. Tom Teach made an error in his income tax calculation and owes the government $745.50 plus exact interest of 18% for the period between April 15 and November 26. How much must he pay in total?

30. Terrific Toys needs to borrow $140,000 to finance an inventory for Christmas sales. The loan will be for 85 days at $11\frac{1}{2}$%. How much more will the interest be if they must pay ordinary, rather than exact, interest?

31. A note for $2,286 is settled after 5 months with a check for $2,362.20. What is the interest rate?

32. Dan Dash needs $168 for the repair of the car that gets him to school. His father agrees to lend him the money, but to discourage him from borrowing he adds $1.54 a month until the debt is paid. What interest rate is he charging?

33. An investment of $1,080 pays $22.68 interest in 72 days. What is the interest rate?

34. Fast Funds Finance Company received payment of a 5-month promissory note. The check was for $3,143.75 and the face value was $3,000. What was the interest rate?

35. Bill Bante deposits $8,000 for 90 days and earns $180 in interest. Find the rate of interest.

36. A loan shark made a loan of $960 and was repaid $1,000 the following month. What rate of interest was he charging?

37. An $1,800 note was repaid on the maturity date. The interest rate charged was 12% and the maturity value was $1,872. What was the term of the note?

38. A $7,200 note with interest at 9% has a maturity value of $7,578. What is the term of the note?

Chapter 7

Name	Date	Class

END-OF-CHAPTER PROBLEMS

39. If $7,000 is invested at 11%, how long will it take before the investment is worth $8,155?

40. How long will it take for $960 to yield $34.40 in interest if the rate is 6%?

41. The Beta Company charges 18% simple interest on overdue accounts. On one account, for $1,500, the penalty was $45. Find how long the account was overdue.

42. How long will it take for $800 to earn $55 interest if the rate is 11%?

43. A savings account paid $22.50 interest in a 1-year period. If the interest rate was $7\frac{1}{2}\%$, how much money was in the account?

44. An investment pays $97.50 in interest every 3 months. If this represents an interest rate of $7\frac{1}{2}\%$, how large is the investment?

45. An investment yields $276 interest payable twice a year. If the interest rate is 9.2%, how large is the investment?

46. Ben loaned Jerry money to pay for a parking sticker at school. Since they were both business students, they agreed to add 9% interest. After 5 months Jerry repaid $20.75 to cover the entire debt. How much did the parking sticker cost?

47. A business must pay a debt of $7,000 in 8 months. Find the amount of money that should be deposited today, at $10\frac{1}{4}\%$ interest, so that enough money will be available to pay the debt.

48. What amount should a lender accept today in payment for a 210-day loan with a maturity value of $8,800 if money has a value of 9%?

49. Mr. A. Turnee accepted a note from a client who could not pay his fee. The term of the note was 6 months and the interest rate was 8%. When the note matures, Mr. Turnee will receive $1,456. How large was his fee?

50. A note dated August 14 was settled with a payment of $1,262.50 on September 28. If the interest rate was 8%, what was the face value of the note?

51. Ms. Reuben needs a new fax machine for her work as a consultant. The machine costs $960, but she will not have the money for 2 months. A friend offers her the money if she will sign a note with maturity value of $973.60 payable in 2 months.

  a. What interest rate is being charged?

  b. If money is worth 8%, is the present value of the note more or less than the face value? (This does not require additional calculation.)

52. A note for $1,800 has a term of 9 months and an interest rate of 11%.

  a. If the money market rate is 11%, what is the present value on the date it is drawn?

  b. If money is worth 12%, what is its present value at the beginning of the term?

53. On November 25, which is his 21st birthday, Larry Lucky will inherit $2,400. However, on August 17 Larry is desperate for money to pay his tuition at Ivy College. If money is worth 9%, what is the cash value of his inheritance on August 17?

54. An $800 note that bears interest at 9% matures in 180 days. What is the present value of the note after three-quarters of the term if money is worth 10%?

55. Rex, Inc., sold a used car for $3,000 and accepted a note with 9.5% interest in exchange. The sale was made on June 26 and payment was due on April 22 of the following year. What was the cash value of the sale if money is worth 10.5%? (The cash value is the present value of the debt at the time of the sale.)

56. A $500 note dated July 21 had a 12% interest rate and matured on October 19. On August 8 the note was sold. What was the value of the note if money was worth 10%?

57. A company accepted a non-interest-bearing note for $900 from a customer. The note was dated January 10 and was due on May 10. On March 11 the company sold it at the money market rate of $11\frac{1}{2}\%$. How much did they get for the note?

Chapter 7

END-OF-CHAPTER PROBLEMS

58. A note for $500 with interest at 8% matures in 72 days. If the payee decides to sell the note 20 days before maturity, what is its value if the money market rate is 9%?

59. Tots Togs offers to buy 7 dozen extra sunsuits if the producer will accept a non-interest-bearing note for $130 maturing in 3 months. The sunsuits are priced at $18.50 a dozen. If money is worth 8%:

  a. Is the present value of the note more or less than the cash value of the sale?

  b. How large is the difference?

60. A retailer who found he could not pay a bill for $1,200 on the day it was due signed a 60-day note with interest at $9\frac{1}{2}$%. Twenty-four days later he found he could settle the debt and paid the present value of the note at the market rate of 10%.

  a. How much did the extension of credit cost him?

  b. If he had paid interest according to the terms of the note for 24 days, how much would the extension of credit have cost him?

SUPERPROBLEM

A retailer has signed two promissory notes. The first, for $1,500, bears interest at 12%, is dated January 15, and is due on October 15. The second has a face value of $3,500, is dated February 10, bears interest at 10%, and is due on July 12.

On April 3 the retailer is notified that he has inherited $5,200 and he decides to pay off his debts. His creditors agree to accept payments as of April 10 at the money market rate of 11%.

  a. What is the difference between the settlement amount and the maturity values of the two notes?

  b. After he pays the bills how much will he have left?

Chapter 7

TEST YOURSELF

Find the missing quantities. Express the time in months.

	Principal	Rate	Time	Interest	Maturity	Section
1.	$3,600	7%	4 months	_____	_____	7.1
2.	$13,200	_____	9 months	$544.50	_____	7.1
3.	$4,500	$7\frac{1}{4}\%$	_____	_____	$5,006.25	7.1
4.	_____	_____	3 months	$255	$17,255	7.1

Find the maturity dates for the following loans dated:

5. July 12 for 180 days. — 7.2

6. February 28 for 8 months. — 7.2

7. Find the exact number of days between March 4 and August 1. — 7.2

8. The Island Wiring Company borrowed $3,200 for 7 months. If the rate of interest was 9%, find: — 7.1

   a. how much interest they will owe.

   b. how much must be repaid at maturity.

9. A loan of $4,200 was repaid with a check for $4,319. If the time was 4 months, what was the interest rate? — 7.1

301

10. An investment pays $5.10 interest every 6 months. If the rate is 8%, how large is the principal? 7.1

11. Judy Burns borrowed $2,500 with interest at 12%. What was the term of her loan if she paid $150 in interest charges? (Assume that time is in months.) 7.1 / 7.3

12. Find the ordinary and exact interest when $400 is borrowed for 80 days at 10% interest. 7.3

13. Tiffany DuPont loaned her boyfriend $800 to repair the truck he uses in his business. She asked him to sign a note that the debt be repaid in 120 days with $5\frac{1}{4}\%$ interest. 7.4

   a. Who is the maker of the note?

   Boyfriend

   b. Who is the payee?

   Tiffany

   c. How much interest is paid?

   $I = P \cdot r \cdot t = 800(.0525) \cdot \frac{120}{360} = \$14$

   d. What is the maturity value?

   $\$814$

14. On May 19 Larry Runner received $1,325.35 in payment of a note dated March 2. He had charged interest at 9%. What was the face value of the note? 7.5

   March 2 = $1300   May 19 — 1,325.35

   $S = P(1 + rt)$

15. A $900 note with interest at 12% has a term of 6 months. What is its present value on the day it is drawn if money is worth 10%? 7.5

   $10800 ? ✗

16. A note with a face value of $3,600 is dated August 20. It is due on October 31 with interest at 11%. After the note is signed the market rate rises to 12%. What is the present value of the note on October 1 at the market rate? 7.5

302

Chapter 7

Name   Date   Class

COMPUTER SPREADSHEET EXERCISES

Find the missing values for the following spreadsheets. Use formulas where possible (see specific instructions below) and $ signs for dollar amounts (round to the nearest cent where necessary). Format your numbers in each column to match those in the spreadsheets, and align the decimal points in all columns where dollar amounts appear. Labels and numbers should appear on your spreadsheet exactly as they are shown here. Assume ordinary interest for all calculations in days. Make sure your spreadsheets fit across the width (or length) of the page before printing.

1. Choose the correct formulas, solve them for the appropriate variables, and translate them to spreadsheet formulas for insertion into the blank cells of the spreadsheet below. Where time is not given, solve for only one of the time periods—years, months, or days. (Do not put values where the dashes appear.) The Drawing toolbar can be used to create the dashed lines as they appear below.

Principal	Rate	Years	Months	Days	Interest	Maturity Value
$4,000	7.50%	---	9	---		
	5.75%	2.0	---	---	$2,875.00	
$7,420		---	---	30		$7,457.10
$800		---	6	---	$34.00	
$2,000	9.25%	---	---			$2,037.00
$1,640	10.00%	---		---	$54.67	
	6.50%	1.5	---	---		$14,816.25

2. Determine the maturity value of the interest-bearing and non-interest-bearing promissory notes described below. Position the column title "Term of Note" across two separate columns, one for the numbers and one for the labels. Then use both in the formula to compute the simple interest, with the label portion as the condition of the IF statement needed to determine the correct calculation.

Your formula might appear as follows:

=IF(TERM="months",x,y)

where x is the simple interest calculation when TERM = "months" is true, and y is the calculation when TERM="months" is false (the term is in days). Be sure to substitute the correct cell address for TERM in the formula.

Face Value	Term of Note	Interest Rate	Interest	Maturity Value
$25,000	3 months	10%		
$5,000	30 days	15%		
$9,000	4 months	16%		
$13,600	60 days	9%		
$3,000	9 months	7%		
$4,000	10 months	0%		
$1,500	60 days	12%		
$2,500	120 days	13%		
$8,500	2 months	8%		
$10,000	73 days	14%		

3. Insert spreadsheet formulas to compute the values in the Interest and Maturity Value columns for the following simple interest loans. The time is given either in years, months, or days. Use an IF function to determine the correct value of time to use in the simple interest formula. For example,

$$\text{=IF(YEARS=0,IF(MONTHS=0,DAYS/360,MONTHS/12),YEARS)}$$

will compute the correct time (*t*) for the formula *I = Prt,* where cell addresses are substituted for the variables in the formula. Use the Borders toolbar to draw the line beneath the column headings.

Principal	Rate	Years	Months	Days	Interest	Maturity Value
$9,000	8.25%	---	8	---		
$17,500	6.75%	3.0	---	---		
$5,460	5.80%	---	---	30		
$790	7.75%	---	5	---		
$2,700	8.50%	---	---	90		
$1,840	7.25%	---	1	---		
$8,750	10.00%	2.5	---	---		

4. You are the payee of each of the promissory notes shown in the following spreadsheet. If the maker wishes to pay off a note before the due date, your policy is to accept the present value of the note at the current market rate as payment in full. Assume that each maker pays early. Determine the correct values for the notes based on the information in the spreadsheet. Also fill in the other missing numbers. Dates must be entered as shown. Subtract the appropriate dates from the due dates to calculate the number of days in the Days and the Days Until Maturity columns.

M.Y. Company - Notes Receivable

Maker	Face Value	Rate on Note	Date of Note	Due Date	Days	Maturity Value	Market Rate	Present Value Date	Days Until Maturity	Present Value
A.V. Consulting	$800.00	11.00%	09-Nov-99	08-Jan-00			13.00%	19-Dec-99		
J. Davis & Co.	$2,500.00	8.00%	02-Feb-00	14-Sep-00			9.25%	26-Jun-00		
Ayers Transmission	$4,250.75	7.75%	27-Dec-00	25-Jun-01			10.00%	22-May-01		
Burson Realty	$990.17	8.50%	07-Apr-99	06-Jul-99			7.75%	07-Jun-99		
CJ Properties	$12,800.00	7.50%	15-Oct-01	23-Jan-02			6.50%	02-Jan-02		
Johnston Video	$14,452.09	10.00%	28-Apr-99	26-Aug-99			6.25%	15-Jul-99		
Krug Furniture	$6,123.00	9.25%	15-Feb-00	15-May-00			7.20%	08-Mar-00		
McNamara Ltd.	$1,907.87	8.25%	11-Dec-98	01-Mar-99			8.40%	29-Dec-98		
Puterbaugh & Sons	$5,321.55	8.30%	29-Feb-00	28-Jul-00			11.50%	28-Jun-00		
Taylor Electronics	$704.50	7.00%	24-Jul-02	04-Oct-02			8.75%	22-Sep-02		

8 ■ Borrowing Money: Bank Discounts

OBJECTIVES

Upon completion of this chapter, the student should be able to:

1. *Use the formulas $D = Sdt$ and $p = S - D$ to find:*

   discount;

   discount rate;

   time; and

   proceeds.

2. *Describe the similarities and differences between simple interest notes and simple discount notes.*

3. *Discount a non-interest-bearing or a simple interest note on any date before the maturity date and find:*

   proceeds;

   amount of interest earned by the first lender; and

   amount of interest earned by the second lender.

4. *Use the formula $S = \dfrac{P}{1 - dt}$ to find the maturity value of a simple discount note.*

5. *Find the savings earned by borrowing with a discount note in order to take advantage of the cash discount on an invoice.*

8.1 SIMPLE INTEREST VERSUS SIMPLE DISCOUNT

Promissory notes in which interest is added to the face value are used between businesspersons, for example, as evidence of a debt for merchandise. The borrower pays the principal and interest at the end of the loan period. When a business borrows money from a bank, however, the debt is usually *discounted;* this means that the interest or discount is collected at the beginning of the loan period instead of at the end. Short-term business borrowing is a common practice, and commercial banks earn a substantial part of their income from the interest charged on this type of loan.

A *simple discount note* has the interest deducted from the face value of the note. *The borrower repays the face value.* This means that the face value and the maturity value of a simple discount note are the same. The amount of interest charged on the note is called the **bank discount,** or just the **discount.** The actual amount of money the borrower receives is called the **proceeds,** which are found by subtracting the discount from the face value (maturity value) of the note. Simple discount notes typically are used for short-term loans of up to 1 year in length. As with a simple interest note, the maturity value is repaid in a single payment at the end of the loan period.

306 CHAPTER 8 BORROWING MONEY: BANK DISCOUNTS

Calculating Discount and Proceeds

The formula for calculating discount is similar to the interest formula. The important difference is that the principal is the maturity value of the debt.

$$\text{Amount of discount} = \text{Maturity value} \times \text{Discount rate} \times \text{Time}$$
$$D = S \times d \times t$$

and

$$\text{Proceeds} = \text{Maturity value} - \text{Amount of discount}$$
$$p = S - D$$

It is important to write this formula accurately so that p (proceeds) will not be confused with P (principal).

■ **Example 1**

A bank loan for $1,200 is discounted at 11% for 72 days. How much does the borrower receive?

1. Determine the amount of discount.

$$D = Sdt$$
$$= 1{,}200 \times 0.11 \times \frac{72}{360}$$

The maturity value is usually called the loan amount, although the borrower does not actually receive this amount.

$$D = \$26.40$$

2. Find the proceeds.

$$p = S - D$$
$$= 1{,}200 - 26.40$$
$$p = \$1{,}173.60$$

The borrower receives $1,173.60 as proceeds from the bank loan.

Check: The proceeds are less than the maturity value.

Note: As was the case for simple interest notes, the banker's rule will be utilized for simple discount notes unless otherwise specified. ■

■ **Example 2**

The owner of an independent supermarket decided that she needed a new, computerized cash register, which would save time for the checkout clerks and result in more accurate inventory control. She needed $3,000 for 90 days and applied to her bank for a loan. When the bank was satisfied that the loan would be repaid as promised, it made the loan at a rate of 9%.

Assume that the borrower signed a 90-day simple discount note with maturity value of $3,000. (See Figure 8-1.) The discount rate is 9%.

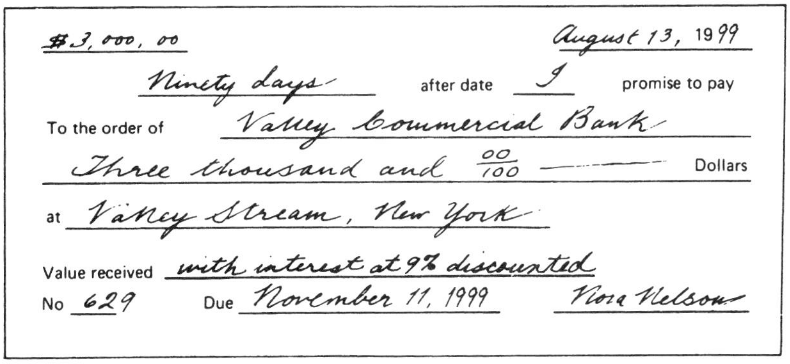
$3,000.00 August 13, 1999
Ninety days after date I promise to pay
To the order of Valley Commercial Bank
Three thousand and 00/100 Dollars
at Valley Stream, New York
Value received with interest at 9% discounted
No 629 Due November 11, 1999 Nora Nelson

FIGURE 8-1 Discounted note.

1. How much did the owner receive?

$$D = Sdt$$
$$= 3{,}000 \times 0.09 \times \frac{90}{360}$$
$$= \$67.50$$
$$p = S - D$$
$$= 3{,}000 - 67.50$$
$$= \$2{,}932.50$$

The owner of the supermarket then had $2,932.50 to use toward the purchase of new registers. This amount is called the **proceeds.** If she needed the full $3,000 as proceeds to buy the cash registers, the note would have had a maturity value larger than $3,000.

Check: The proceeds are less than the maturity value. ■

■ **Example 3**

The similarities and differences between the note of Example 2, which has a face value of $3,000, discounted at 9% for 90 days, and a simple interest note with a face value of $3,000 with an interest rate of 9% for 90 days, are summarized as follows.

	DISCOUNT NOTE FACE VALUE (S) = $3,000	INTEREST NOTE FACE VALUE (P) = $3,000
Interest charged	$D = Sdt$	$I = Prt$
	$D = 3{,}000 \times 0.09 \times \frac{90}{360}$	$I = 3{,}000 \times 0.09 \times \frac{90}{360}$
	$D = \$67.50$	$I = \$67.50$
Maker receives	$p = S - D$	
	$= 3{,}000 - 67.50$	
	$p = \$2{,}932.50$	$P = \$3{,}000$ (face value)
Maker repays		$S = P + I$
		$= 3{,}000 + 67.50$
	$S = \$3{,}000$ (face value)	$S = \$3{,}067.50$

The stated interest rate (9%) and the amount of interest ($67.50) are the same for both notes. But the discounted note gives the borrower only $2,932.50 compared with the $3,000 that the maker of an interest note receives. Thus, *based on the total actual amount of money the borrower has available to spend, the actual interest rate is higher in the discounting procedure.* The borrower is really paying slightly more than 9% for the money received because 9% of $2,932.50 for 90 days would be only $65.98 interest, not the $67.50 that was paid.

To find the actual interest rate, the following procedure (discussed in Chapter 7) is used:

$$I = Prt$$

$$67.50 = 2{,}932.50 \times r \times \frac{90}{360}$$ *The borrower receives only $2,932.50, and true interest is calculated on this amount.*

$$67.50 = 733.125 r$$

$$\frac{67.50}{733.125} = \frac{733.125 r}{733.125}$$ *Divide both sides by the coefficient of r.*

$$r = 0.0920 = 9.2\%$$

For the discounted loan the true interest rate paid is 9.2%, not 9% as stated.

Because of the possible confusion resulting from the different ways of calculating interest charges, the federal government passed the **Truth in Lending Act** in 1969. The law does not regulate rates but requires that the lender reveal both the amount of interest and the annual percentage rate charged on the money received, correct to the nearest $\frac{1}{4}$%. We will cover this in Chapter 9 when we study installment buying. ■

■ **Example 4**

A discount of $180 is charged for a 54-day bank loan on August 13. If the discount rate is 15%:

1. How much must be repaid at maturity?
   The amount to be repaid is the maturity value (*S*) in the discount formula.

$$D = Sdt$$

$$180 = S \times 0.15 \times \frac{54}{360}$$

$$180 = S(0.0225)$$

$$\frac{180}{0.0225} = \frac{S(0.0225)}{0.0225}$$ *Divide both sides by the coefficient of S.*

$$\$8{,}000 = S$$

The borrower will repay $8,000 when the loan is due.

2. How much did the borrower receive?

$$p = S - D$$
$$= 8{,}000 - 180$$
$$p = \$7{,}820$$

The borrower received $7,820 as proceeds and paid $180 in interest charges.

Check: In a discounted loan, *p* (proceeds) is always less than *S* (maturity value).

3. When must the loan be repaid?
   The maturity date is 54 days after August 13.

   August 13:   225 (from the calendar table)
              + 54
              ―――
                279

   The 279th day of the year is October 6, and the loan must be repaid on that date. ■

■ **Example 5**   Mr. Lowland signed a $7,000 note that his bank discounted. The note was dated April 26 and matured on July 7. The proceeds were $6,874, just enough to pay an invoice due on April 26. What was the discount rate?

$$D = Sdt$$

$$126 = 7{,}000 \times d \times \frac{72}{360}$$ *Discount is $7,000 − $6,874 = $126. Use the calendar table to find t.*

$$126 = 1{,}400d$$

$$\frac{126}{1{,}400} = \frac{1{,}400d}{1{,}400}$$ *Divide both sides by the coefficient of d.*

$$d = 0.09 = 9\%$$

The discount rate was 9%. ■

■ **Example 6**   On July 8 Mr. Lowland signed another note for $6,000. The discount rate had risen to 10%. If he received $5,600 to pay another invoice, what was the time of the note? (Assume that time is in months.)

$$D = 6{,}000 - 5{,}600$$ *Subtract the proceeds from the maturity value.*

$$D = \$400$$

$$D = Sdt$$

$$400 = 6{,}000 \times 0.10 \times \frac{t}{12}$$

$$400 = 50t$$

$$\frac{400}{50} = \frac{50t}{50}$$ *Divide both sides by the coefficient of t.*

$$t = 8 \text{ months}$$

Check: The time is less than 1 year. ■

Chapter 8

Name _____ Date _____ Class _____

TRY THESE PROBLEMS (Set I)

Find the missing quantities for the following loans.

	MATURITY VALUE	TERM	RATE	DISCOUNT	PROCEEDS
1.	$4,500	60 days	12%		
2.	$2,800	90 days	$9\frac{3}{4}$%		
3.	$900	210 days	11%		
4.		120 days	$7\frac{1}{2}$%	$212.50	
5.		60 days	12%	$200	
6.		18 days		$3.40	$796.60
7.		150 days		$125	$2,875

8. A borrower signs a 150-day note for $6,000. If the bank discounts the note at 15%, what will be the amount received from the bank?

9. The First National Bank is currently charging a 16% discount rate. Find the discount and the proceeds on a $6,240 note for 30 days.

10. Bill Jenkins signs a $2,000, 90-day note at the bank and receives proceeds of $1,925. What is the discount rate?

11. The proceeds of a 60-day note were $879. If the discount was computed on a maturity value of $900, what was the discount rate?

12. A note having a face value of $8,000 was discounted at 12%. If the discount was $320, find the term of the loan. (Assume that time is in months.)

13. Jim Jones signs a $5,400 discount note. The bank charges a 10% discount rate and the proceeds are $5,355. Find the time of the note. (Assume that time is in months.)

14. Mary Land borrowed $910 for 3 months from a bank that discounted the loan at 12%. When the transaction was completed, she received $882.70 from the bank.

   a. What were the proceeds?

   b. What was the maturity value of the loan?

   c. How much discount was charged?

15. If Mary Land (Problem 14) had signed a simple interest note for $882.70 at the same interest rate of 12% and for the same time of 3 months, how much would she have repaid?

8.2 DISCOUNTING A PROMISSORY NOTE

One characteristic of a promissory note that makes it a valuable business instrument is that it is *negotiable;* it can be sold. The new owner then becomes a payee. The sale does not affect the maker of the note except that he or she now pays the amount due to the new payee, which could be a bank, a business, or an individual. The maturity value and maturity date for the note remain unchanged.

For instance, a manufacturer may deliver merchandise to a retailer and not demand payment for several months. Instead the retailer might be required to sign a promissory note. If cash is needed before the note becomes due, the manufacturer can sell it to a bank. The bank will give the manufacturer the maturity value of the note, less a bank discount. Typically, the note is sold **with recourse,** meaning that if the retailer does not pay, the bank can collect from the manufacturer.

The bank buying the note is, in effect, granting a loan equal to the maturity value of the note for the time it must hold the note until it receives payment. The term of the note is split into two parts.

If the note is with interest, the original payee *earns* interest for the time he holds the note, from the original date to the date he sells it. He *pays* interest for the time the bank holds the note, from the date of discount to the maturity date, the **discount period.** The interest rate and discount rate are usually not the same. (See Figure 8-2.)

Discounting is simpler for a non-interest-bearing note and is illustrated first.

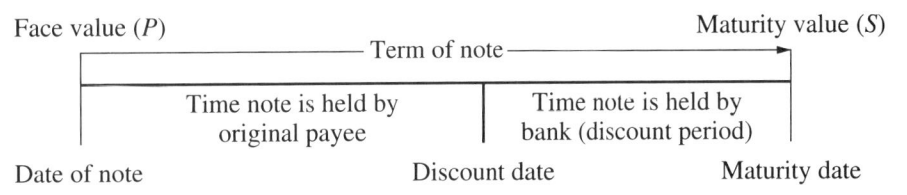

FIGURE 8-2 Time diagram for discounting a promissory note: the basic form.

■ **Example 1**

Mr. W. Atlas, an engineer, holds a non-interest-bearing note for $21,000 from an architect who could not pay his bill. The note had a 6-month term. Unfortunately, other clients also were late in paying and Mr. Atlas decided to discount the note at his bank in order to meet his payroll. Two months after he accepted the note he sold it to his bank, which discounted it at 12.5%. How much did Mr. Atlas get for the note?

Before beginning a note-discounting problem, draw the time diagram (Figure 8-3) and determine the discount period.

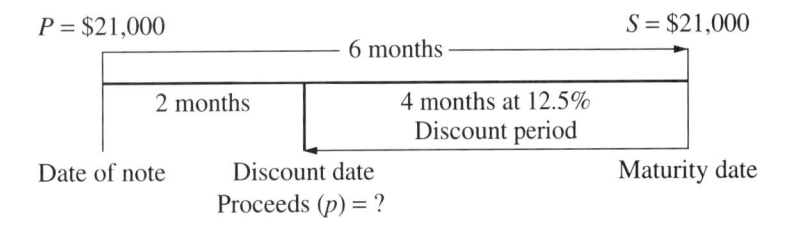

FIGURE 8-3 Time diagram for discounting a promissory note: Example 1.

$D = Sdt$ $d = $ *discount rate of* $12.5\%;$
$t = 4$ *months. (Mr. Atlas held the note 2 months, leaving 4 months to maturity.)*

$$= 21{,}000 \times 0.125 \times \frac{4}{12}$$

$D = \$875$

$p = S - D$

314 CHAPTER 8 BORROWING MONEY: BANK DISCOUNTS

$$= 21{,}000 - 875$$
$$p = \$20{,}125$$

1. The architect still must pay $21,000 on the due date, but now he pays it to the bank.
2. Mr. Atlas gets $20,125 from the bank on the discount date instead of $21,000 from the architect on the maturity date.
3. The bank holds the note for 4 months and earns $875 in interest. ■

Interest-Bearing Notes

Before attempting to find the proceeds when a business discounts an interest-bearing note, draw the time diagram and record the relevant facts of the problem. The diagram can be essential in diagnosis and solution. As each step of the calculation is completed, place the appropriate information on the diagram. The entire process should be done in four steps:

a. Find the interest and maturity value. (Discount is always based on maturity value.)
b. Find the discount period (from the date of discount to the date of maturity).
c. Calculate the discount (based on the maturity value, discount rate, and discount period).
d. Find the proceeds (by subtracting discount from maturity value).

■ **Example 2**

New Walls Co. sold $5,400 worth of wallpaper to Permagant, a retailer of paints and wallpaper. Because of a fire in the store, Permagant was unable to pay the bill when it came due, and New Walls Co. accepted a 120-day note with interest at 12%. Forty days before the note was due, New Walls Co. needed cash and discounted the note at City National Bank. The discount rate was 14%.

1. How much did New Walls Co. get for the note? (Another way of asking this question is "How large were the proceeds?")

    a. Find the interest and maturity value.

$$I = Prt$$
$$= 5{,}400 \times 0.12 \times \frac{120}{360}$$

r and t used to find the maturity value are part of the note.

$$I = \$216$$
$$S = P + I$$
$$= 5{,}400 + 216$$
$$S = \$5{,}616$$

The maturity value of the note is $5,616. This amount will be used to find the discount.

    b. Find the discount period. The discount period is given in the problem as "forty days before the note was due."

    c. Calculate the discount.

$$D = Sdt$$
$$= 5{,}616 \times 0.14 \times \frac{40}{360}$$

d is the discount rate, 14%; t is the discount period, 40 days.

$$D = \$87.36$$

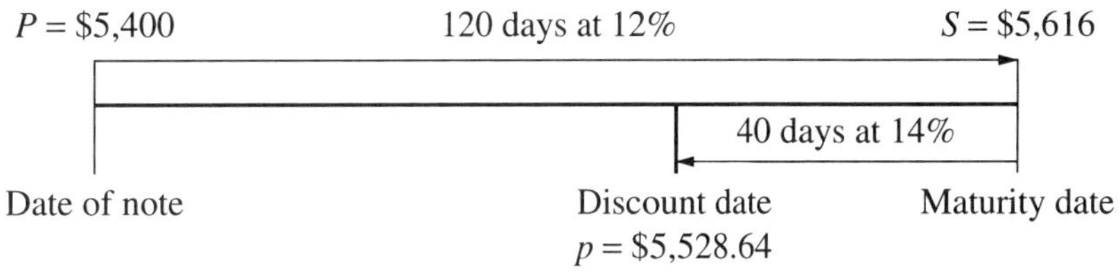

```
P = $5,400         120 days at 12%         S = $5,616
|————————————————————————————————|
        |——————————————————————|
                              40 days at 14%
Date of note       Discount date          Maturity date
                   p = $5,528.64
```

FIGURE 8-4 Time diagram with note values: Example 2.

d. Find the proceeds.

$$p = S - D$$
$$= 5{,}616 - 87.36$$
$$p = \$5{,}528.64$$

New Walls Co. received $5,528.64 as proceeds for the note. Figure 8-4 shows the completed time diagram.

2. How much did New Walls Co. earn in interest?

The proceeds, $5,528.64, are $128.64 more than the face value of the note ($5,528.64 − $5,400.00 = $128.64). New Walls thus earned interest of $128.64. However, they did not earn the full 12% on the note for the time they held it because the discount rate they paid was higher than the interest rate they charged.

3. How much would New Walls Co. have earned in interest if they had held the note to maturity?

They would have earned $216, the interest on the face value at 12% for the term of the note.

4. How did the sale of the note affect Permagant?

Permagant, the maker of the note, was unaffected by the sale except that they will pay $5,616 (the maturity value) to City National Bank instead of to New Walls Co.

5. How much did the bank earn in interest?

When the note is paid the bank will have earned $87.36, the amount charged for discounting the note and holding it for 40 days until maturity.

Check: The interest on the note that New Walls Co. would have earned by holding the note to maturity, less the discount paid to the bank, should equal the actual interest the company received.

$216.00 (answer to part 3)
− 87.36 (answer to part 5)
$128.64 (answer to part 2) ■

■ **Example 3**   A toy store, stocking up for Christmas, offered to buy $6,120 worth of additional merchandise from Q.P. Doll Co. if the company would accept an 8% note at the end of its regular credit period. The company agreed and the note was signed on November 6, 1999 to mature on February 14, 2000. On January 5, 2000, Q.P. Doll Co. discounted the note at First Valley Bank and paid a 10% discount rate.

316 CHAPTER 8 BORROWING MONEY: BANK DISCOUNTS

1. What were the proceeds of the note?

   a. Find the maturity value.

   The term of the note is from November 6, 1999, to February 14, 2000.

   The time is:

   $$
   \begin{aligned}
   \text{December 31:} &\quad 365 \\
   \text{November 6:} &\quad -310 \\
   \hline
   \text{Days left in 1999:} &\quad 55 \\
   \text{Days in 2000 to February 14:} &\quad 45 \\
   \hline
   \text{Term of note:} &\quad 100 \text{ days}
   \end{aligned}
   $$

   $$I = Prt$$
   $$= 6{,}120 \times 0.08 \times \frac{100}{360}$$
   $$I = \$136$$
   $$S = P + I$$
   $$= 6{,}120 + 136$$
   $$S = \$6{,}256$$

   The maturity value is $6,256.

   b. Find the discount: Use the time diagram as shown in Figure 8-5.

   $$D = Sdt$$
   $$= 6{,}256 \times 0.10 \times \frac{40}{360}$$

   d is the discount rate = 10%; t is the discount period = 40 days from the discount date to maturity.

   $$D = \$69.51.$$

   The discount was $69.51.

   c. Find the proceeds.

   $$p = S - D$$
   $$= 6{,}256 - 69.51$$
   $$p = \$6{,}186.49$$

   The proceeds from the sale of the note were $6,186.49.

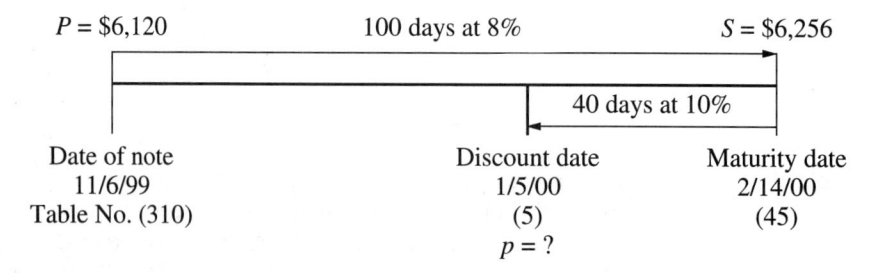

```
P = $6,120            100 days at 8%              S = $6,256
   |──────────────────────────────────────────────────|
                              |─── 40 days at 10% ───|
Date of note              Discount date         Maturity date
  11/6/99                    1/5/00                2/14/00
Table No. (310)                (5)                   (45)
                              p = ?
```

+FIGURE 8-5 Time diagram with note values: Example 3.

2. Was this an advantageous transaction for Q.P. Doll Co.? The company received $66.49 ($6,186.49 − $6,120) more for the note than it would have received for the merchandise, but they had money tied up in the note. However, they did satisfy a customer and increase sales.

3. How much did First Valley Bank earn?
The bank earned $69.51 for holding the note 40 days.

Check:

Interest to maturity:	$136.00
Discount:	69.51
Earned by first payee:	$ 66.49 ∎

Chapter 8

Name _____ Date _____ Class _____

TRY THESE PROBLEMS (Set II)

Find the discount and proceeds for the following non-interest-bearing notes that are discounted before maturity.

	FACE VALUE	DISCOUNT DATE	MATURITY DATE	DISCOUNT RATE
1.	$5,300	July 5	September 8	9%
2.	$3,640	September 12	December 26	10.5%
3.	$25,000	June 10	July 22	12%

Find the maturity value and the proceeds for the following interest-bearing notes that are discounted before maturity.

	FACE VALUE	INTEREST RATE	TERM	DISCOUNT RATE	DISCOUNT PERIOD
4.	$1,200	9%	4 months	10%	1 month
5.	$4,000	10%	180 days	14%	30 days
6.	$1,140	12%	5 months	18%	36 days

Find the maturity date and the term of discount in each of the following.

	DATE OF NOTE	TIME OF NOTE	DATE OF DISCOUNT
7.	May 24	30 days	May 31
8.	January 19	1 month	February 3
9.	November 15	75 days	December 18
10.	June 16	80 days	July 3

Find the maturity value and the proceeds for the following interest-bearing notes that are discounted before maturity.

	PRINCIPAL	RATE OF INTEREST	TIME OF NOTE	DATE OF NOTE	DATE OF DISCOUNT	DISCOUNT RATE
11.	$3,600	11%	150 days	April 13	June 12	$12\frac{1}{2}$%
12.	$40,000	15%	180 days	July 1	November 6	13.5%
13.	$2,200	12%	45 days	October 2	October 31	14%

14. A non-interest-bearing note with a face value of $1,600 is discounted at 9% on September 30. The note is dated March 20 and matures on November 9. What are the proceeds?

15. A 150-day note has a maturity value of $8,100. If it is discounted at $10\frac{1}{2}$%, 100 days after it is drawn, find:

   a. The discount.

   b. The proceeds.

16. An 80-day note with a face value of $450 bears interest at 8%. It is discounted 50 days before maturity at 9%. Find:

   a. The maturity value.

   b. The discount.

   c. The proceeds.

17. The payee of a 90-day promissory note discounted it at her bank after holding it for 30 days. The note has a face value of $6,400 and bears interest at 12%. The discount rate is 15%. Find:

   a. How much the payee would have earned in interest if she had held the note to maturity.

   b. How much the bank earned for discounting the note.

   c. How much she received from the bank as proceeds.

   d. How much the original payee earned on the entire transaction.

18. Artsware, Inc., accepted a $7,000 note on July 8. The terms of the note were 8% for 90 days. Artsware discounted the note on August 24 at the Washington Bank at 9%. What proceeds did Artsware receive?

8.3 MATURITY VALUE OF A DISCOUNTED LOAN

A business borrowing money from a bank in the form of a simple discount note will usually know how much it needs as proceeds rather than the maturity value of the note, particularly if the note is needed to cover a specific debt. For this type of situation, the maturity value of the note can be found by using the following formula, which combines two that have been used earlier.*

$$p = S(1 - dt)$$

The formula can also be written as

$$S = \frac{p}{1 - dt}$$

Either version may be used to find maturity value when p, d, and t are known. However, the last form is more convenient since it is set up in solution form.

■ **Example 1**

What is the maturity value of a note that is discounted at 9% for 40 days if the maker receives $10,890 as proceeds?

$$S = \frac{p}{1 - dt}$$

S is unknown; p, d, and t are given.

$$= \frac{10{,}890}{1 - 0.09 \times \frac{40}{360}}$$

$$= \frac{10{,}890}{1 - 0.01}$$

Complete the arithmetic in the denominator first.

$(1 - 0.01 = 0.99)$

$$= \frac{10{,}890}{0.99}$$

$$S = \$11{,}000$$

The maker would repay $11,000. The loan cost him $110, the difference between what he repays ($11,000) and what he receives as proceeds for his use ($10,890). ($11,000 − $10,890 = $110.)

Check: The maturity value is somewhat higher than the proceeds. A more complete check can be made by doing the problem in reverse—starting with the maturity value and finding the proceeds.

$$D = Sdt \qquad\qquad p = S - D$$

$$= 11{,}000 \times 0.09 \times \frac{40}{360} \qquad = 11{,}000 - 110$$

$$D = \$110 \qquad\qquad p = \$10{,}890$$

■

---

*The formula combines $D = Sdt$ and $p = S - D$. Therefore, $p = S - Sdt$ or $p = S(1 - dt)$.

322 CHAPTER 8 BORROWING MONEY: BANK DISCOUNTS

CALCULATOR HINT

You can use the memory keys on your calculator to solve Example 1.

ENTER	DISPLAY
0.09 ☒ 40 ➗ 360 M−	0.01 and the letter M
1 M+	1
10,890 ➗ MR ＝	11,000

The maturity value is $11,000.

Note: When you press MR, 0.99 is displayed. This shows that 0.01 was subtracted from 1.

■ **Example 2**

The owner of a small business must pay a $7,200 installment on her taxes on September 15 and must also pay a bill for $4,700. She needs a loan until October 15 and borrows the required amount from her bank at a discount rate of 12%.

1. How much must she repay on October 15?

   The maturity value of the loan is required. The proceeds must cover both the tax payment ($7,200) and the bill ($4,700).

   $p = 7,200 + 4,700 = \$11,900$

   $S = \dfrac{P}{1 - dt}$ *S must be found; p, d, and t are given.*

   $= \dfrac{11,900}{1 - 0.12 \times \dfrac{30}{360}}$ *There are 30 days between Sept. 15 and Oct. 15.*

   $= \dfrac{11,900}{1 - 0.01}$ *Complete the arithmetic in the denominator.*

   $= \dfrac{11,900}{0.99}$ *Divide*

   $S = \$12,020.20$

   The amount to be repaid is $12,020.20.

2. How much did the loan cost?

   $D = S - p$
   $= 12,020.20 - 11,900$
   $D = \$120.20$

   The discount is $120.20, which is the cost of the loan.

Check: The maturity value exceeds the proceeds by a reasonable amount. **Numerical check:**

$D = Sdt$ $p = S - D$

$D = 12,020.20 \times 0.12 \times \dfrac{30}{360}$ $= 12,020.20 - 120.20$

$D = \$120.20$ $p = \$11,900$ ■

8.4 BORROWING TO ANTICIPATE AN INVOICE

A business may find that they can save money by borrowing from a bank in order to take advantage of cash discounts offered by their suppliers. Such loans are often from banks in the form of simple discount notes. Typically, the notes are for very short periods and may be made informally when the firm borrows frequently using the same bank. The business will save money provided the cash discount exceeds the bank discount they must pay. On a single bill the savings may be small, but on a large volume the amount saved can be substantial.

■ **Example 1**

A company can deduct 2% from a bill for merchandise amounting to $3,197.96 if they pay early. To do this they must borrow the money from a bank for 15 days at a discount rate of 12%.

1. What will be the maturity value of the note that the company signs?

   The company needs to borrow the amount of the merchandise minus the cash discount. This amount will be the proceeds of the bank note. The cash discount can be found as follows:

   $$\%D \times Cr = \$D$$

   For a review of cash discounts, see Chapter 4.

   $$0.02 \times 3{,}197.96 = \$D$$

   $$\$63.96 = \$D$$

   Now determine the proceeds of the bank note. The amount required to pay the bill is:

   $3,197.96 for merchandise
   −63.96 less cash discount
   Total: $3,134.00

   The proceeds of the note must be $3,134. To find the maturity value:

   $$S = \frac{P}{1 - dt}$$

   S must be found; p, d, and t are given.

   $$= \frac{3{,}134}{1 - 0.12 \times \frac{15}{360}}$$

   There are 15 days in the discount period.

   $$= \frac{3{,}134}{1 - 0.005}$$

   Complete the arithmetic in the denominator.

   $$= \frac{3{,}134}{0.995}$$

   Divide

   $$S = \$3{,}149.75$$

   The company will pay the bank $3,149.75 when the note is due.

2. How much is saved by borrowing to take advantage of the cash discount?

   There will be a savings if the cost of the loan is less than the cash discount. The cost of the loan (the bank discount) is the maturity value less the proceeds.

   $$D = S - p = 3{,}149.75 - 3{,}134.00 = \$15.75$$

   The business can save $63.96 in cash discount and will pay $15.75 in bank discount. *The difference,*

   $$\$63.96 - 15.75 = \$48.21$$

   is the net amount saved by taking the loan and anticipating the bill.

324 CHAPTER 8   BORROWING MONEY: BANK DISCOUNTS

Check: The maturity value is moderately higher than the proceeds. The bank discount ($15.75) should not be large for a 15-day note under $5,000. ∎

■ **Example 2**

The Glamour Advertising Agency has found it very profitable to borrow money to take advantage of cash discounts. The magazine in which it places advertisements for its clients requires payment by the fifth of the month if a cash discount is taken. Their clients usually pay them by the twentieth of the month. Because they maintain a large account in the bank and have established their credit over several years, the bank discounts their loans at 6%, which is less than the bank's regular rate.

On March 5 the agency paid a bill for $16,000 and took a cash discount of 4%. On the same day they borrowed the money from the bank at 6% for 15 days. How much did they save?

1. Determine the proceeds of the bank loan. The advertising agency needs only enough to pay the bill *after the cash discount has been deducted.*

$$\%Pd \times Cr = \$Pd$$

$$0.96 \times 16{,}000 = \$Pd$$

% paid is the complement of 4% cash discount.
(100% − 4% = 96%)

$$\$15{,}360 = \$Pd$$

The agency needs $15,360 to pay the bill. Therefore, $p = \$15{,}360$.

2. Find the maturity value of the loan.

$$S = \frac{P}{1 - dt}$$

$$S = \frac{15{,}360}{1 - 0.06 \times \frac{15}{360}}$$

Values of d (6%) and t (15 days) are given in the problem.

$$= \frac{15{,}360}{1 - 0.0025}$$

Complete the arithmetic in the denominator.

$$= \frac{15{,}360}{0.9975}$$

Divide

$$S = \$15{,}398.50$$

Check: The proceeds ($15,360) are less than the maturity value ($15,398.50).

3. Find how much was saved.
A bill for $16,000 was settled by paying $15,398.50 (the maturity value of the loan). Therefore, the amount saved is

$$\$16{,}000 - 15{,}398.50 = \$601.50$$

The agency saved $601.50 by taking the bank loan to anticipate the bill.

Check: Cash discount (saved) = 0.04 × 16,000 = $640.00
Bank discount (paid) = 15,398.50 − 15,360 =   38.50
Amount saved = $601.50   ∎

■ **Example 3**

The Yarn Bazaar would like to anticipate a bill but needs a short-term bank loan. The bill totals $1,560.46 and includes $37.50 for freight. The credit terms are 2/10, n/30. If the bank discounts the loan at 9%, how much will the store save?

1. Determine the proceeds of the loan. The cash discount applies only to merchandise, and the freight must be subtracted before the discount is calculated.

$$\text{Merchandise cost} = 1{,}560.46 - 37.50 = \$1{,}522.96$$

The 2% discount is taken on the cost of merchandise, $1,522.96.

$\%D \times Cr = \$D$ *The arithmetic is simpler if the discount is found first.*

$0.02 \times 1{,}522.96 = \$D$

$\$30.46 = \$D$

Amount paid $= 1{,}522.96 - 30.46 = \$1{,}492.50$

The amount required to pay the bill is:

$$\begin{array}{rl} \$1{,}492.50 & \text{for merchandise} \\ +37.50 & \text{for freight} \\ \hline \text{Total: } \$1{,}530.00 & \end{array}$$

The proceeds of the bank loan must be $1,530.

2. Find the maturity value of the loan.

Notice that the term of the loan is not stated in the problem. When this situation occurs, we will make some assumptions *using the credit terms* (2/10, *n*/30) of the invoice. We will assume that the money will be borrowed on the *last possible date* that the cash discount may be taken, the 10th day after the invoice date, and that it will be repaid on the 30th day after the invoice date, *when the bill is due*. The time between the 10th and 30th days is 20 days.

$S = \dfrac{P}{1 - dt}$ *S is to be found; p, d, and t are known.*

$= \dfrac{1{,}530}{1 - 0.09 \times \dfrac{20}{360}}$

$= \dfrac{1{,}530}{1 - 0.005}$

$= \dfrac{1{,}530}{0.995}$

$S = \$1{,}537.69$

The Yarn Bazaar will pay the bank $1,537.69 when the loan is due.

3. Find the amount saved.

$$\begin{array}{rl} \text{Invoice amount:} & \$1{,}560.46 \\ \text{Maturity value} & -1{,}537.69 \\ \hline \text{Amount saved:} & \$22.77 \end{array}$$

The store will save $22.77 by borrowing to anticipate the bill. ∎

8.5 COMPARING SIMPLE INTEREST AND SIMPLE DISCOUNT

The following review should help to distinguish between the simple interest promissory notes discussed in Chapter 7 and the simple discount notes discussed in this chapter.

Both kinds of notes have two important similarities.

1. The time is generally less than 1 year.
2. They are repaid with a single payment at the end of the time period.

A comparison of the two notes is shown in Table 8-1.

TABLE 8-1 COMPARING SIMPLE INTEREST AND SIMPLE DISCOUNT NOTES

SIMPLE INTEREST NOTE	SIMPLE DISCOUNT NOTE
I = interest	D = discount
P = principal (face value)	p = proceeds (amount the borrower receives)
r = rate of interest	d = discount rate
t = time in years or part of a year	t = time in years or part of a year
S = maturity value (amount repaid)	S = maturity value (amount repaid)
Face value: stated on note	Face value: same as maturity value
Amount of interest: $I = Prt$	Amount of interest: $D = Sdt$
Maturity value: $S = P + I$	Maturity value: same as face value
Amount received by borrower: principal (or face value)	Amount received by borrower: proceeds $(S - D)$
Distinguishing phrases:	Distinguishing phrases:
Interest at a certain rate	Discounted at a certain rate
A maturity value greater than face value	A maturity value equal to face value
	Proceeds

As listed above, the FACE VALUE has different meanings for a simple interest and a simple discount note. The face value of a simple interest note is the principal or the amount borrowed. The face value of a simple discount note is the maturity value or the amount repaid.

The MATURITY VALUE is always the amount repaid, but a borrower who uses a simple interest note repays the principal plus the interest ($S = P + I$), while a borrower who uses a simple discount note repays the face value.

The amount the borrower (maker) receives is the principal (face value) of a simple interest note. With a discount note the maker receives the proceeds ($p = S - D$).

Now go to Try These Problems (Set III).

NEW FORMULAS IN THIS CHAPTER

- $D = Sdt$ (Amount of discount = Maturity value × Discount rate × Time)

- $p = S - D$ (Proceeds = Maturity value − Amount of discount)

- $S = \dfrac{p}{1 - dt}$ or $p = S(1 - dt)$

Chapter 8

TRY THESE PROBLEMS (Set III)

Find the maturity value of the following discounted notes.

	PROCEEDS	DISCOUNT RATE	TERM
1.	$1,176	8%	3 months
2.	$1,314.35	9%	4 months
3.	$474	10%	45 days
4.	$1,900	$7\frac{1}{2}\%$	120 days

5. An 80-day note was discounted at 9%. If the borrower received $600, how much must he repay?

6. The proceeds of a 5-month note were $3,108. If the bank discount had been 8%, what was the face value of the note?

7. Finer Feathered Friends has an opportunity to buy some merchandise from a bankrupt supplier at a very low price, but they must pay cash. The owner selects merchandise costing $716 and obtains a loan from the bank to pay the bill. The purchase is made on July 25 and the debt matures on August 14. The discount rate is 10%.

   a. How much will the bank receive on the due date?

   b. How much does the merchandise cost the store?

8. A company can save $82.64 by paying a bill early. However, they must borrow the money from a bank in order to do this. If the bank charges $27.90 for borrowing, how much would the company save?

9. A company can deduct 2% from a bill for merchandise amounting to $3,414.29 if they pay early. To do this they must borrow money from a bank for 15 days at a discount rate of 10%.

   a. What will be the face value of the note?

   b. How much will be saved by taking out the note?

10. Great Gardens, Inc., owes $609.95 for its spring supply of seeds. The bill has credit terms 2/15, *n*/30. A bank note was discounted at 9% in order to take advantage of the cash discount.

   a. What is the maturity value of the note?

   b. How much will Great Gardens save by borrowing to take advantage of the cash discount?

11. A bank note is discounted at 12% in order to take advantage of the cash discount on an invoice for $656.82, with terms of 2/10, *n*/30.

   a. What is the face value of the note?

   b. How much is saved by borrowing to take advantage of the cash discount?

12. The Value Discount Store has an invoice that shows $3,745.50 for merchandise and freight charges of $40. The terms are 4/10, 3/15, *n*/30. To get the largest cash discount the store borrows the required amount from a bank at a discount rate of 12%. They expect to be able to repay the loan after 18 days.

   a. How much is saved by borrowing from the bank?

   b. How much does the bank earn in interest?

13. The face values of both a simple interest and a simple discount note are $3,250. Both notes have interest rates of 10% for 72 days.

   a. How much interest is charged for each?

   b. How much does the maker receive on each?

   c. How much does the maker repay on each?

14. The maturity values of two 6-month notes are $650.

   a. Compare the present value at 12% interest with the proceeds at 12% discount.

   b. What amount does the maker receive from each note?

   c. How much interest is paid on each note?

Chapter 8

END-OF-CHAPTER PROBLEMS

Fill in the missing items for the following bank loans. Assume that the term is in months for Problem 4 and in years for Problem 6.

	TERM	DISCOUNT RATE	MATURITY VALUE	PROCEEDS	AMOUNT OF DISCOUNT
1.	210 days	7%	$ 874		
2.	75 days	$9\frac{1}{2}$	4,000		
3.	1 year		3,060	$2,723.40	
4.		9		3,556.02	$109.98
5.	6 months		1,600		116
6.		$12\frac{1}{2}$	5,000	4,375	
7.	100 days	12		348	
8.	90 days	11		7,780	

Find the term and the discount period for the following notes. State all answers in days.

	DATE OF NOTE	DISCOUNT DATE	MATURITY DATE
9.	March 6	April 25	September 2
10.	April 8	June 20	August 8

329

	DATE OF NOTE	DISCOUNT DATE	MATURITY DATE
11.	September 12	November 4	December 23
12.	March 16	July 5	October 25
13.	September 25	November 10	December 3

Find the maturity value and the proceeds for the following interest-bearing notes that are discounted before maturity.

	FACE VALUE	INTEREST RATE	TERM	DISCOUNT RATE	DISCOUNT PERIOD
14.	$2,100	9%	90 days	15%	30 days
15.	1,500	8	72 days	10	40 days
16.	740	9	45 days	$10\frac{1}{4}$	30 days
17.	600	10	150 days	12	10 days
18.	3,960	10	6 months	$10\frac{1}{2}$	4 months

19. A bank loan of $980 is discounted at $9\frac{1}{2}$%. If the loan is due in 90 days, what are the proceeds?

20. A 120-day note for $5,300 is discounted at 12%. How much does the borrower receive?

21. Tiny Togs Co., manufacturers of children's clothes, borrowed $3,350 from a bank on May 15. The loan was discounted at 10% and was to be repaid on July 26.

   a. How much did the company receive from the bank?

   b. How much did the loan cost?

Chapter 8

END-OF-CHAPTER PROBLEMS

22. On June 27 Jill Shawn borrowed $8,200 from a bank to modernize her store. The debt was to be repaid on September 25 and the bank charged $11\frac{1}{2}$% discount.

    a. How much did she receive from the bank?

    b. How much did the loan cost her?

23. A jewelry store owner owes $1,079.45 for improvements made to the burglar alarm system. If he signs a 60-day note for $1,200 discounted at 12%, will he have enough to pay the bill?

24. A note dated March 21 matures on May 8. The maturity value is $405 and the proceeds are $398.79. Find the discount rate.

25. A note with a face value of $850 was discounted at 12%. If the discount was $34, find the term of the loan.

26. The face value of a note was $2,538 and the proceeds were $2,504.16. If the term of the note was 40 days, what was the discount rate?

27. A note with a face value of $5,256 was discounted at 8%. If the discount was $315.36, find the time of the loan in days.

28. Bill Grey received $940.80 as proceeds on a 3-month loan. If he repaid $960, what was the discount rate?

29. The owner of the Health Hut decided to install new equipment to sell fresh fruit juices. On August 15 she borrowed $1,600, which was discounted at 10%. She received $1,580 from the bank as proceeds of the note.

    a. What was the term of the note?

    b. When must the note be repaid?

30. What is the face value of a note discounted at 12% for 90 days if the borrower receives $814.80 as proceeds?

31. A 4-month note is discounted at $9\frac{1}{2}$%. If the proceeds are $2,324, how much must be repaid on the note?

32. Sheri Turner signed a 72-day note with a discount rate of 8% and received $7,714. How much must she repay?

33. The face values of both a simple interest and a simple discount note are $5,000. Both notes have interest rates of 9% for 60 days.

   a. How much interest is charged for each type of note?

   b. How much would each borrower receive?

   c. How much would each payee receive at maturity?

34. Suppose the notes in Problem 33 have a face value of $2,500 and the rate is 10% for 90 days.

   a. How much interest is charged for each?

   b. How much does each borrower now receive?

   c. How much is repaid at maturity?

35. Two notes both have maturity values of $954 after 270 days.

   a. Compare the principal at 8.51% interest with the proceeds at 8% discount.

   b. What amount does the maker receive from each note?

   c. How much interest is paid on each note?

Chapter 8

Name _____ Date _____ Class _____

END-OF-CHAPTER PROBLEMS

36. Two 9-month notes both have maturity values of $1,204. The first is a simple interest note and the second is a simple discount note. The rate for both is 10%.

   a. How much does the maker have to spend for each note?

   b. What amount of interest is paid on each?

37. Fancy Furniture Co. owed $876.20 for merchandise plus $10.30 for insurance. The bill was due on May 6 and the owner found he was unable to pay it. To keep a good credit rating, he decided to borrow the necessary amount from a bank that charged 9% discount. The maturity date for the loan was July 5.

   a. How much must he repay the bank?

   b. What was the cost of borrowing?

38. A note dated June 5 is discounted on October 3 at 11%. The face value of the note is $960, the interest rate is $8\frac{1}{2}\%$, and the maturity date is November 2. Find:

   a. The maturity value.

   b. The discount.

   c. The proceeds.

   d. The amount of interest earned by the first payee.

39. Mr. A. Frend owned a 90-day non-interest-bearing note for $8,500. He intended to hold the note to maturity when he loaned the money, but he suddenly needed a new car. He decided to discount the note 24 days before it was due and paid a discount rate of $12\frac{1}{2}\%$. What is the most expensive car he can buy if he uses the proceeds to pay for the car?

40. Plush Pets accepted a note for $690 with interest at $9\frac{1}{2}\%$ for 60 days. They immediately discounted it at a bank that charged them 10%. How much less than $690 did the company receive from the bank?

41. Fresh Flowers Co. had a failure in its heating system, and the stock in the hothouse was destroyed. Until the insurance claim was settled they were short of cash and discounted a customer's note. The note was for $7,200 with interest at 6% dated January 5 for 120 days. It was discounted on March 21 at 11%.

a. How much more cash does the company have after discounting the note?

b. How much interest would they have earned on the note if they had held it to maturity?

c. How much did the bank make on the transaction?

42. A $1,728 note was discounted on December 30, 1998. The note included interest at 7%, was dated November 5, 1998, and was due on January 19, 1999. How much more than the face value of the note was received from the bank if the discount rate was 10%?

43. Fine Fabrics Shop owes a mill $650 for merchandise. The bill is dated July 24, with credit terms 4/15, n/45 E.O.M. If the store borrows at 12% discount, how much is saved by borrowing and paying the bill on August 15?

44. An invoice for $978.35 has terms 3/10, 2/20, n/30. If a bank note is discounted at 9% to take advantage of the 3% discount, how much will be saved?

45. Mr. Bookwurm owed $1,217.35 for a shipment of books. The invoice had credit terms 2/10, n/30, and he wanted to pay the bill on time to deduct the discount. He borrowed the necessary amount from a bank, which discounted the loan at $10\frac{1}{2}$%. How much did he save?

46. Fenster and Fenetre, Inc., purchased window glass priced at $902.63 less 10%. In addition to the trade discount, the supplier offered credit terms of 3/10, n/30. When the bill came, there was an additional charge of $8 for special handling. To pay the bill and take advantage of the cash discount, the company discounted a loan at their bank at 9%.

a. How much would they have had to pay without the cash discount?

b. How much must they repay the bank?

c. How much did they save by borrowing?

Chapter 8

SUPERPROBLEM

The Stationery Store bought:

   15 electric pencil sharpeners @ $20 less 10%;
   200 fancy pens @ $2.50 less 8/2; and
   5 globes @ $17.

The bill had terms 2/10, *n*/30 and was dated May 18. On May 28 the store made a partial payment of $300, but on June 17, when the remainder of the debt was due, there was no cash available. Instead, the supplier accepted a 60-day note with a low interest rate of 6%. On June 20 the supplier discounted the note at 12%. How much did the supplier lose by accepting and discounting the note?

Chapter 8

TEST YOURSELF

Find the missing quantities for the following bank loans. Assume that the term is in days for Problem 3.

	Term	Discount Rate	Maturity Value	Proceeds	Amount of Discount	Section
1.	30 days	8%	$7,500	_____	_____	8.1
2.	10 months	_____	$510	$471.75	_____	8.1
3.	_____	$9\frac{1}{4}\%$	_____	$4,541	$259	8.1

Find the term and the discount period for the following note. State both answers in days.

	Date of Note	Discount Date	Maturity Date	Term	Discount Period	
4.	March 14	April 27	June 2	_____	_____	8.2

5. The Back Bay Bank is currently discounting notes at 7%. A borrower signs a 40-day note for $3,150.   8.1

   a. How much is the discount?

   b. What are the proceeds?

   c. How much must be repaid at maturity?

6. Abby Adams received $962.50 after signing a 90-day, $1,000 note. What discount rate was being charged?   8.1

					Section

7. A note having a face value of $4,000 was discounted at 12%. If the discount was $160, what was the time of the loan? (Assume that time is in months.)   8.1

8. The discount was $425 on a 120-day note. If the rate was $7\frac{1}{2}\%$:   8.1

   a. What was the face value of the loan?

   b. How much were the proceeds?

Find the maturity value and proceeds for the following interest-bearing note that is discounted before maturity.

	Face Value	Interest Rate	Term	Discount Rate	Discount Period	Maturity Value	Proceeds	
9.	$11,000	7%	180 days	10%	45 days	_____	_____	8.2

10. A 180-day note has a maturity value of $3,250. If it is discounted at $10\frac{1}{2}\%$, 120 days after it is drawn, find: 8.2

   a. The discount.

   b. The proceeds.

11. Smiling Sunny's Toys accepted a $2,000 note on July 1. The terms of the note were 9% interest for 90 days. Smiling Sunny discounted the note on August 10, when the discount rate was 12%.   8.2

   a. What proceeds were received from the bank?

   b. How much did the bank earn?

   c. How much did Smiling Sunny earn on the entire transaction?

Chapter 8

TEST YOURSELF

	Section
12. The proceeds of a 5-month note were $7,315. If the discount rate was 9%, what was the face value of the note?	8.3
13. A bank note is discounted at 12% in order to take advantage of the cash discount on an invoice for $985.23. The terms are 2/10, *n*/30.	8.4
a. What is the face value of the note?	
b. How much is saved by borrowing to take advantage of the cash discount?	

Use the following information for Problems 14–16.

    The face values of both a simple interest and a simple discount note are $6,000. Both notes have interest rates of $8\frac{1}{2}\%$ for 4 months.

14. How much interest is charged for each type of note?	8.5
15. How much would each borrower receive?	8.5
16. How much is repaid at maturity?	8.5

Chapter 8

COMPUTER SPREADSHEET EXERCISES

Find the missing values for the following spreadsheets. Use formulas to compute the missing values, with $ signs for dollar amounts. (Where necessary, round to the nearest cent.) Format your numbers in each column to match those in the spreadsheets and align the decimal points for dollar amounts. Labels and numbers should appear on your spreadsheet exactly as they are shown here. Make sure your spreadsheet fits across the width or length of the page before printing. Assume ordinary interest for all calculations where time is in days.

1. Use formulas to compute the missing quantities for the following discounted bank loans.

Maturity Value	Term (in Days)	Discount Rate	Amount of Discount	Proceeds
$1,020	90	8.00%		
$8,000	60	7.50%		
$7,120	210			$6,870.80
		10.50%	$228.20	$9,551.80
$5,900	360		$560.50	
$9,200		12.75%		$8,965.40
	120	11.25%		$1,239.70
	45	9.00%		$1,977.50

2. Find the proceeds of the following interest-bearing promissory notes. Use formulas to obtain the correct values. Position the column title "Term" across two separate columns, one for the numbers and one for the labels. Do the same with the column title "Discount Period." Then use the number and label column entries in the appropriate formulas. You will need to use the IF function to test whether the term and discount period are in months or days. If "months" use 12 in the time fraction part of the simple interest and bank discount formulas; if "days" use 360.

Face Value	Interest Rate	Term	Interest	Maturity Value	Discount Rate	Discount Period	Bank Discount	Proceeds
$1,200	10.50%	90 days			9.00%	30 days		
$1,500	8.00%	72 days			8.50%	1 month		
$18,500	6.70%	4 months			12.25%	42 days		
$11,275	6.95%	10 months			9.75%	3 months		
$960	7.75%	180 days			11.00%	5 months		
$9,900	12.00%	5 months			8.30%	2 months		
$8,000	7.50%	45 days			9.25%	30 days		
$14,780	8.80%	9 months			7.75%	4 months		
$600	8.25%	150 days			12.00%	120 days		
$4,400	10.25%	2 months			7.24%	10 days		

3. Find the maturity date and maturity value for each of the following interest-bearing notes that are discounted before maturity. Then find the proceeds. To compute the maturity date, add the number of days in the term to the date of the note. This can be accomplished by adding the appropriate cells. Dates must be entered as shown, or in one of the other acceptable Microsoft Excel date formats. Assume ordinary interest in all calculations. Align the decimal points in columns containing dollar amounts.

Maker	Face Value	Rate on Note	Date of Note	Term in Days	Maturity Date	Maturity Value	Discount Date	Discount Rate	Proceeds
Bradburn & Sons	$5,750.00	7.75%	04-Nov-99	75			19-Dec-99	7.25%	
Hetzer Jewelry	$3,490.00	10.75%	21-Mar-00	120			02-Jun-00	9.70%	
Jackson Meats	$1,872.50	8.00%	15-Jul-98	100			26-Sep-98	8.15%	
JP Autobody	$788.00	8.25%	10-Jun-99	90			07-Aug-99	6.35%	
Kraft Counseling	$8,675.25	6.50%	25-Aug-00	150			03-Jan-01	10.00%	
Meyer Consulting	$2,340.00	11.50%	15-Sep-00	110			15-Oct-00	12.25%	
Reed Insurance	$15,125.00	8.90%	12-May-01	60			28-May-01	7.99%	
Taylor Plumbing	$1,995.80	7.70%	27-Jun-00	72			17-Aug-00	6.85%	
Vincent Realty	$12,525.00	9.60%	07-Dec-98	180			14-Mar-99	7.80%	
Williams Corp.	$910.35	5.45%	17-Jan-02	75			31-Jan-02	8.40%	

9 ■ Consumer Credit

OBJECTIVES

Upon completion of this chapter, the student should be able to:

1. *Compute, for an installment plan purchase:*
   the cost of credit;
   the down payment;
   the monthly payment; and
   the annual percentage rate.
2. *Compute, for charge accounts:*
   the monthly interest charge according to the United States rule;
   the new balance due; and
   the amount needed to close the account.
3. *Use the rule of 78 to determine interest saved when an installment plan is prepaid.*
4. *Determine the balance due when prepaying an installment loan.*

Consumer credit is available in many forms, from many sources, for many purposes, and at different costs. For many consumers, credit can be more difficult to avoid than to get, since credit-extending companies oftentimes will aggressively recruit applicants using tools such as direct mailings, special promotions, and preapproved credit lines. Credit extended to consumers has two advantages. It increases sales because many people buy merchandise and services on credit when they do not have cash available. In addition, the interest paid on the debt is income for the companies extending credit.

The simple interest and simple discount notes discussed in the preceding chapters are normally repaid with a single payment at the end of the loan period. A borrower, however, may prefer to repay part of the debt before the loan is due. Most lenders offer consumer credit through some type of **installment plan**—whereby a customer can take possession of a purchase and pay for it later, with interest, by a series of regular periodic payments.* People typically pay high rates for the privilege of this type of borrowing. This chapter discusses various types of installment buying and also covers revolving credit cards.

9.1 CREDIT PLANS

There are many variations in credit plans. The following discussion describes some general features. Among the most familiar instruments of consumer credit are the general credit cards like MasterCard, Visa, and Discover Card, and the charge accounts of department stores. Available in addition are loans to finance auto and boat purchases, college tuition, home improvements, and purchases of appliances, furniture and clothing, as well as loans for nonspecific uses—money when it is needed from bank credit lines or personal loans. Issuing the credit are banks, retail outlets, finance companies, savings and loan associations, and credit unions.

*In this chapter, "installment plan" will be used to describe high-interest plans that run for a limited time period. Chapter 11 includes longer-term loans where a comparatively lower rate is used.

Costs vary. Some credit plans are without charge to the borrower if the debt is paid on time. These include the credit extended by retail outlets, utility companies, and some of the general credit cards issued by banks and financial service companies. If the bill is paid within a specified time, usually between 10 and 30 days, there is no charge to the buyer. When a general credit card is used, however, the retailer or restaurant from which the purchase is made does pay a small percent of the bill to the issuing company, and this cost is reflected in slightly higher prices. Charges to the buyer for other types of credit vary.

Before consumer protection legislation was enacted in the late 1960s, it was legal for an installment loan to be advertised, for example, at 5% interest when the true interest rate was nearer to 10%. It was also legal to quote the terms of the agreement in such a way as to minimize or disguise the true amount or percent of interest charged. Interest at $1\frac{1}{2}$% a month sounds much lower than 18% a year, although they are the same. A $25 down payment with installments of $50 a month does not seem expensive if the buyer fails to consider for how many months the payments continue or makes no comparison with the cash price of the merchandise.

These practices are no longer legal. In 1969 Congress passed the **Truth in Lending Act,** which requires that the lender disclose all details of the transaction, including the cash price, amount of down payment, and amount financed. Most important is the requirement of disclosure of the **finance charge** (the extra cost for buying on credit), which, with minor exceptions, must be shown as an **annual percentage rate.** The law also specifies how the interest rate is to be calculated and what costs must be included as finance charges. As a result of the law, all periodic account statements must include a long list of details. (See Figure 9-1)

It is important to understand what is *not* included in the Truth in Lending Act as well as what is included. The law provides only for disclosure, so that every consumer using credit has correct knowledge with which to determine whether the credit is worth the extra cost. It does not regulate how much may be charged for credit, and finance charges legally can be the same as before the law was passed. In addition, the law is a consumer protection act and *does not apply* to business loans.

A consumer buying on credit should know the dollar cost of the loan, the rate of interest charged, and the amount and number of payments to be made. The dollar cost is the difference between the cash price for the merchandise and the total cost on the installment plan. For a consumer who borrows money, the dollar cost is the difference between what he gets from the lending institution and what he repays.

Cost of Credit

■ **Example 1**

A credit union permits its members to borrow sums up to $10,000 and to repay the debt in 18 equal installments. If a member borrowed $4,000 and repaid it in installments of $235.60, how much is the finance charge?

There is no down payment in a money loan, and the cost of the installment plan is the total of the monthly payments. The amount financed by the borrower is $4,000. To find the finance charge the following formula is used:

$$\text{Finance charge} = \text{Total of monthly payments} - \text{Amount financed}$$

Therefore,

$$\text{Finance charge} = (235.60 \times 18) - 4{,}000$$
$$= 4{,}240.80 - 4{,}000$$
$$= \$240.80$$

CREDIT PLANS 345

PLEASE WRITE YOUR ACCOUNT NO ON YOUR CHECK AND ENCLOSE THIS PORTION WITH YOUR PAYMENT SO ADDRESS ON REVERSE APPEARS IN WINDOW. Do not Fold, Staple Or Clip.
IF YOU HAVE ANY QUESTIONS ABOUT YOUR ACCOUNT, PLEASE WRITE OR CALL US BETWEEN 9 A.M. AND 9 P.M.
IF YOU TELEPHONE YOUR INQUIRY, YOU DO NOT PRESERVE YOUR RIGHTS UNDER FEDERAL LAW. SEND INQUIRIES TO:

PHONE NO.
1-800-942-1977

ACCOUNT TYPE	ACCOUNT NUMBER	CREDIT LINE	AVAILABLE CREDIT	BILLING DATE	PAYMENT DUE DATE ⑦
MASTERCARD		7,000 ⑧		06/06/00	06/28/00

DATE POSTED	DATE TRANS	REFERENCE NUMBER	TRANSACTION DESCRIPTION			DEBITS/CREDITS (-)
05/10	04/26	75263008130472066501306	CONTINENT0058444601698	PORTLAND	OR	2,861.70
05/10	04/26	75263008130472066501306	CONTINENT0058444601695	PORTLAND	OR ②	2,861.70
05/13	05/11	75402218133000000004253	HYATT KINGSGATE ACCO	AUCKLAND	NZL	177.54
05/17	04/18	75485418137001301000752	PETITE CORNER REST	ELMONT	NY	60.73
05/17	05/06	75485308135772036757339	STOUFFER CORPORATION	LOS ANGELES	CA	79.68
05/17	05/13	75402218137000000004226	HYATT KINGSGATE ACCO	ROTORUA	NZL	293.29
05/18	04/26	75414038138138202263609	AIR NEW ZEALAND-740	LOS ANGELES	CA	74.65
05/18	04/26	75414038138138202263708	AIR NEW ZEALAND-740	LOS ANGELES	CA	74.65
05/19	05/17	75402218139000000004596	PLIMMER TOWER HOTEL	WELLINGTON	NZL	161.68
05/24	05/23	75353108144901389301179	SHERATON HOTELS	BRISBANE	AUS ②	62.63
06/02	06/02	75353108154901374300753	SHERATON HOTELS	TOWNSVILLE	AUS	120.39
06/06			LATE CHARGE			10.00

ACCOUNT OVERLIMIT. TO AVOID ADDITIONAL OVERLIMIT FEES REMIT $41.05

SUMMARY	PREVIOUS BALANCE	PURCHASES ADVANCES AND OTHER DEBITS	PAYMENTS AND OTHER CREDITS	FINANCE CHARGE At Periodic Rate	Transaction Fees	NEW BALANCE	AMOUNT PAST DUE INCLUDED IN MINIMUM PAYMENT	⑨ YOUR MINIMUM PAYMENT ▼
	①	②	③	④		⑥		
PURCHASES	108.01	6,838.64	.00	94.40		7,041.05		
ADVANCES	.00	.00	.00	.00	.00	.00		
TOTALS	108.01	6,838.64	.00	94.40	.00	7,041.05	10.00	128.00

* YOU CAN AVOID ANY FINANCE CHARGES ON *
* PURCHASES IF YOU PAY YOUR PURCHASES *
* BALANCE IN FULL EVERY MONTH. *

RATES APPLIED TO BALANCES ⑤				
TRANSACTIONS	BALANCE SUBJECT TO FINANCE CHARGE	MONTHLY PERIODIC RATE	NOMINAL A.P.R.	ANNUAL PERCENTAGE RATE
PURCHASES	6,671.52	1.415%	16.980%	16.980%
ADVANCES	.00	1.415%	16.980%	16.980%

1. Previous balance
2. Purchases since last bill
3. Amounts credited since last bill
4. Details of finance charges.
5. Interest rates and balances to which they are applied
6. New balance
7. Date by which payment must be made
8. Maximum credit allowed for this customer
9. Minimum payment required

FIGURE 9-1 Monthly statement for a general charge account.

346 CHAPTER 9 CONSUMER CREDIT

The borrower would repay $240.80 more than she received.

Check: The total repaid must be more than the amount received. ∎

■ **Example 2**

A self-cleaning oven is priced at $629 and can be bought on installments with a down payment of $100 and 24 equal monthly payments of $26 each. How much does the use of credit cost the customer?

The amount financed must be found first.

Amount financed = Cash price − Down payment

Amount financed = 629 − 100

= $529

Finance charge = Total of monthly payments − Amount financed

= (26 × 24) − 529

= 624 − 529

= $95

A customer using the installment plan would pay $95 more than if he bought the oven for cash. This $95, regardless of whether it is called interest, service charge, insurance, or something else is the finance charge under the law. ∎

■ **Example 3**

A retailer who sells furniture on the installment plan adds 12% simple interest to the debt to cover clerical work, interest, and insurance. A $600 couch is sold with a down payment of $75, with the remainder and interest to be paid in nine monthly installments. What is the finance charge for the transaction?

The retailer may, for his convenience, use the 12% interest rate to determine the finance charge. He may not, however, say to the customer that the annual rate is 12%. It is a **nominal rate** (a rate in name only), but it is not the true rate of interest. The retailer will have to disclose a much higher rate to the customer; the reason will be explained a little later in this chapter.

The interest charge is calculated on the debt remaining after the down payment is deducted from the cash price; in this problem the interest is charged on $525 ($600 − $75). The time is the period during which payments are made. There are nine monthly installments, so the time is nine months.

$$I = Prt$$
$$= 525 \times 0.12 \times \frac{9}{12}$$
$$= \$47.25$$

A finance charge of $47.25 will be added to the debt. ∎

9.2 DETERMINING THE PERIODIC PAYMENT

When the finance charge determined by the lender is added to the debt, the payment each period can be found by dividing the total owed by the number of payments.

DETERMINING THE PERIODIC PAYMENT 347

■ **Example 1**

A watch priced at $175 may be bought on credit. The required down payment is 20% of the cash price, and a service charge of $15 is added. If the debt is repaid in 10 equal monthly payments, what is the amount of each payment?

$$\text{Down payment} = 0.20 \times 175 = \$35$$

$$\text{Amount financed} = \text{Cash price} - \text{Down payment}$$

Amount financed = 175 − 35 $140

Add service charge   + 15

Total debt $155

$$\text{Each payment} = \frac{\text{Total debt}}{\text{Number of payments}}$$

$$= \frac{155}{10}$$

Each payment = $15.50

The debt will be repaid in 10 monthly installments of $15.50 each. ■

■ **Example 2**

A color television set can be bought for $450 cash. On the installment plan a down payment of $50 is required. If simple interest at 15% is added as a finance charge and the debt is paid off in 18 equal monthly installments:

1. What is the additional cost for credit?

$$\text{Amount financed} = \text{Cash price} - \text{Down payment}$$

Amount financed = 450 − 50 = $400

$$I = Prt$$
$$= 400 \times 0.15 \times \frac{18}{12}$$
$$I = \$90$$

The customer would pay $90 more than the cash price if she bought on the installment plan.

2. How much would each monthly payment be?

The monthly payments must cover the amount of the debt ($400) and the interest charge ($90). The total of $490 ($400 + $90) is divided into 18 equal payments.

$$\text{Each payment} = \frac{\text{Total debt}}{\text{Number of payments}} = \frac{490}{18}$$

Each payment = $27.22*

The total debt would be paid in 18 installments of $27.22 each.

A second method of determining the periodic payment is more complex and requires some knowledge of annuities. It is discussed in Chapter 11. ■

---

*The last payment would be $27.26 to compensate for rounding.

Chapter 9

Name _____ Date _____ Class _____

TRY THESE PROBLEMS (Set I)

Find the cost of credit for the loans in Problems 1 to 3.

	LOAN AMOUNT	NUMBER OF PAYMENTS	PAYMENT AMOUNT
1.	$4,200	18	$250
2.	3,000	24	135
3.	1,800	12	174.50

4. An installment purchase is made with a down payment of $90 and 10 monthly payments of $30 each. The cash price of the merchandise was $325. What was the finance charge?

5. A new car was purchased for $25,000. The trade-in allowance was $4,000 and the remainder was financed with a 5-year loan at 10% simple interest. What was the finance charge?

6. If the entire debt for the car in Problem 5 is repaid in equal monthly installments, how large is each payment?

Find the amount of each monthly payment for the installment purchases in Problems 7–9.

	CASH PRICE	DOWN PAYMENT	FINANCE CHARGE	NUMBER OF PAYMENTS
7.	$500	20%	$25	10
8.	650	25%	9% simple interest	36
9.	320	None	14% simple interest	12

10. A personal computer priced at $1,875 includes the monitor but not the printer. The printer costs an additional $316. A customer buys both on the installment plan with a down payment of 10% of the cash price plus a finance charge of 8% simple interest on the remaining debt. How large is the finance charge if the debt is paid in 10 months?

11. A refrigerator-freezer is priced at $900 cash. If bought on the installment plan, a 15% down payment is required and 10% simple interest is added. The entire debt is repaid in 18 equal monthly installments.

   a. How much is the finance charge?

   b. How large is each payment?

12. The selling price of a microwave oven is $239. When sold on the credit plan, a down payment of 15% is required and a $30 finance charge is added to the cost. If the debt is to be repaid in 12 equal monthly installments:

   a. What is the total cost on the installment plan?

   b. How large is the monthly payment?

9.3 UNITED STATES RULE

Example 3 from the previous section used a simple interest rate to determine the finance charge on an installment loan. The stated rate of 12% (also referred to as the *nominal rate*) was taken on the original loan balance for the entire loan period. However, 12% is not the true annual rate of interest (**annual percentage rate**) and cannot be legally quoted according to the Truth in Lending Act.

To illustrate, assume that $1,000 is borrowed for one year and a finance charge of $100 is paid. In this case, the simple interest rate would be 10%. The annual percentage rate would also be 10% if the loan was not paid back until the end of the one-year loan period. (The borrower has use of an average amount of $1,000 over the entire loan period.) Now assume the same loan and finance charge are paid off in 12 monthly installments. Each payment reduces the amount of money that the borrower has use of, but interest is still being paid at the rate of 10% on the entire amount. This means that the annual percentage rate of interest is higher than 10%. Applied to installment loans, the annual percentage rate is the annual simple interest rate that is *actually* being paid *on the outstanding balance of the loan* at the time each payment is made. (We will compute annual percentage rates in Section 9.5.)

When a loan payment is made or an entire loan is paid off before it is due, the **United States rule,** or **U.S. rule,** is used by the federal government as well as most courts, states, and financial institutions to calculate the proper interest credit for the borrower and to determine how much more is owed on the debt. It gets its name from the fact that a U.S. Supreme Court ruling in the nineteenth century upheld this method of applying payments to a loan and its interest. The rule states that:

- each payment must first be used to pay any interest owed;
- the balance of the payment is then used to decrease the debt; and
- interest is calculated only on the *outstanding balance* of the loan.

Example 1 illustrates how the rule would work. The details of the loan in the example will be shown in an **amortization schedule,** a table that gives the amount of principal and interest and the remaining balance for each payment.

■ **Example 1**

A 12-month installment loan of $1,000 has an annual percentage rate of 10%. Use the United States rule to prepare an amortization schedule for this loan.

According to the rule, interest is calculated each month at 10% on the *outstanding balance,* not on $1,000. The borrower pays interest only on what is owed, the unpaid balance of the debt. The monthly payment is $87.92. (The last payment is $87.87 to adjust for rounding.)

The calculations for the first 2 months are given in detail. The procedure is exactly the same for the remaining 10 periods.

End of first month: Interest owed = *Prt*

$$= 1{,}000 \times 0.10 \times \frac{1}{12}$$

Interest owed = $ 8.33

Payment = $87.92

To pay interest   − 8.33

To reduce principal $79.59

New balance = 1,000 − 79.59 = $920.41

End of second month: Interest owed = $920.41 \times 0.10 \times \dfrac{1}{12}$

$$\begin{aligned}
\text{Interest owed} &= \$\ 7.67 \\
\text{Payment} &= \$87.92 \\
\text{To pay interest} &\ \underline{-7.67} \\
\text{To reduce debt} &\ \ \$80.25
\end{aligned}$$

New balance = 920.41 − 80.25 = $840.16

As shown in Table 9-1, with a debt of $1,000 the total amount of interest is reduced from $100 to $54.99 when the United States rule is used and 10% interest is calculated only on unpaid balances.

TABLE 9-1 DEBT AMORTIZATION BY UNITED STATES RULE

END OF MONTH	UNPAID BALANCE	PAYMENT	TO PAY INTEREST	TO REDUCE DEBT	NEW BALANCE
1	$1,000.00	$87.92	$8.33	$79.59	$920.41
2	920.41	87.92	7.67	80.25	840.16
3	840.16	87.92	7.00	80.92	759.24
4	759.24	87.92	6.33	81.59	677.65
5	677.65	87.92	5.65	82.27	595.38
6	595.38	87.92	4.96	82.96	512.42
7	512.42	87.92	4.27	83.65	428.77
8	428.77	87.92	3.57	84.35	344.42
9	344.42	87.92	2.87	85.05	259.37
10	259.37	87.92	2.16	85.76	173.62
11	173.62	87.92	1.45	86.47	87.14
12	87.14	87.87	0.73	87.14	0.00
		$1,054.99	$54.99	$1,000.00	

The United States rule is applied when payments are made on real estate mortgages, consumer bank loans, and charge accounts. Mortgages are frequently very long term debts extending 25 or 30 years. The monthly payments are equal and are determined on the basis of the original loan, the interest rate, and time. At the end of the term the debt is reduced to zero. The same is true for shorter-term installment debts like bank loans and auto financing. Charge accounts, however, give the customer more choice in how payments are made.

9.4 CHARGE ACCOUNTS

Interest rates permitted on charge accounts vary among the states. The bill shown in Figure 9-2 was issued in a state that permits a rate of 21%, but the store that issued the bill charges an annual rate of 19.8% or 1.65% per month. There is a 50¢ minimum finance charge.

This bill shows a payment and credit of $322.51, $10 less than the total amount owed. The finance charge, calculated as 1.65% of the average daily balance, is $3.18. The payment first covers the finance charge, and the remainder is applied to reduce the debt.

Payment	$322.51	
Finance charge	3.18	
To reduce debt	$319.33	
Amount still owed	$ 13.18	(332.51 − 319.33)
Add new purchases	21.11	(11.37 + 9.74)
New balance	$ 34.29	

CHARGE ACCOUNTS   353

```
BILL CLOSING DATE 07/23/01      TEL# 212-355-5900      ACCOUNT NUMBER
```

DATE	STORE	REFERENCE NUMBER	DEPT.	DESCRIPTION	AMOUNT
06/27	FM	07174445 00550057	420-00	CREDIT BETTER TRADITIONAL SHOES	-64.93
07/12	FM	00280028	155-20	ARPEL	② 9.74
07/14		02030047		*** PAYMENT - THANK YOU ***	-257.58
07/22	FM	00840085	131-00	CLINIQUE COSMETICS	② 11.37

WITH OUR COMPUTERIZED REGISTRY
YOU MAY SELECT YOUR GIFTS
FROM OUR FABULOUS ASSORTMENTS.
THE REGISTRY IN N.Y. 705-2800
AND ALL OUR STORES

PREVIOUS BALANCE ①	PAYMENTS & CREDITS ③	PURCHASES & OTHER CHARGES	AVERAGE DAILY BALANCE ⑦	FINANCE CHARGE ⑧	NEW BALANCE ⑤	AMOUNT NOW DUE ⑨
332.51	322.51	21.11	192.86	3.18	34.29	20.00

④ THE ANNUAL PERCENTAGE RATE OF FINANCE CHARGE IS 19.8%(1.65% PERIODIC RATE) OF THE TOTAL OF THE AVERAGE DAILY BALANCE AND   %(% PERIODIC RATE) OF ANY AMOUNT IN EXCESS OF $
ANY 50¢ FINANCE CHARGE IS A MINIMUM CHARGE AS PERMITTED BY STATE LAW. THE AMOUNT NOW DUE INCLUDES MISSED PAYMENTS IF ANY. TO AVOID AN ADDITIONAL FINANCE CHARGE THE NEW BALANCE MUST BE RECEIVED
⑥ IN FULL WITHIN 27 DAYS OF THE BILL CLOSING DATE ABOVE. YOU NEED NOT PAY ANY AMOUNT YOU HAVE QUESTIONED PENDING OUR COMPLIANCE WITH REG Z SEC.226.13   NOTICE: SEE REVERSE SIDE FOR IMPORTANT INFORMATION

1. Opening balance
2. Purchases since last bill
3. Amounts credited since last payment
4. Interest rates
5. Closing balance
6. Last date for payment to avoid finance charge
7. Average daily balance on which finance charge is calculated
8. Finance charge
9. Minimum payment required

FIGURE 9-2 Charge account statement.

If this bill is paid in full within 27 days of the billing date, there will be no further finance charge. The customer will have received credit with no interest charge from the time the new purchases were made. The next bill will show a zero balance unless additional purchases are made during the month.

Many customers in this situation use **revolving credit,** which allows charges up to a stated maximum, and a minimum monthly payment is usually required based on the amount owed. Assume this store requires a minimum payment of $20. This would leave a balance of $14.29. If no additional purchases are made, the new bill will show a debt of $14.79 (14.29 + 0.50). The minimum finance charge would apply because the calculated charge of 1.65% on $14.29 is only 24¢.

CHAPTER 9 CONSUMER CREDIT

■ **Example 1**

A revolving charge account at People's Department Store permits the customer to pay a minimum of $15 a month with a $1\frac{1}{2}$% per month finance charge on any unpaid balance. A customer receives a bill for $240 carried over from the previous month plus a finance charge of $3.60. Find the total interest charged if the following payments are made:

1. Two $50 payments.
2. One $75 payment.
3. One payment just large enough to close the account.

	Unpaid Balance	+	Finance Charge	=	Amount Owed	Payment	To Pay Interest	To Reduce Debt	New Balance
1.	$240.00		$3.60	=	$243.60	$50.00	$3.60	$46.40	$193.60
	193.60		(193.60 × 0.015 = $2.90)	=	196.50	50.00	2.90	47.10	146.50
2.	146.50		(146.50 × 0.015 = 2.20)	=	148.70	75.00	2.20	72.80	73.70
3.	73.70		(73.70 × 0.015 = 1.11)	=	74.81	74.81	1.11	73.70	0.00
						$249.81	$9.81	$240.00	

The total interest charge is $9.81.

Check: Total payments less original debt for merchandise should equal the sum of the interest column ($249.81 − $240 = $9.81). ■

We will assume in all problems that the finance charge is calculated only on the overdue balance and not on new purchases made before the next billing date. This is not true for all charge accounts. In some, if the amount of the bill is not paid in full by the required date, the new finance charge is based on the average daily balance outstanding,* including new purchases. This means that a customer who does not pay in full by the required date loses the privilege of the initial credit period for new purchases.

■ **Example 2**

Mr. Dresser opened a revolving charge account in a men's furnishings store and bought $92 worth of shirts and ties during the month. The account required a minimum monthly payment of $20, and a finance charge of $1\frac{1}{2}$% per month on unpaid balances was added. When he received his first bill, he sent a check for $20. During the next month he bought another $40 worth of merchandise and put it on his account. He then paid another $20. If he makes no additional purchases, what balance should appear on his third bill?

1. The first bill is for $92 with no finance charge. The payment of $20 is applied to the principal and reduces the debt to $72 ($92 − $20).

2. The second bill includes three parts.

    The unpaid balance $72.00

    Interest on the unpaid balance for one month at $1\frac{1}{2}$%.

    $I = 72 \times 0.015 =$ $1.08

    Additional purchases   40.00

    Total $113.08

    Payment   − 20.00

    New balance $93.08

---

*Average daily balance is determined by adding the outstanding balance on each day in the billing cycle and dividing the total by the number of days in the billing cycle.

3. The third bill includes two parts.

$$\text{The unpaid balance} \quad \$93.08$$

Interest on the unpaid balance for one month at $1\frac{1}{2}\%$.

$$I = 93.08 \times 0.015 = \quad \underline{1.40}$$

$$\text{Total} \quad \$94.48$$

The third bill for Mr. Dresser's account should be for $94.48, which includes a finance charge of $1.40. ∎

Chapter 9

TRY THESE PROBLEMS (Set II)

For charge accounts in Problems 1 and 2, prepare a table showing unpaid balance, finance charge, amount owed, payment, interest, principal and new balance until the account is paid off. There is no finance charge with the first payment. If the amount owed is within $3 of the monthly payment, pay the entire amount owed.

	UNPAID BALANCE	MONTHLY PAYMENT	INTEREST PER MONTH
1.	$112	$24	$1\frac{1}{2}\%$
2.	90	15	1%

3. A watch priced at $300 is purchased on an easy payment plan. The cost is paid off at $48 a month starting 1 month from the purchase date. After that, interest will be charged at 18% a year on the unpaid balance with no minimum finance charge. Set up the table showing unpaid balance, finance charge, amount owed, payment, interest, principal, and new balance until the account is paid off. When the account balance drops below $80, pay the entire amount.

For Problems 4 through 11, use credit terms of $1\frac{1}{2}\%$ interest per month on all unpaid balances and a minimum interest charge of 50¢. Payment must be within 30 days to avoid the finance charge.

4. A charge account customer had a balance for merchandise of $86 carried over from the previous month. She bought $42.68 worth of new purchases and added it to her account. What total, including interest, will her next bill show?

For Problems 5 through 7, a charge account statement is dated May 4 and has a balance for merchandise of $189.49 carried over from the previous month.

5. What additional charge will be shown on the bill?

6. If a $25 payment is made and there are no new purchases during May, what total should the June statement show?

7. If a payment of $30 is made on the May bill and a new purchase of $35.96 is added, what charges should the June statement show?

8. A balance of $589 remains on a revolving charge account after a payment is made. How large a finance charge will be added to the next bill?

9. A charge account statement lists $84.46 for merchandise and a finance charge for 1 month. If a payment of $15 is made, how much will be used to reduce the merchandise debt?

10. Mae Muzik opened a charge account at People's Department Store and used it when she bought a VCR for $299. When she received the bill, she sent a check for $35.

    a. By how much was her debt reduced?

    b. The following month she sent a check for $20. By how much was the merchandise debt reduced?

11. A charge account customer received a bill showing a previous balance of $27.02 and a finance charge of 50¢. By the required payment date he sent a check for $15. Before the next billing period he bought $18.28 worth of merchandise and put it on the account. What finance charge and total should be shown on the next bill?

9.5 ANNUAL PERCENTAGE RATE

The Federal Truth in Lending Act requires the lender to tell the borrower in writing the true annual rate of interest or **annual percentage rate (APR)**. It is the rate that is actually being paid on the balance due at the time of each installment payment. Like the monthly payment, it can be determined from annuity tables, but a much more convenient table is available from the Federal Reserve Board for the use of retailers and lending institutions that extend consumer credit.

To illustrate use of the table, consider the following example:

■ **Example 1**

Wire City sold a desktop computer for $1,200. They agreed to accept payment in 12 equal monthly payments and added 9% simple interest on the original amount. What annual percentage rate is charged?

Table 9-2 shows part of the APR table. The instructions given for finding the annual percentage rate apply to this problem. The complete table (Table C-4) begins on page 597.

TABLE 9-2 ANNUAL PERCENTAGE RATE FOR MONTHLY PAYMENT PLANS

NUMBER OF PAYMENTS	14.00%	14.25%	14.50%	14.75%	15.00%	15.25%	15.50%	15.75%	16.00%	16.25%	16.50%	16.75%	17.00%	17.25%	17.50%	17.75%
					(Finance Charge per $100 of Amount Financed)											
1	1.17	1.19	1.21	1.23	1.25	1.27	1.29	1.31	1.33	1.35	1.37	1.40	1.42	1.44	1.46	1.48
2	1.75	1.78	1.82	1.85	1.88	1.91	1.94	1.97	2.00	2.04	2.07	2.10	2.13	2.16	2.19	2.22
3	2.34	2.38	2.43	2.47	2.51	2.55	2.59	2.64	2.68	2.72	2.76	2.80	2.85	2.89	2.93	2.97
4	2.93	2.99	3.04	3.09	3.14	3.20	3.25	3.30	3.36	3.41	3.46	3.51	3.57	3.62	3.67	3.73
5	3.53	3.59	3.65	3.72	3.78	3.84	3.91	3.97	4.04	4.10	4.16	4.23	4.29	4.35	4.42	4.48
6	4.12	4.20	4.27	4.35	4.42	4.49	4.57	4.64	4.72	4.79	4.87	4.94	5.02	5.09	5.17	5.24
7	4.72	4.81	4.89	4.98	5.06	5.15	5.23	5.32	5.40	5.49	5.58	5.66	5.75	5.83	5.92	6.00
8	5.32	5.42	5.51	5.61	5.71	5.80	5.90	6.00	6.09	6.19	6.29	6.38	6.48	6.58	6.67	6.77
9	5.92	6.03	6.14	6.25	6.35	6.46	6.57	6.68	6.78	6.89	7.00	7.11	7.22	7.32	7.43	7.54
10	6.53	6.65	6.77	6.88	7.00	7.12	7.24	7.36	7.48	7.60	7.72	7.84	7.96	8.08	8.19	8.31
11	7.14	7.27	7.40	7.53	7.66	7.79	7.92	8.05	8.18	8.31	8.44	8.57	8.70	8.83	8.96	9.09
12	7.74	7.89	8.03	8.17	8.31	8.45	8.59	8.74	8.88	9.02	9.16	9.30	9.45	9.59	9.73	9.87
13	8.36	8.51	8.66	8.81	8.97	9.12	9.27	9.43	9.58	9.73	9.89	10.04	10.20	10.35	10.50	10.66
14	8.97	9.13	9.30	9.46	9.63	9.79	9.96	10.12	10.29	10.45	10.67	10.78	10.95	11.11	11.28	11.45
15	9.59	9.76	9.94	10.11	10.29	10.47	10.64	10.82	11.00	11.17	11.35	11.51	11.71	11.88	12.06	12.24
16	10.20	10.39	10.58	10.77	10.95	11.14	11.33	11.52	11.71	11.90	12.09	12.28	12.46	12.65	12.84	13.03
17	10.82	11.02	11.22	11.42	11.62	11.82	12.02	12.22	12.42	12.62	12.83	13.03	13.23	13.43	13.63	13.83
18	11.45	11.66	11.87	12.08	12.29	12.50	12.72	12.93	13.14	13.35	13.57	13.78	13.99	14.21	14.42	14.64
19	12.07	12.30	12.52	12.74	12.97	13.19	13.41	13.64	13.86	14.09	14.31	14.54	14.76	14.99	15.22	15.44
20	12.70	12.93	13.17	13.41	13.64	13.88	14.11	14.35	14.59	14.82	15.06	15.30	15.54	15.77	16.01	16.25
21	13.33	13.58	13.82	14.07	14.32	14.57	14.82	15.06	15.31	15.56	15.81	16.06	16.31	16.56	16.81	17.07
22	13.96	14.22	14.48	14.74	15.00	15.26	15.52	15.78	16.04	16.30	16.57	16.83	17.09	17.36	17.62	17.88
23	14.59	14.87	15.14	15.41	15.68	15.96	16.23	16.50	16.78	17.05	17.32	17.60	17.88	18.15	18.43	18.70
24	15.23	15.51	15.80	16.08	16.37	16.65	16.94	17.22	17.51	17.80	18.09	18.37	18.66	18.95	19.24	19.53
25	15.87	16.17	16.46	16.76	17.06	17.35	17.65	17.95	18.25	18.55	18.85	19.15	19.45	19.75	20.05	20.36
26	16.51	16.82	17.13	17.44	17.75	18.06	18.37	18.68	18.99	19.30	19.62	19.93	20.24	20.56	20.87	21.19
27	17.15	17.47	17.80	18.12	18.44	18.76	19.09	19.41	19.74	20.06	20.39	20.71	21.04	21.37	21.69	22.02
28	17.80	18.13	18.47	18.80	19.14	19.47	19.81	20.15	20.48	20.82	21.16	21.50	21.84	22.18	22.52	22.86
29	18.45	18.79	19.14	19.49	19.83	20.18	20.53	20.88	21.23	21.58	21.94	22.29	22.64	22.99	23.35	23.70
30	19.10	19.45	19.81	20.17	20.54	20.90	21.26	21.62	21.99	22.35	22.72	23.08	23.45	23.81	24.18	24.55
31	19.75	20.12	20.49	20.87	21.24	21.61	21.99	22.37	22.74	23.12	23.50	23.88	24.26	24.64	25.02	25.40
32	20.40	20.79	21.17	21.56	21.95	22.33	22.72	23.11	23.50	23.89	24.28	24.68	25.07	25.46	25.86	26.25
33	21.06	21.46	21.85	22.25	22.65	23.06	23.46	23.86	24.26	24.67	25.07	25.48	25.88	26.29	26.70	27.11
34	21.72	22.13	22.54	22.95	23.37	23.78	24.19	24.61	25.03	25.44	25.86	26.28	26.70	27.12	27.54	27.97
35	22.38	22.80	23.23	23.65	24.08	24.51	24.94	25.36	25.79	26.23	26.66	27.09	27.52	27.96	28.39	28.83
36	23.04	23.48	23.92	24.35	24.80	25.24	25.68	26.12	26.57	27.01	27.46	27.90	28.35	28.80	29.25	29.70
37	23.70	24.16	24.61	25.06	25.51	25.97	26.42	26.88	27.34	27.80	28.26	28.72	29.18	29.64	30.10	30.57
38	24.37	24.84	25.30	25.77	26.24	26.70	27.17	27.64	28.11	28.59	29.06	29.53	30.01	30.49	30.96	31.44
39	25.04	25.52	26.00	26.48	26.96	27.44	27.92	28.41	28.89	29.38	29.87	30.36	30.85	31.34	31.83	32.32
40	25.71	26.20	26.70	27.19	27.69	28.18	28.68	29.18	29.68	30.18	30.68	31.18	31.68	32.19	32.69	33.20
41	26.39	26.89	27.40	27.91	28.41	28.92	29.44	29.95	30.46	30.97	31.49	32.01	32.52	33.04	33.56	34.08
42	27.06	27.58	28.10	28.62	29.15	29.67	30.19	30.72	31.25	31.78	32.31	32.84	33.37	33.90	34.44	34.97
43	27.74	28.27	28.81	29.34	29.88	30.42	30.96	31.50	32.04	32.58	33.13	33.67	34.22	34.76	35.31	35.86
44	28.42	28.97	29.52	30.07	30.62	31.17	31.72	32.28	32.83	33.39	33.95	34.51	35.07	35.63	36.19	36.76
45	29.11	29.67	30.23	30.79	31.36	31.92	32.49	33.06	33.63	34.20	34.77	35.35	35.92	36.50	37.08	37.66
46	29.79	30.36	30.94	31.52	32.10	32.68	33.26	33.84	34.43	35.01	35.60	36.19	36.78	37.37	37.96	38.56
47	30.48	31.07	31.66	32.25	32.84	33.44	34.03	34.63	35.23	35.83	36.43	37.04	37.64	38.25	38.86	39.46
48	31.17	31.77	32.37	32.98	33.59	34.20	34.81	35.42	36.03	36.65	37.27	37.88	38.50	39.13	39.75	40.37
49	31.86	32.48	33.09	33.71	34.34	34.96	35.59	36.21	36.84	37.47	38.10	38.74	39.37	40.01	40.65	41.29
50	32.55	33.18	33.82	34.45	35.09	35.73	36.37	37.01	37.66	38.30	38.94	39.59	40.24	40.89	41.55	42.20
51	33.25	33.89	34.54	35.19	35.84	36.49	37.15	37.81	38.46	39.12	39.79	40.45	41.11	41.78	42.45	43.12
52	33.95	34.61	35.27	35.93	36.60	37.27	37.94	38.61	39.28	39.96	40.63	41.31	41.99	42.67	43.36	44.04
53	34.65	35.32	36.00	36.68	37.36	38.04	38.72	39.41	40.10	40.79	41.48	42.17	42.87	43.57	44.27	44.97
54	35.35	36.04	36.73	37.42	38.12	38.82	39.52	40.22	40.92	41.63	42.33	43.04	43.75	44.47	45.18	45.90
55	36.05	36.76	37.46	38.17	38.88	39.60	40.31	41.03	41.74	42.47	43.19	43.91	44.64	45.37	46.10	46.83
56	36.76	37.48	38.20	38.92	39.65	40.38	41.11	41.84	42.57	43.31	44.05	44.79	45.53	46.27	47.02	47.77
57	37.47	38.20	38.94	39.68	40.42	41.16	41.91	42.65	43.40	44.15	44.91	45.66	46.42	47.18	47.94	48.71
58	38.18	38.93	39.68	40.43	41.19	41.95	42.71	43.47	44.23	45.00	45.77	46.54	47.32	48.09	48.87	49.65
59	38.89	39.66	40.42	41.19	41.96	42.74	43.51	44.29	45.07	45.85	46.64	47.42	48.21	49.01	49.80	50.60
60	39.61	40.39	41.17	41.95	42.74	43.53	44.32	45.11	45.91	46.71	47.51	48.31	49.12	49.92	50.73	51.55

- Find the finance charge. For the $1,200 loan, the finance charge is:

$$I = Prt$$
$$= 1,200(0.09)(\frac{12}{12})$$
$$= \$108$$

- Find the amount financed. For the $1,200 loan, the total is the amount financed. When a down payment is required, it must first be deducted from the cash price.

- Find the finance charge per $100. Use the following formula:

$$\text{Finance charge per \$100} = \frac{\text{Finance charge}}{\text{Amount financed}} \times 100$$

For the $1,200 loan,

$$\text{Finance charge per \$100} = \frac{108}{1,200} \times 100$$

$$= 0.09 \times 100$$

Finance charge per $100 = $9

(A second method of finding the charge per $100 is shown in Example 2.)

- Use the APR table to find the annual percentage rate. In Table 9-2, find the number of payments in the left-hand column. Start at the beginning of the table. In the row corresponding to the number of payments (12), move across until you find the charge per $100, in this problem $9. The exact amount is not usually an entry in the table. There is no 9 entry, but it can be located between 8.88 and 9.02. Choose the closer one, 9.02, and move upward to read the percent at the top of the column: 16.25%. If the value to be located is exactly halfway between two entries, use the larger one.

The 16.25% found in the table is the annual percentage rate or true interest rate that must be disclosed to the borrower if $108 is charged on a loan of $1,200 to be repaid in 12 equal monthly installments. The *nominal interest rate* is 9%. Note that the APR is almost twice the nominal rate. This relationship can be used to check answers in many problems.

■ **Example 2**

Redi Cash Co. adds a finance charge of $28 to a $200 loan to be repaid in 24 equal monthly installments. What is the annual percentage rate?

Finance charge = $28

Amount financed = $200

$$\text{Finance charge per \$100} = \frac{\text{Finance charge}}{\text{Amount financed}} \times 100$$

$$= \frac{28}{200} \times 100$$

Finance charge per $100 = $14

Number of payments = 24

In the row for 24 payments in Table 9-3, 14 is between 13.82 and 14.10, being closer to 14.10. The percent at the top of the column is 13.00%. A finance charge of $28 on $200 repaid in 24 installments is at the annual percentage rate of 13.00%.

TABLE 9-3 ANNUAL PERCENTAGE RATE FOR MONTHLY PAYMENT PLANS

ANNUAL PERCENTAGE RATE

NUMBER OF PAYMENTS	10.00%	10.25%	10.50%	10.75%	11.00%	11.25%	11.50%	11.75%	12.00%	12.25%	12.50%	12.75%	13.00%	13.25%	13.50%	13.75%
(Finance Charge per $100 of Amount Financed)															
1	0.83	0.85	0.87	0.90	0.92	0.94	0.96	0.98	1.00	1.02	1.04	1.06	1.08	1.10	1.12	1.15
2	1.25	1.28	1.31	1.35	1.38	1.41	1.44	1.47	1.50	1.53	1.57	1.60	1.63	1.66	1.69	1.72
3	1.67	1.71	1.76	1.80	1.84	1.88	1.92	1.96	2.01	2.05	2.09	2.13	2.17	2.22	2.26	2.30
4	2.09	2.14	2.20	2.25	2.30	2.35	2.41	2.46	2.51	2.57	2.62	2.67	2.72	2.78	2.83	2.88
5	2.51	2.58	2.64	2.70	2.77	2.83	2.89	2.96	3.02	3.08	3.15	3.21	3.27	3.34	3.40	3.46
6	2.94	3.01	3.08	3.16	3.23	3.31	3.38	3.45	3.53	3.60	3.68	3.75	3.83	3.90	3.97	4.05
7	3.36	3.45	3.53	3.62	3.70	3.78	3.87	3.95	4.04	4.12	4.21	4.29	4.38	4.47	4.55	4.64
8	3.79	3.88	3.98	4.07	4.17	4.26	4.36	4.46	4.55	4.65	4.74	4.84	4.94	5.03	5.13	5.22
9	4.21	4.32	4.43	4.53	4.64	4.75	4.85	4.96	5.07	5.17	5.28	5.39	5.49	5.60	5.71	5.82
10	4.64	4.76	4.88	4.99	5.11	5.23	5.35	5.46	5.58	5.70	5.82	5.94	6.05	6.17	6.29	6.41
11	5.07	5.20	5.33	5.45	5.58	5.71	5.84	5.97	6.10	6.23	6.36	6.49	6.62	6.75	6.88	7.01
12	5.50	5.64	5.78	5.92	6.06	6.20	6.34	6.48	6.62	6.76	6.90	7.04	7.18	7.32	7.46	7.60
13	5.93	6.08	6.23	6.38	6.53	6.68	6.84	6.99	7.14	7.29	7.44	7.59	7.75	7.90	8.05	8.20
14	6.36	6.52	6.69	6.85	7.01	7.17	7.34	7.50	7.66	7.82	7.99	8.15	8.31	8.48	8.64	8.81
15	6.80	6.97	7.14	7.32	7.49	7.66	7.84	8.01	8.19	8.36	8.53	8.71	8.88	9.06	9.23	9.41
16	7.23	7.41	7.60	7.78	7.97	8.15	8.34	8.53	8.71	8.90	9.08	9.27	9.46	9.64	9.83	10.02
17	7.67	7.86	8.06	8.25	8.45	8.65	8.84	9.04	9.24	9.44	9.63	9.83	10.03	10.23	10.43	10.63
18	8.10	8.31	8.52	8.73	8.93	9.14	9.35	9.56	9.77	9.98	10.19	10.40	10.61	10.82	11.03	11.24
19	8.54	8.76	8.98	9.20	9.42	9.64	9.86	10.08	10.30	10.52	10.74	10.96	11.18	11.41	11.63	11.85
20	8.98	9.21	9.44	9.67	9.90	10.13	10.37	10.60	10.83	11.06	11.30	11.53	11.76	12.00	12.23	12.46
21	9.42	9.66	9.90	10.15	10.39	10.63	10.88	11.12	11.36	11.61	11.85	12.10	12.34	12.59	12.84	13.08
22	9.86	10.12	10.37	10.62	10.88	11.13	11.39	11.64	11.90	12.16	12.41	12.67	12.93	13.19	13.44	13.70
23	10.30	10.57	10.84	11.10	11.37	11.63	11.90	12.17	12.44	12.71	12.97	13.24	13.51	13.78	14.05	14.32
24	10.75	11.02	11.30	11.58	11.86	12.14	12.42	12.70	12.98	13.26	13.54	13.82	14.10	14.38	14.66	14.95
25	11.19	11.48	11.77	12.06	12.35	12.64	12.93	13.22	13.52	13.81	14.10	14.40	14.69	14.98	15.28	15.57
26	11.64	11.94	12.24	12.54	12.85	13.15	13.45	13.75	14.06	14.36	14.67	14.97	15.28	15.59	15.89	16.20
27	12.09	12.40	12.71	13.03	13.34	13.66	13.97	14.29	14.60	14.92	15.24	15.56	15.87	16.19	16.51	16.83
28	12.53	12.86	13.18	13.51	13.84	14.16	14.49	14.82	15.15	15.48	15.81	16.14	16.47	16.80	17.13	17.46
29	12.98	13.32	13.66	14.00	14.33	14.67	15.01	15.35	15.70	16.04	16.38	16.72	17.07	17.41	17.75	18.10
30	13.43	13.78	14.13	14.48	14.83	15.19	15.54	15.89	16.24	16.60	16.95	17.31	17.66	18.02	18.38	18.74
31	13.89	14.25	14.61	14.97	15.33	15.70	16.06	16.43	16.79	17.16	17.53	17.90	18.27	18.63	19.00	19.38
32	14.34	14.71	15.09	15.46	15.84	16.21	16.59	16.97	17.35	17.73	18.11	18.49	18.87	19.25	19.63	20.02
33	14.79	15.18	15.57	15.95	16.34	16.73	17.12	17.51	17.90	18.29	18.69	19.08	19.47	19.87	20.26	20.66
34	15.25	15.65	16.05	16.44	16.85	17.25	17.65	18.05	18.46	18.86	19.27	19.67	20.08	20.49	20.90	21.31
35	15.70	16.11	16.53	16.94	17.35	17.77	18.18	18.60	19.01	19.43	19.85	20.27	20.69	21.11	21.53	21.95
36	16.16	16.58	17.01	17.43	17.86	18.29	18.71	19.14	19.57	20.00	20.43	20.87	21.30	21.73	22.17	22.60
37	16.62	17.06	17.49	17.93	18.37	18.81	19.25	19.69	20.13	20.58	21.02	21.46	21.91	22.36	22.81	23.25
38	17.08	17.53	17.98	18.43	18.88	19.33	19.78	20.24	20.69	21.15	21.61	22.07	22.52	22.99	23.45	23.91
39	17.54	18.00	18.46	18.93	19.39	19.86	20.32	20.79	21.26	21.73	22.20	22.67	23.14	23.61	24.09	24.56
40	18.00	18.48	18.95	19.43	19.90	20.38	20.86	21.34	21.82	22.30	22.79	23.27	23.76	24.25	24.73	25.22
41	18.47	18.95	19.44	19.93	20.42	20.91	21.40	21.89	22.39	22.88	23.38	23.88	24.38	24.88	25.38	25.88
42	18.93	19.43	19.93	20.43	20.93	21.44	21.94	22.45	22.96	23.47	23.98	24.49	25.00	25.51	26.03	26.55
43	19.40	19.91	20.42	20.94	21.45	21.97	22.49	23.01	23.53	24.05	24.57	25.10	25.62	26.15	26.68	27.21
44	19.86	20.39	20.91	21.44	21.97	22.50	23.03	23.57	24.10	24.64	25.17	25.71	26.25	26.79	27.33	27.88
45	20.33	20.87	21.41	21.95	22.49	23.03	23.58	24.12	24.67	25.22	25.77	26.32	26.88	27.43	27.99	28.55
46	20.80	21.35	21.90	22.46	23.01	23.57	24.13	24.69	25.25	25.81	26.37	26.94	27.51	28.08	28.65	29.22
47	21.27	21.83	22.40	22.97	23.53	24.10	24.68	25.25	25.82	26.40	26.98	27.56	28.14	28.72	29.31	29.89
48	21.74	22.32	22.90	23.48	24.06	24.64	25.23	25.81	26.40	26.99	27.58	28.18	28.77	29.37	29.97	30.57
49	22.21	22.80	23.39	23.99	24.58	25.18	25.78	26.38	26.98	27.59	28.19	28.80	29.41	30.02	30.63	31.24
50	22.69	23.29	23.89	24.50	25.11	25.72	26.33	26.95	27.56	28.18	28.80	29.42	30.04	30.67	31.29	31.92
51	23.16	23.78	24.40	25.02	25.64	26.26	26.89	27.52	28.15	28.78	29.41	30.05	30.68	31.32	31.96	32.60
52	23.64	24.27	24.90	25.53	26.17	26.81	27.45	28.09	28.73	29.38	30.02	30.67	31.32	31.98	32.63	33.29
53	24.11	24.76	25.40	26.05	26.70	27.35	28.00	28.66	29.32	29.98	30.64	31.30	31.97	32.63	33.30	33.97
54	24.59	25.25	25.91	26.57	27.23	27.90	28.56	29.23	29.91	30.58	31.25	31.93	32.61	33.29	33.98	34.66
55	25.07	25.74	26.41	27.09	27.77	28.44	29.13	29.81	30.50	31.18	31.87	32.56	33.26	33.95	34.65	35.35
56	25.55	26.23	26.92	27.61	28.30	28.99	29.69	30.39	31.09	31.79	32.49	33.20	33.91	34.62	35.33	36.04
57	26.03	26.73	27.43	28.13	28.84	29.54	30.25	30.97	31.68	32.39	33.11	33.83	34.56	35.28	36.01	36.74
58	26.51	27.23	27.94	28.66	29.37	30.10	30.82	31.55	32.27	33.00	33.74	34.47	35.21	35.95	36.69	37.43
59	27.00	27.72	28.45	29.18	29.91	30.65	31.39	32.13	32.87	33.61	34.36	35.11	35.86	36.62	37.37	38.13
60	27.48	28.22	28.96	29.71	30.45	31.20	31.96	32.71	33.47	34.23	34.99	35.75	36.52	37.29	38.06	38.83

When only the nominal rate and time are given, we can find the finance charge per $100 by using the formula:

> Charge per $100 = Nominal rate × Time of loan in years × 100

As an illustration, assume that Example 2 was restated as follows:

Redi Cash Co. charges 7% on loans to be repaid in 24 monthly installments. What is the annual percentage rate?

Charge per $100 = Nominal rate × Time of loan in years × 100

$$\text{Charge per } \$100 = 0.07 \times \frac{24}{12} \times 100 = \$14, \text{ the same value found by the first method.}$$

Note that the amount of debt financed was not needed for this problem. ∎

■ **Example 3**

Two retailers offer the same color television set on credit plans. One offers the set at $425 cash and sells it on installments with a down payment of $75 and 24 monthly payments of $18 each. The second offers a longer time for payment, 36 months, and calculates the carrying charge at a nominal rate of 10%. Which retailer offers the lower annual percentage rate?

First retailer

$$\text{Amount financed} = \text{Cash price} - \text{Down payment}$$
$$= 425 - 75$$
$$\text{Amount financed} = \$350$$

$$\frac{\text{Finance}}{\text{charge}} = \frac{\text{Total of monthly}}{\text{payments}} - \frac{\text{Amount}}{\text{financed}}$$
$$= (18 \times 24) - 350$$
$$= 432 - 350$$
$$= \$82$$

$$\text{Charge per } \$100 = \frac{\text{Finance charge}}{\text{Amount financed}} \times 100$$
$$= \frac{82}{350} \times 100$$

Charge per $100 = $23.43

Appendix Table C-4 value for 24 payments and $23.43 per $100 = 21.00%.

Second retailer

Only the nominal rate and time are given.

$$\text{Charge per } \$100 = \text{Nominal rate} \times \text{Time in years} \times 100$$
$$= 0.10 \times 3 \times 100$$

Charge per $100 = $30

Appendix Table C-4 value for 36 payments and $30 per $100 = 18.00%.
The second retailer is offering a lower annual percentage rate.

Check: The fact that the two rates are close is a first check. Both rates can be checked by approximation (≅ means "is approximately equal to").

First retailer:

$$I = Prt, \text{ used to find the nominal interest rate}$$
$$82 = 350 \times r \times 2$$
$$82 = 700r$$
$$r = \frac{82}{700} = 0.117 = 12\% \text{ (to the nearest percent)}$$

The APR ≅ 2 × 12% = 24%, compared with the calculated result of 21.00%.

Second retailer: The nominal rate is given in the problem. The APR ≅ 2 × 10% = 20%, compared with the calculated result of 18.00%. ∎

Chapter 9

Name	Date	Class

TRY THESE PROBLEMS (Set III)

Find the annual percentage rate charged for the following installment loans.

	AMOUNT FINANCED	FINANCE CHARGE	MONTHS
1.	$120	$10	12
2.	$520	$80	24
3.	—	12% simple interest	36

In Problems 4 and 5, find the annual percentage rate charged for each loan.

	AMOUNT FINANCED	MONTHLY PAYMENT	TIME
4.	$1,200	$74.10	$1\frac{1}{2}$ years
5.	$3,800	135.00	3 years

6. A $1,200 loan requires payments of $58 each month for 24 months.

   a. What is the finance charge on the loan?

   b. What annual percentage rate must be revealed according to the Truth in Lending Act?

7. A loan of $900 is to be repaid in 15 monthly installments of $67.50 each. Find the annual percentage rate.

8. Fast Funds, Inc., advertised a reduced rate on its 60-month auto loans. A loan of $20,000 could be repaid with installments of $432 a month. What was the APR for the loan?

9. The finance charge on a loan is calculated at 7% simple interest. If it is repaid in 24 monthly payments, what is the annual percentage rate?

10. Roy Racer bought a new car for $19,400 and traded his old one in for $6,000. He financed the remaining debt with a loan he repaid in monthly installments of $450 over 3 years. What annual percentage rate did he pay?

11. A refrigerator priced at $428 is purchased on a credit plan with a down payment of 15% and a service charge of 9% simple interest. The buyer is to make equal payments each month for $1\frac{1}{2}$ years.

   a. What is the finance charge on the refrigerator?

   b. How much is the monthly payment?

   c. What is the total cost on the installment plan?

   d. What interest rate will the seller have to disclose to the buyer?

12. A computer printer is priced at $149. On the credit plan a down payment of $25 is required and a finance charge of $15.50 is added. The total is then paid in 12 equal monthly payments.

   a. What is the nominal interest rate?

   b. What is the APR?

9.6 RULE OF 78

It is not unusual for an installment loan to be paid off before it is due. A person might find that he has some extra money or is able to borrow at a cheaper rate and pay off a more expensive loan.

When a loan is paid off early, the borrower is entitled to a reduction in the finance or interest charge because of the shorter time the debt has run. The United States rule, studied earlier in this chapter, is the method used by the U.S. government as well as most states and financial institutions to determine the balance and interest refunded on a debt. According to the rule, the finance charge is not the same for each month but declines as the debt matures. Therefore, the amount of interest that would have been paid had the loan continued to the end of the term must be determined by calculation. The **rule of 78** is a variation of the U.S. rule, and is still used by many lenders to calculate the amount by which the finance charge is reduced. Lenders often use the rule to protect against early payoffs on small loans, since it allows more of the finance charge to be earned during the early months of the loan.

Using the rule of 78, the numbers of the months (*1*st, *2*nd, *3*rd, *4*th, etc.) in an installment plan are totaled. In a 12-month plan, the numbers 1 through 12 add up to 78. (This is how the method obtained its name.) The total interest is divided so that $\frac{12}{78}$ is paid in the first month, $\frac{11}{78}$ in the second, $\frac{10}{78}$ in the third, and so on, until $\frac{1}{78}$ of the total interest is paid in the final month of a 12-month plan. Thus, if the loan is paid off in full after nine payments have been made, the borrower has paid $\frac{72}{78}$ of the total interest charge. The remaining $\frac{3}{78} + \frac{2}{78} + \frac{1}{78} = \frac{6}{78}$ of the total is the amount of interest saved by paying early (the fraction $\frac{6}{78}$ is called the **refund fraction**). If the borrower pays the last three installments in full, the bank will refund this amount of interest.

The method is similar if the loan is for more or less than a year. For example, for a 24-month loan, instead of the sum of the integers 1 through 12, or 78, the denominator is the sum of the integers 1 through 24, or 300. This total is found easily by using the formula for the sum of the first *n* consecutive integers:

$$\frac{n(n+1)}{2} = \frac{24(25)}{2} = 300$$

The numerator of the refund fraction is the total of the numbers of the remaining months. If the 24-month loan were paid off after 19 payments, with 5 payments remaining, the numerator would be:

$$\frac{n(n+1)}{2} = \frac{5(6)}{2} = 15$$

The rebate or reduction in the interest charge would be:

$$\frac{15}{300} \times \text{Finance charge}$$

To compute the amount of finance charge to be refunded on any installment loan use:

> Interest refund = Refund fraction × Finance charge

where the sum of the numbers of the remaining months is the numerator of the refund fraction, and the sum of the numbers of the total months is the denominator.

■ **Example 1**

A 12-month installment loan with a finance charge of $234 is repaid in full after the eighth payment. Find the refund on the finance charge.

$$12 - 8 = 4 \qquad \text{\textit{There are 4 payments remaining.}}$$

$$\frac{n(n+1)}{2} = \frac{4(5)}{2} = 10 \qquad \text{\textit{Sum the numbers of remaining payments.}}$$

$$\frac{n(n+1)}{2} = \frac{12(13)}{2} = 78 \qquad \text{\textit{Sum the numbers of all the payments.}}$$

$$\frac{10}{78} \times 234 = \$30$$

Multiply the finance charge by $\frac{10}{78}$ to find the rebate.

The reduction in the interest charge is $30. ∎

■ **Example 2** A 24-month auto loan for $6,750 had monthly payments of $330. The loan was paid in full after the ninth installment.

1. How much interest was to be paid on the loan?
2. How much interest was saved by paying the loan off early?
3. How much must be paid after the ninth installment to cancel the loan?

1. $330 \times 24 = \$7{,}920$ *Find the total of the monthly payments.*

    $7{,}920 - 6{,}750 = \$1{,}170$ *Finance charge = total of payments − amount financed.*

2. $24 - 9 = 15$ *There are 15 payments remaining.*

    $\frac{n(n+1)}{2} = \frac{15(16)}{2} = 120$ *Sum the numbers of the remaining payments.*

    $\frac{n(n+1)}{2} = \frac{24(25)}{2} = 300$ *Sum the numbers of all the payments.*

    $\frac{120}{330} \times 1{,}170 = \$468$ *Multiply the finance charge by $\frac{120}{300}$ to find the interest saved.*

3. $330 \times 15 = \$4{,}950$ *Multiply the monthly payment by the number of payments remaining to find the amount still due.*

    $4{,}950 - 468 = \$4{,}482$ *Subtract the interest saved from the amount still due to find the payment needed to cancel the loan.* ∎

Now go to Try These Problems (Set IV).

NEW FORMULAS IN THIS CHAPTER

- Amount financed = Cash price − Down payment
- Finance charge = Total of monthly payments − Amount financed
- Each periodic payment = $\dfrac{\text{Total debt}}{\text{Number of payments}}$
- Finance charge per \$100 = $\dfrac{\text{Finance charge}}{\text{Amount financed}} \times 100$
- Finance charge per \$100 = Nominal rate × Time in years × 100
- Sum of the first n consecutive integers = $\dfrac{n(n+1)}{2}$
- Rule of 78 interest refund = Refund fraction × Finance charge

Chapter 9

| | Name | Date | Class |

TRY THESE PROBLEMS (Set IV)

Find the interest saved under the rule of 78 for the following installment loans that are paid off before maturity.

	FINANCE CHARGE	TOTAL NUMBER OF PAYMENTS	NUMBER OF PAYMENTS REMAINING
1.	$ 450	12	4
2.	$1,700	48	14
3.	$1,218	36	7

4. Andrea Rice took out a college loan with 36 monthly payments. If the total interest was $2,210.50 and she paid off the loan 12 months early, how much interest was refunded to her?

5. Chris De Witt purchased a camcorder by paying $100 down and agreeing to make 24 monthly payments of $33.70. The cash price was $700. After making 15 payments he decided to pay the loan in full.

   a. What was the finance charge on the original purchase?

   b. How much of this charge would be refunded to him?

   c. What amount is needed to pay the loan in full after 15 payments?

6. Tommy Tuner purchases a $5,400 used car on the installment plan. He agrees to make 18 monthly payments, which will include a finance charge at 12% simple interest.

   a. How much interest does Tommy expect to pay?

   b. What monthly payment is needed to finance the car?

   c. After the 12th payment he decides to pay the balance due on the car. How much did he save in interest?

   d. How much was needed to pay the loan in full at this time?

Chapter 9

Name	Date	Class

END-OF-CHAPTER PROBLEMS

Find the cost of credit in each of the following loans.

	AMOUNT OF LOAN	NUMBER OF PAYMENTS	MONTHLY PAYMENT
1.	$ 268	15	$ 20.50
2.	5,000	24	240.00
3.	595	10	65.00
4.	2,225	36	97.50
5.	3,000	18	175.00

Find the amount of each monthly payment in the following installment purchases.

	CASH PRICE	DOWN PAYMENT	FINANCE CHARGE	NUMBER OF PAYMENTS
6.	$285	$50	$22	12
7.	995	None	82.50	15
8.	490	20% of cash price	12% simple interest	20
9.	650	$75	9% simple interest	12
10.	125	25% of cash price	11% simple interest	9

For Problems **11–15**, find the annual percentage rates for the installment purchases in Problems 6–10.

16. A color television can be bought for 15% down and $40 a month for 10 months. If the cash price is $395, what is the charge for the use of the installment plan?

17. A couch is priced at $930 if bought for cash. On the store's installment plan, a down payment of $100 and 24 payments of $40 each are required. What is the finance charge?

18. Kate Kleen bought a new washer and dryer that were priced at $450 if bought for cash. She decided to use the credit plan and paid 15% down and $35.50 a month for 14 months.

   a. How much did the appliances cost on the installment plan?

   b. What was the cost of credit?

19. Fred Fenster needed to replace all the windows in his house and obtained a loan from a finance company to cover the cost. He borrowed $900 and repaid the debt in 36 equal installments of $30.25 each. How much was the finance charge?

20. A typewriter sells for either $260 cash or 15% down, a carrying charge of $25, and the balance in 12 installments. What is the monthly installment payment?

21. A wireless speaker system is priced at $199.95. It can be bought with a 25% down payment and an additional carrying charge of $90. The debt must be repaid in 15 equal monthly installments.

   a. How large is the monthly payment?

   b. How much does the speaker system cost on the installment plan?

22. Jack Jillson borrowed $2,400 from Easy Money, Inc. The company added $285.50 as a finance charge and required the debt to be repaid in 30 monthly installments. How large was the monthly payment?

Chapter 9

| Name | Date | Class |

END-OF-CHAPTER PROBLEMS

23. A laptop computer can be bought for $1,545 cash or 10% down and 24 equal payments that cover the remaining debt and a finance charge calculated as 10% simple interest.

    a. How large is each monthly payment?

    b. What annual percentage rate is charged?

24. Rex, Inc., sold a used car for $1,800. They agreed to accept payment for the car in 15 equal monthly payments and added 11% simple interest on the original amount.

    a. How much is each payment?

    b. Find the APR.

25. A stereo component system is priced at $850 for cash sales. On the "easy payment plan" the down payment required is 15% of the cash price. Twelve percent simple interest is added to cover financing costs. The entire debt is to be repaid in 18 equal monthly payments.

    a. How large is each payment?

    b. What is the interest rate that must be quoted to the buyer?

26. Gil Gordon purchased a portable color television set for $225. He decided to charge the set and make payments of $26.75 a month for 9 months. Find the annual percentage rate for this loan.

27. Caron Cooke is paying for the remodeling of her kitchen by an initial check for $300 and 36 monthly installments of $100 each. If she had had sufficient cash to pay for the entire job upon completion, it would have cost $3,000.

    a. How much does the job cost on the installment plan?

    b. What is the cost of credit?

    c. What nominal (simple) interest rate is charged?

    d. What is the APR?

For Problems 28 through 33, use credit terms of 1% interest per month on all unpaid balances and a minimum interest charge of 50¢. Payment must be within 25 days to avoid the finance charge. For revolving credit, the minimum payment required is $20.

28. A customer who has been using the revolving charge plan and paying $20 a month has an outstanding balance of $65.70 for merchandise. If he again pays $20, how much merchandise debt will be shown on the next bill?

29. A customer owes the store $39.65 for merchandise carried over from the previous month. If the bill is to be paid in full, what total must be paid?

30. A customer with a balance for merchandise of $215.03 from the previous month has been paying $55 a month. If the usual payment is made, what total, including interest, will appear on the next bill?

31. Find total interest paid if the customer in Problem 30 continues making payments until the account is closed. Start with $215.03 and continue the calculation until the amount owed is less than $56. Then pay the entire remaining debt.

32. A customer opened an account on February 18 and bought a sink that cost $55.98, a ceiling fixture for $28.99, and an electric cooker for $31.79. The bill arrived on March 3, and on March 17 she sent a check for the minimum payment. If she buys nothing else before the next billing date, what total will the new bill show?

33. A customer with no outstanding balance bought $172 worth of merchandise. She paid $25 each month and made no new purchases until the account had no outstanding balance. When the amount owed was less than $36, she paid the entire amount.

    a. Prepare a table showing the interest charged, the amount by which the debt is reduced each month, and the new balance.

    b. How much interest did she pay in total?

Chapter 9

END-OF-CHAPTER PROBLEMS

34. Universal Department Store offered its charge account customers the following terms. There would be no charge on purchases paid within 30 days of the billing date. After that, there would be a $1\frac{1}{2}\%$ per month finance charge on any unpaid balance with a minimum finance charge of 50¢ per month. A monthly minimum payment of $10 is required on the revolving plan.

Mrs. Eve Adams purchased $42.14 worth of merchandise during the first month her account was opened. When her bill came, she paid the minimum amount. The next month she bought a dress for $53.47 and housewares totalling $29.95.

a. What will be the finance charge on her next bill?

b. What will be the total amount owed?

35. Mrs. Adams (Problem 34) decided to pay $22 a month and make no new purchases until her account had a zero balance. Set up the schedule showing unpaid balance, finance charge, amount owed, payment, interest, principal, and new balance until the account is paid off.

36. A $600 bank loan is repaid in 12 monthly installments of $54 each.

a. What is the finance charge?

b. What is the annual percentage rate?

37. An air conditioner priced at $495 can be bought on the credit plan for 15% down and 18 equal monthly payments. A carrying charge of 12% simple interest is included in the payments.

a. How large is the finance charge?

b. How large is each payment?

c. What interest rate must be disclosed to the buyer?

373

38. Lisa Adams borrowed $5,000 for college expenses and agreed to repay the loan after graduation in 36 monthly payments at 11% simple interest.

  a. How much interest is included in the payments?

  b. How much does Lisa pay each month?

  c. If Lisa paid off the rest of the loan after her 26th payment, how much interest did she save? (Use the rule of 78.)

  d. What amount did Lisa need to cancel the loan after the 26th payment?

39. If a loan of $1,200 is repaid in 15 monthly installments of $87:

  a. What is the nominal or simple interest rate?

  b. What is the APR?

40. An automobile loan of $6,000 is repaid in 36 months with monthly payments of $202. Find the finance charge and the interest rate that must be disclosed to the borrower.

41. A credit union offers its members low interest loans with long repayment periods. A loan of $2,000 can be repaid in 24 equal monthly payments of $93.

  a. How much interest is paid in the two years?

  b. What annual percentage rate is charged?

42. Don Driver bought a used car and needed a loan of $4,000 to finance it. He had a choice of two plans. Bank A offered a 24-month loan with payments of $191. Bank B offered a 36-month loan with payments of $133. Which bank offered the lower APR?

Chapter 9

END-OF-CHAPTER PROBLEMS

43. A student loan of $6,000 is to be repaid in 36 monthly payments of $192.50 each. After 24 payments the borrower decides to pay the remaining debt in full.

    a. How much interest is saved? (Use the rule of 78.)

    b. How large is the final payment?

    c. What is the total amount paid to cancel the debt?

SUPERPROBLEM

Interest of 1.5% per month was charged on a 12-month loan for $600.

    a. How much was the finance charge?

    b. After eight payments a check was sent to cancel the loan. Using the rule of 78, how much interest was saved? What was the amount of the check?

    c. If the loan had originally been for 8 months at the same interest rate, how much would the finance charge have been?

Chapter 9

	Name	Date	Class

TEST YOURSELF

Section

Find the cost of credit in each of the following loans.

	Amount of Loan	Number of Payments	Monthly Payment	
1.	$5,000	12	$465	9.1
2.	$1,725	36	$61.50	9.1

Find the amount of each monthly payment in the following installment purchases.

	Cash Price	Down Payment	Finance Charge	Number of Payments	
3.	$2,399	$100	$148	10	9.2
4.	$780	20% of cash price	9% simple interest	24	9.2

5. At the E-Z Finance Co. a borrower can get a loan of $500 and can repay it in 12 equal monthly payments of $50. How large is the finance charge? 9.1

6. Furniture with a cash price of $650 can be bought on installment by paying $90 down and a finance charge of 16% simple interest. If 15 monthly payments are made, find:

    a. The finance charge. 9.1

    b. The monthly payment. 9.2

    c. The total cost on the installment plan. 9.1

    d. The interest rate the seller will have to reveal to the buyer. 9.5

For Problems 7–11, use credit terms of $1\frac{1}{2}\%$ per month on all unpaid balances and a minimum interest charge of 50¢. Payment must be within 30 days to avoid the finance charge.

7. A charge account customer received a bill showing a previous balance of $918 and a finance charge of $13.77. By the required payment date he sent a check for $150. Before the next billing period he bought $205 of new merchandise and put it on the account. What balance, including interest, will the next bill show?   9.4

8. Anna Thacker opened a charge account at Goodstuff Department Store and used it when she bought a new suit for $199. When she received the bill she sent a check for $50. By how much is her debt reduced?   9.4

For Problems 9–11, a charge account statement is dated September 3 and has a balance for merchandise of $172.29 carried over from the previous month.

9. What total would be shown on the September bill?   9.4

10. If a $25 payment is made in September and a purchase of $26.50 is added to the account during the month, what charges will appear on the October bill?   9.4

11. If the October total is paid off with monthly $50 payments (or less for the final payment) with no new purchases being made, set up the table showing finance charge, payments, interest, and new balance until the entire amount is paid.   9.4

12. A table tennis set is on sale for $199 if bought for cash. If bought on the installment plan no down payment is required, but a finance charge of $20 is added. The total is then paid in 12 equal monthly payments.

   a. How large is the monthly payment?   9.2

   b. What annual percentage rate is charged?   9.5

Chapter 9

TEST YOURSELF

Section

13. An outdoor gas grill can be bought on the credit plan for $336, which includes a service charge of $36. If the cost is repaid in 15 equal monthly installments:

    a. What is the nominal or simple interest rate?

    b. What is the APR?

9.5

14. Misty Day bought an automobile on an installment plan that required monthly payments of $173.82. The finance charge on the 48-month loan was $3,496. She decided to pay off the loan 12 months early. Find:

    a. The interest refunded using the rule of 78.

    b. The amount needed to pay the loan in full.

9.6

15. A stereo system is priced at $999 when bought for cash. On the installment plan a down payment of $100 is required and the remainder is paid in 18 monthly payments of $65 each. After 9 payments, a customer decides to pay the remaining debt in full.

    a. How much of the finance charge is saved? (Use the rule of 78.)

    b. How large is the final payment?

9.6

Chapter 9

COMPUTER SPREADSHEET EXERCISES

Complete the following spreadsheets. Use formulas where possible and $ signs for dollar amounts. Align decimal points for dollar amounts. All columns and numbers in your spreadsheets should appear as shown. If your spreadsheet is too wide to fit across the width of the page, print lengthwise using the Landscape orientation and the Fit Columns to Page scaling option in File Page Setup.

1. Compute the finance charge and the size of each payment. The down payment is based on the cash price. Assume equal monthly installments.

Cash Price	Down Payment	Add-on Rate	Number of Payments	Finance Charge	Monthly Payment
$4,000	10%	15.00%	15		
$1,125	25%	19.80%	9		
$375	20%	20.00%	18		
$8,230	15%	10.00%	6		
$957	5%	22.50%	12		
$1,204	30%	12.50%	10		
$533	20%	8.95%	15		
$6,970	10%	18.40%	18		

2. Fill in the missing amounts for the last three columns of the following spreadsheet. An accurate way to compute the annual percentage rate for computer applications is to use the formula

$$APR = \frac{72I}{3P(n+1) + I(n-1)}$$

where:

I represents the interest or finance charge;

P is the principal; and

n is the number of payments.

Use this formula in the rightmost column of your spreadsheet. Round APRs to the nearest hundredth of a percent. Assume equal monthly installments.

Amount Financed	Number of Payments	Add-on Rate	Finance Charge	Monthly Payment	APR
$5,400	18	10.0%			
$930	6	9.0%			
$2,410	10	8.0%			
$16,880	36	11.0%			
$12,900	24	9.5%			
$3,256	12	12.0%			
$8,200	15	14.0%			
$7,214	24	8.5%			
$9,865	10	7.5%			
$37,625	30	13.0%			

3. Jewelry priced at $879 is purchased on the easy payment plan. The cost is paid off at $88 a month starting 1 month from the purchase date. After the first payment, interest will be charged at 15% a year on the unpaid balance with no minimum finance charge. Complete the columns in the following table based on this information. When the account balance drops below $140, pay the entire amount (use the IF function to

381

determine the correct payment). Give totals for the Payment, To Pay Interest, and To Reduce Debt columns.

Unpaid Balance	Finance Charge	Amount Owed	Payment	To Pay Interest	To Reduce Debt	New Balance
$879.00	---	$879.00	$88.00	---		

10 ■ Compound Interest

OBJECTIVES

Upon completion of this chapter, the student should be able to use appropriate tables or a calculator to compute:

1. *Compound interest and compound amount when:*

    there is a single deposit;

    an additional deposit is made during the time period;

    the number of compounding periods is larger than the last entry in the table; and *the rate changes during the time period.*

2. *Effective rate.*

3. *Present value.*

For most short-term loans, interest is computed only once, using the simple interest formula. The simple interest calculation is based on the principal alone, and not on any past interest earned. For longer-term financial arrangements, however, interest is commonly compounded. In *compounding,* the interest is added to the original principal at stated periods; the total (principal plus interest) then becomes the principal for the next interest calculation. **Compound interest** is the total interest that accumulates over the life of the loan or investment. All financial institutions in the U.S. pay compound interest on savings account deposits; it is also paid on money-market accounts, certificates of deposit, installment loans, credit cards, and insurance and retirement plans.

10.1 SIMPLE VERSUS COMPOUND INTEREST

On August 29, 1814, New York City lent the Federal Government $1 million to defend the city against British attacks. In 1975, there was a question of whether the debt had been repaid. It was. It was settled on June 15, 1815, for $1,028,183.75, including interest. But, "if the debt were still owed . . . principal and interest would total $11 billion."

This excerpt from a *New York Times* article of November 19, 1975, can be used to illustrate the differences between simple interest, studied in Chapter 7, and compound interest, the subject of this chapter. For a quick calculation will show that for $1 *million* to accumulate to $11 *billion*, even in 161 years, there must be more than simple interest involved. By the simple interest formula using 9%:

$$I = Prt$$
$$= 1,000,000 \times 0.09 \times 161$$
$$I = \$14,490,000$$

The debt would be only $15,490,000, or approximately $15.5 *million*, instead of the indicated $11 *billion*.

$$S = P + I$$
$$= 1{,}000{,}000 + 14{,}490{,}000$$
$$S = \$15{,}490{,}000$$

Although the rate used to reach the total of $11 billion is not given, there is no doubt that the interest has been **compounded;** that is, at regular periods the interest has been added to the principal, and *interest has been earned on interest*. Even at 6%, if the interest was added to the principal each year, and thus became part of the principal for the next year, the debt would reach $11 billion in 161 years.

So far we have performed numerous simple interest calculations—each of which was based on the principal alone. We now know that compound interest calculations differ from those at simple interest in that the principal is increased periodically by adding the past interest earned. An illustration using a principal of $1,000 invested at 9% for one year will compare the two methods and show how increases in compounding frequency affect the final amount.

Simple Interest

After one year, a principal of $1,000 at *9% simple interest* becomes:

$$I = Prt$$
$$= 1{,}000 \times 0.09 \times 1 = \$90$$
$$S = P + I = \$1{,}090$$

Interest at 9% Compounded Quarterly

To compute the final amount at *9% compounded quarterly:*

The simple interest is calculated for 3 months—a quarter of a year—and is added to the principal. This new amount becomes the principal for the next 3-month period. Thus $1,000 at 9% compounded quarterly for 1 year will yield $93.08 in interest and will accumulate to $1,093.08. (See Table 10-1.)

TABLE 10–1 $1,000 AT 9% INTEREST COMPOUNDED QUARTERLY

$I = Prt$	INTEREST	AMOUNT
First period: $I = 1{,}000 \times 0.09 \times \frac{3}{12}$		
$I = \$22.50$	$22.50	
$S = 1{,}000 + 22.50 = \$1{,}022.50$		$1,022.50
Second period: $I = 1{,}022.50 \times 0.09 \times \frac{3}{12} = \$23.01$	23.01	
$S = 1{,}022.50 + 23.01 = \$1{,}045.51$		1,045.51
Third period: $I = 1{,}045.51 \times 0.09 \times \frac{3}{12} = \$23.52$	23.52	
$S = 1{,}045.51 + 23.52 = \$1{,}069.03$		1,069.03
Fourth period: $I = 1{,}069.03 \times 0.09 \times \frac{3}{12} = \$24.05$	24.05	
$S = 1{,}069.03 + 24.05 = \$1{,}093.08$	———	$1,093.08
	$93.08	

Interest at 9% Compounded Monthly

For the $1,000 at *9% compounded monthly:*

The procedure is the same as for compounding quarterly, but the interest is calculated 12 times a year, each time for 1 month, or $\frac{1}{12}$ of a year. Thus $1,000 at 9% compounded monthly for 1 year will yield $93.80 in interest and will accumulate to $1,093.80. (See Table 10-2.)

TABLE 10-2 $1,000 AT 9% INTEREST COMPOUNDED MONTHLY

MONTH	INTEREST CALCULATION	INTEREST	AMOUNT
1	$1,000 \times 0.09 \times \frac{1}{12}$	$ 7.50	$1,007.50
2	$1,007.50 \times 0.09 \times \frac{1}{12}$	7.56	1,015.06
3	$1,015.06 \times 0.09 \times \frac{1}{12}$	7.61	1,022.67
4	$1,022.67 \times 0.09 \times \frac{1}{12}$	7.67	1,030.34
5	$1,030.34 \times 0.09 \times \frac{1}{12}$	7.73	1,038.07
6	$1,038.07 \times 0.09 \times \frac{1}{12}$	7.79	1,045.86
7	$1,045.86 \times 0.09 \times \frac{1}{12}$	7.84	1,053.70
8	$1,053.70 \times 0.09 \times \frac{1}{12}$	7.90	1,061.60
9	$1,061.60 \times 0.09 \times \frac{1}{12}$	7.96	1,069.56
10	$1,069.56 \times 0.09 \times \frac{1}{12}$	8.02	1,077.58
11	$1,077.58 \times 0.09 \times \frac{1}{12}$	8.08	1,085.66
12	$1,085.66 \times 0.09 \times \frac{1}{12}$	8.14	$1,093.80
		$93.80	

Three Results

Several facts about compound interest are illustrated by the calculation. (See Table 10-3.)

- Compound interest is based on simple interest.
- Simple interest is computed once for the entire period. When interest is compounded, it is computed periodically by the simple interest formula and is added to the principal to get a new principal for the next period.
- The more frequent the compounding, the larger the amount of interest.

TABLE 10-3 COMPARISON OF SIMPLE VERSUS COMPOUND INTEREST ON $1,000 FOR ONE YEAR

9% RATE	INTEREST	AMOUNT
Simple interest	$90.00	$1,090.00
Quarterly compounding	93.08	1,093.08
Monthly compounding	93.80	1,093.80

A practical consideration is also apparent. Even for 1 year the calculation is long and tedious, and for a much longer period it would be prohibitive. In practice, compound interest is calculated using computers. However, we will use tables and/or calculators for all computations.

Financial institutions commonly pay interest compounded daily (based on a 365-day year), so that the interest is credited every day that the money is on deposit. But, as can be seen from Table 10-4, the advantage of more frequent compounding becomes very small.

TABLE 10-4 COMPARISON OF SIMPLE VERSUS COMPOUND INTEREST ON $1,000 FOR ONE YEAR

9% RATE	INTEREST	AMOUNT
Simple interest	$90.00	$1,090.00
Compounded 4 times a year	93.08	1,093.08
Compounded 12 times a year	93.80	1,093.80
Compounded 24 times a year	93.99	1,093.99
Compounded 72 times a year	94.11	1,094.11
Compounded 144 times a year	94.14	1,094.14
Compounded 365 times a year	94.16	1,094.16

10.2 FINDING COMPOUND AMOUNT BY TABLE

The following terms will be used in discussing compound interest.

- The **compound amount** is the final sum at the end of the term.

- **Compound interest** is the difference between compound amount and the original principal, if no additional deposits are made during the period.

- The **conversion period** (or compounding period) is the stated, regular interval (such as annually, semiannually, quarterly) for which the interest is computed and then added to the principal.

- The **nominal rate** is the stated annual rate. For example, if a deposit earns 9% compounded monthly, the nominal rate is 9% a year.

Two values must be determined to use the compound interest tables: the interest rate per period (i) and the total number of periods (n). In the following formulas, special care must be taken to distinguish between *conversion periods in 1 year* and *total number of conversion periods*.

$$\text{Interest rate per period } (i) = \frac{\text{Nominal interest rate}}{\text{Number of conversion periods in 1 year}}$$

$$\text{Total number of conversion periods } (n) = \text{Number of conversion periods in 1 year} \times \text{Number of years}$$

These formulas should be understood rather than memorized. The first means, for example, that if you get 12% interest for a whole year, you will get 6% for a half-year (two compounding periods), 3% for a quarter of a year (four compounding periods), and so on.

The second means, for example, that if there are 2 compounding periods in 1 year, there are 2 × 3 = 6 in 3 years. Or, if there are 12 compounding periods in 1 year, there are 12 × 3 = 36 in 3 years.

Find i and n for the following investments.

FINDING COMPOUND AMOUNT BY TABLE 387

■ **Example 1** 9% compounded annually, for 10 years:

$$i = \frac{\text{nominal rate}}{\text{number of conversion periods in 1 year}}$$

$$i = \frac{9\%}{1} = 9\%$$

Compounded annually means that interest is calculated once a year and added to the principal.

n = number of conversion periods in one year × number of years

$n = 1 \times 10 = 10$ periods

For 9% compounded annually for 10 years, i (rate per period) = 9% and n (total number of periods) = 10. ■

■ **Example 2** 14% compounded quarterly for 6 years:

$$i = \frac{\text{nominal rate}}{\text{number of conversion periods in 1 year}}$$

$$i = \frac{14\%}{4} = 3\tfrac{1}{2}\%$$

Quarterly compounding means 4 conversion periods per year.

n = number of compounding periods per year × number of years

$n = 4 \times 6 = 24$ periods

For 14% compounded quarterly for 6 years, $i = 3\tfrac{1}{2}\%$ and $n = 24$ periods. ■

■ **Example 3** 8% compounded monthly for 3 years:

$$i = \frac{8\%}{12}$$

Monthly compounding means 12 conversion periods per year.

$$i = \tfrac{2}{3}\%$$

Reduce to lowest terms.

$n = 12 \times 3 = 36$ periods

For 8% compounded monthly for 3 years, $i = \tfrac{2}{3}\%$ and $n = 36$ periods. ■

■ **Example 4** $5\tfrac{1}{2}\%$ compounded semiannually for $3\tfrac{1}{2}$ years.

$$i = \frac{5\tfrac{1}{2}\%}{2} = 5\tfrac{1}{2} \div 2$$

$$= \frac{11}{2} \times \frac{1}{2}$$

$$= \frac{11}{4} = 2\tfrac{3}{4}\%$$

$$n = 2 \times 3\tfrac{1}{2} = 2 \times \tfrac{7}{2} = 7 \text{ periods}$$

For $5\tfrac{1}{2}\%$ compounded semiannually for $3\tfrac{1}{2}$ years, $i = 2\tfrac{3}{4}\%$ and $n = 7$ periods. ■

Chapter 10

TRY THESE PROBLEMS (Set I)

Find the rate per period (*i*) and the total number of periods (*n*) for the following investments.

1. 8% compounded annually for 6 years.

2. 7% compounded semiannually for 15 years.

3. 5% compounded annually for 4 years.

4. 6% compounded semiannually for 8 years.

5. 10% compounded quarterly for 11 years.

6. $8\frac{3}{4}$% compounded semiannualy for 12 years.

7. $8\frac{1}{2}$% compounded monthly for 9 years.

8. $6\frac{1}{2}$% compounded quarterly for $3\frac{1}{2}$ years.

9. 11% compounded monthly for 5 years.

10. 7% compounded quarterly for $6\frac{1}{2}$ years.

11. 8% compounded quarterly for 3 years and 3 months.

12. 9% compounded monthly for 5 years and 7 months.

Using the Tables

Bank deposits in the U.S. earn compound interest. There are many types of accounts available, often at different rates and under different conditions. In a **savings account** or **passbook account**, money can be deposited or withdrawn anytime, with no penalty. Historically, rates on these accounts have varied from $2\frac{1}{2}\%$ to 6%. Most financial institutions give a higher interest rate for a **certificate of deposit (CD)**, where the depositor agrees to leave a minimum amount of money for a specific period of time. **Money-market accounts** pay interest comparable to CDs and, unlike CDs, offer penalty-free withdrawal of funds and limited check writing privileges. Interest rates on money-market accounts can change frequently, whereas rates on certificates of deposit remain the same throughout the term.

Calculation of a compound amount can be difficult, especially if the number of periods is large. In order to simplify the computation, a compound interest table is often used. Based on the values of rate per period (i) and number of periods (n), the table gives the values to which $1 will increase. The table value is then multiplied by the principal to find the compound amount. To illustrate, consider the following:

> Find the compound amount and the compound interest for a principal of $1,000 that is invested at 9% compounded quarterly for 1 year.

Recall that compound amount and interest for this same investment were calculated at the beginning of the chapter without the table (see Table 10-1).

To use the table, find the values of i and n.

$$i = \frac{9\%}{4} = 2\frac{1}{4}\%$$

$$n = 4 \times 1 = 4 \text{ periods}$$

Table C-5,* Amount of $1 at Compound Interest, begins on page 602. Table 10-5 shows one page of the table. Locate $2\frac{1}{4}\%$ in the column headings and go down to the row corresponding to four periods. The amount is 1.09308332. This means that $1 left at $2\frac{1}{4}\%$ per period for four periods will increase to $1.09 (rounded). Since the principal in this problem is $1,000 and not $1, multiply the table value by $1,000 to get the compound amount. In general,

---

Compound amount = Principal $\times$ Table C-5 value

For our illustration,

$$\text{Compound Amount} = 1{,}000 \times 1.09308332$$
$$= \$1{,}093.08$$

Note: It is not necessary to use all the decimal places in the table. Use one more decimal place, without rounding, as there are digits in the principal, including cents. For this problem seven decimal places for the six digits in $1,000.00 would be adequate.

$$1{,}000 \times 1.0930833 = \$1{,}093.08$$

The compound amount is found directly from the table. Compound interest is found by subtracting the original principal from the compound amount.

Compound interest = Compound amount − Original principal

---

*The formula for compound amount is $S = P(1 + i)^n$. The table has been calculated for $P = 1$ and different values of i and n.

TABLE 10-5 AMOUNT OF 1 AT COMPOUND INTEREST $s = (1 + i)^n$

n	2%	2¼%	2½%	2¾%	n
1	1.0200 0000	1.0225 0000	1.0250 0000	1.0275 0000	1
2	1.0404 0000	1.0455 0625	1.0506 2500	1.0557 5625	2
3	1.0612 0800	1.0690 3014	1.0768 9063	1.0847 8955	3
4	1.0824 3216	1.0930 8332	1.1038 1289	1.1146 2126	4
5	1.1040 8080	1.1176 7769	1.1314 0821	1.1452 7334	5
6	1.1261 6242	1.1428 2544	1.1596 9342	1.1767 6836	6
7	1.1486 8567	1.1685 3901	1.1886 8575	1.2091 2949	7
8	1.1716 5938	1.1948 3114	1.2184 0290	1.2423 8055	8
9	1.1950 9257	1.2217 1484	1.2488 6297	1.2765 4602	9
10	1.2189 9442	1.2492 0343	1.2800 8454	1.3116 5103	10
11	1.2433 7431	1.2773 1050	1.3120 8666	1.3477 2144	11
12	1.2682 4179	1.3060 4999	1.3448 8882	1.3847 8378	12
13	1.2936 0663	1.3354 3611	1.3785 1104	1.4228 6533	13
14	1.3194 7876	1.3654 8343	1.4129 7382	1.4619 9413	14
15	1.3458 6834	1.3962 0680	1.4482 9817	1.5021 9896	15
16	1.3727 8571	1.4276 2146	1.4845 0562	1.5435 0944	16
17	1.4002 4142	1.4597 4294	1.5216 1826	1.5859 5595	17
18	1.4282 4625	1.4925 8716	1.5596 5872	1.6295 6973	18
19	1.4568 1117	1.5261 7037	1.5986 5019	1.6743 8290	19
20	1.4859 4240	1.5605 0920	1.6386 1644	1.7204 2843	20
21	1.5156 6634	1.5956 2066	1.6795 8185	1.7677 4021	21
22	1.5459 7967	1.6315 2212	1.7215 7140	1.8163 5307	22
23	1.5768 9926	1.6682 3137	1.7646 1068	1.8663 0278	23
24	1.6084 3725	1.7057 6858	1.8087 2595	1.9176 2610	24
25	1.6406 0599	1.7441 4632	1.8539 4410	1.9703 6082	25
26	1.6734 1811	1.7833 8962	1.9002 9270	2.0245 4575	26
27	1.7068 8648	1.8235 1588	1.9478 0002	2.0802 2075	27
28	1.7410 2421	1.8645 4499	1.9964 9502	2.1374 2682	28
29	1.7758 4469	1.9064 9725	2.0464 0739	2.1962 0606	29
30	1.8113 6158	1.9493 9344	2.0975 6758	2.2566 0173	30
31	1.8475 8882	1.9932 5479	2.1500 0677	2.3186 5828	31
32	1.8845 4059	2.0381 0303	2.2037 5694	2.3824 2138	32
33	1.9222 3140	2.0839 6034	2.2588 5086	2.4479 3797	33
34	1.9606 7603	2.1308 4945	2.3153 2213	2.5152 5626	34
35	1.9998 8955	2.1787 9356	2.3732 0519	2.5844 2581	35
36	2.0398 8734	2.2278 1642	2.4325 3532	2.6554 9752	36
37	2.0806 8509	2.2779 4229	2.4933 4870	2.7285 2370	37
38	2.1222 9879	2.3291 9599	2.5556 8242	2.8035 5810	38
39	2.1647 4477	2.3816 0290	2.6195 7448	2.8806 5595	39
40	2.2080 3966	2.4351 8897	2.6850 6384	2.9598 7399	40
41	2.2522 0046	2.4899 8072	2.7521 9043	3.0412 7052	41
42	2.2972 4447	2.5460 0528	2.8209 9520	3.1249 0546	42
43	2.3431 8936	2.6032 9040	2.8915 2008	3.2108 4036	43
44	2.3900 5314	2.6618 6444	2.9638 0808	3.2991 3847	44
45	2.4378 5421	2.7217 5639	3.0379 0328	3.3898 6478	45
46	2.4866 1129	2.7829 9590	3.1138 5086	3.4830 8606	46
47	2.5363 4351	2.8456 1331	3.1916 9713	3.5788 7093	47
48	2.5870 7039	2.9096 3961	3.2714 8956	3.6772 8988	48
49	2.6388 1179	2.9751 0650	3.3532 7680	3.7784 1535	49
50	2.6915 8803	3.0420 4640	3.4371 0872	3.8823 2177	50

For our investment, this gives

$$\text{Compound interest} = 1{,}093.08 - 1{,}000$$

$$= \$93.08$$

$1,000 left for 1 year at 9% compounded quarterly will grow to $1,093.08 and will yield $93.08 in interest. This is the same result as found by the long method.

■ **Example 5** Using Table C-5, find the compound amount and compound interest for $1,000 at 9% compounded monthly for 1 year.

$$i = \frac{9\%}{12} = \frac{3}{4}\%$$

$$n = 12 \times 1 = 12$$

Table C-5 value (seven decimal places) = 1.0938069

Compound amount = Principal × Table C-5 value

= 1,000 × 1.0938069

Compound amount = $1,093.81

Compound interest = Compound amount − Principal

= 1,093.81 − 1,000

Compound interest = $93.81

$1,000 left at 9% compounded monthly will yield $93.81 in interest and accumulate to $1,093.81 after 1 year. This is the same result obtained by the long method of calculation. The difference of 1¢ is due to rounding in the long calculation. ∎

■ **Example 6** Find the compound interest and the compound amount if $3,000 is left on deposit at 8% for 12 years with interest compounded quarterly.

$$i = \frac{8\%}{4} = 2\%$$ *The nominal rate is 8%; there are four conversion periods per year.*

$n = 12 \times 4 = 48$ periods *There are 12 years and four conversions periods per year.*

Table C-5 value = 2.5870703 *Use seven decimal places for six digits in principal 3,000.00.*

Compound amount = Principal × Table C-5 value

= 3,000 × 2.5870703

Compound amount = $7,761.21

Compound interest = Compound amount − Principal

= 7,761.21 − 3,000

Compound interest = $4,761.21

If $3,000 is left at 8% compounded quarterly for 12 years, it will amount to $7,761.21, including $4,761.21 in compound interest.

Two points should be noted in this illustration.

1. Regardless of the wording of the problem, which may ask for the interest before the amount, the compound amount is always found before compound interest when the table is used.

2. The interest earned is more than the original principal. This occurred because the interest rate was high and the time on deposit was long. This relationship has been used in bank advertisements that invite you to double your money in 10 to 15 years, depending on the interest rate. ∎

394 CHAPTER 10   COMPOUND INTEREST

■ **Example 7**   Find the compound amount and the compound interest on $600 at 10% compounded semiannually after 2 years and 6 months.

$$i = \frac{10\%}{2} = 5\%$$

$$n = 2 \times 2\frac{6}{12} = 2 \times 2\frac{1}{2} = 5 \text{ periods}$$

DOUBLING YOUR MONEY

If you wish to approximate how long it will take to double your money you can use the same method that bankers and accountants use. They know that money doubles in approximately $\frac{72}{r}$, where r is the nominal rate.*

Consider the following rates:

  3% (approximates the returns on passbook savings accounts)

  $5\frac{1}{2}\%$ (typical of the rates given on certificates of deposit)

  10% (the average annual return on common stocks since 1926)

At 3% money will double in approximately 24 years. $\left(\frac{72}{3}\right)$

At $5\frac{1}{2}\%$ money will double in approximately 13 years. $\left(\frac{72}{5.5}\right)$

At 10% money will double in approximately 7 years. $\left(\frac{72}{10}\right)$

*Based on annual compounding.

Table C-5 value = 1.276281   *Use six decimal places for five digits in principal (600.00).*

Compound amount = 600 (1.276281)

Compound amount = $765.77

Compound interest = 765.77 − 600

Compound interest = $165.77

At the end of 2 years and 6 months, $600 at 10% compounded semiannually will have earned $165.77 in compound interest and will amount to $765.77. ■

■ **Example 8**   Find the maturity value of a 5-year note bearing interest at 12% compounded monthly if the face value of the note is $1,500. How much interest will be earned?

$$i = \frac{12\%}{12} = 1\%$$

$$n = 12 \times 5 = 60$$

Table C-5 value = 1.8166967

Compound amount = 1,500 × 1.8166967

Compound amount = $2,725.05

Compound interest = 2,725.05 − 1,500

Compound interest = $1,225.05

The maturity value of the note is $2,725.05. This includes $1,225.05 compound interest. ∎

Compounding also applies to situations other than those involving loans or investments.

■ **Example 9**

Stephanie Jones begins a job with a weekly salary of $920 and is given a 3% raise each year. What is her weekly salary after 5 years?

The weekly salary after 5 years is a compound amount.

$$i = \frac{3\%}{1} = 3\%$$

$$n = 1 \times 5 = 5$$

Table C-5 value = 1.159274

Compound amount = 920 × 1.159274

Compound amount = $1,066.53

If a weekly salary of $920 increases at 3% compounded annually for 5 years, it will grow to $1,066.53.

Chapter 10

TRY THESE PROBLEMS (Set II)

Find the compound amount and the compound interest for Problems 1–6.

1. $525 at 7% compounded annually for 8 years.

2. $620 at 10% compounded semiannually for 10 years.

3. $950 at 12% compounded quarterly for 8 years.

4. $1,200 at $10\frac{1}{2}$% compounded monthly for 3 years and 4 months.

5. $400 at 7% compounded quarterly for 4 years.

6. $400 at 7% compounded annually for 4 years.

7. $300 is invested in a passbook savings account earning $3\frac{1}{2}$% compounded monthly. How much will the investment be worth after $7\frac{1}{2}$ years?

8. $10,000 remains in a savings account paying 4% compounded quarterly for 5 years and 3 months. How much interest will be earned?

9. Tara Doby begins a job with an annual salary of $49,200 and is given a $3\frac{1}{2}$% raise each year. What is her annual salary after 4 years?

10. Van McGowan purchased a $2,000 2-year CD paying 5% compounded monthly.

    a. How much will the CD be worth at the end of the period?

    b. How much interest will be earned?

11. Tammy Towe purchased a $2,000, 2-year CD earning 5% compounded quarterly.

    a. What is the CD worth at maturity?

    b. How much interest will be earned?

12. Compare your answers for Problems 5 and 6. Also compare your answers for Problems 10 and 11. What conclusions can you draw?

13. How much interest will be earned on a $3,000, 6-month CD that earns 6% compounded monthly?

14. An account paying 6% compounded monthly is opened with $500. How much will there be in the account after 2 years and 5 months?

15. $900 is invested at 8% compounded quarterly for 3 years and 9 months.

    a. How much will the investment be worth at the end of the period?

    b. How much of the total will be interest?

16. Dorothy Desire deposits $4,300 in the bank, which pays interest at $7\frac{1}{2}$% compounded monthly. How much will Dorothy have in this account at the end of 3 years?

10.3 USING A FORMULA TO FIND COMPOUND AMOUNT

The formula for compound amount,

$$S = P(1+i)^n$$

can be evaluated using your calculator. Let's redo Example 7 this way.

To use the formula, first find i and change it to a decimal.

$$i = \frac{10\%}{2} = 5\% = 0.05$$

Then find n.

$$n = 2 \times 2\frac{1}{2} = 5$$

Substitute these numbers into the formula.

$$S = P(1+i)^n$$
$$S = 600(1+0.05)^5$$
$$S = 600(1.05)^5 \qquad (1.05^5 \text{ means multiply } 1.05 \text{ by itself five times.})$$

Enter the following sequence on your calculator to get the compound amount:

Enter	Display
1.05 ☒ 1.05 ☒ 1.05 ☒ 1.05 ☒ 1.05 ☐	1.2762815 (*This is the value in Table* C-5.)
☒ 600 ☐	765.7689 (*This is rounded to* $765.77.)

The compound interest is found by subtracting.

$$765.77 - 600 = \$165.77$$

If your calculator has a "y^x" key or an "x^y" key it is even easier to solve this kind of problem. This key takes a number and raises it to a power. Using the y^x key to find the value of 1.05^5, the entire procedure becomes:

ENTER	DISPLAY
1.05 y^x 5 ☐ ☒ 600 ☐	765.7689 ($765.77)

■ **Example 1**

Find the compound amount and compound interest on $450 at 12% compounded quarterly for 1 year.

$$i = \frac{12\%}{4} = 3\% = 0.03$$
$$n = 4 \times 1 = 4$$
$$S = P(1+i)^n$$
$$S = 450(1.03)^4$$

400 CHAPTER 10   COMPOUND INTEREST

ENTER	DISPLAY
1.03 y^x 4 = × 450 =	506.4789 ($506.48)

The compound amount is $506.48.
The compound interest is $56.48 (506.48 − 450). ■

You *must* have a calculator with the y^x key to work a problem with daily compounding.

■ **Example 2**   A deposit of $500 is made in an account that earns 6.5% interest compounded daily. Find the value of this account in 2 years.

$$i = \frac{6.5\%}{365} = 0.0178082\% = 0.000178082$$

$$n = 365 \times 2 = 730$$

$$S = P(1 + i)^n = 500(1.000178082)^{730}$$

ENTER	DISPLAY
1.000178082 y^x 730 = × 500 =	569.4076 ($569.41)

The value of this account in 2 years will be $569.41. ■

■ **Example 3**   Linda Green begins a job at a salary of $36,000 a year and is given a $3\frac{1}{2}\%$ raise each year. What is her annual salary after 5 years?

$$i = \frac{3\frac{1}{2}\%}{1} = 3\frac{1}{2}\% = 0.035$$

$$n = 1 \times 5 = 5$$

$$S = P(1 + i)^n = 36{,}000(1.035)^5$$

ENTER	DISPLAY
1.035 y^x 5 = × 36000 =	42756.707 ($42,756.71)

Linda's annual salary in 5 years will be $42,756.71. ■

Chapter 10

TRY THESE PROBLEMS (Set III)

In the following problems, use the formula $S = P(1 + i)^n$ to find the compound amount.

Find the compound amount and the compound interest for Problems 1–6.

1. $500 at 7% compounded annually for 3 years.

2. $2,500 at 10% compounded semiannually for 2 years.

3. $3,000 at 9% compounded monthly for 10 years.

4. $840 at 5.5% compounded daily for 1 year.

5. $7,000 at 9% compounded daily for 3 years.

6. $12,000 at 8% compounded quarterly for 4 years and 9 months.

7. Howard Griffin begins a job at a salary of $780 a week and is given a 4% raise each year. What is his weekly salary after 3 years?

8. $2,500 is invested in a savings account earning $3\frac{1}{2}\%$ compounded monthly. How much will the investment be worth after 4 years?

9. A deposit of $9,000 remains in a savings account paying 4% compounded daily for 2 years. How much interest will be earned?

10. Sharonda Layton purchased a $3,000, 5-year CD earning $5\frac{1}{4}\%$ compounded daily.

    a. How much will the CD be worth at the end of the period?

    b. How much interest will be earned?

11. Ravi Iyer purchased a $500, 6-month CD earning $4\frac{1}{2}\%$ compounded quarterly.

   a. What is the CD worth at maturity?

   b. How much interest will be earned?

12. An account paying $5\frac{1}{2}\%$ compounded monthly is opened with $1,200. How much will there be in the account after 6 years?

10.4 RATE CHANGES DURING PERIOD

Thus far we have considered only the case in which the interest rate remains unchanged throughout the period and no additional deposits are made. When such changes occur, the problem can be worked out in the same way as before, except that it is now broken into parts and treated like a separate problem each time a change occurs.

■ **Example 1**

$5,000 was invested in an account for 6 years. When the deposit was first made, the account was paying 8% compounded quarterly. After $2\frac{1}{2}$ years, the rate was changed to 8% compounded monthly. Find the compound amount and compound interest at the end of the 6 years.

The 6-year period is broken into two parts, the first $2\frac{1}{2}$ years when the conversion period was 3 months, and the next $3\frac{1}{2}$ years when the conversion period was 1 month. The compound amount found for the first part will be the principal for the second.

First part: $5,000 invested for $2\frac{1}{2}$ years at 8% compounded quarterly.

$$i = \frac{8\%}{4} = 2\%$$

$$n = 4 \times 2\frac{1}{2} = 10 \text{ periods}$$

Table C-5 value = 1.2189944

Compound amount = 5,000 × 1.2189944

Compound amount = $6,094.97

Second part: $6,094.97 invested for $3\frac{1}{2}$ years at 8% compounded monthly.

$$i = \frac{8\%}{12} = \frac{2}{3}\%$$

$$n = 12 \times 3\frac{1}{2} = 42 \text{ periods}$$

Table C-5 value = 1.3219009

Compound amount = 6,094.97 × 1.3219009

Compound amount = $8,056.95

Compound interest = Compound amount − Original principal

= 8,056.95 − 5,000

Compound interest = $3,056.95

The compound interest for the entire period is $3,056.95. The compound amount at the end of the period is $8,056.95. ■

10.5 ADDITIONAL DEPOSIT MADE DURING PERIOD

If an additional deposit is made during the period, the problem is again broken into parts. Each time a deposit is made, it must be added to the compound amount at the time of deposit to get the new principal. The interest is the final amount less the original principal *plus any deposits made during the period.*

404 CHAPTER 10 COMPOUND INTEREST

■ **Example 1**

On July 1, 1996, a deposit of $400 was made in a bank that paid 6% interest compounded quarterly. On January 1, 1998, an additional deposit of $200 was made in the same account. What will be the balance in the account on January 1, 2001? How much of this balance is compound interest?

First part: $400 at 6% compounded quarterly for $1\frac{1}{2}$ years, from July 1, 1996, to January 1, 1998.

$$i = \frac{6\%}{4} = 1\frac{1}{2}\%$$

$$n = 4 \times 1\frac{1}{2} = 6 \text{ periods}$$

Table C-5 value = 1.093443

Compound amount = 400 × 1.093443

Compound amount = $437.38

The balance in the account plus the $200 deposit is the principal for the second part of the problem.

$$437.38 + 200 = \$637.38$$

Second part: $637.38 at 6% compounded quarterly for 3 years, from January 1, 1998, to January 1, 2001.

$$i = \frac{6\%}{4} = 1\frac{1}{2}\%$$

$$n = 4 \times 3 = 12 \text{ periods}$$

Table C-5 value = 1.195618

Compound amount = 637.38 × 1.195618

Compound amount = $762.06

Compound interest = Compound amount − (Original principal + Deposits)

= 762.06 − (400 + 200)

= 762.06 − 600

Compound interest = $162.06

The balance in the account will be $762.06; $162.06 will be interest. ■

10.6 WHEN *n* IS LARGER THAN LAST ENTRY IN TABLE

The same method, breaking the problem into parts, can be used to solve problems in which the total number of periods is more than the largest number of periods found in the table.

■ **Example 1**

When Junior was 3 years old his parents realized that college costs were going up and they decided to make some provisions in advance for his education. They deposited $10,000 in an account paying interest at 5% compounded monthly. If they leave the account untouched until he is 18 years old and if the interest remains the same, how much will there be in the account when he is ready for college?

$$i = \frac{5\%}{12} = \frac{5}{12}\%$$

$$n = 12 \times 15 = 180 \text{ periods}$$

Since the table gives only 100 periods, the problem is now broken into two parts, 100 periods and 80 periods.

First part:

$$\text{Table C-5 value } (i = \frac{5}{12}\%, n = 100) = 1.51558426$$

$$\text{Compound amount} = 10{,}000 \times 1.51558426$$

$$\text{Compound amount} = \$15{,}155.84$$

Second part:

$$\text{Table C-5 value } (i = \frac{5}{12}\%, n = 80) = 1.39464627$$

$$\text{Compound amount} = \$15{,}155.84 \times 1.39464627$$

$$\text{Compound amount} = \$21{,}137.04$$

The account will have a balance of $21,137.04 when Junior is 18 years old. ∎

10.7 EFFECTIVE RATE

The terms **effective rate** and **effective annual yield** appear frequently in bank advertisements. Figure 10-1 shows the rates offered by one bank for different minimum deposits and different time periods. In each case the effective annual yield is higher than the nominal rate. For example, 5.85% compounded daily is equivalent to an annual yield of 6.02%.

Deposit	Annual Yield	Annual Rate
$1,000–$9,999	5.97%	5.80%
$10,000–$24,999	6.02%	5.85%
$25,000–$49,999	6.08%	5.90%
Term: 6 months, 4 days. Daily Compounding		

FIGURE 10-1 Comparison of bank rates.

In this illustration all the rates are compounded daily, and they can be compared directly. However, when an investor has a choice between different rates that also are compounded differently, the effective rate is used to make the comparison and to decide which rate is more profitable.

The **effective rate** is the simple interest rate that would give the same return in 1 year as the compound interest rate. It answers the question, What nominal rate compounded annually would give the same return?

Using a principal of $1 at 9% compounded monthly for 1 year:

$$i = \frac{9\%}{12} = \frac{3}{4}\%$$

$$n = 12 \times 1 = 12$$

$$\text{Table C-5 value} = 1.09380$$

$$\text{Compound amount} = 1 \times 1.09380$$

$$\text{Compound amount} = \$1.0938$$

$$\text{Compound interest} = 1.0938 - 1$$

$$\text{Compound interest} = \$.0938$$

By the simple interest formula:

$$I = Prt$$

$$0.0938 = 1 \times r \times 1$$

$$r = 0.0938 = 9.38\%$$

The effective rate is 9.38%. This means that 9.38% compounded annually is equivalent to 9% compounded monthly. The annual return will be the same.

Although this may appear to be a special case because the principal was $1, the result would be the same if any other principal was used, or if the principal was represented by P in the calculation.

Calculation of the effective rate can be greatly simplified by observing that the same result is found by subtracting 1 from the compound amount of $1 in 1 year and converting the result to a percent. In other words,

$$\text{Effective rate} = (1 + i)^N - 1$$

For this calculation i has its usual meaning and value, but we will use N instead of n for the number of compounding periods in 1 year when finding the effective rate. As before, the $(1 + i)^N$ value can be computed on a calculator or found in Table C-5.

■ **Example 1**

Find the effective rate of interest if the nominal rate is 10% compounded quarterly.

$$i = \frac{10\%}{4} = 2\frac{1}{2}\% = 0.025$$

$$N = 4$$

For the effective rate, N is the number of conversion periods in 1 year.

$$\text{Effective rate} = (1 + i)^N - 1$$

$$= (1 + 0.025)^4 - 1 = (1.025)^4 - 1$$

$$= 1.10381 - 1 = 0.10381 = 10.38\%$$

Check: The effective rate is equal to the nominal rate if compounding is annual. Otherwise, the effective rate is moderately higher than the nominal rate—10.38%, as compared with 10%. ■

Chapter 10

TRY THESE PROBLEMS (Set IV)

Find the compound amount and compound interest for the following investments. The interest rate was changed during the term.

1. $500; 8% compounded annually for 2 years, then 10% compounded quarterly for $1\frac{1}{2}$ years.

2. $3,000; 8% compounded monthly for 1 year, then 9% compounded semiannually for 2 years.

3. $400; 6% compounded monthly for 2 years, then 7% compounded monthly for 3 years.

4. $100; $8\frac{1}{2}$% compounded annually for 3 years, then 9% compounded quarterly for 4 years.

Find the compound amount and the compound interest for the following investments.

5. $850 at 7% compounded semiannually for 3 years, with an additional deposit of $250 at that time. The total then remains invested for an additional 6 years.

6. $1,000 at 6% compounded quarterly for 2 years. An additional $300 is then deposited, and the total remains for $1\frac{1}{2}$ more years.

7. $300 at $10\frac{1}{2}$% compounded monthly for 5 years. The amount of $150 is deposited after 5 years, and the total remains for an additional 3 years.

Find the effective rate for each of the nominal rates in Problems 8 through 11. Round to the nearest hundredth of a percent.

8. 9% compounded quarterly.

9. 9% compounded semiannually.

10. 8% compounded monthly.

11. 8% compounded annually.

407

12. Which yields more interest, 5% compounded monthly or 5.12% compounded annually? (*Hint:* Compare effective rates.)

13. $4,000 is invested at 9% compounded monthly for 8 years.

  a. Find the value of the investment at the end of the period.

  b. How much interest is earned?

  c. What is the effective rate of interest?

14. A deposit of $400 is made in a bank that pays $5\frac{1}{2}$% compounded quarterly. The money remains in the account for 4 years, then the interest rate is changed to 6% compounded monthly and the money remains on deposit for 3 more years.

  a. What will be the balance in the account 7 years after the initial deposit?

  b. How much interest will be earned?

  c. What are the effective rates for the two periods?

15. If an investment pays 7% compounded quarterly, find the balance on January 1, 2004, in an account with the following deposits:
First deposit: July 1, 2000, $3,000.
Second deposit: July 1, 2003, $2,000.

10.8 PRESENT VALUE

The present value of an investment at compound interest has the same meaning as the present value at simple interest. It is the principal that must be invested now at a given rate to accumulate to a given amount at the end of a definite period. Present value answers questions like, How much must be deposited today at 9% compounded quarterly to accumulate a balance of $5,000 in 3 years?

Present-value problems can be done with the compound-amount table,* but this involves division, which is avoided by using Table C-6, Present Value of 1 at Compound Interest, starting on page 618, where the division has already been done. No new methods are needed; just use the correct table.

> Present value = Compound amount × Table C-6 value

It is also possible to find the present value by using a calculator and the formula $P = \dfrac{S}{(1+i)^n}$, which can also be written as†

$$P = S(1+i)^{-n}$$

■ **Example 1**

What sum must be invested now at 9% compounded monthly to accumulate to $8,000 in 6 years? How much interest will be earned?

$$i = \frac{9\%}{12} = \frac{3}{4}\%$$

$$n = 12 \times 6 = 72 \text{ periods}$$

Table C-6 value = 0.5839236

Present value = Compound amount × Table C-6 value

= 8,000 × 0.5839236

Present value = $4,671.39

Check: The present value must be less than the compound amount given in the problem. When doing present-value problems, be sure the table value is less than 1.

Compound interest = Compound amount − Present value

= 8,000 − 4,671.39

Compound interest = $3,328.61 ■

---

*The formula for compound amount, $S = P(1+i)^n$, is solved for P and gives $P = \dfrac{S}{(1+i)^n} = S\left(\dfrac{1}{(1+i)^n}\right) = S\left(\dfrac{1}{\text{Table C-5 value}}\right) = S \text{ (Table C-6 value)}$.

†Recall that a negative exponent indicates a fraction. Thus, $\dfrac{S}{(1+i)^n} = S(1+i)^{-n}$.

> **CALCULATOR HINT**
>
> If your calculator has a power key, y^x or x^y, you can calculate any of the values in Table C-6. The table is generated using the formula $P = S(1 + i)^{-n}$, where S is $1.
>
> For Example 1, where
>
> $$i = \frac{3}{4}\% = 0.0075 \text{ and } n = 72$$
>
> the Table C-6 value would be calculated as
>
> $$P = \$1 (1 + 0.0075)^{-72} = (1.0075)^{-72}$$
>
> | ENTER | DISPLAY |
> |---|---|
> | 1.0075 y^x 72 $+/-$ $=$ | 0.58392363 |
>
> 0.58392363 is the value in Table C-6

■ **Example 2** Mr. Bigbucks wants to provide a $10,000 wedding gift for his daughter. She is now 9 years old, and he would like the fund to be available by the time she is 20. He decides on an investment that pays 8% compounded semiannually. How large must the deposit be? How much of the final $10,000 will be interest?

$$i = \frac{8\%}{2} = 4\%$$

$$n = 2 \times 11 = 22 \text{ periods}$$

Table C-6 value = 0.42195538

Present value = Compound amount $\times$ Table C-6 value

= 10,000 $\times$ 0.42195538

Present value = $4,219.55

Compound interest = Compound amount − Present value (Principal)

= 10,000 − 4,219.55

Compound interest = $5,780.45

A deposit of $4,219.55 at 8% compounded semiannually will yield $10,000 at the end of 11 years. Of this, $5,780.45 will be interest. ■

■ **Example 3** Mr. Speedy has been financing his cars and realizes that this adds substantially to the cost. He decides to provide funds in advance so that he can buy his next car for cash and avoid the finance charge. He expects to need a new car in 5 years and to pay about $24,000 for it. He can put a sum of money in:

1. A safe investment at 5% compounded quarterly; or
2. A more risky investment at 9% compounded monthly.

How much must he invest now to give him $24,000 in 5 years in each plan?

Plan 1

$$i = \frac{5\%}{4} = 1\frac{1}{4}\%$$

$$n = 4 \times 5 = 20 \text{ periods}$$

Table C-6 value = 0.78000854

Present value = 24,000 × 0.78000854 = $18,720.20

Plan 2

$$i = \frac{9\%}{12} = \frac{3}{4}\%$$

$$n = 12 \times 5 = 60 \text{ periods}$$

Table C-6 value = 0.63869970

Present value = 24,000 × 0.6386997 = $15,328.79

Mr. Speedy will have $24,000 at the end of 5 years if he invests $18,720.20 at 5% compounded quarterly or $15,328.79 at 9% compounded monthly. The decision on which investment is better depends on whether Mr. Speedy is also Mr. Cautious. ■

Now go to Try These Problems (Set V).

NEW FORMULAS IN THIS CHAPTER

- Interest rate per period (i) = $\dfrac{\text{Nominal interest rate}}{\text{Number of conversion periods in 1 year}}$

- Total number of conversion periods (n) = $\begin{Bmatrix}\text{Number of conversion} \\ \text{periods in 1 year}\end{Bmatrix} \times \begin{Bmatrix}\text{Number of} \\ \text{years}\end{Bmatrix}$

- Compound amount = Principal × Table C-5 value or $S = P(1 + i)^n$

- Compound interest = Compound amount − (Principal + Deposits)

- Effective rate = $(1 + i)^N - 1$, where N is the number of conversion periods in 1 year

- Present value = Compound amount × Table C-6 value or $P = S(1 + i)^{-n}$

Chapter 10

Name　　　　　　　　Date　　　　　　　　Class

TRY THESE PROBLEMS (Set V)

Find the present value and compound interest in Problems 1 through 6:

	MATURITY VALUE	TIME	INTEREST RATE
1.	$ 600	3 years	10% compounded quarterly
2.	1,400	15 years	7% compounded semiannually
3.	7,200	10 years	8% compounded annually
4.	2,500	7 years	9% compounded monthly
5.	10,000	5 years	$7\frac{1}{2}$% compounded monthly
6.	4,000	6 years	11% compounded quarterly

7. How much must be invested at 4% compounded monthly to amount to $20,000 in 7 years? How much interest will be earned?

8. Kate Richards promised her nephew that she would give him $3,000 when he graduates from high school in 4 years. If money is worth 9% compounded quarterly, how much must she invest today so that the money will be available when he graduates?

9. How much must be invested at 6% compounded quarterly to cover the down payment of $17,000 on a house that the Davenports plan to buy in 9 years? How much interest will they earn?

10. Rocco Boyd loaned $21,000 to the owner of a new restaurant. He will be repaid at the end of 6 years with interest at 10% compounded semiannually. Find how much he will be repaid.

11. Sue Smith wants to have $20,000 in savings when her daughter reaches college age in 10 years. If her bank will pay 7% compounded semiannually on long-term deposits:

   a. What single deposit made now will achieve Sue's goal?

   b. How much interest will the investment earn during this time?

12. Joyce Hammel wants to buy new furniture in 5 years at an estimated cost of $5,400. If she invests $2,750 now at 8% compounded semiannually, will she have enough money to buy the furniture? (*Hint:* Find the present value of $5,400 or the maturity value of $2,750.)

13. A developer can buy some property from a landowner for $86,000. The landowner claims that in one year the same property will be selling for $93,000. If money is worth 7% compounded monthly, should the developer buy the land now or in one year? (*Hint:* Find the present value of $93,000 or the maturity value of $86,000.)

Chapter 10

END-OF-CHAPTER PROBLEMS

Find the interest rate per period (*i*) and the total number of conversion periods (*n*) for the following investments.

1. 8% compounded quarterly for 9 years.

2. 7% compounded semiannually for 14 years.

3. $5\frac{1}{2}$% compounded annually for 4 years.

4. 12% compounded monthly for $11\frac{1}{2}$ years.

5. $4\frac{1}{2}$% compounded monthly for 3 years and 9 months.

6. 13% compounded quarterly for 6 years and 3 months.

For Problems **7–11,** find the effective rates for the investments in Problems 1–5. Round to the nearest hundredth of a percent. Find the compound amount and the compound interest for Problems 12 through 16.

12. **a.** $5,000 at 7% compounded annually for 10 years.

    b. $5,000 at 7% compounded monthly for 10 years.

13. $8,300 at 6% compounded quarterly for 9 years.

14. $6,000 at 5% compounded monthly for 11 years.

15. $700 at 7% compounded annually for 6 years and then at a new interest rate of 9% compounded monthly for an additional 3 years.

16. $800 at 8% compounded semiannually for 10 years. A deposit of $100 is then added and the total remains for another 3 years.

Find the present value that would yield the following maturity values and also find the interest earned.

	MATURITY VALUE	RATE	TIME
17.	$ 8,000	10% compounded quarterly	3 years
18.	15,000	13% compounded semiannually	15 years
19.	300	12% compounded quarterly	13 years

Fill in the missing quantities for each of the following:

	MATURITY VALUE	PRESENT VALUE	RATE	YEARS	INTEREST
20.	$2,000		6% compounded semiannually	8	
21.		$1,246.33	6% compounded semiannually	8	
22.	500		12% compounded quarterly	9	
23.		400	9% compounded semiannually	24	

24. Brian Sprankle deposited $3,000 in a savings account that paid $5\frac{1}{2}$% compounded quarterly.

   a. How much would he have in the account after 10 years?

   b. How much interest has been earned?

Chapter 10

END-OF-CHAPTER PROBLEMS

25. How much interest will be earned by an investment of $6,000 if the interest rate is 10% compounded semiannually and the term of the investment is 16 years?

26. Which would be a more profitable investment, $8\frac{1}{2}$% compounded annually or 8% compounded quarterly? (Compare the effective rates.)

27. Which would yield more interest in 1 year, 8% simple interest or compound interest with an effective rate of 8%?

28. In which bank should you deposit your money for the highest interest rate: Bank A ($6\frac{1}{4}$% compounded annually) or Bank B (6% compounded monthly)?

29. An investment of $4,000 earns interest of 11% compounded semiannually.

   a. How much will it be worth after $8\frac{1}{2}$ years?

   b. How much interest is included in the total?

   c. What is the effective interest rate?

30. If Steve Welsh invests $2,200 at $7\frac{1}{2}$% compounded monthly, how much interest will he earn in:

   a. 8 years?

   b. 12 years?

   c. What is the effective interest rate?

417

31. A deposit of $900 is made in a savings account that pays $5\frac{1}{2}\%$ compounded quarterly. It remains for $3\frac{1}{2}$ years. At that time the interest rate is changed to 6% compounded quarterly.

  a. What is the balance in the account after an additional 2 years?

  b. How much interest does the balance include?

32. What is the final balance if $6,000 is left on deposit at 8% compounded quarterly for 13 years and then the rate is changed to 8% compounded monthly for the next 6 years? How much interest is earned?

33. $5,000 is deposited in an account paying 10% compounded semiannually and left for 8 years. At that time an additional $2,000 is deposited.

  a. How much is in the account after 8 years without the new deposit?

  b. How much will be in the account after an additional 6 years?

  c. How much interest will be earned in the 14 years?

34. A deposit of $2,000 is made for a 5-year-old child. It remains at 7% compounded annually for 10 years.

  a. How much is there in the account at the end of 10 years?

  b. If an additional deposit of $500 is made after 10 years and the rate is changed to 8% compounded monthly, how much will there be in the account when the child is 19 years old?

35. A bank balance was $4,000 on July 1, 1999. The account had been opened on July 1, 1997, and the interest was 9% compounded quarterly. On July 1, 1999, an additional deposit of $1,000 was made.

  a. If the interest rate remains the same, how much will be in the account on January 1, 2002?

  b. What was the original deposit?

Chapter 10

END-OF-CHAPTER PROBLEMS

36. Mandy Moore purchased a $5,000, 1-year CD earning $5\frac{1}{2}$% compounded quarterly.

    a. What is the maturity value of the investment?

    b. If money is worth 6% compounded quarterly, what is the present value of the investment?

37. A 6-year, $20,000 investment is made at 10% compounded quarterly.

    a. What is the maturity value of the investment?

    b. If money is worth 9% compounded semiannually, what is the present value of this investment?

38. A note is due in 3 years with simple interest at 9%. The face value of the note is $25,000.

    a. Find the maturity value of the note.

    b. For how much can the note be sold if the money can be deposited at 12% compounded semiannually?

39. Which yields more interest for Siri Products: $6,000 at 12% simple interest for 7 years, or $5,000 at 12% compounded semiannually for 7 years? How much more interest is obtained?

40. A simple interest note has a face value of $2,000, bears interest at 8%, and matures in 9 months. How much would have to be invested for the same time at 6% compounded monthly to yield the same maturity value?

41. A debt of $3,000 is to be paid in 3 years with interest at 9% compounded quarterly.

    a. How much interest will be added to the debt?

    b. What simple interest rate would give the same amount of interest in 3 years?

    c. What is the effective interest rate?

419

42. A retailer is offered merchandise for $500 cash or $520 due in 6 months. If money is worth 8% compounded monthy, which is the cheaper price?

43. A builder can buy some property from Land-Space, Inc., for $47,000. Land-Space claims that in one year the same property will be selling for $50,000. If money is worth 8% compounded quarterly, should the builder buy the land now or in one year?

44. Find the interest earned on $10,000 for 4 years at 8% compounded: (a) yearly, (b) semiannually, (c) quarterly, (d) monthly; (e) find the simple interest.

45. Find the interest earned on $10,000 at 8% compounded quarterly for: (a) 5 years, (b) 10 years, and (c) 20 years.

46. Will an investment double in value if it earns interest at 9% compounded monthly for 8 years?

47. If you started a job at a salary of $1,000 a week and were given an 8% raise each year, how much would you be earning after 5 years?

48. A man borrowed $25,000 to pay for his son's college education. The debt is due in 6 years, and interest is charged at an effective rate of $8\frac{1}{2}$% (i.e., $8\frac{1}{2}$% compounded annually).

   a. How much will he repay when the debt is due?

   b. How much would he have had to deposit 15 years ago at the same interest rate to have the required $25,000 today?

49. Modern Furniture would like to purchase new computers for their office showroom in 4 years. They estimate the cost at $15,000. If they invest their Money at 12% compounded monthly:

   a. What single deposit should Modern make to have the needed funds in 4 years?

   b. How much interest will they earn?

   c. What is the effective rate of interest?

Chapter 10

END-OF-CHAPTER PROBLEMS

50. How much must be invested now in order to have $4,500 in 10 years if money is worth 10% compounded quarterly?

51. Richard Right must repay $5,000 in 8 years. If money is worth 9% compounded monthly,

    a. What would be the cash settlement if the debt were paid now?

    b. How much interest is included in the maturity value?

52. How much must be invested now at $8\frac{1}{2}\%$ compounded annually to amount to $50,000 in 20 years? How much interest will be earned?

53. $1,000 is invested at 7% compounded monthly to pay a debt in 5 years. When the debt is due, there are other funds available and the investment remains at the same interest rate for an additional $6\frac{1}{2}$ years. How much interest is earned during the entire period?

54. How much must be invested now at 8% compounded annually to have $10,000 at the end of 20 years? If the investment then remains at the same interest rate for an additional 5 years, how much will it be worth?

55. Myra Meeks wants to attend State University and will need to have $15,000, 5 years from today. If her bank pays 7% compounded quarterly, how much must she deposit today so that she will have the required amount in 5 years?

56. Don Jackson loaned his friend $9,000 to help him open a new art gallery. The debt will be repaid at the end of 4 years with interest at 12% compounded semiannually.

    a. How much will Don receive at the end of 4 years?

    b. What is the effective rate of interest?

57. A 10-year-old child will inherit $20,000 when she reaches her 18th birthday.

   a. If money is worth 8% compounded monthly, what is the present value of her inheritance?

   b. If she doesn't take the money, and the interest rate remains unchanged, how much will she have on her 21st birthday?

58. An account was opened for a newborn child with a deposit of $5,000. The interest rate was 5% compounded semiannually. It was hoped that when the child reached her eighteenth birthday and was ready for college, the balance would cover the first 2 years' tuition. If the college she chose was charging $8,500 a year:

   a. How many years would the account actually cover?

   b. What opening balance would have been equal to the tuition for 2 years?

SUPERPROBLEM

One partner in a new business invested $20,000. At the end of 3 years the business was not doing well and he decided to take his money out. It was agreed that his share was worth $22,000 and he accepted this amount.

a. If he had left his money in a bank account at a fixed rate of 7% compounded monthly for the same period, how much more would he have had?

b. If he had put just enough in the same bank at the same rate to have $22,000 at the end of 3 years and had loaned the rest of the $20,000 to a friend with a $5\frac{1}{2}$% simple interest note, how much would he have had at the end of 3 years?

c. If, after 3 years, he added the maturity value of the note to the amount on deposit (part b), how much would he have had at the end of an additional year if the interest rate increased to 9% compounded monthly?

Chapter 10

TEST YOURSELF

Find the rate per period (*i*) and the number of compounding periods (*n*) for the following investments.

1. 6% compounded quarterly for 15 years. 10.2

2. $3\frac{1}{2}$% compounded monthly for 2 years and 10 months. 10.2

Find the compound amount and the compound interest for Problems 3 and 4.

3. $4,400 at 7% compounded annually for 3 years. 10.3

4. $900 at $4\frac{1}{2}$% compounded semiannually for 8 years. 10.3

5. Alice Adams invests $3,000 at 9% compounded semiannually for 8 years. 10.2

   a. How much will she have after 8 years?

   b. How much interest will she have earned?

6. $600 is deposited in a savings bank that pays $5\frac{1}{2}$% compounded quarterly. 10.2

   a. How much will the account contain after 9 years and 3 months?

   b. How much interest will be earned?

7. Brandy Nunn begins a job at a salary of $750 per week and is given a 3% raise each year. What is her weekly salary after 4 years? 10.2

8. Carrie Bonilla opened a new passbook savings account. She deposited $2,000 at 4% compounded quarterly. At the end of 4 years she deposited an additional $1,000. What is the balance in Carrie's account at the end of 6 years? 10.5

9. Ray Winters opened a savings account with a deposit of $300. The bank paid 6% compounded quarterly. After 2 years the rate changed to 7% compounded quarterly. How much did Ray have in his account at the end of 5 years? 10.4

10. Find the effective rate for 6% compounded monthly. 10.7

11. In which bank should you deposit your money for the highest rate: Bank A paying $4\frac{1}{4}\%$ compounded annually or Bank B paying 4% compounded monthly? (Compare the effective rates.) 10.7

12. Reese Arnold wants to have $20,000 in 10 years. Interest can be earned at 8% compounded quarterly. Find: 10.8

    a. The sum he must invest.

    b. The compound interest earned.

13. A business can buy some property from a landowner for $160,000. The landowner claims that in one year the same property will be selling for $170,000. If money is worth 7% compounded semiannually, should the business buy the land now or in one year? 10.8

Chapter 10

COMPUTER SPREADSHEET EXERCISES

Find the missing values for the following spreadsheets. Use formulas where possible and $ signs for dollar amounts. Align decimal points in columns where they appear. All columns and numbers in your spreadsheets should appear as shown. If your spreadsheet is too wide to fit comfortably across the width of the page, use the File Page Setup Landscape orientation and the Fit Columns to Page scaling option to print lengthwise.

1. Complete the spreadsheet below. Express Rate per Period as a percent rounded to two decimal places.

Interest Rate	Term of Investment (in Years)	Compounding Frequency (per Year)	Rate per Period (i)	Total Number of Periods (n)
10.50%	3.75	4		
7.25%	9.50	2		
8.00%	2.00	1		
9.75%	8.25	12		
13.50%	4.00	12		
11.00%	5.20	365		

2. Find the compound amount and compound interest. Use the formula

$$S = P(1+i)^n$$

for the compound amount. Denote exponentiation by using the $\wedge$ symbol in your spreadsheet formula.

Amount Invested	Interest Rate	Term of Investment (Years)	Compounding Frequency (per Year)	Compound Amount	Compound Interest
$1,400	6.50%	5	2		
$24,000	12.75%	6	12		
$15,500	8.00%	15	365		
$8,950	13.75%	3	1		
$20,000	12.25%	9	4		
$7,500	7.75%	4	12		
$19,500	6.00%	20	2		
$2,300	10.50%	11	365		
$5,000	9.25%	13	4		
$3,068	11.00%	7	2		

3. Fill in the missing values. Use the following nested IF statement to determine the compounding frequency per year:

=IF(TYPE="annually",1, IF(TYPE="semiannually",2,IF(TYPE= "quarterly",4,IF(TYPE="monthly",12,365))))

where the cell addresses in the "Type of Compounding" column are substituted for TYPE in the formula. The formula for effective rate is

$$\text{Effective rate} = (1+i)^N - 1$$

where i is the rate per period and N is the number of times compounded per year. Denote exponentiation in your spreadsheet formula by using the $\wedge$ symbol. Display the effective rates as percents and round to the nearest hundredth of a percent. Position the column title "Time" across two separate columns, one for the numbers and one for the labels. Then use the number column entries in the appropriate formulas. The formula for present value is

$$P = S(1+i)^{-n}$$

Be sure to make the exponent negative in your spreadsheet formula.

Maturity Value	Interest Rate	Type of Compounding	Compounding Frequency (per Year)	Effective Rate	Time	Present Value	Compound Interest
$1,600	7.00%	annually			3 years		
$2,000	4.50%	semiannually			5.5 years		
$25,350	10.75%	monthly			7 years		
$1,140	5.25%	daily			16 years		
$86,300	8.25%	quarterly			12.75 years		
$4,590	9.30%	daily			21.2 years		
$12,250	13.65%	semiannually			4 years		
$17,880	11.80%	daily			5.8 years		

4. Many people dream of accumulating $1,000,000 in savings. This exercise computes the number of years necessary to do this based on annual or daily compounding of various amounts at various rates of interest. The Microsoft Excel function

=NPER(rate per period, payment, present value, future value)

computes the number of compounding periods required for a present value to grow to a given maturity value, earning a fixed interest rate per compounding period. For example

=NPER(10%/12,0,−10000,20000)

returns 83.52, which tells you that it will take 83.52 months to double your $10,000. To find the number of years, divide by 12. In this example, like the spreadsheet below, we do not add to our beginning investment, so the value entered for payment is 0. Also note that the number entered for present value is negative. Of course, in your spreadsheet you will have to replace the values in the example with the appropriate cell addresses and numbers. Place $1,000,000 in a separate cell and use the cell address in the function (instead of the number). Use relative row and column addresses where necessary. Round calculations for the number of years to the nearest tenth.

Click on the Drawing button on the Excel toolbar and select the rectangle to begin enclosing your table within boxes as shown below. Move the mouse pointer to where you want to begin drawing a box, click, and drag across the worksheet. Release the mouse button when the box is the size you want.

Looking at the completed worksheet, how many years will it take $5,000 to accumulate to $1,000,000 at: (a) 14% compounded annually? (b) 14% compounded daily? (c) Compare your answers to (a) and (b). Explain why the answers are not the same.

Number of Years Needed to Accumulate $1,000,000									
Annual Compounding						Daily Compounding			
	Interest Rate						Interest Rate		
Beginning Investment	5%	8%	11%	14%	Beginning Investment	5%	8%	11%	14%
$500					$500				
$1,000					$1,000				
$5,000					$5,000				
$10,000					$10,000				
$25,000					$25,000				
$50,000					$50,000				

11 ■ Annuities

OBJECTIVES

Upon completion of this chapter, the student should be able to:

1. *Use appropriate tables to compute:*
   amount of an annuity;
   present value of an annuity; and
   rent:
   present value known; and
   maturity value known.
2. *Determine the total amount received or paid during the term of an annuity and the interest that is included.*
3. *Prepare an amortization schedule showing how a loan is paid off.*
4. *Prepare a sinking fund schedule to verify that the required maturity value will be obtained.*
5. *Distinguish among the six different types of compound interest and annuity problems.*

11.1 WHAT IS AN ANNUITY?

We are all familiar with annuities, although we may know them only by specific names: rent, salaries, social security payments, installment payments, mortgage payments, life insurance premiums, and others. What these all have in common is that a series of payments, usually equal, is made at regular intervals. That is the definition of an **annuity**—a series of payments, usually equal, made periodically.

Under the general heading of annuity, however, there are many differences. The annuity may run for a definite time or it may be indefinite, as in the case of a life insurance policy. Payments may be made at the beginning of each period or at the end. Rent is paid in advance; salaries are paid at the end of the period. There may or may not be compound interest involved. The interest conversion period and payment period may be the same or different.

The type of annuity that we will study is an investment annuity that includes compound interest. It is also **ordinary, simple,** and **certain:**

- *ordinary,* because payments are made at the end of each payment period;
- *simple,* because the interest conversion period coincides with the payment period;
- *certain,* because the annuities begin and end on definite dates.

Following is a brief illustration of how money accumulates in an ordinary, simple, annuity certain. We will find the total for deposits of $100 made at the end of each quarter, for one year, in an account paying 6% compound quarterly. The calculations here are done manually. Later we will explain how to use tables that simplify calculating annuities.

428 CHAPTER 11   ANNUITIES

			TOTAL ON DEPOSIT
March 31:	Deposit $100		$100.00
June 30:	Add interest for one quarter on $100 at 6%	$ 1.50	
	Add deposit of $100	100.00	
	Total added	$101.50	$201.50
Sept. 30:	Add interest for one quarter at 6% on $201.50	$ 3.02	
	Add deposit of $100	100.00	
	Total added	$103.02	$304.52
Dec. 31:	Add interest for one quarter at 6% on $304.52	$ 4.57	
	Add deposit of $100	100.00	
	Total added	$104.57	$409.09

At the end of the year, after the fourth deposit, the account balance is $409.09. A total of $400 has been deposited (four deposits of $100 each), and $9.09 interest has been credited to the account. During the first year there are four deposits but three conversions of interest. If the annuity continues, there will be four deposits and four conversions each year.

The following terms are used in discussing annuities.

- **Rent** is the periodic payment.
- **Term** is the length of time the annuity continues.
- **Amount** is the final value at the end of the term.
- **Payment interval** is the time between two successive payments. It is also the conversion period for a simple annuity.

Annuity calculations are really complex compound interest calculations, and they will all be done with tables similar to the compound interest tables. The values of i and n are the same as for compound interest, and the table values are located in the same way. The essential part of the problem is to understand what is required and then to use the correct tables.

11.2 AMOUNT OF AN ANNUITY

The amount (or maturity value) of an annuity depends on the size of the periodic payment, the term, and the interest rate. The formula for the amount of an annuity when the periodic payment is $1 is:

$$S_{\overline{n}|i} = \frac{(1+i)^n - 1}{i}$$

This formula is not needed if Table C-7, Amount of an Annuity of 1 Per Period, is used. Table C-7 gives values for $S_{\overline{n}|i}$ (read "S angle n at i") for various values of i and n assuming a periodic payment of $1. For any size payment:

> Amount of an annuity = Size of each payment × Table C-7 value

In symbols, the formula becomes

$$S_n = R \times S_{\overline{n}|i}$$

where S_n is used for the amount of an annuity to distinguish it from the amount S at compound interest. R is the periodic payment or rent.

■ **Example 1**

Denise Downs deposits $2,000 every 6 months in an account paying 6% compounded semiannually. She makes these deposits for 2 years.

1. What is the amount of the annuity?
2. How much will Denise have deposited herself?
3. How much interest has accumulated?

1.
$$i = \frac{6\%}{2} = 3\%$$

$$n = 2 \times 2 = 4 \text{ periods}$$

Table C-7 value $(S_{\overline{n}|i}) = 4.1836270$

Use one more decimal place as there are digits in rent, including cents.

Amount of annuity (S_n) = Rent × Table C-7 value

$$= 2,000 \times 4.1836270$$

Amount of annuity $(S_n) = \$8,367.25$

Denise's account will contain $8,367.25 after 2 years.

2. There will be four deposits of $2,000 each.

$$4 \times 2,000 = \$8,000$$

Denise will deposit $8,000.

3. The difference between the amount and the total deposited is the accumulated interest.

Compound interest = Amount of annuity − Total deposits

$$= 8,367.25 − 8,000$$

Compound interest = $367.25

The annuity will earn $367.25 interest during the 2 years. ■

Note in this problem that there are four deposits but only three conversions of interest. Since the first deposit is made at the end of the first semiannual compounding period, this account earns interest for three periods. The second deposit earns interest for two periods, the third for one period, and the fourth earns no interest at all. The diagram below illustrates the activity in the account.

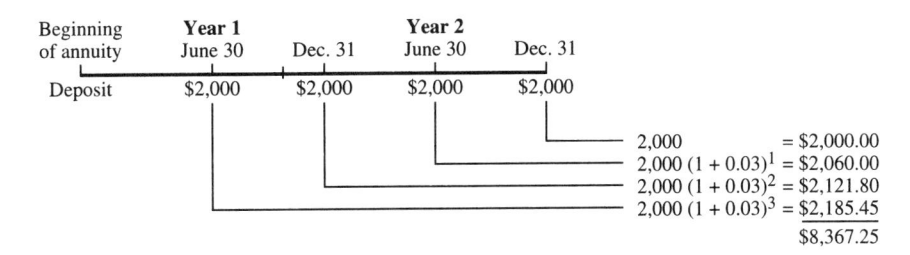

Beginning of annuity	**Year 1** June 30	Dec. 31	**Year 2** June 30	Dec. 31
Deposit	$2,000	$2,000	$2,000	$2,000

$2,000 = \$2,000.00$
$2,000 (1 + 0.03)^1 = \$2,060.00$
$2,000 (1 + 0.03)^2 = \$2,121.80$
$2,000 (1 + 0.03)^3 = \$2,185.45$
$\$8,367.25$

> **CALCULATOR HINT**
>
> Example 1 can be solved using your calculator and the formula for S_n. The value of the fraction represents the table value.
>
> $$S_n = \text{Rent} \times \frac{(1+i)^n - 1}{i}$$
>
> $$S_n = 2{,}000 \times \frac{(1+0.03)^4 - 1}{0.03}$$
>
> First, do the addition inside the parentheses.
>
> $$S_n = 2{,}000 \times \frac{(1.03)^4 - 1}{0.03}$$
>
> Then evaluate the fraction by following the order of operations, and multiply by the rent.
>
> | ENTER | DISPLAY | |
> |---|---|---|
> | 1.03 y^x 4 = | 1.1255088 | |
> | − 1 = | 0.1255088 | |
> | ÷ 0.03 = | 4.183627 | *This is the Table C-7* |
> | × 2000 = | $8,367.25 | *value.* |
>
> Denise's account will contain $8,367.25 after 2 years.

■ **Example 2**

Back in 1981, a new father decided to put $500 a year in an investment paying 8% compounded annually. He believed that when the eighteenth payment was made there would be enough in the fund to pay for his child's college education with enough left over for a new car. By 1999, when the last payment was made, the college of his choice was charging $9,200 a year for tuition. For how many years of college would the fund pay? How much of the fund was accumulated interest?

$$i = \frac{8\%}{1} = 8\%$$

$$n = 1 \times 18 = 18 \text{ periods, from 1981 to 1999}$$

Table C-7 value ($S_{\overline{n}|i}$) = 37.450243

Amount (S_n) = Rent × Table C-7 value

= 500 × 37.450243

Amount (S_n) = $18,725.12

The fund would pay for about 2 years of college tuition. Forget the car.

Compound interest = Amount of annuity − Total deposits

= 18,725.12 − (500 × 18)

= 18,725.12 − 9,000

Compound interest = $9,725.12

The father had deposited $9,000. The annuity had earned $9,725.12 in interest. ■

Example 3

A man plans to provide a fund to make it possible for him to retire before social security age. He decides to deposit $200 every quarter at 8% compounded quarterly. He continues these deposits for 11 years and then finds that he can no longer afford the deposits. However, he can leave the fund untouched. How much will he have toward retirement at the end of another 10 years? How much of the fund is interest?

This is a two-part problem. First, there is an annuity for 11 years, and then the fund earns compound interest for the next 10 years. (See Figure 11-1.)

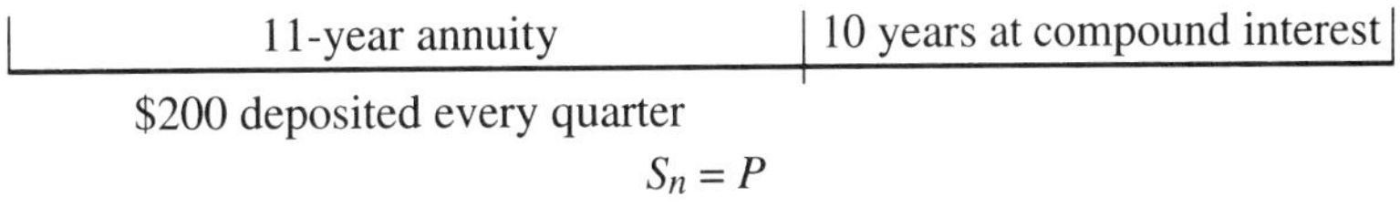

11-year annuity	10 years at compound interest

$200 deposited every quarter
$$S_n = P$$

FIGURE 11-1 Retirement fund.

First part: the annuity

$$i = \frac{8\%}{4} = 2\%$$

$$n = 4 \times 11 = 44 \text{ periods}$$

Table C-7 value = 69.502657

Amount (S_n) = Rent × Table C-7 value

= 200 × 69.502657

Amount (S_n) = $13,900.53

Check: $200 deposited 44 times with no interest would give $200 × 44 = $8,800. The amount of the annuity should be much more, since it includes interest.

Second part: compound amount
The amount of the annuity (S_n) is the principal left at compound interest.

$$i = \frac{8\%}{4} = 2\%$$

$$n = 4 \times 10 = 40 \text{ periods}$$

Table C-5 value = 2.20803966

Compound amount = Principal × Table C-5 value

= 13,900.53 × 2.20803966

Compound amount = $30,692.92

The final account balance at the end of 21 years would be $30,692.92.

Compound interest = Amount − Deposits

= 30,692.92 − 8,800

Compound interest = $21,892.92

The annuity and compound interest account would earn $21,892.92 in interest, almost three times the amount deposited. ■

Chapter 11

TRY THESE PROBLEMS (Set I)

Find the amount of the annuity, the total deposited, and the interest earned for each of the following.

	RENT	COMPOUND INTEREST RATE	TIME
1.	$600	8% annually	8 years
2.	$450	7% semiannually	12 years
3.	$200	9% quarterly	11 years
4.	$300	12% monthly	4 years

Find the maturity value and the interest earned.

	PERIODIC PAYMENT	COMPOUND INTEREST RATE	TIME
5.	$5,000	5% monthly	$2\frac{1}{2}$ years
6.	$2,600	11% quarterly	10 years
7.	$700	8% semiannually	5 years and 6 months
8.	$1,000	9% monthly	3 years and 6 months

9. Find the maturity value if $300 is deposited every 6 months in an account paying 9% compounded semiannually. The term is 12 years.

10. A deposit of $350 is made each quarter in an account paying interest at 8% compounded quarterly.

   a. How much will there be in the account after $6\frac{1}{2}$ years?

   b. What will be the total amount deposited?

   c. How much interest will be earned?

11. Ridge Products is investing $20,000 every 6 months with the intention of purchasing new machinery in 5 years. The money is being invested in a mutual fund paying 6% compounded semiannually.

   a. How much will Ridge invest during the 5 years?

   b. What will be the amount in their account?

   c. How much interest will the account have earned?

12. A "new car fund" is established by depositing $380 a month in an account paying 7% compounded monthly.

   a. How expensive a car will the fund provide after 4 years?

   b. How much interest is included in the total amount?

13. $300 is deposited in an account each month for $2\frac{1}{2}$ years. The account earns 12% compounded monthly. After this time payments stop.

   a. How much will there be in the account when the payments stop?

   b. If the accumulated amount is then left for an additional 3 years at the same interest rate, how much will there finally be in the account?

   c. How much interest will be earned during the entire period?

14. Each month Nancy Taylor deposits $50 into her savings and loan account, which pays 6% compounded monthly. After 4 years the payments stop, but the account is left untouched for another 3 years.

   a. How much is in the account at the end of 7 years?

   b. How much interest is earned?

11.3 PRESENT VALUE OF AN ANNUITY

Thus far we have considered annuities with rents paid *in,* with the largest value of the annuity at the end of the term. The reverse condition is also possible. Rents may be paid *from* an annuity, with the largest value at the beginning of the term.

Retirement annuities are of this type. A fund is built up over a period of years and then, on retirement, the annuity pays out a fixed amount at regular intervals. The fixed series of withdrawals is the annuity and the sum required at the beginning of the term when the payouts begin is the **present value** of the annuity. It is the lump sum that, when invested or deposited today, at a given rate, will provide a specific series of equal periodic withdrawals in the future.

Present value also refers to the value of an investment at any time before maturity. The **cash equivalent** of an installment plan is the present value of an annuity at the beginning of the term. It is found when a problem asks, How much money can I pay *now* instead of making a series of periodic payments?

To calculate the present value of an annuity, use the formula

> Present value of an annuity = Size of each payment × Table C-8 value

or, in symbols,

$$A_n = R \times a_{\overline{n}|i}$$

where

A_n is the present value of the annuity;

R is the rent or periodic payment; and

$a_{\overline{n}|i}$ is the present value of the annuity when the rent is $1 (Table C-8, Present Value of Annuity of 1 per Period).

■ **Example 1**

What is the present value of an annuity if it pays $200 at the end of each quarter for 10 years and the interest is 9% compounded quarterly?

$$i = \frac{9\%}{4} = 2\frac{1}{4}\%$$

$$n = 4 \times 10 = 40 \text{ periods}$$

Table C-8 value $(a_{\overline{n}|i}) = 26.193522$

Present value (A_n) = Rent × Table C-8 value

$= 200 \times 26.193522$

Present value $(A_n) = \$5,238.70$

The present value of the annuity is $5,238.70. This means that $5,238.70 deposited at 9% compounded quarterly would yield 40 payments of $200 each, or a total of $8,000. The difference between the principal on deposit at the beginning of the term (the present value) and the amount paid out is the compound interest.

Compound interest = Amount paid out − Present value

= 8,000 − 5,238.70

Compound interest = $2,761.30 ■

■ **Example 2**

An alumna gave her alma mater a one-time gift to provide an award of $1,000 a year for 20 years. The money was invested at 8% compounded annually. How large was the gift? How much of the total awarded during the 20 years will be interest?

The money is to be paid out of an existing fund; the present value is required.

$$i = \frac{8\%}{1} = 8\%$$

$$n = 1 \times 20 = 20 \text{ periods}$$

Table C-8 value = 9.8181474

Present value (A_n) = Rent × Table C-8 value

$$= 1{,}000 \times 9.8181474$$

Present value (A_n) = $9,818.15

Interest = Amount paid out − Present value

$$= (20 \times 1{,}000) - 9{,}818.15$$

$$= 20{,}000 - 9{,}818.15$$

Interest = $10,181.85

The original gift was $9,818.15. Of the $20,000 to be paid as awards, more than half, $10,181.85, will be from compound interest.

Another example of the present value of an annuity is the cash equivalent of a set of equal payments made at regular intervals, as in a car or education loan or in the payment of a home mortgage. The **cash equivalent** is the sum that can be invested *today* that would have the same maturity value as would the down payment and periodic payment of an annuity. ■

■ **Example 3**

Jeff Jones would like to receive payments of $4,000 each quarter for 10 years after he retires. Money is worth 8% compounded quarterly.

1. How much money must he have on deposit at retirement in order to receive these payments?

2. If Jeff will retire in 15 years, what single deposit should he make now so that he will have the necessary funds at retirement?

3. How much will he receive in total payments?

4. How much interest will the single deposit earn before the payments end?

This problem consists of two investments. One is the annuity for 10 years and the other is a single deposit made 15 years before the annuity begins. (See Figure 11-2.)

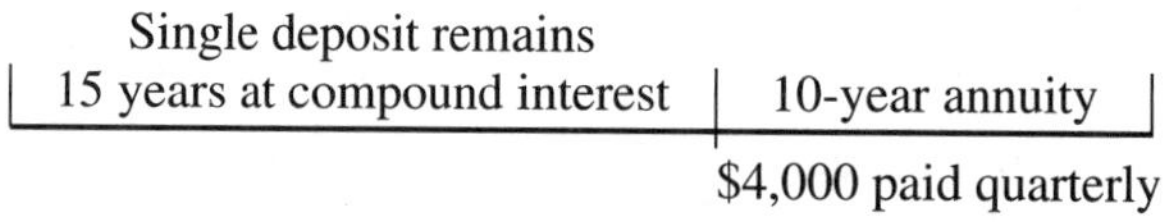

```
         Single deposit remains
    | 15 years at compound interest | 10-year annuity |
                                    $4,000 paid quarterly
```

FIGURE 11-2 Retirement payments.

1. The money he must have on deposit at retirement is the *present value of an annuity*.

$$i = \frac{8\%}{4} = 2\%$$

$$n = 4 \times 10 = 40$$

Table C-8 value = 27.355479

$$\text{Present value } (A_n) = \text{Rent} \times \text{Table C-8 value}$$
$$= 4{,}000 \times 27.3554792$$
$$\text{Present value } (A_n) = \$109{,}421.92$$

Jeff must have $109,421.92 on deposit when he retires.

Check: $4,000 withdrawn 40 times with no interest would give $4,000 × 40 = $160,000. The present value of the annuity ($109,421.92) should be less than the total of the withdrawals because interest continues to be added after payments from the fund begin.

2. The *single deposit* he should make now is the *present value at compound interest* of the amount needed at retirement in 15 years.

   Because only one deposit is involved, the formula

   $$\text{Present value} = \text{Compound Amount} \times \text{Table C-6 value}$$

   is used. Recall that this formula gives the present value at compound interest of a given maturity value.* For our example:

   $$i = \frac{8\%}{4} = 2\%$$
   $$n = 4 \times 15 = 60$$
   $$\text{Table C-6 value} = 0.30478227$$
   $$\text{Present value} = \text{Compound amount} \times \text{Table C-6 value}$$
   $$= 109{,}421.92 \times 0.30478227$$
   $$\text{Present value} = \$33{,}349.86$$

   Jeff must deposit $33,349.86 15 years before retirement. This deposit will have a maturity value of $109,421.92.

Check: With interest compounded at 8%, money will double in approximately 9 years $\left(\frac{72}{8}\right)$. Therefore, in 9 years Jeff will have more than $66,000. An additional 7 years will add large amount of interest, although the deposit will not double again. $109,421.92 is a reasonable total.

3. To find the total received:

   $$\text{Total received} = \text{Rent} \times \text{Number of payments}$$
   $$= 4{,}000 \times 40$$
   $$\text{Total received} = \$160{,}000$$

   Jeff will receive $160,000 in payments.

4. The interest earned is the difference between the total received from the annuity and the original investment.

   $$\text{Interest} = \text{Total received} - \text{Original investment}$$
   $$= 160{,}000 - 33{,}349.86$$
   $$\text{Interest} = \$126{,}650.14$$

This may seem like a large amount of interest, but remember the time period is 25 years, 15 years for the single deposit and 10 years for the annuity. ■

*Present value can also be found using $P = S(1+i)^{-n}$

438 CHAPTER 11   ANNUITIES

Cash Equivalent Price

■ **Example 4**

An auto dealer offered a customer a used car for $1,000 down and 24 monthly payments of $180 each. What is the cash equivalent of this offer if money is worth 9% compounded monthly?

The cash equivalent is the down payment plus the present value of the annuity at the given interest rate.

> Cash equivalent price = Down payment + Present value of annuity

$$i = \frac{9\%}{12} = \frac{3}{4}\%$$

$$n = 24 \text{ periods}$$

Table C-8 value = 21.889146

Present value (A_n) = Rent × Table C-8 value

= 180 × 21.889146

Present value (A_n) = $3,940.05

Cash equivalent = 1,000 + 3,940.05

Cash equivalent = $4,940.05

The dealer would get an equivalent price for the car if he accepted $4,940.05 in cash instead of $5,320 over 2 years ($1,000 + 24 × $180). ■

■ **Example 5**

A color television set is offered by two stores on the installment plan. Store A's plan requires a down payment of $75 with 8 monthly payments of $50 each. Store B requires a down payment of $85 with 16 monthly installments of $25 each. Which store has the better plan if they both include interest at 12% compounded monthly?

First, we must decide what "better" means, and it may have different meanings for different customers.

1. For a customer who cannot afford $50 a month, the second plan is better simply because the monthly payment is much smaller.

2. For another who does not want to be bothered with payments for more than a year, the first plan with the shorter payment period may be better.

3. A third, who knows a little business mathematics, might calculate quickly that there is more money paid in the second plan, and therefore the first is better.

   Cost = Down payment + Installment payments

   Plan A = 75 + (8 × 50) = 75 + 400 = $475

   Plan B = 85 + (16 × 25) = 85 + 400 = $485

4. A fourth customer, who covered annuities in business mathematics, decides that the only correct way to compare the plans is to find the cash equivalent values. In each case this will be the present value of the annuity plus the down payment.

Plan A

$$i = \frac{12\%}{12} = 1\%$$

$$n = 8 \text{ periods}$$

Table C-8 value = 7.65167

Present value = 50 × 7.65167

Present value = $382.58

Cash equivalent price = Down payment + Present value of annuity

= 75 + 382.58

Cash equivalent price = $457.58

Plan B

$$i = \frac{12\%}{12} = 1\%$$

$$n = 16 \text{ periods}$$

Table C-8 value = 14.71787

Present value = 25 × 14.71787

Present value = $367.95

Cash equivalent price = Down payment + Present value of annuity

= 85 + 367.95

Cash equivalent = $452.95

On the basis of the cash prices equivalent to the installment payment plans, Plan B is better. The extended time allowed for payment more than makes up for the fact that the actual money paid is somewhat higher than in Plan A.

Lottery payments are another good example of the need to find the present value of an annuity. ■

■ **Example 6** A state lottery advertised a $1,000,000 prize to be paid in 20 yearly installments of $50,000 each. If money can be invested at $8\frac{1}{2}\%$ compounded annually, how much money must be placed in a fund to cover the payments?

$$i = 8\frac{1}{2}\%$$

$$n = 20$$

Table C-8 value = 9.46333661

Present value = 50,000 × 9.46333661

= $473,166.83

If the state invests $473,166.83 (less than half of the $1,000,000 they have to pay) they will have enough money to make payments of $50,000 a year for 20 years. ■

Chapter 11

Name   Date   Class

TRY THESE PROBLEMS (Set II)

Find the present value required in order to receive each of the following annuities, the total amount each annuity would pay, and the interest included in each.

	PERIODIC PAYMENT	COMPOUND INTEREST RATE	TIME
1.	$600	8% annually	10 years
2.	$500	7% semiannually	20 years
3.	$200	10% quarterly	12 years

	RENT	COMPOUND INTEREST RATE	TIME
4.	$350	$10\frac{1}{2}$% monthly	4 years
5.	$100	11% quarterly	8 years
6.	$3,000	9% quarterly	9 years

7. Rose Anthony would like to receive $200 a month during the 4 years she is in college. Her mutual fund pays 8% compounded monthly.

   a. How much would Rose need on deposit when she starts college in order to withdraw $200 a month?

   b. What will be the total amount of Rose's withdrawals?

   c. How much of the amount withdrawn will be interest?

441

8. How large a fund is required to pay $4,000 annually for 25 years if money can be invested at 8% compounded annually?

9. A debt is repaid with 12 quarterly installments of $100 each. If money is worth 10% compounded quarterly, what is the cash equivalent price?

10. An industrial warehouse was purchased for $10,000 down and monthly payments of $900 for 4 years.
    a. What was the total cost to the buyer?

    b. If money is worth 9% compounded monthly, what is the cash equivalent price?

11. What is the cash value of a $200-a-month salary increase during the first year if money is worth 7% compounded monthly?

12. Roberta Greene would like to set up an annuity that will pay her $500 each quarter for 12 years after she retires.
    a. If money is worth 12% compounded quarterly, how much must be in the fund at retirement to provide this annuity?

    b. If Roberta will retire in 20 years, how much should she deposit today in order to have the necessary amount at retirement?

    c. How much interest would be earned if she made this deposit today and then collected the annuity after retirement?

13. Mike Stevens will retire in 12 years. He wants to receive $2,000 each 6 months for 15 years after retirement.
    a. If money is worth 10% compounded semiannually, how much must he have on deposit at retirement in order to receive the payments?

    b. What amount deposited today will provide the necessary funds for the annuity?

    c. How much interest will be earned?

11.4 FINDING RENT: PRESENT VALUE OF ANNUITY KNOWN

A loan is **amortized** if both the principal and interest are paid off by a series of equal periodic payments. A car loan paid with a series of equal monthly payments is an example of a loan that is amortized. The present value of the debt is the amount borrowed, and the series of payments that reduces the debt to zero is the rent.

Another familiar example of amortizing a long-term debt is the series of payments made on a home mortgage, which frequently runs for 15 or 30 years. The payments are usually made each month. Part of each payment is first used to pay the interest due, and the remainder is deducted from the balance (principal) of the debt (United States rule). The interest rate on a home mortgage is always quoted as a percent per year and it is also an accurate, or annual percentage, rate.*

Amortization is a variation of the present value of an annuity procedure already studied because the entire debt exists at the beginning of the time period, and the problem is to determine what periodic payment is needed to reduce the debt to zero at the end of the term. The amount owed is the present value of the annuity, and the payment may be found by using the formula for present value: (A_n) = Rent × Table C-8 value. This formula can be solved for the rent, R;† however, to avoid division we will use a different table. Table C-9, Periodic Rent of Annuity Whose Present Value Is 1, gives the periodic payment (rent) of an annuity when the present value is $1.

To find the periodic payment when the present value of an annuity is known, multiply the present value by the appropriate value from Table C-9.

> Rent = Present value × Table C-9 value

The total amount paid to amortize a loan is found by multiplying the payment (rent) by the number n of payments made.

Total paid = Rent × Number of payments

The interest is the difference between the total paid and the original loan.

Interest = Total paid − Original loan

■ **Example 1**

Nicole King borrowed $20,000 to build an addition to her home. She will repay the loan in monthly payments for 4 years. Interest is at 9% compounded monthly.

1. How much is the monthly payment?
2. What is the total amount paid?
3. How much interest is included in the payments?

1.
$$i = \frac{9\%}{12} = \frac{3}{4}\%$$

$$n = 12 \times 4 = 48 \text{ periods}$$

Table C-9 value = 0.02488504

Rent = Present value × Table C-9 value

= 20,000 × 0.02488504

Rent = $497.70

---

*Annual percentage rates are discussed in Chapter 9.

†$A_n = R a_{\overline{n}|i}$

$R = \dfrac{A_n}{a_{\overline{n}|i}} = A_n \times \dfrac{1}{a_{\overline{n}|i}}$

$R = A_n \times$ Table C-9 value

444 CHAPTER 11 ANNUITIES

The loan will be amortized by monthly payments of $497.70 for 4 years.

2.
$$\text{Total paid} = \text{Rent} \times \text{Number of payments}$$
$$= 497.70 \times 48$$
$$= \$23,889.60$$

There will be 48 payments of $497.70. The total paid (principal plus interest) will be $23,889.60.

3. Interest will be computed each month at $\frac{3}{4}$% on the balance still owed. The total interest will be

$$\text{Interest} = \text{Total paid} - \text{Original loan}$$
$$= \$23,889.60 - 20,000$$
$$= \$3,889.60 \quad \blacksquare$$

■ **Example 2**   Realty Co. sold a piece of property for $60,000. A down payment of $10,000 was made, and the remainder is to be paid in equal semiannual installments, the first due 6 months after the date of sale. The interest is 8% compounded semiannually, and the debt is to be amortized in 5 years.

1. What semiannual payment is required?
2. What will be the total amount of the payments?
3. How much interest will be paid?
4. What is the total cost of the property?

1. The debt of $50,000 ($60,000 − $10,000) is the present value of an annuity for which the rent must be found. Since the first payment is due at the end of 6 months, this is an ordinary annuity.

$$i = \frac{8\%}{2} = 4\%$$

$$n = 2 \times 5 = 10$$

Table C-9 value = 0.12329094

$$\text{Rent} = \text{Present value} \times \text{Table C-9 value}$$
$$= 50,000 \times 0.12329094$$
$$\text{Rent} = \$6,164.55$$

The debt will be paid off in 5 years by semiannual installments of $6,164.55.

2.
$$\text{Total paid} = \text{Rent} \times \text{Number of payments}$$
$$= \$6,164.55 \times 10$$
$$= \$61,645.50$$

The total amount of the payments will be $61,645.50.

3.
$$\text{Interest} = \text{Total paid} - \text{Original loan}$$
$$= 61,645.50 - 50,000$$
$$= \$11,645.50$$

The interest paid will be $11,645.50.

4. The total cost of the property is found by adding the down payment to the total paid to amortize the loan.

$$\text{Total cost} = 10{,}000 + 61{,}645.50$$
$$= \$71{,}645.50$$

The total cost of the property is $71,645.50

Persons making payments to amortize a loan are often given an **amortization schedule** that shows how much of each payment goes toward the principal, how much is interest, and the principal still owed. Table 11-1 illustrates an actual amortization schedule of a $134,000 loan at $7\frac{1}{2}\%$ interest for 25 years. The monthly payment is $990.25. It is interesting to note that a total of $297,074.45 will be repaid. This includes the original principal of $134,000 and interest of $163,074.45.

TABLE 11-1 AMORTIZATION SCHEDULE

RATE: 7.5%   PAYMENT: $990.25   LOAN: $134,000

PAYMENT NUMBER	PAYMENT ON INTEREST	PRINCIPAL	BALANCE OF LOAN
1	$837.50	$152.75	$133,847.25
2	$836.55	$153.70	$133,693.55
3	$835.58	$154.66	$133,538.89
4	$834.62	$155.63	$133,383.26
5	$833.65	$156.60	$133,226.65
61	$768.26	$221.99	$122,699.63
62	$766.87	$223.38	$122,476.25
63	$765.48	$224.77	$122,251.48
64	$764.07	$226.18	$122,025.31
65	$762.66	$227.59	$121,797.72
181	$521.40	$468.85	$82,954.35
182	$518.46	$471.78	$82,482.57
183	$515.52	$474.73	$82,007.83
184	$512.55	$477.70	$81,530.13
185	$509.56	$480.68	$81,049.45
231	$350.02	$640.22	$55,363.75
232	$346.02	$644.22	$54,719.52
233	$342.00	$648.25	$54,071.27
234	$337.95	$652.30	$53,418.97
235	$333.87	$656.38	$52,762.59
291	$59.82	$930.43	$8,639.99
292	$54.00	$936.25	$7,703.74
293	$48.15	$942.10	$6,761.64
294	$42.26	$947.99	$5,813.66
295	$36.34	$953.91	$4,859.74
296	$30.37	$959.87	$3,899.87
297	$24.37	$965.87	$2,933.99
298	$18.34	$971.91	$1,962.08
299	$12.26	$977.99	$984.10
300	$6.15	$984.10	$0.00
Total	$163,074.45	$134,000.00	

446 CHAPTER 11 ANNUITIES

The amount of interest, which is more than the original loan, is not unusual in long-term loans. ■

■ **Example 3** Prepare an amortization schedule for Example 2, in which a semiannual payment of $6,164.55 will amortize a $50,000 loan at 8% for 5 years.

The *interest for the period* is found by using the simple interest formula. For payment 1,

$$I = Prt$$
$$= 50,000 \times 0.08 \times \frac{1}{2}$$
$$I = \$2,000$$

The *amount to principal* is found by subtracting the interest for the period from the amount of payment.

$$\text{Amount to principal} = \text{Payment} - \text{Interest for Period}$$
$$= 6,164.55 - 2,000$$
$$\text{Amount to principal} = \$4,164.55$$

The *principal at the end of the period* is found by subtracting the amount to principal from the previous principal at the end of the period.

$$\text{Principal at end of period} = \text{Previous principal} - \text{Amount to principal}$$
$$= 50,000 - 4,164.55$$
$$= \$45,835.45$$

For payment 2,

$$I = 45,835.45 \times 0.04 = \$1,833.42$$
$$\text{Amount to principal} = 6,164.55 - 1,833.42 = \$4,331.13$$
$$\text{Principal at end of period} = 45,835.45 - 4,331.13 = \$41,504.32$$

Since interest is rounded to the nearest cent for each of these periods, it is not unusual to find that the totals are off by a few cents. Lenders typically adjust the final payment in order to have the correct amounts.

PAYMENT NUMBER	AMOUNT OF PAYMENT	INTEREST FOR PERIOD	AMOUNT OF PRINCIPAL	PRINCIPAL AT END OF PERIOD
0	—	—	—	$50,000.00
1	$6,164.55	$2,000.00	$4,164.55	45,835.45
2	6,164.55	1,833.42	4,331.13	41,504.32
3	6,164.55	1,660.17	4,504.38	36,999.94
4	6,164.55	1,480.00	4,684.55	32,315.39
5	6,164.55	1,292.62	4,871.93	27,443.45
6	6,164.55	1,097.74	5,066.81	22,376.64
7	6,164.55	895.07	5,269.48	17,107.16
8	6,164.55	684.29	5,480.26	11,626.89
9	6,164.55	465.08	5,699.47	5,927.42
10	6,164.55	237.10	5,927.45	(0.03)
		$11,645.47	$50,000.03	

11.5 FINDING RENT: AMOUNT OF ANNUITY KNOWN

A frequent reason for annuities is to establish a fund that will have a given value at the end of the term. For example, suppose some new parents decide that they want to put aside enough each year so that at the end of 18 years there would be $50,000 in a college fund for their child. The problem is now to find the rent of the annuity, the sum that must be deposited each period to give the required $50,000 at the end of the term.

It can be shown mathematically that the values in Table C-9 that were used to find rent when the *present value was known* can also be used to find rent when the *amount of the annuity is known*. The value of i is subtracted from the Table C-9 value.

> Rent = Amount of annuity × (Table C-9 value − Interest rate per period)

■ **Example 1**

How much must be deposited each 6 months in an investment earning 9% compounded semi-annually to have $50,000 at the end of 18 years?

$$i = \frac{9\%}{2} = 4\frac{1}{2}\% = 0.045$$

$$n = 2 \times 18 = 36$$

Table C-9 value − i = 0.05660578 − 0.045

Table C-9 value − i = 0.01160578

Rent = 50,000 × 0.01160578

Rent = $580.29

$580.29 must be deposited twice each year to have $50,000 at the end of 18 years.

When a corporation or government uses an annuity to build up a fund to pay a debt when it is due, the fund is called a **sinking fund**. A sinking fund also may be established to replace machinery or equipment at the end of the depreciation period. ■

■ **Example 2**

A corporation will need to replace a $60,000 machine after 8 years of use. Allowing for inflation, the management decides to establish a sinking fund of $75,000 to replace the equipment at the end of its useful life.

1. How much must they deposit each quarter in an account paying 9% compounded quarterly?
2. How much will the total amount of deposits be?
3. How much interest will the sinking fund earn?

1. The amount of the annuity is $75,000. The rent must be found.

$$i = \frac{9\%}{4} = 2\frac{1}{4}\% = 0.0225$$

$$n = 4 \times 8 = 32 \text{ periods}$$

Rent = Amount of annuity × (Table C-9 value − i)

= 75,000 × (0.04417415 − 0.0225)

= 75,000 × 0.02167415

Rent = $1,625.56

The corporation will have to deposit $1,625.56 each quarter to have the necessary $75,000 at the end of 8 years.

448 CHAPTER 11 ANNUITIES

2.
$$\text{Total deposits} = \text{Rent} \times \text{Number of payments}$$
$$= \$1{,}625.56 \times 32$$
$$= \$52{,}017.92$$

There will be 32 deposits of $1,625.56 each. Therefore, the total amount deposited is $52,017.92

Check: Since each payment is earning interest until the end of the term, the total deposits should be smaller than the amount of the annuity.

3. The interest earned is found by subtraction.
$$\text{Interest} = \text{Amount of annuity} - \text{Total deposits}$$
$$= 75{,}000 - \$52{,}017.92$$
$$= \$22{,}982.08$$

While a sinking fund is in progress, a **sinking fund table** may be prepared to keep track of the amount in the fund. This table shows how much interest the fund has earned during each period and the current balance. ∎

■ **Example 3**

How much must be deposited each year in an account paying 8% compounded annually to have $2,000 at the end of 5 years?

$$i = \frac{8\%}{1} = 8\% = 0.08$$
$$n = 1 \times 5 = 5$$
Table C-9 value $- i = 0.2504564 - 0.08$
Table C-9 value $- i = 0.1704564$
$$\text{Rent} = 2{,}000 \times 0.1704564$$
$$\text{Rent} = \$340.91$$

$340.91 must be deposited each year to have $2,000 at the end of 5 years.

A sinking fund table for these deposits is shown below.

PAYMENT NUMBER	INTEREST EARNED	AMOUNT OF PAYMENT	YEARLY INCREASE	TOTAL IN ACCOUNT
1	$ 0	$ 340.91	$340.91	$ 340.91
2	27.27	340.91	368.18	709.09
3	56.73	340.91	397.64	1,106.73
4	88.54	340.91	429.45	1,536.18
5	122.89	340.91	463.80	1,999.98
	$295.43	$1,704.55		

The interest earned for each period is found by using the simple interest formula. For example, for payment 3:
$$I = Prt = 709.09 \times 0.08 \times 1 = \$56.73$$

Note that the final total in the account is $.02 less than the $2,000 that is needed because of rounding error. ∎

Relationship Among Annuity Values

We have considered several values of annuities in this chapter and compared magnitudes to check calculated results. These relationships can now be summarized in the following inequality:

Present value < Total payments < Amount (< means "is less than.")

Chapter 11

TRY THESE PROBLEMS (Set III)

Find the rent for the following annuities and calculate the total payments and the interest earned.

	PRESENT VALUE	TERM	INTEREST RATE
1.	$3,000	4 years	12% compounded monthly
2.	25,000	10 years	10% compounded quarterly
3.	5,000	13 years	$8\frac{1}{2}$% compounded annually
4.	60,000	25 years	6% compounded semiannually

	MATURITY VALUE	TERM	INTEREST RATE
5.	$10,000	15 years	5% compounded semiannually
6.	4,500	6 years	11% compounded quarterly
7.	60,000	45 years	7% compounded annually
8.	7,000	$3\frac{1}{2}$ years	9% compounded monthly

9. A retired person has $50,000 in an investment paying 10% compounded semiannually. If he wishes the fund to provide payments twice a year for 20 years, how much can he withdraw each time?

10. A $9,000 used car is sold with a down payment of $1,000 and quarterly payments for 4 years. If interest at 12% compounded quarterly is included, how much is each payment? How much interest is paid?

11. a. How much must be deposited each month at 9% compounded monthly to have $15,000 at the end of 4 years?

   b. What is the total amount deposited?

   c. How much interest is earned?

12. A company must pay an outstanding debt of $50,000, 10 years from now.

   a. How much must they deposit each quarter at 9% compounded quarterly to have the required amount in 10 years?

   b. How much interest will be earned?

13. Sally Small borrowed $10,000 to remodel her home. She will repay the loan in quarterly payments for 11 years. Interest is 8% compounded quarterly.

   a. What is Sally's quarterly payment?

   b. How much interest will be paid?

11.6 REVIEW OF COMPOUND INTEREST AND ANNUITIES

The following review should help to distinguish among the different types of problems presented in Chapter 10, "Compound Interest," and Chapter 11, "Annuities." Although all annuities we have studied receive compound interest, for purposes of differentiation we will use *compound interest* to classify one type of problem and *annuities* to classify another.

First decide whether the problem is a compound interest problem or an annuity problem. A compound interest problem involves only *one payment or deposit;* an annuity problem involves *a series of payments or deposits.* Which table to use depends on what is given in the problem and what is to be found. The following discussion and Figure 11-3 should clarify this procedure.

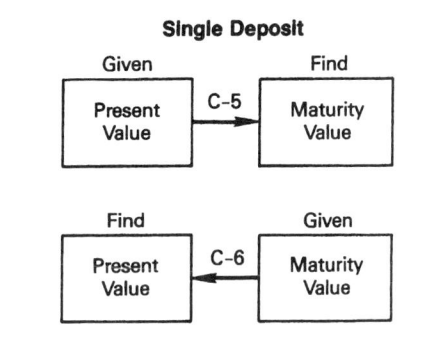

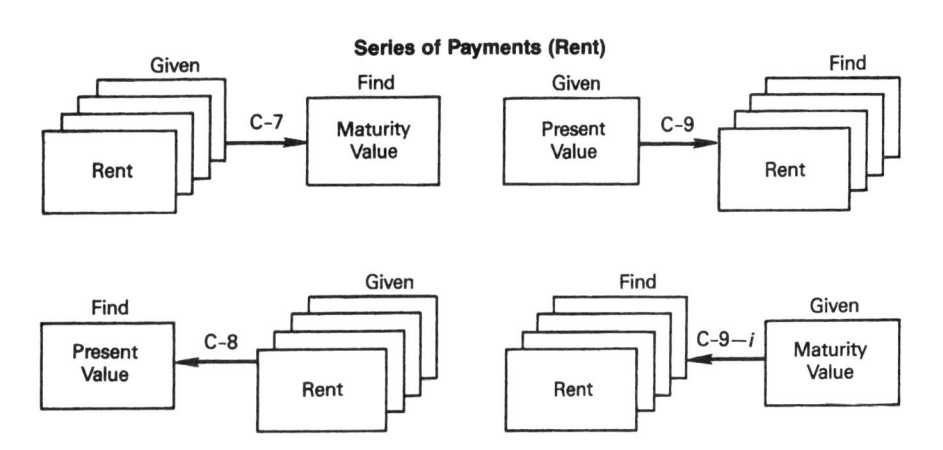

FIGURE 11-3 Single deposit and series of deposits.

Compound Interest Problems

If a *single deposit* is involved, determine whether the maturity value (the amount at the end of the investment) or the present value (the amount at the beginning of the investment) is given.

1. Use Table C-5 if the amount at the beginning of the period is given and the maturity value is to be found. For example, if $1,000 is deposited at 9% compounded monthly, how much will it be worth in 10 years?

2. Use Table C-6 if the maturity value is given and the present value must be found. For example, what amount must be invested at 9% compounded monthly in order to have $1,000 in 10 years?

Annuity Problems

If a *series of payments* is involved, determine first whether the present value, maturity value, or rent is given in the problem.

1. Use Table C-7 if the periodic payment (rent) is given and you wish to find the maturity value of a series of payments. For example, $1,000 is deposited each month in an

account paying 9% compounded monthly. What will be the value of the account after 10 years?

2. Use Table C-8 if the periodic payment is given and you wish to find the amount needed at the beginning of the period to provide these payments. For example, the Smiths want to withdraw $1,000 a month for 10 years after their retirement. How much must they have in their account at retirement if interest is 9% compounded monthly?

3. Use Table C-9 when you know the value at the beginning of the period and you want to find the periodic payment that will reduce this amount to zero at the end of the period. This procedure is used to amortize a debt. For example, Bob Smith borrowed $9,000 and wishes to repay the debt in monthly payments for 10 years. If interest is 9% compounded monthly, how large must each payment be?

4. Use (Table C-9 – i) when you are given the maturity value needed at end of the period and you wish to find the regular payment needed to reach this goal. The business application of this procedure is the sinking fund. For example, Bob Smith wishes to have $9,000 in 10 years. If money is worth 9% compounded monthly, what monthly payment must he make in order to reach his goal?

Now go to Try These Problems (Set IV).

NEW FORMULAS IN THIS CHAPTER

- Amount of an annuity = Rent × Table C-7 value

$$S_n = R \times s_{\overline{n}|i}$$

- Present value of an annuity = Rent × Table C-8 value

$$A_n = R \times a_{\overline{n}|i}$$

- Cash equivalent price = Down payment + Present value of annuity

- When present value is known:

$$\text{Rent} = \text{Present value} \times \text{Table C-9 value}$$

- When amount is known:

$$\text{Rent} = \text{Amount of annuity} \times (\text{Table C-9 value} - i)$$

Chapter 11

TRY THESE PROBLEMS (Set IV)

1. If money is worth 9% compounded semiannually:

   a. How much must be invested today so that you can receive $900 twice a year for the next 5 years?

   b. What will investments of $900 twice a year be worth after 5 years?

   c. What semiannual payments must be made so that you will have $900 in 5 years?

   d. What semiannual payments must be made to pay off a loan of $900 in 5 years?

   e. What amount must be invested today in order to have $900 in 5 years?

   f. What will a $900 investment be worth in 5 years?

   g. Find the interest for parts a through f.

2. The Taylor family is considering various investment possibilities to help finance their children's college educations. They asked a friend who works in a bank to answer the questions that follow. The bank pays 8% compounded quarterly.

   a. If they deposit $10,000 at the bank today, how much will they have in 12 years?

   b. If they want to have $20,000 in the bank at the end of 12 years:

      (i) How much must they deposit today?

      (ii) How much must they deposit each quarter?

   c. If they deposit $250 each quarter, how much will they have in 12 years?

   d. What quarterly payment must they make if they borrow $20,000 to meet their expenses and amortize the debt in 12 years?

   e. If their oldest child were to begin college today, how much must be on deposit for her to receive $800 each quarter for the 10 years it will take her to become a doctor?

Chapter 11

Name _____ Date _____ Class _____

END-OF-CHAPTER PROBLEMS

Find the missing quantities for each of the following:

	RENT	COMPOUND INTEREST RATE	TIME	AMOUNT	INTEREST
1.	$140	10% quarterly	12 years		
2.		7% monthly	$2\frac{1}{2}$ years	$6,500	
3.	$300	11% semiannually	15 years		
4.	$90	9% monthly	$3\frac{3}{4}$ years		
5.		8% annually	50 years	$60,000	
6.		9% quarterly	10 years	$4,000	

Find the missing quantities for each of the following:

	PERIODIC PAYMENT	COMPOUND INTEREST RATE	TIME	PRESENT VALUE	INTEREST
7.	$200	5% monthly	3 years		
8.		8% quarterly	8 years	$3,000	
9.	$400	10% quarterly	9 years		
10.	$600	9% semiannually	10 years		
11.		8% monthly	3 years	$16,000	
12.		11% quarterly	$7\frac{1}{2}$ years	$22,000	

455

13. A deposit of $300 is made each quarter in an account paying 7% compounded quarterly. If the deposits continue for 7 years:

    a. How much will there be in the account at the end of the period?

    b. How much of the total will be interest?

14. A company executive invests 20% of his $3,000 monthly salary each month for 4 years. If the money earns 8% interest compounded monthly, how much will he have in the account at the end of the term?

15. A college fund is established for a young child. Deposits of $400 are made quarterly for 12 years, and interest is paid at 11% compounded quarterly. The child doesn't actually start college or touch the fund for another 2 years.

    a. What is the amount in the fund?

    b. How much is interest?

16. A college account was established for a child when she was 10 years old. A deposit of $500 was made twice a year in an investment that paid 8% compounded semiannually. When the child was 18 years old the deposits were stopped, but she decided not to go to college until the following year. The fund remained untouched at the same interest rate for an additional year. How much did she have when she started college? How much interest was earned?

17. A city puts $40,000 a year in a special fund paying $8\frac{1}{2}\%$ interest compounded annually.

    a. How much will the fund contain at the end of 20 years?

    b. How much of the total will be interest?

18. Layton Industries is investing $3,000 each quarter to modernize their equipment in 10 years. Their investment earns 9% compounded quarterly.

    a. How much will they have in 10 years?

    b. How much interest will have been earned?

Chapter 11

Name	Date	Class

END-OF-CHAPTER PROBLEMS

19. Louis Max is 40 years old. He deposits $2,000 each year into a retirement account that pays 8% compounded annually.

    a. How much will there be in the account when Mr. Max reaches the age of 65 and is ready to retire?

    b. How much interest will have been earned?

20. An insurance company must settle a claim by paying $7,500 a year for the next 20 years. If money can be invested at $8\frac{1}{2}\%$ compounded annually, how much must they deposit today so they can make the payments?

21. Stan Gordon would like to receive $10,000 annually for 15 years after he retires. He can invest in an account paying $8\frac{1}{2}\%$ compounded annually.

    a. How much must Stan's account contain when he retires in order for him to receive these payments?

    b. How much will he receive after retirement?

    c. How much interest will the account be earning while he is receiving these payments?

22. A debt is to be paid off with 15 monthly payments of $110 each. If money is worth 12% compounded monthly, how much cash would give the lender the equivalent of the annuity?

23. Jim Atkins will enter college in 5 years. At that time his parents want him to be able to withdraw $3,000 each 6 months for 4 years. They can invest money at 7% compounded semiannually.

    a. What amount must be on deposit when he starts college so that he can receive these payments?

    b. How much should his parents deposit today so that he can receive these payments?

    c. How much interest will this deposit earn by the time Jim finishes college?

24. As part of her contract with Ace Sales, the new sales manager is guaranteed an annual retirement bonus of $2,000 every quarter for 8 years. Her retirement will start in 6 years. The company can invest money at 12% compounded quarterly.

   a. How much must be on deposit when she retires so she can receive her bonus?

   b. How much must Ace Sales invest today to provide this retirement fund?

   c. How much interest will be earned by the time the annuity ends?

25. Linda Lucky won a Two Million Dollar Lottery. She is to be paid a $50,000 installment every 6 months for 20 years. If money can be invested at 6% compounded semiannually, how much should the lottery agency invest today to provide sufficient funds for these payments?

26. Billy Bully signs a professional sports contract. As a signing bonus, he is given an option of $120,000 cash now or a bonus of $11,000 a year for 12 years. If money is worth 8% compounded annually:

   a. Which offer is the best?

   b. How much difference is there between the two options?

27. Sid Saver deposited $30,000 in an account paying 8% compounded annually. He left the money for 10 years and then put the money in a new account paying 9% compounded quarterly. If he plans to liquidate the account in 11 years, how much can he draw from it quarterly?

28. Bob Coker bought a $132,000 co-op apartment by paying $12,000 down and signing a 25-year mortgage at 9% compounded semiannually.

   a. How much is each payment?

   b. How much interest will Bob pay?

   c. What will the total cost of the apartment be?

Chapter 11

END-OF-CHAPTER PROBLEMS

29. A delivery truck cost $30,000. After a down payment of $8,000 the balance is financed with monthly payments at 9% compounded monthly for 4 years. How large is each monthly payment?

30. A debt of $5,000 is to be paid off in 3 years with equal monthly payments. If the interest charged is $7\frac{1}{2}\%$ compounded monthly, how large must the payments be?

31. A couple's retirement plan includes an annual withdrawal from a bank account. The account balance is $80,000 at the beginning of the retirement period, and the interest rate is 8% compounded annually. At the time of retirement they plan on using up the money completely in 15 years. How much can they draw from it each year?

32. A house is bought with a down payment of $30,000 and a $90,000 mortgage with 7% interest. The mortgage is to be amortized by quarterly payments for 10 years. How large should the payments be?

33. A farm was bought for a down payment of $40,000 and a mortgage of $200,000, which will be amortized by quarterly payments for 12 years. If the payments include interest at 7% compounded quarterly:

   a. How large is each payment?

   b. How much is paid altogether?

   c. How much is interest?

34. A company will need to replace a machine at the end of 8 years. They expect the machine to cost about $62,000, and they decide to establish a sinking fund to cover the cost. If they can get 6% compounded quarterly, how much must they deposit each quarter to have the money to replace the machine?

35. How much must be deposited each quarter for 4 years at 10% compounded quarterly to amount to $8,000?

36. Phyllis and Harold Winkler want to have $10,000 when they retire in 7 years to take a trip around the world.

   a. How much must they save each quarter to reach their goal if money is worth 9% compounded quarterly?

   b. How much interest will they earn on their savings?

37. Irene David wants to save money to buy a $20,000 car 4 years from now. How much money should she deposit each month in an account paying 7% compounded monthly so that she will have the $20,000 for the car? How much interest will she earn?

38. Cracker Films wants to buy a new high definition TV set 3 years from now. They believe the set will cost $2,500 and the bank pays interest at 8% compounded quarterly. The company wants to deposit an equal amount at the end of each quarter for the next 3 years in order to accumulate enough money to buy the TV. How large should each deposit be?

39. Hadley and Padley expect to need $15,000 in 5 years to purchase new office equipment. Their bank pays 10% compounded quarterly.

   a. What quarterly payment should be made so that they will have the necessary funds in 5 years?

   b. What will their total deposits be?

   c. How much interest will they have earned?

40. A fund is established to provide an annuity of $500 twice a year for 10 years. Payments into the fund are made semiannually for 8 years to reach the required amount. If money is worth 10% compounded semiannually:

   a. How much must the fund have in it when payments begin?

   b. How large is the periodic payment needed to establish the fund?

41. Lisa Turner invested $5,000 when her son was born. How much will be available when her son is 21 years old if the investment pays 7% compounded annually? How much interest will have been earned?

42. First City Bank pays 8% interest, compounded quarterly, to investors. If Rita Rogers invests $7,500, how much will she have at the end of $6\frac{1}{2}$ years?

Chapter 11

END-OF-CHAPTER PROBLEMS

43. Willy Gamble is scheduled to receive $8,500 in 11 years. If money is worth $8\frac{1}{2}\%$ compounded annually, how much is this inheritance worth today?

44. A man owns a note for $1,000 due in 5 years. What should a buyer pay for this note if money is worth 8% compounded quarterly?

45. Susie Cohen wants to add an extra room to her house in 5 years. She estimates it will cost $30,000.

  a. How much should she deposit if she wants to have the needed amount available in 5 years? Money is worth 5% compounded quarterly.

  b. How much interest will her deposit earn?

46. Buddy Mentin invested $8,000 in a fund that was paying $5\frac{1}{2}\%$ compounded semiannually.

  a. How much will his investment be worth after 10 years?

  b. How much interest will he have earned?

Find the periodic payment necessary to amortize the following loans. Then prepare an amortization schedule showing how each payment is divided between interest and outstanding principal.

	PRESENT VALUE	TERM	INTEREST RATE
47.	$5,000	2 years	10% compounded semiannually
48.	$1,000	4 years	8% compounded annually

461

Find the periodic payment needed to finance the following sinking funds. Then prepare a sinking fund table showing that these payments will have the desired maturity value.

	MATURITY VALUE	TERM	INTEREST RATE
49.	$2,000	5 years	$8\frac{1}{2}$% compounded annually
50.	$4,000	2 years	12% compounded quarterly

SUPERPROBLEM

Pete Planwell started saving for retirement when he was 30 years old. He thought that $40,000 a year for 10 years would be enough, with social security and his pension, to permit him to live comfortably. He planned to retire at 67. Interest rates were low when he started and he could not forecast what would happen so far in advance, but he thought 7% compounded annually was a good estimate. If he started the fund with a deposit of $5,000, how much would he have to deposit each year until he reached 67 to provide enough money for his retirement?

Chapter 11

TEST YOURSELF

Find the amount of the annuity and the interest earned.

	RENT	COMPOUND INTEREST RATE	TIME		
1.	$500	$6\frac{1}{2}\%$ annually	9 years		11.2

Find the present value required in order to receive the following annuity and the interest included.

	RENT	COMPOUND INTEREST RATE	TIME		
2.	$1,800	9% monthly	3 years		11.3

Find the rent for the following annuities and calculate the interest earned.

	PRESENT VALUE	TERM	INTEREST RATE		
3.	$80,000	6 years	7% compounded quarterly		11.4

	MATURITY VALUE	TERM	INTEREST RATE		
4.	$27,000	$5\frac{1}{2}$ years	5% compounded semiannually		11.5

5. On her tenth birthday, Lisa's grandfather put $1,000 in an account that earns 5% compounded quarterly. If the principal and interest were left to accumulate, what was the value of the account on Lisa's 21st birthday? How much interest had been earned?   10.2

6. To provide for the eventual retirement of outstanding bonds, Hopewell Corporation invests $20,000 in a special fund each year. If interest is at 8% compounded annually: 11.2

   a. How much will be in the fund at the end of 15 years?

   b. How much will Hopewell have deposited?

   c. How much interest will be earned?

463

	Section

7. Aceto Research Foundation wants to deposit a sufficient sum of money at 10% compounded quarterly in order to be able to withdraw $5,000 every 3 months for the next 5 years.   11.3

   a. What amount should be deposited?

   b. How much will Aceto's total withdrawals be?

   c. How much interest will be earned?

8. The Electric Power and Light Company must repay an outstanding debt of $900,000, 20 years from now. If the interest rate is 6% compounded semiannually: 11.5

   a. How much should the company invest every 6 months?

   b. How much will they deposit in 20 years?

   c. How much interest will the company earn?

9. Harold High borrowed $15,000 and agreed to repay it in equal quarterly installments over a $7\frac{1}{2}$-year period. If the interest rate is 7% compounded quarterly: 11.4

   a. How large is each payment?

   b. How much interest is paid?

10. Judy Bardy plans to modernize her heating system in 4 years at a cost of $4,000. If her savings and loan bank pays 7% compounded quarterly, how much should she deposit now to have $4,000 in 4 years? How much interest will be earned?   10.8

11. Leo Lundy deposited $200 each month in an account paying $7\frac{1}{2}$% compounded monthly. After $3\frac{1}{2}$ years the deposits stopped but Leo forgot about the account until 5 years later.   11.2

   a. How much was in the account when deposits stopped?

   b. What was the value of the account 5 years later?

   c. How much interest had been earned?

12. At retirement, Juan Martinez wishes to deposit enough money so that he can withdraw $15,000 a year for the next 10 years. If money earns $8\frac{1}{2}$% compounded annually: 11.3

   a. How much must he have on deposit at retirement?

   b. He estimates that he will work for the next 15 years. How much should he deposit now so that he will have the necessary funds at retirement?

   c. How much interest will be earned if he makes this deposit and then collects the annuity after retirement?

Chapter 11

Name	Date	Class

COMPUTER SPREADSHEET EXERCISES

Many computer spreadsheet functions have been created to perform the complex calculations associated with annuities. The annuity functions introduced in the following spreadsheets are those used in Microsoft Excel. Be sure to use formulas to compute the missing values and $ signs for $ amounts. Align decimal points in all columns where they are used. All columns and numbers in your spreadsheets should appear as shown. If your spreadsheet is too wide to fit across the width of the page, print lengthwise using the File Page Setup Landscape orientation and the Fit Columns to Page scaling option.

1. Find the missing quantities. The function

    =FV(rate per period,number of compounding periods,payment)

    determines the amount of an ordinary annuity based on a series of equal payments (rent), each earning a specified rate per period over the number of compounding periods in the term. For example, assume you deposit $1,000 every 6 months for the next 20 years into a bank account paying 5% interest, compounded semiannuallly.

    =FV(5%/2,40,−1000)

    returns $67,402.55, the value of your account at the end of 20 years. (Note that the payment must be entered as a negative value.) The calculation assumes that the interest is paid on the last day of each period and you make each payment on the last day of the period.

Rent	Interest Rate	Compounding Frequency (per Year)	Term	Amount of Annuity	Total Deposited	Interest Earned
$5,500	9.00%	1	8.00 years			
$800	7.50%	12	6.75 years			
$200	8.00%	2	12.50 years			
$1,500	6.00%	4	24.25 years			
$650	10.00%	2	10.00 years			
$2,430	5.25%	1	29.00 years			
$1,845	11.75%	4	2.75 years			
$790	6.65%	12	9.40 years			

2. Persons making payments to amortize a loan are often given an amortization schedule that shows how much of each payment goes toward the principal, how much is interest, and the principal still owed. You are to complete the following amortization schedule for a $25,000 loan repaid with monthly payments for 4 years with interest at 8.5% compounded monthly. Installment loans are computed like ordinary annuities in that payments are made at the end of each payment period. Use the PMT function to compute the amount of the monthly payment. The function takes the form

    =PMT(rate per period,number of compounding periods,present value)

    For example,

    =PMT(8%/12,360,−50000)

    returns the value $366.88, which is the monthly payment on a $50,000 mortgage with an interest rate of 8% and a 30-year term. (Don't forget to use a minus sign in front of present value.)

    Be sure to put the values for Rate, Payment, and Loan in separate cells so the cell addresses can be used in formulas. The interest for the period is found by multiplying the loan balance at the end of the previous period by the interest rate per period.

Compute the principal by subtracting the interest for the period from the monthly payment. Give totals for the Interest and Principal columns. For your spreadsheet to have the appearance of the one seen here, some column titles will have to be placed over more than one column. Include the horizontal lines at the top and bottom of your spreadsheet.

```
                        AMORTIZATION SCHEDULE
              Rate: 8.5%    Payment:        Loan: $25,000
                            Payment on
   Payment Number   Interest    Principal    Balance of Loan
         1
         2
         3
         4
         5
         6
         7
         8
         9
        10
        11
        12
        13
        14
        15
        16
        17
        18
        19
        20
        21
        22
        23
        24
        25
        26
        27
        28
        29
        30
        31
        32
        33
        34
        35
        36
        37
        38
        39
        40
        41
        42
        43
        44
        45
        46
        47
        48
      Total
```

3. The function

=PV(rate per period,number of compounding periods,payment)

determines the present value for a series of equal payments, each of a specified amount, discounted at a periodic interest rate, over the number of compounding periods in the term. The payments are made at the end of each compounding period.

Chapter 11

Name | Date | Class

COMPUTER SPREADSHEET EXERCISES

For example, assume you have just won a million-dollar lottery prize. You can either receive 20 annual payments of $50,000 each (a total of $1,000,000 over 20 years) at the end of each year or a lump-sum payment of $400,000 now. You want to find out which option is worth more in today's dollars. If you were to accept the annual payments of $50,000, assume that you could invest the money at a rate of 9% compounded annually. The formula

=PV(9%,20,–50000)

returns $456,427.28. This tells you that the $1,000,000 paid over 20 years is worth $456,427.28 in today's dollars, which is more than the lump-sum payment.

The individuals in the spreadsheet below have won various lottery prizes. You are to fill in the missing numbers and determine the better payout option. The leftmost column shows the names of the lottery winners; however, these names are not in the correct order. Put the names in alphabetical order, but *leave the other columns unchanged.* The Microsoft Excel Sort Ascending button can be used to do this. Divide the lottery amount by the term to find the annual payment. Use the PV function to find the present value of each annuity. You are to write an IF statement to place the words "Lump-sum" or "Annuity" in the Better Payout Option column. Choose "Lump-sum" if the lump-sum payout is greater than or equal to the present value of the annuity. Choose "Annuity" if the opposite is true. Assume that all annuities are ordinary and that the assumed rate of return is an annual compounded rate.

Name	Lottery Amount	Lump-sum Payment	Annual Payment	Term (in Years)	Assumed Rate of Return on Annual Payments	Present Value of Annuity	Better Payout Option
Ungar, F.	$30,000,000	$13,052,600		20	10.00%		
Kocher, B.	$2,000,000	$875,900		25	7.50%		
Bradburn, J.	$12,000,000	$3,097,332		40	9.25%		
Powell, C.	$27,000,000	$9,255,014		30	8.75%		
Quinlan, B.	$11,000,000	$6,133,291		10	12.40%		
Drazda, W.	$60,000,000	$34,897,717		15	7.05%		
Keller, C.	$1,000,000	$309,899		20	14.90%		
Price, L.	$88,000,000	$28,642,865		25	11.50%		
Welsh, E.	$47,000,000	$16,132,442		40	6.75%		
Battle, K.	$5,000,000	$2,077,181		25	8.30%		

12 ■ Understanding Statistics

OBJECTIVES

Upon completion of this chapter, the student should be able to:

1. *Interpret the following types of graphs:*
   bar graph;
   line graph;
   pictograph;
   pie graph;
   component bar graph; and
   statistical map.
2. *Compute the mean, median, and mode.*
3. *Compute the range and standard deviation.*
4. *Compute and interpret simple index numbers.*

Statistics can generally be thought of as the science of collecting, organizing, analyzing, presenting, and interpreting numerical data. Statistical thinking determines the management of the economy and the operations of business to a greater degree than is generally recognized. An understanding of important statistical concepts is essential today for making important business decisions. Statistics are used to inform or misinform us, to persuade or dissuade us, to get us to buy a particular product or to analyze trends. In economics, business, health, politics, education, and practically every other field of human activity, information is often given in numbers and in the form of graphs and charts.

Consider the advantage of a statistical description (one based on numbers) compared with the word descriptions we all use. A student says, "I expect to earn a high salary when I graduate and start to work full-time."

Does this statement have the same meaning for all the student's classmates? How high is a "high" salary? The answers would vary considerably.

But if the statement were "I expect to earn about $1,000 a week when I graduate and start to work full-time," there would be no question of its meaning.

Similarly, we could ask: How tall is a "tall" woman? How large is a "large" class? What time is an "early" start for a day?

If we specify that the woman is 5 feet 9 inches, that the class has 40 students, and that the early start for the day is 6 A.M., we all begin at the same point in any further discussion. The statements are not necessarily true because they contain numbers, but now we all give the same meaning to each statement.

470 CHAPTER 12   UNDERSTANDING STATISTICS

Only a few statistical ideas can be included in one chapter of a textbook, and only those most often used in business and general publications will be included here.

12.1 PRESENTATION OF NUMERICAL FACTS

Read the following paragraphs.

Yields on checking accounts, certificates of deposit, and money market accounts were down for the 6-month period ended yesterday.

The Bate Rate Monitor's national index reported that the average yield for checking accounts was down 22 basis points to 1.26 percent. (A basis point is one-hundredth of a percentage point.) Certificates of deposit yields were all down. A 6-month CD fell 68 basis points, to 4.17 percent, while the $2\frac{1}{2}$-year and 5-year fell 79 and 71 basis points respectively to 4.43 and 4.42 percent.

As of yesterday's Treasury bill auction, . . . the average yield for the 3-month bill was down 65 basis points for the 6-month period, to 4.36 percent, and the rate on 6-month bills was down 66 basis points, to 4.27 percent.

Elsewhere in the funds market, the Bank Rate Monitor said that the average yield for money market accounts fell 23 basis points to 2.76 percent.

If you were bored or confused, and perhaps stopped in the middle, you have discovered why statistical material is presented to only a limited extent in this text form. Now look at Table 12-1 and Figure 12-1 and observe how much clearer and more compact the report is.

A table gives the data more exactly than a graph, and more data can be included without causing confusion. But a graph is more attractive and is used more often when the presentation is for readers who are not at ease with numbers and who are not likely to take much time to examine a table.

Tables

Data are available on the population and economy of the United States from many sources, the most important of which is the federal government. A good reference, with which every business student should be familiar, is the *Statistical Abstract of the United States,* an extensive set of tables that gives basic statistics on every phase of life in the United States and also some international comparisons. Tables 12-2, 12-3, and 12-4 are brief summaries of some of the data published by the federal government.

TABLE 12-1   WEEKLY YIELDS ON CHECKING ACCOUNTS, BANK CDS, TREASURY BILLS, AND MONEY MARKET ACCOUNTS. AUGUST 5, 1998 AND FEBRUARY 5, 1999

	YIELD		
	WEEK ENDING AUGUST 5, 1998	WEEK ENDING FEBRUARY 5, 1999	BASIS POINT DECREASE*
	In Percent		
Checking Accounts	1.48	1.26	22
Certificates of Deposit			
6-month CD	4.85	4.17	68
$2\frac{1}{2}$-year CD	5.22	4.43	79
5-year CD	5.13	4.42	71
Treasury Bills			
3-month	5.01	4.36	65
6-month	5.01	4.27	66
Money Market Accounts	2.99	2.76	23

*A basis point is one one-hundredth of a percentage point
Source: Data from *Bank Rate Monitor* and *The New York Times*

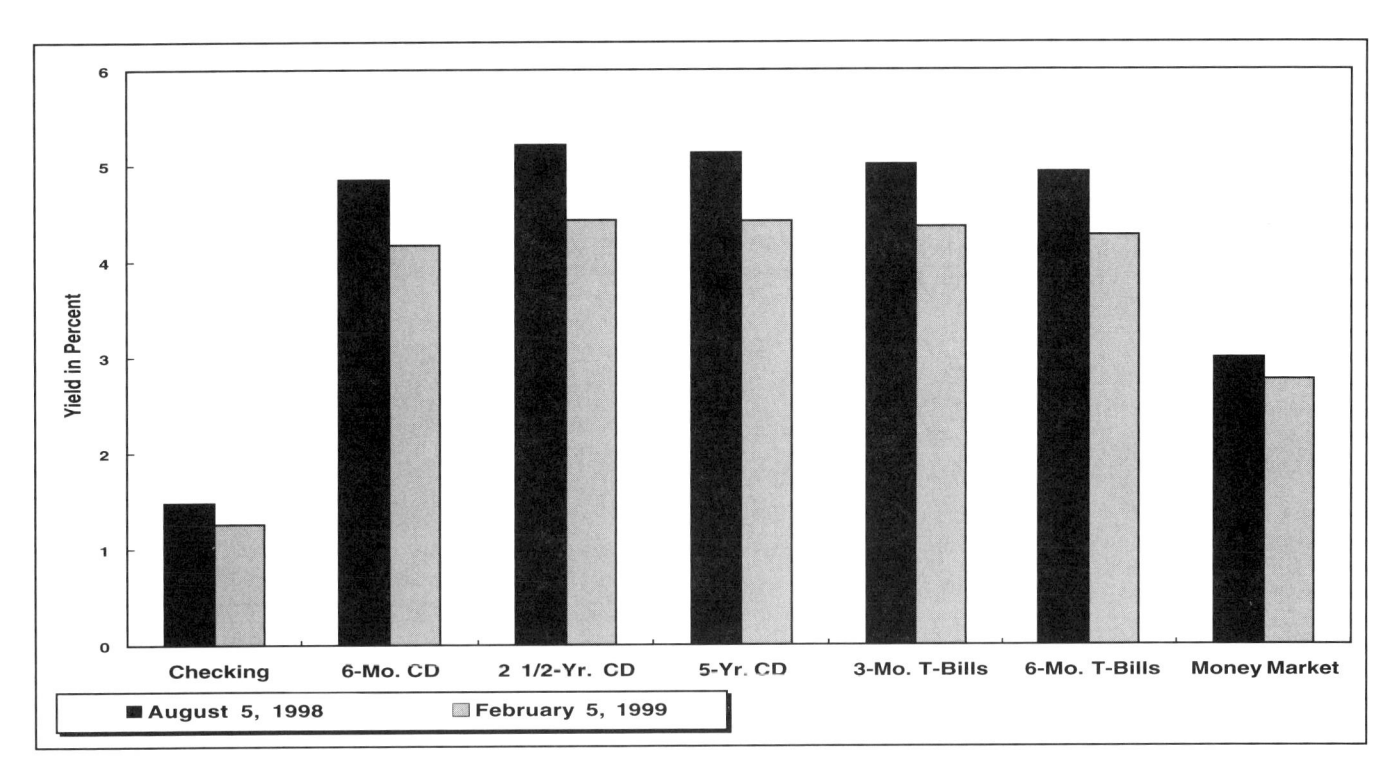

FIGURE 12-1 Weekly yields on checking accounts, bank CDs, Treasury bills, and money market accounts. August 5, 1998 and February 5, 1999. (*Source:* Data from *Bank Rate Monitor* and the *New York Times*.)

TABLE 12-2 POPULATION BY AGE, 1980–1997 (PERCENT OF TOTAL)

AGE	1980	1990	1997
Under 5 years	7.2%	7.5%	7.2%
5–14	15.4%	14.2%	14.4%
15–24	18.7%	14.9%	13.6%
25–34	16.4%	17.3%	14.8%
35–44	11.4%	15.1%	16.5%
45–54	10.1%	10.0%	12.6%
55–64	9.5%	8.5%	8.2%
65 and over	11.3%	12.5%	12.7%
	100.0%	100.0%	100.0%

Source: U.S. Bureau of Census.

TABLE 12-3 PERSONAL CONSUMPTION EXPENDITURES

	BILLION DOLLARS		PERCENT OF TOTAL	
	1990	1997	1990	1997
Durable goods	477	659	12.4%	12.0%
Nondurable goods	1,245	1,592	32.4%	29.0%
Services	2,118	3,235	55.2%	59.0%
Total	3,840	5,486	100.0%	100.0%

Source: U.S. Bureau of Economic Analysis.

Table 12-2, which shows the population of the United States by age groups and the corresponding percents of the total, has implications for business and government. It illustrates, for example, why the social security system has needed to increase the payment rate and the wage ceiling. The proportion of the population 65 years old and over has increased from 11.3% to 12.7% in 17 years, and this trend is expected to accelerate. At the other end of the scale, there has been a sharp drop in the proportion of those 24 years old and younger—from 41.3% to 35.2%. Thus, there will be a smaller proportion of the population contributing to the system and a larger proportion drawing benefits.

For the business community, the population data imply less expansion in markets for baby, youth, teenage, and young-adult products, and more for items such as medicines and retirement homes. For politicians and advertisers, the changing age composition may require changes in methods of appeal. In 1980, the average age of the population was 30.0 years, whereas in 1997 the average had increased to 34.9.

Table 12-3 shows how we spend our money. The proportion spent for durable goods (like cars, refrigerators, and television sets) has changed only a little in the 7 years recorded in the table. The proportion spent on services (e.g., schools, lawyers, doctors, and auto mechanics), however, has increased mostly at the expense of money spent for nondurable goods (food, clothing, and other products that do not last long).

TABLE 12-4 FASTEST GROWING OCCUPATIONS, BY PERCENT

OCCUPATION	PERCENT INCREASE
Database administrators, computer specialists	118%
Computer engineers	109%
Systems analysts	103%
Personal and home care aides	85%
Physical and corrective therapy assistants	79%
Home health aides	76%
Medical assistants	74%
Desktop publishing specialists	74%
Physical therapists	71%
Occupational therapy assistants	69%
Paralegals	68%
Occupational therapists	66%
Teachers, special education	59%
Human services workers	55%
Data processing equipment repairers	52%

Source: U.S. Bureau of Labor Statistics

Line Graphs

Table 12-4 shows changes in job opportunities expected by the year 2006. This list includes only those occupations that are expected to show the fastest growth. Most of the occupations that are projected to undergo a large percent increase are associated with computers or health care.

Figure 12-2 illustrates a simple **line graph.** It shows how many imported cars were sold in the U.S. each year from 1987 through 1997. The horizontal scale has the years marked, and the vertical scale is in millions of cars. It shows that sales of imported passenger cars have steadily decreased since 1987, with only a slight increase in 1997 disrupting the trend.

To "read" the graph—that is, to find the sales of imported cars for a particular year:

- Locate the year on the horizontal axis.
- Move up vertically until you meet the graph.
- Go across to the vertical axis.
- Read the value from the scale.

The broken lines in the graph illustrate how to find the sales total for 1995. The point giving this value falls between 1.5 and 2 and can be approximated at 1.6 million cars.

Now look at Figure 12-3. Although it looks different, it is based on the same information. The fall in imported car sales looks sharper because the bottom of the graph has been cut off and the scale starts at 1 instead of at zero. This is a common error that tends to distort the data. Sometimes this error is made just to save space by eliminating the unused portion of the graph, but sometimes it is done with the intention of visually distorting the data to produce a particular effect. The accepted solution to the frequent need to save space is shown in Figure 12-4. The break in the scale says to the reader, "A part has been omitted; read carefully."

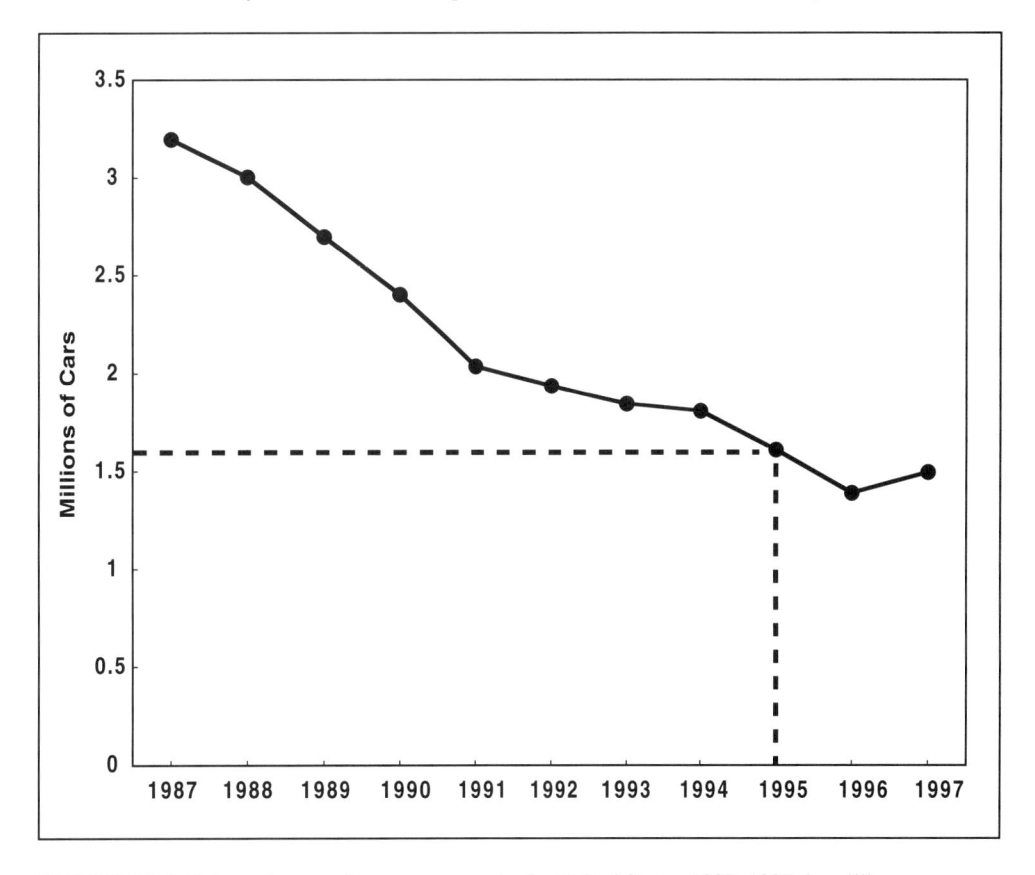

FIGURE 12-2 Sales of imported passenger cars in the United States, 1987–1997, in millions. (*Source:* Automotive News Data Center.)

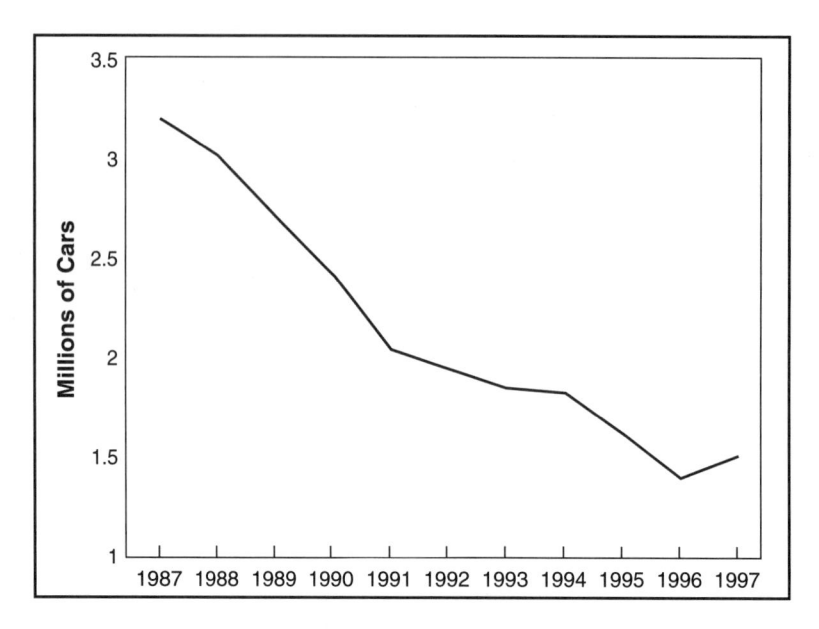

FIGURE 12-3 Sales of imported passenger cars in the United States, 1987–1997, in millions. (*Source:* Automotive News Data Center.)

■ **Example 1**

Two or more lines may be drawn on the same graph if the graph remains clear and easily read. Figure 12-5 shows average expenditures for domestic cars and for imported cars from 1987 through 1997. The years are shown on the horizontal axis (*x*-axis) and expenditures are indicated on the vertical axis (*y*-axis). Two general conclusions can be drawn on the basis of the graph.

1. Since 1987, the average expenditure for imported cars has been greater than the average expenditure for domestic cars.

2. The difference in the two averages has been increasing.

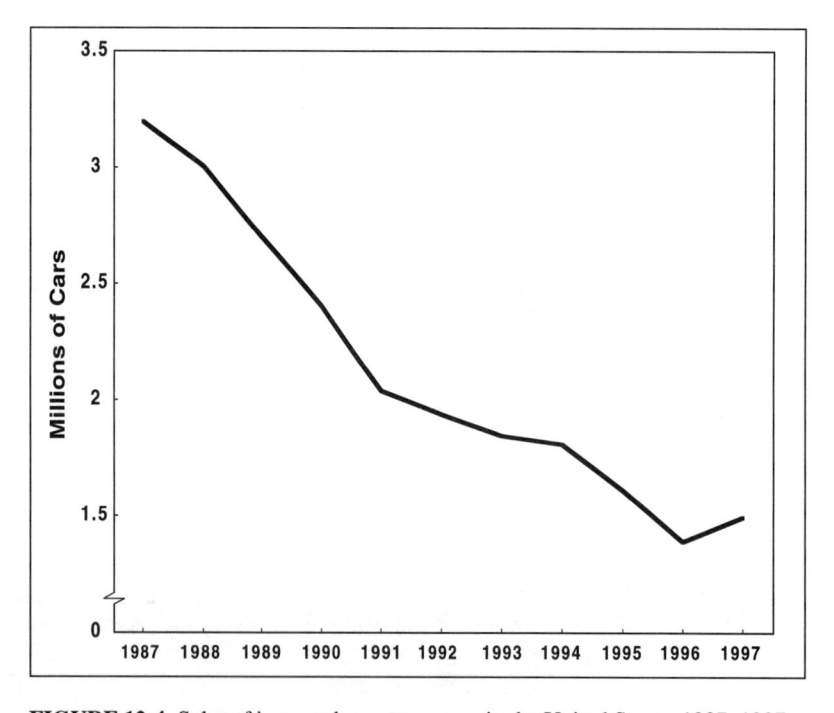

FIGURE 12-4 Sales of imported passenger cars in the United States, 1987–1997, in millions. (*Source:* Automotive News Data Center.)

PRESENTATION OF NUMERICAL FACTS **475**

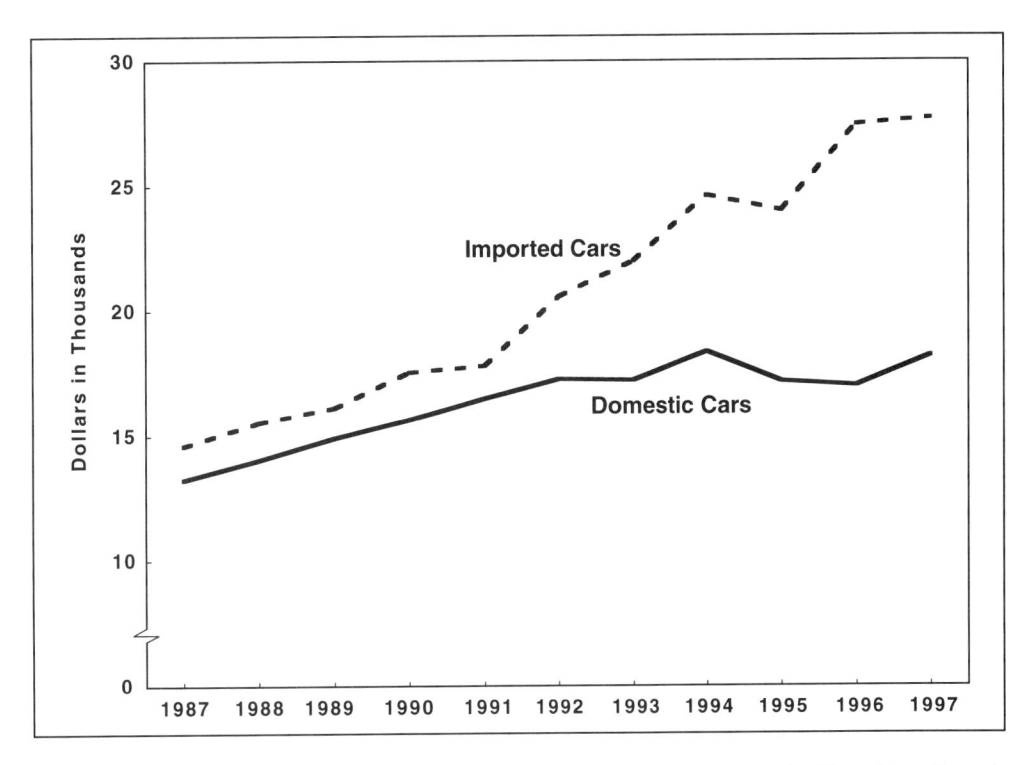

FIGURE 12-5 Average expenditure for new cars, 1987–1997. (*Source:* Automotive News Data Center.)

Numerical information is readily estimated from the graph. When you "read" a graph, you should not expect to get an exact answer, only a reasonable approximation. Following are some of the questions that can be answered.

1. What were the average expenditures for domestic cars and for imported cars in 1989? How large was the difference?

    The average expenditure for imported cars was about 16.0 thousand dollars, or $16,000. (The expenditure scale is in thousands of dollars.) The average expenditure for domestic cars was about 15.0 thousand dollars, or $15,000. The difference was $1,000.

2. What were the average expenditures for domestic cars and for imported cars in 1995? How large was the difference?

The line for imported cars can be read at about $24,000. The line for domestic cars reaches a little less than halfway between $15,000 and $20,000 and can be approximated as $17,000. In 1995 the average expenditure for imported cars was about $7,000 higher than for domestic cars. ∎

Bar Graphs

A second type of graph used frequently is a **bar graph.** The simplest, with single bars, is shown in Figure 12-6. (Figure 12-1 is also a bar graph, with two bars for each category.) The length of the bar gives a visual impression of the value it represents, and it is best not to break the scale. Figure 12-6 shows the production of passenger automobiles in the United States for 1997 by company and indicates clearly the dominance of General Motors. The bars may be drawn horizontally, as shown, or vertically.

∎ **Example 2**

From Figure 12-6, estimate the total production of passenger autos for 1997.

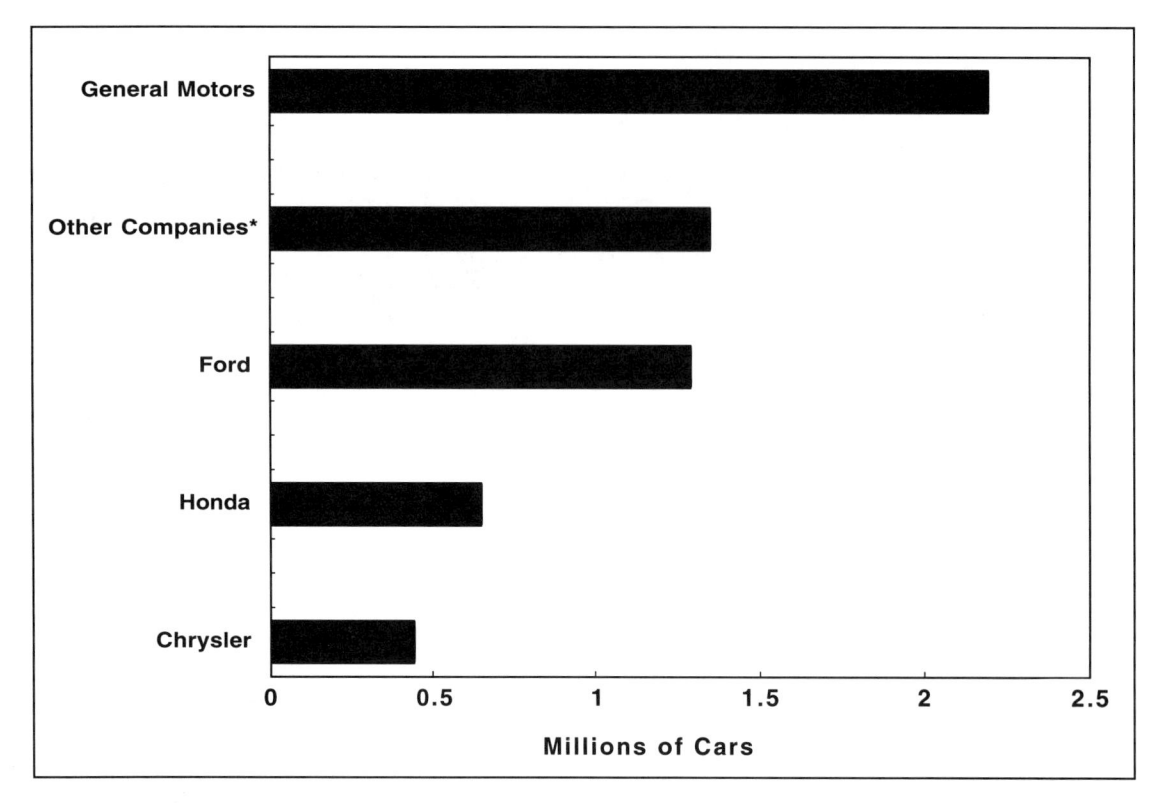

FIGURE 12-6 U.S. car production, 1997, in millons. (*Source:* Automotive News Data Center). *Other companies are BMW, Mazda, Mitsubishi, Nissan, Nummi, Suburu-Isuzu, and Toyota.

The graph is read in the same way as a line graph. In this case, however, since the company names are on the vertical axis we start there, follow the bar to the end, and then go down to the horizontal scale, which is in millions of cars. This gives approximately:

Chrysler	0.4 million
Honda	0.6 million
Ford	1.3 million
Other Companies	1.4 million
General Motors	2.2 million
Total	5.9 million

A variation of a bar graph, called a *pictograph,* is frequently used in nontechnical publications. Figure 12-7 is based on the same data as Figure 12-6 and also shows the production of passenger cars in 1997 by company. Each of the little autos represents 200,000 cars, and for quantities smaller than 200,000 the auto is cut proportionately. The only advantage of a pictograph is that it is more attractive than a simple bar graph for many readers. But it is also less accurate. One type of pictograph can be very misleading. Figure 12-8 shows the increase in the number of men and women in the labor force between 1980 and 1998. Pictures of a man and a woman are used instead of a bar or a line of small figures. The height of the figures corresponds to the top of a bar and gives the value that should be read from the graph. For men, this would be 62 million in 1980 and 72 million in 1998, an increase of 16%. For women, the values are 46 million in 1980 and 61 million in 1998, an increase of 33%. However, the reader tends to respond to the area of the figures instead of the height and may then conclude that the number of women in the labor force nearly doubled during the 18-year period.

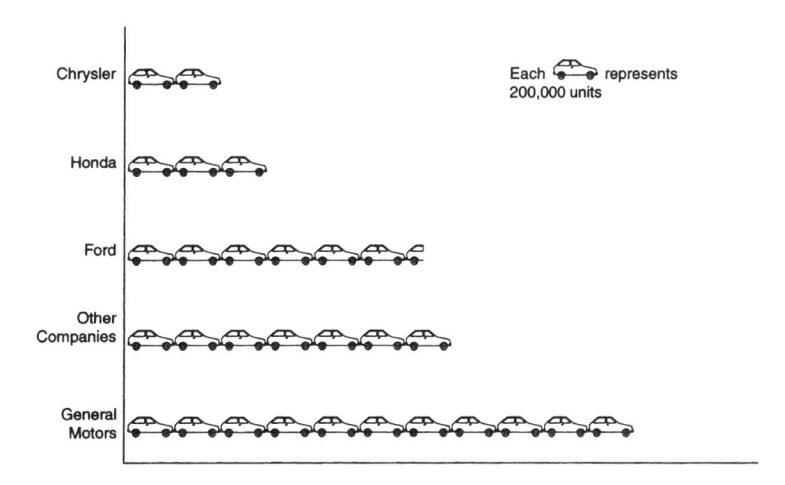

FIGURE 12-7 U.S. car production, 1997. (*Source:* Automotive News Data Center.)

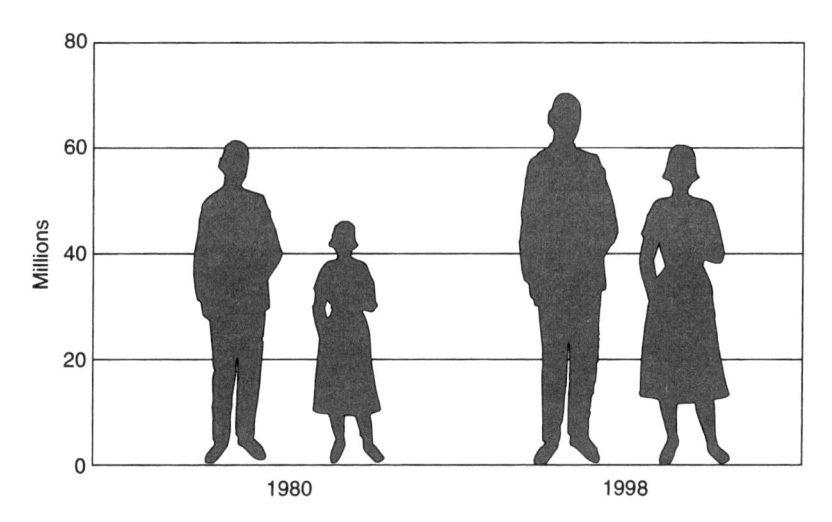

FIGURE 12-8 Number of men and women in the labor force (in millions). (*Source:* Bureau of Labor Statistics.)

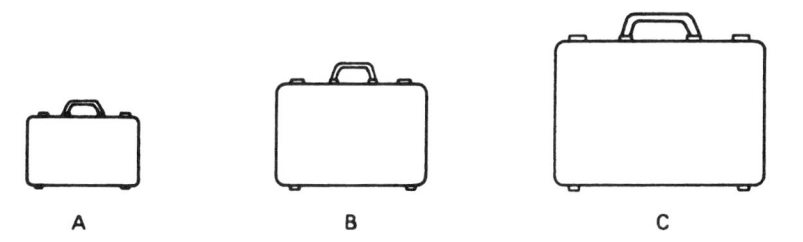

FIGURE 12-9 Growth of managerial and supervisory staff in Cleanaire Corporation.

This type of graph can be used accurately if the area, rather than one dimension, represents the data. Figure 12-9 shows the increase in managerial and supervisory personnel in a growing company for 3 years. An attaché case is used to represent the total for each year. Since this figure is essentially a rectangle, it is not difficult to change the dimensions so that the area accurately represents the new total. In this case the number of employees doubled (B) and then doubled again (C). ■

Chapter 12

TRY THESE PROBLEMS (Set I)

Problems 1 through 6 are based on Figure 12-10.

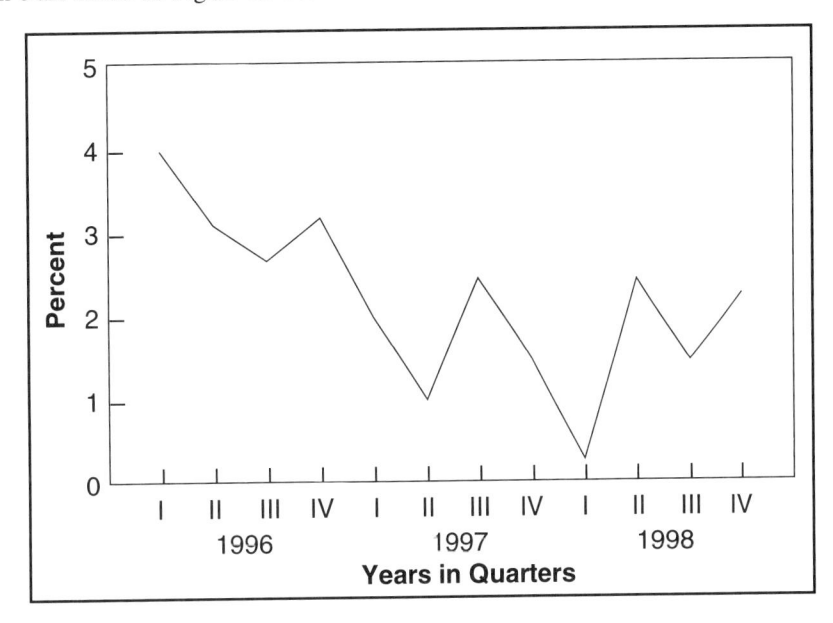

FIGURE 12-10 Annualized quarterly percent change in the consumer price index. (*Source:* U.S. Department of Labor.) (Figures are seasonally adjusted.)

1. Figure 12-10 is a _____ graph.

2. What was the percent change in the index for the first quarter of 1996?

3. What was the percent change in the index for the second quarter of 1998?

4. What quarter had the smallest percent change in the index?

5. How large was the difference between the quarterly percent changes for the second and third quarters of 1997?

6. From the graph, what general conclusion can be stated about the quarterly percent changes in the Consumer Price Index?

Problems 7 through 10 are based on Figure 12-11.

7. Figure 12-11 is a _____ graph.

8. What was the employment in 1992?

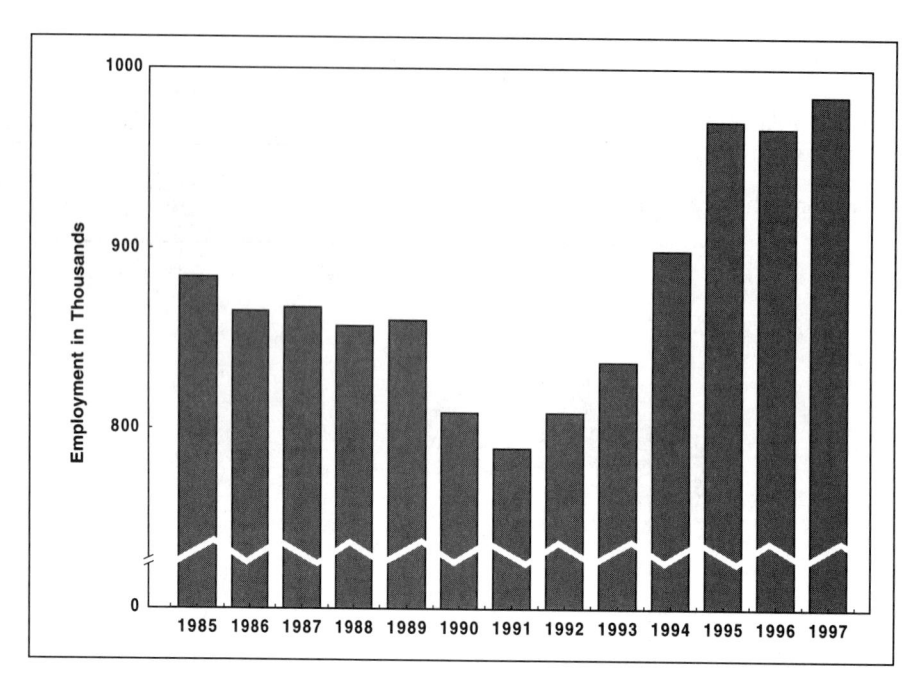

FIGURE 12-11 Employment in manufacturing U.S. Motor Vehicles and Equipment, 1988–1997. (Annual Average.) (*Source:* U.S. Department of Commerce, Bureau of Labor Statistics.)

TABLE 12-5   MOTHERS PARTICIPATING IN THE LABOR FORCE
(PERCENT OF TOTAL)

	MOTHERS WITH CHILDREN UNDER 6 YEARS OLD	MOTHERS WITH CHILDREN 6–13 YEARS OLD
1980	45.3%	62.6%
1985	53.7	68.1
1990	58.8	72.3
1997	63.6	76.5

Source: U.S. Bureau of Labor Statistics.

9. Which year had the largest numbers of people employed?

10. How large was the difference between employment in 1994 and 1995?

Problems 11 through 13 are based on Table 12-5.

11. What general trend is shown in the table?

12. Which of the two groups has shown the larger percent increase in labor force participation between 1980 and 1997?

13. What service industry would you expect to profit most from the indicated change?

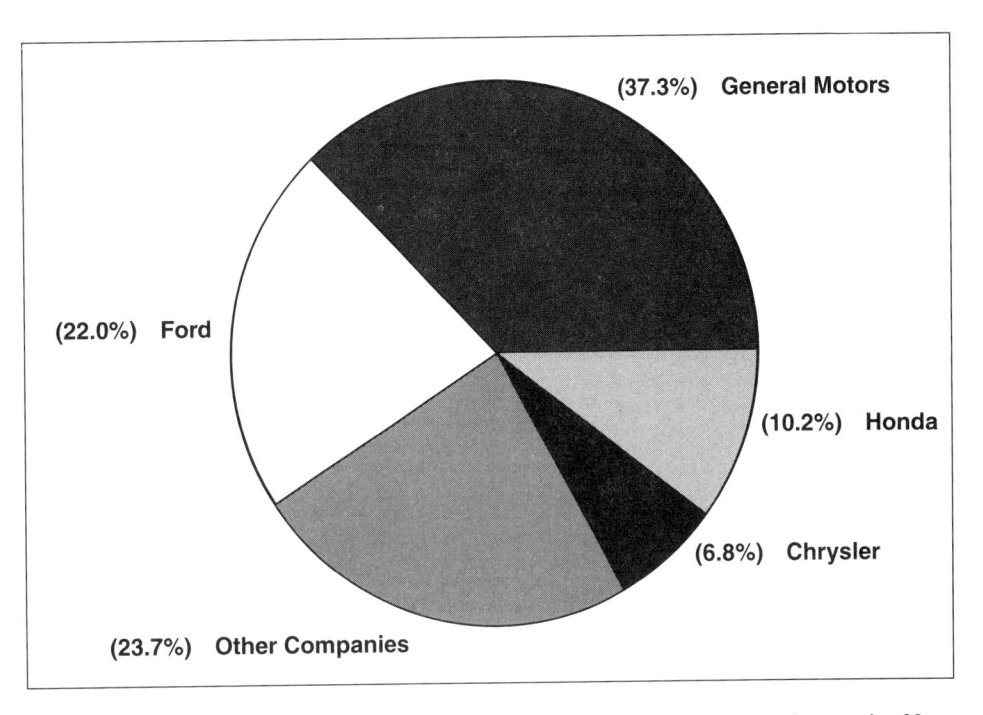

FIGURE 12-12 Car production in the United States, 1997, by company. (*Source:* Automotive News Data Center.) *Other companies include BMW, Mazda, Mitsubishi, Nissan, Nummi, Subaru-Isuzu, and Toyota.

Pie Graphs

A third type of graph, a **pie chart** or **circle graph,** is used to show the breakdown of a whole unit into its parts. It is often used to show the sources of a government's or company's income or the purposes for which it spends money. The data are graphed in percents of total, but the numerical values may be included. Figure 12-12 shows car production for 1997 by company in this form.

In Example 2 of this section the production totals were read from Figure 12-6, showing the production of passenger cars in the United States in 1997 (in millions of cars).

Chrysler	0.4 million
Honda	0.6
Ford	1.3
Other Companies	1.4
General Motors	2.2
Total	5.9 million

To convert each company's production to a percent of total, use the basic percent equation. For example, for General Motors:

$$R \times B = P$$

$$R(5.9) = 2.2 \qquad \text{The base is the total production; the rate is unknown.}$$

$$\frac{R(5.9)}{5.9} = \frac{2.2}{5.9} \qquad \text{Divide both sides by the coefficient of } R.$$

$$R = 0.3728 = 37.3\%$$

The percents for different companies are found the same way.

PRODUCTION OF PASSENGER CARS IN THE UNITED STATES, 1997 (PERCENT OF TOTAL)

Chrysler	6.8%
Honda	10.2
Ford	22.0
Other Companies	23.7
General Motors	37.3
Total	100.0%

■ **Example 3** Use Figure 12-13 to answer the following questions.

1. Which company in 1996 has the largest percent of sales in the computer and office equipment market?

   IBM, with 41.2% of total sales, is the dominant company.

2. Which two companies have almost the same market share? What total percent of the market do these two control?

   Xerox (10.3%) and Compaq (10.1%) have a similar share of the market. Together they make up 20.4% (10.3 + 10.1) of the market.

3. If sales of computers and office equipment were about 184.5 billion in 1997, how much revenue was generated by IBM?

$$R \times B = P$$

$$0.412 \times 184.5 = P$$

The base is 184.5 billion. It is unnecessary to write out $184.5 billion provided the word "billion" is included in the answer.

$$76.01 = P$$

IBM's computer and office equipment sales were approximately $76.01 billion.

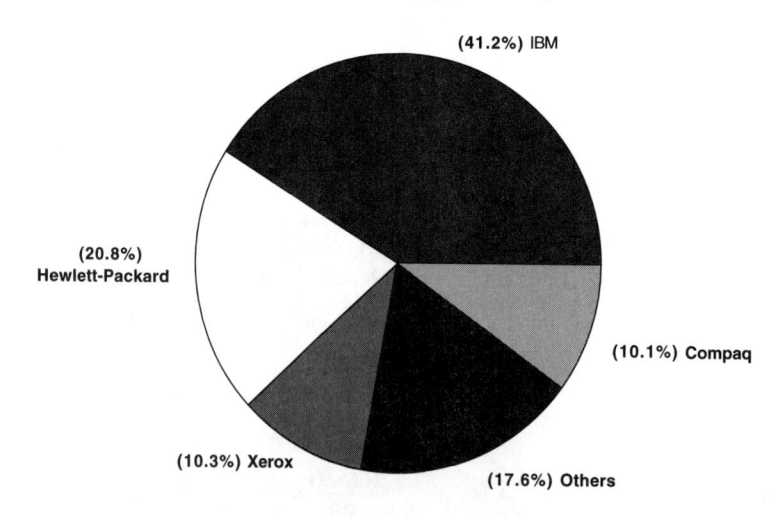

FIGURE 12-13 Leading U.S. businesses, computer and office equipment, 1997. (*Source:* FORTUNE magazine.)

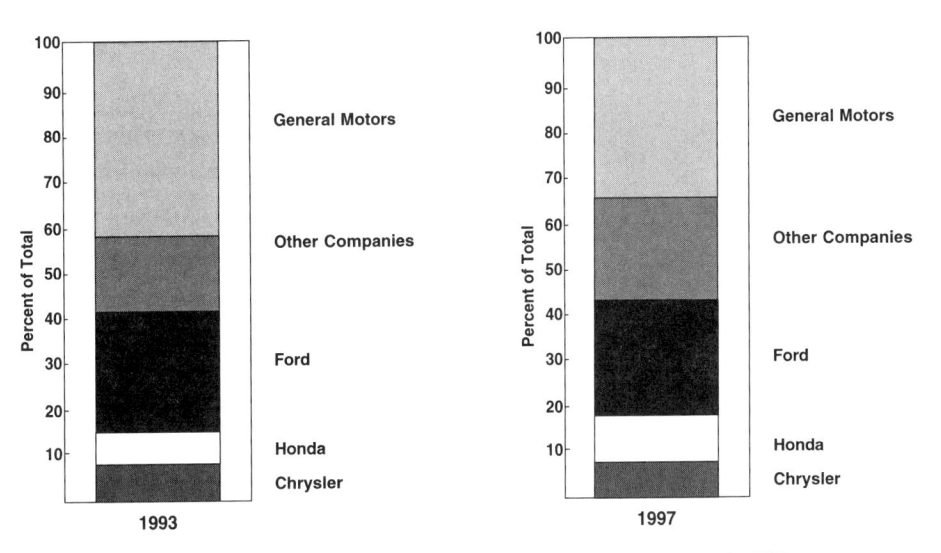

FIGURE 12-14 Production of passenger cars in the United States, 1993 and 1997 (percent of total). (*Source:* Automotive News Data Center.)

Component Bar Graph

The breakdown of a total into its parts can be shown in a *component* bar graph as well as in circular form. The graph may be based on numerical data or percent of total. If more than one year's data are to be shown, two or more bars may be used. Figure 12-14 shows a comparison between automobile production for 1993 and 1997. It indicates the increased proportion of the passenger car market for Honda and Other Companies at the expense of General Motors and Chrysler. Ford's proportion stayed about the same.

Statistical Map

The **statistical map** uses shading or colors and a key to show how a particular characteristic (population, number of cars owned, income, etc.) is distributed over a geographical area. Figure 12-15 shows the disposable personal income per person by state in 1997. To read the graph, match the shading of the state to the key in the upper right-hand corner, and read the percent from the key. In this map the area just below the key contains additional information on the national disposable personal income that cannot be seen from the graph.

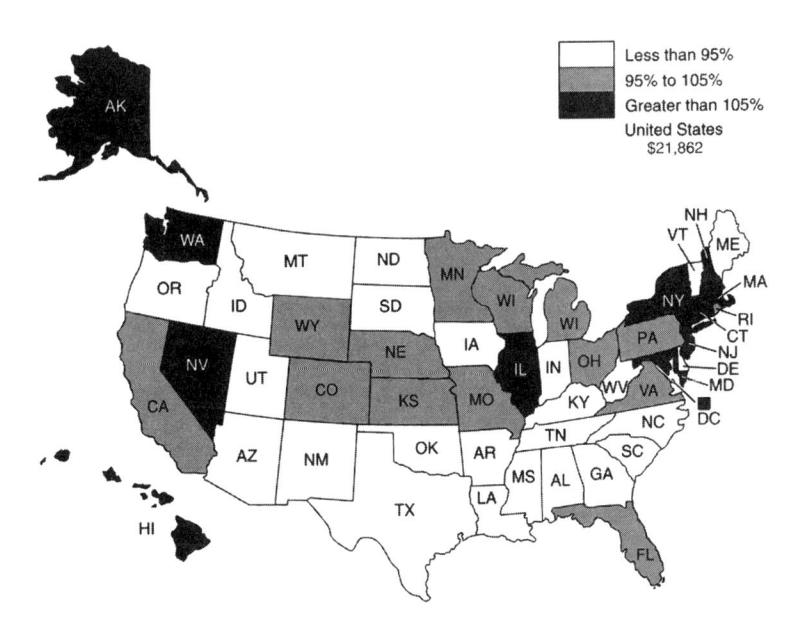

FIGURE 12-15 Disposable personal income per person: 1997 (percent of national average). (*Source:* U.S. Bureau of Census.)

■ **Example 4**

1. Using Figure 12-15, compare the disposable personal incomes for California, Kansas, and New York.

    California and Kansas have the same shading on the map and according to the key had disposable personal incomes per person between 95% and 105% of the national average. The New York income was higher, being greater than 105% of the national average.

2. Locate Illinois on the map and find the disposable personal income.

    Illinois has a disposable personal income greater than 105% of the national average. ■

Chapter 12

TRY THESE PROBLEMS (Set II)

Problems 1 through 5 are based on Figure 12-16.

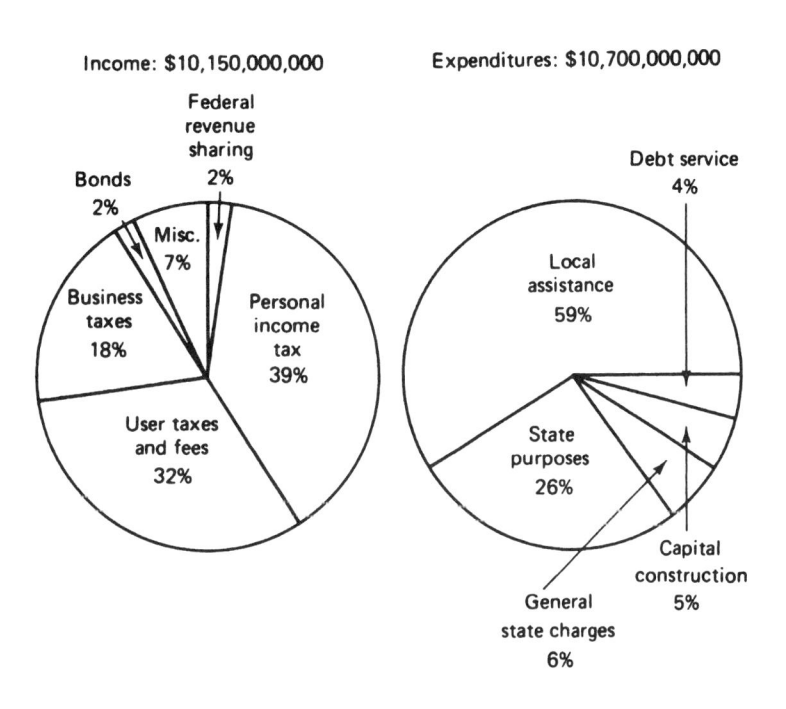

Income: $10,150,000,000
- Federal revenue sharing 2%
- Bonds 2%
- Misc. 7%
- Business taxes 18%
- Personal income tax 39%
- User taxes and fees 32%

Expenditures: $10,700,000,000
- Debt service 4%
- Local assistance 59%
- State purposes 26%
- Capital construction 5%
- General state charges 6%

FIGURE 12-16 Proposed budget for Good State, 2003–2004.

1. What percent of Good State's expected income will come from taxes?

2. What part of this state's expenditures is not for local assistance?

3. What source of income yields the most money?

4. How much of the state's expenditures is for capital construction?

5. How much of the state's income is contributed by the federal government?

485

Problems 6 through 13 are based on Figure 12-17.

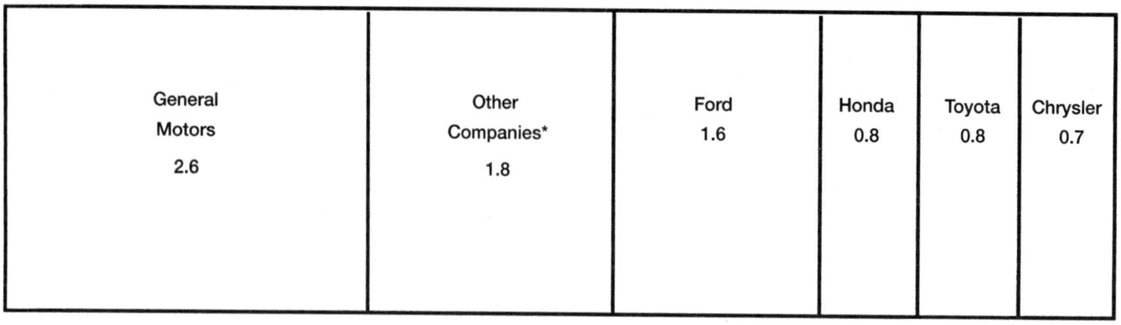

| General Motors 2.6 | Other Companies* 1.8 | Ford 1.6 | Honda 0.8 | Toyota 0.8 | Chrysler 0.7 |

*BMW, Mazda, Mitsubishi, Nissan, and Subaru-Isuzu, Volkswagen, and Volvo.

FIGURE 12-17 U.S. new car sales, 1997 calendar year (in millions). Total = 8.3 million. (*Source:* Automotive News Data Center.)

6. This type of graph is called a _____.

7. If this type of data were shown on a pie chart, what change would have to be made?

8. How many new cars sold in the United States in 1997 were sold by General Motors, Ford, and Chrysler?

9. How many new cars were sold by Toyota, Honda, and Other Companies?

10. What proportion of new cars sold in the United States in 1997 were sold by General Motors, Ford, and Chrysler?

11. What proportion were sold by Toyota, Honda, and Other Companies?

12. What was Honda's share of the 1997 U.S. new car market?

13. What was Ford's share of the market?

12.2 AVERAGES

One number is often needed to summarize a large group of numbers. It is important that this number be representative of the entire group of numbers. It should be a number around which the other numbers seem to cluster.

Measures of central tendency, or **averages,** locate the center of a group of numbers. The three most commonly used measures of central tendency are the **arithmetic mean** (usually called "the **mean**"), the **median,** and the **mode.**

Mean

College students are generally familiar with the term "average" and know how it is calculated. If an examination is returned in class, there is always the question, How did the rest of the class do? What was the class average? What is asked for is a single number to represent the entire list of class grades. If the teacher says the class average was 70, it will be assumed that all the grades were added and the total divided by the number in the class to obtain 70.

In statistics this average is called the **mean.** It is only one of several "averages" used for different kinds of problems.

■ Example 1

A class with five students had the following scores on an examination:

$$98, 80, 92, 90, 20$$

Find the mean score.

To find the mean, use the formula

$$\text{Mean} = \frac{\text{Sum of values}}{\text{Number of values}}$$

In other words, add the grades and divide by the number of grades.

$$\text{Mean} = \frac{98 + 80 + 92 + 90 + 20}{5}$$

$$= \frac{380}{5} = 76$$

The mean of the grades is 76.

The student with 80 on the test might be pleased that his grade is "above average." But the 76 average does not represent the grades well at all. Four of the five are higher and only one is lower, so much lower in fact that it has distorted the average. ■

Median

A better "average" in this case, and for any set of numbers where there are unusual or extreme values like the 20, is the **median,** or middle. *Median* has the same meaning used in fields other than statistics. On the highway the sign that says "Do not cross median" means do not cross the divider in the *middle* of the road that separates two-way traffic.

To find the median, or middle value,

1. Arrange the values in order of size.

2. Determine the median value by applying one of the following rules:

   a. The median is the middle number if there is an odd number of values.

   b. The median is the average of the middle two numbers if there is an even number of values.

For the test scores in Example 1,

$$\left.\begin{array}{r}98\\92\end{array}\right\} \quad \textit{Two grades are higher}$$

$$90 \quad \leftarrow \textit{Middle grade or median}$$

$$\left.\begin{array}{r}80\\20\end{array}\right\} \quad \textit{Two grades are lower}$$

The median grade is 90, and the teacher could report that the class average was 90. There are two grades higher and two lower. It no longer is important that there is a 20 in the group. The lowest grade could be 60, for example, and the median would still be 90.

All the median tells you is that 90 is the *middle* grade. Now the student with a grade of 80 would know that his grade is "below average." This is a better evaluation of a grade of 80 on this test, since it was the second lowest in the class.

■ **Example 2**   Find the median grade of the following set of test scores: 42, 88, 72, 93, 85, 67.

To find the median grade, order the numbers from lowest to highest (or highest to lowest).

$$\begin{array}{c}42\\67\\72\\85\\88\\93\end{array}$$

Notice that there is no grade that separates the three lowest grades from the three highest. Therefore, to find the median we must take the *average of the two middle numbers*.

$$\frac{72 + 85}{2} = 78.5$$

The median is 78.5. Half of the grades are lower than 78.5 and half are higher.

The mean and median are the two most widely used averages in statistical reporting. The mean is used, for example, in reporting average hours of work per week, average age of factory equipment, and average wages in all factories in the United States. However, when the government reported that the average income for a household in the United States in 1997 was $37,005 a year, it was using the median. This is because there are extreme values in household incomes, ranging up into the millions. These few, very high incomes distort the mean, making it higher than the median, just as the very low grade of 20 made the mean of the grades lower than the median. ■

Mode

A third "average" is the **mode,** the value that occurs most often. In the list of five grades there is no mode, since each grade occurs only once. However, if the student who got 90 complained about the grading and had his mark changed to 92, the list would become

$$98, 92, 92, 80, 20$$

and the mode would be 92.

The mode would be useful, for example, to a shoe manufacturer who needs to know the most commonly occurring shoe size. It would also be useful in a marketing project to know the modal or most common income in an area.

Note: If all values occur the same number of times, the data are said to have no mode. When two values occur most often, the data are said to be *bimodal.*

Example 3

A consumer checking the price of an 8-ounce jar of jelly found the following prices in seven stores in her area.

$$\$0.79, \$0.89, \$0.79, \$0.93, \$0.85, \$0.82, \$0.77$$

Find the mean, median, and mode of the prices.

$$\text{Mean} = \frac{\text{Total of values}}{\text{Number of values}}$$

$$\text{Mean} = \frac{79 + 89 + 79 + 93 + 85 + 82 + 77}{7}$$

$$\text{Mean} = \frac{584}{7} = \$0.83 \text{ (rounded)}$$

For the median calculation, arrange the values in order of size.

$$\left.\begin{array}{r}93\\89\\85\end{array}\right\} \textit{Three values are higher}$$

$$82 \quad \leftarrow \textit{Median or middle value}$$

$$\left.\begin{array}{r}79\\79\\77\end{array}\right\} \textit{Three values are lower}$$

The mode is determined by inspection, since only one value, $0.79, occurs more than once. The mean price is $0.83. The median price is $0.82. The modal price is $0.79. ■

Weighted Mean

In the previous illustrations of the calculation of the mean we assumed, without stating it, that all the individual values were equally important. This is a valid assumption. But there are other cases where this assumption cannot be made. A weighted mean is used when the values are not of equal importance. The most familiar illustration for a college student is probably the grade point average.

Example 4

Shown below are Mary's grades and credits for each course for her first semester at Morefun College. If each A is worth 4 points; each B, 3 points; each C, 2 points; and each D, 1 point, compute Mary's grade point average for the semester.

COURSE	GRADE	NUMBER OF CREDITS
Mathematics	A	4
Art appreciation	B	3
Science	C	3
English	D	3
Physical education	F	2

If we take the simple mean of the grades without regard for the credits, we have:

$$\frac{4 + 3 + 2 + 1 + 0}{5} = \frac{10}{5} = 2.0$$

But this is not the way grade point averages are calculated. The result is not accurate because her five courses are not of equal importance from a credit standpoint. We must weigh the grade for each course according to the number of credits granted for the course.

COURSE	GRADE	NUMBER OF CREDITS	GRADE × CREDITS
Mathematics	A	4	16
Art appreciation	B	3	9
Science	C	3	6
English	D	3	3
Physical education	F	2	0
		15	34

Each grade is multiplied by the number of credits, the total is found, and it is then divided by the sum of the weights. This gives:

$$\text{Grade point average} = \frac{34}{15} = 2.3$$

Mary's grade point average is 2.3. The weighted mean is higher than the simple mean because the better grades occurred in the larger credit courses.

The general formula for the weighted mean can be stated as follows:

$$\text{Weighted mean} = \frac{\text{Sum of (values} \times \text{weights)}}{\text{Sum of weights}}$$

We will now apply this formula to calculate a weighted mean for a business problem. ■

■ **Example 5**

The Nut House, Inc., packages and sells mixed nuts, which include peanuts, cashews, and hazelnuts. They pay 40¢ a pound for peanuts, 80¢ a pound for the cashews, and 70¢ a pound for hazelnuts. What is the average (mean) cost per pound?

If the company used equal amounts of all three types, it would be correct to take a simple average and say the mean cost per pound is 63¢.

$$\text{Mean} = \frac{40 + 80 + 70}{3} = \frac{190}{3} = \$0.63$$

But the company uses a much larger proportion of peanuts than of either of the other two; in fact, in a production run of 500 pounds, they will use 250 pounds of peanuts, 100 pounds of cashews, and 150 pounds of hazelnuts. Using these weights to give each type of nut its actual importance, we have:

TYPE	VALUE	WEIGHT	VALUE × WEIGHT
Peanuts	0.40	250	$100
Cashews	0.80	100	80
Hazelnuts	0.70	150	105
		500	$285

Thus, the weighted mean (average cost per pound) can be found as follows:

$$\text{Average cost per pound} = \frac{285}{500} = \$0.57$$

The weighted mean is lower than the simple mean because there is a larger proportion of the cheaper peanuts than the more expensive types of nuts. ■

> **CALCULATOR HINT**
>
> You can use the memory keys on your calculator to solve Example 2.
>
> | ENTER | DISPLAY |
> |---|---|
> | 0.40 [×] 250 [M+] | 100 |
> | 0.80 [×] 100 [M+] | 80 |
> | 0.70 [×] 150 [M+] | 105 |
> | [MR] [÷] 500 [=] | 0.57 |
>
> The weighted mean is $0.57.

12.3 DISPERSION

Averages are one way of describing a set of numbers, but they are useful only to a limited extent. You might know, for example, that the average temperature in an area for a particular date is 74°F, but that would not tell you in advance whether you would need a coat that day. It could be much hotter or much colder than 74°. You can get an average of 60 on six quizzes by getting

$$50, 60, 60, 60, 70, 60 \quad \text{or} \quad 100, 40, 50, 70, 20, 80$$

The difference between the two sets of grades is the scatter, or distance of the individual values from the mean of 60. In statistics this characteristic is called **dispersion.** We will consider here only two measures of dispersion, the range and the standard deviation.

Range

The **range** is the simplest measure of dispersion to calculate. It is the difference between the highest and lowest value in the group.

$$\text{Range} = \text{Highest value} - \text{Lowest value}$$

The range for the first set of quizzes is $70 - 50 = 20$. For the second set, the range is $100 - 20 = 80$.

The range is used in reporting stock market prices when the daily highs and lows are given. It is also used in reporting the weather when the historical high and low for the data are included.

Standard Deviation

The usefulness of the range as a measure of dispersion is limited since it considers only the highest and lowest values and ignores the other numbers in the group. A more widely used measure of dispersion that takes into account how *all* the numbers are dispersed is the **standard deviation.** It tells us an average distance of each value from the mean of group.

Before illustrating the calculation of the standard deviation, it is necessary to review a concept from elementary algebra. If a number is squared, it is multiplied by itself. The operation is indicated by an exponent, a small 2 to the upper right of the number. Thus, 7^2 means multiply 7 by 7, and $(-2)^2$ means multiply -2 by -2.

$$7^2 = 7 \times 7 = 49$$
$$(-2)^2 = (-2)(-2) = 4$$

The inverse operation is taking the square root of a number, that is, finding the number that, when multiplied by itself, will give the original number. The operation is indicated by the symbol $\sqrt{}$. Thus, $\sqrt{36}$ means find the number that, when multiplied by itself, will give 36.

$$\sqrt{36} = 6, \text{ because } 6 \times 6 = 36$$
$$\sqrt{16} = 4, \text{ because } 4 \times 4 = 16$$

When calculating the standard deviation you will need to find both the square and square root of various numbers. Pressing the square root ($\sqrt{}$) key on your calculator after entering the number will give you the square root.

■ **Example 1**

Sally Student received the following grades on a set of quizzes.

$$6, 3, 1, 8, 7$$

Find the mean and the standard deviation.

To find the mean, add the values and divide by the number of values.

$$\text{Mean} = \frac{\text{Total of values}}{\text{Number of values}} = \frac{6+3+1+8+7}{5} = \frac{25}{5} = 5$$

To compute standard deviation use the following procedures:

1. Subtract the mean from each value to obtain the distance from the mean. (The sum of these distances will always be 0.)

2. Square each distance.

3. Add the squared results.

4. Divide by the number of values and take the square root.

VALUES	d (DISTANCE FROM MEAN)	d^2 (DISTANCE SQUARED)
6	6 − 5 = 1	1^2 = 1
3	3 − 5 = −2	$(-2)^2$ = 4
1	1 − 5 = −4	$(-4)^2$ = 16
8	8 − 5 = 3	3^2 = 9
7	7 − 5 = 2	2^2 = 4
		Total = 34

The formula for standard deviation is:

$$\text{Standard deviation} = \sqrt{\frac{\text{Sum of squared distances}}{\text{Number of values}}}$$

or, in symbols,

$$\text{Standard deviation} = \sqrt{\frac{\Sigma d^2}{N}}$$

where

Σ = "the sum of"

d^2 = distance squared

N = number of values

$\sqrt{}$ = "the square root of"

For this problem:

$$\text{Standard deviation} = \sqrt{\frac{\Sigma d^2}{N}} = \sqrt{\frac{34}{5}} = \sqrt{6.8} = 2.6$$

■ **Example 2** Compare Sally's grades with those of Peter Pupil, who received the following grades on the quizzes.

$$9, \ 1, \ 1, \ 8, \ 6$$

Find the mean and standard deviation:

$$\text{Mean} = \frac{9 + 1 + 1 + 8 + 6}{5} = \frac{25}{5} = 5$$

VALUES	d (DISTANCE FROM MEAN)	d^2 (DISTANCE SQUARED)
9	9 − 5 = 4	$4^2 = 16$
1	1 − 5 = −4	$(-4)^2 = 16$
1	1 − 5 = −4	$(-4)^2 = 16$
8	8 − 5 = 3	$3^2 = 9$
6	6 − 5 = 1	$1^2 = 1$
		Total = 58

$$\text{Standard deviation} = \sqrt{\frac{\Sigma d^2}{N}} = \sqrt{\frac{58}{5}} = \sqrt{11.6} = 3.4$$

The mean of Peter's grades is 5 and the standard deviation is 3.4.

Sally and Peter both had the same mean grade. However, the standard deviation of Sally's grades (2.6) is smaller than the standard deviation of Peter's grades (3.4).

Since the standard deviation is a measure of the average distance of the grades from the mean, it can be used to compare the dispersion of two different groups of numbers from the same mean. The smaller standard deviation indicates that the numbers are less spread from the mean. The larger standard deviation means that the numbers are more dispersed.

Thus, although Sally and Peter both had the same mean grade, Sally was more consistent in her test results. Peter's grades were farther from the mean. ■

The standard deviation is used in many areas of business statistics. In a production process, for example, an increase in the standard deviation of a set of measurements may indicate that a machine needs adjustment, that the products show too much variation. A sales manager might check which salespeople are most consistent in their week-to-week sales totals, that is, which sales records have the smallest standard deviation. Other important uses of the standard deviation are in probability and sampling, which are not within the scope of this text.

Chapter 12

TRY THESE PROBLEMS (Set III)

Find the mean, median, mode, range, and standard deviation for the following.

1. 2, 1, 4, 28, 4, 3

2. 8, 10, 16, 21, 27

3. $4, $10, $6, $8, $7

A college student working part-time for 7 weeks earned the following amounts per week: $100, $126, $110, $116, $116, $112, $104.

4. What was the mean salary?

5. What was the median salary?

6. What was the mode?

7. Find the range.

8. Find the standard deviation. (Round to the nearest cent.)

9. If the student earned the same amount every week, what would the standard deviation be?

10. Another student had a mean salary of $112 with a standard deviation of $18. Compare the dispersion of salaries of the two students.

Find the grade point average for each of the following students. Assume that A = 4, B = 3, C = 2, D = 1, F = 0.

11. CREDITS	GRADE
4	B
2	F
5	C
1	A
3	B

12. CREDITS	GRADE
3	C
3	D
4	B
2	A
4	B

13. Metro Appliance employs 10 persons whose titles and salaries are listed below. Compute the mean salary of the employees.

POSITION	NUMBER OF EMPLOYEES	WEEKLY SALARY
Owner	1	$1,600
Sales	5	$ 700
Stock	2	$ 300
Office	2	$ 400

The Citrus Co. cans a combination of orange and grapefruit sections. By the time the fruit is ready for the can, the oranges cost the company 86¢ a pound and the grapefruit costs 76¢ a pound. The mixture is made up of 40% oranges and 60% grapefruit.

14. What is the cost of the combined fruit? (The percents may be used as weights.)

15. There is a price change to 76¢ for oranges and 74¢ for grapefruit, and the company changes the mix to 50%–50%. What is the new cost per pound?

16. With the 50%–50% mix, is the weighted mean different from the simple mean? Why?

12.4 SAMPLING

In a survey of 2,041 adults who were asked whether the Earth goes around the Sun or the Sun goes around the Earth, 21 percent replied incorrectly; 7 percent said they did not know. *

Have you ever wondered how the ratings of television programs are determined? Most of us can say, "They didn't ask me what I watch." How can they make a judgment about which programs all the television sets in the United States are tuned to? Or how can any pre-election survey show what proportion of the voting population will be for candidate X? Again, they never asked most of us for whom we would vote.

These conclusions about whole populations (all the television sets, all the voters) are based on a statistical process called **sampling** in which a conclusion about the whole population is made on the basis of a sample. The word **sample** as used in statistics means the same as when it is used in any other context—a sample is simply a part of the whole. Other more familiar samples are a small piece of fabric or carpeting, a taste of the soup being cooked, and an examination. On the basis of a sample, we judge whether the fabric or carpeting will look good where we need it, whether the soup tastes right, and whether the student knows enough to pass the course.

Sampling is widely used in business—in the grading of grain and cotton; in accounting, particularly auditing; to check on whether machines are producing accurate products; to determine whether or not a new product will sell well; and so on. It would be impossible or too time consuming and expensive to check every bale of cotton, to ask every bank depositor to check his balance, to measure every product as it came off the machine, or to ask every potential customer whether he or she would buy the new product.

The theory of statistical sampling is much too complex to include here. But a few facts about samples should be understood if you are to accept or reject, on an informed basis, much of the information given by polls, advertising, and government statistics.

- *A good statistical sample is representative of the whole population.* If the soup has been well stirred, a spoonful will represent accurately the taste of the whole potful. But if an examination includes many obscure, unimportant points, it is a poor sample of the course material. If a company advertises that four out of five doctors recommend a certain product for cold relief, maybe they asked only doctors who own stock in the company.

- *The size of the sample alone does not determine whether the results are reliable.* The political polls, often based on approximately 1,000 voters, are generally accepted as reliable even though these 1,000 voters are used to represent approximately 96 million people who actually may vote. The Nielsen television ratings, also accepted as reliable, are based on TV viewing in 4,000 households, although there are about 100 million households in the United States, almost all of which own television sets.

- *A sample can indicate conditions only at the time the survey is made.* If a political poll shows that candidate X will receive 46% of the vote, it indicates only that if the election were held when the poll was taken candidate X would receive 46% of the votes.

- *Conclusions based on a sample are not really as definite and as certain as they appear to be.* If the poll gave candidate X 46% of the vote, we are pretty sure, *but not certain*, that *approximately* 46% of the votes would be for him if the election were held then. There is still a chance—small, but it exists—that the true percent is not near 46%.

12.5 INDEX NUMBERS

An **index number** is a special form of ratio, used to show changes over a period of time. A quantity (sales, price, production, etc.) is compared with the corresponding value in some earlier period, which is called the **base.** The ratio is then multiplied by 100.

*From the *New York Times.*

$$\text{Index} = \frac{\text{Given year's value}}{\text{Base year's value}} \times 100$$

Advantages of Index Numbers

An index number has several advantages compared with the original data.

An Index Shows Percent Change From the Base Year. If there were no change, the numerator and denominator would be the same. The ratio would be 1.00 and the index would be 100. Therefore, any difference from 100 indicates change. If the index is greater than 100, there has been some percent increase. If it is less than 100, there has been a percent decrease. *The percent change is the difference from* 100. For example, in Table 12-6,

TABLE 12-6 CALCULATING A SALES INDEX

YEAR	RATIO	CHANGED TO A DECIMAL	MULTIPLIED BY 100 = INDEX
1995	$\frac{200{,}000}{200{,}000} =$	1.00	100
1996	$\frac{250{,}000}{200{,}000} =$	1.25	125
1997	$\frac{200{,}000}{200{,}000} =$	1.00	100
1998	$\frac{190{,}000}{200{,}000} =$	0.95	95
1999	$\frac{220{,}000}{200{,}000} =$	1.10	110

- The index for 1996 (125) indicates a 25% increase in sales compared with 1995 (125 − 100).
- The index for 1997 (100) means that sales were the same in 1997 as in 1995.
- The index for 1998 (95) indicates a 5% decrease in sales compared with 1995 (100 − 95).

■ **Example 1**

A retailer has recorded the following annual sales. Calculate a sales index with 1995 as the base year.

YEAR	SALES
1995	$200,000
1996	250,000
1997	200,000
1998	190,000
1999	220,000

If we use 1995 as the base year, the ratio of sales for each year compared with 1995 is changed to a decimal and multiplied by 100 (see Table 12-6).

The index of sales, on a 1995 base (compared with 1995), is the last column.

One Index Number Can Represent Changes in Many Quantities. The most familiar and most widely used composite index number is the *Consumer Price Index.* This single number tells us each month what is happening to the inflation rate, that is, how much more or less we are paying for the goods and services we buy than we did in the past.

The Consumer Price Index, published monthly by the Bureau of Labor Statistics, is a measure of the average change in prices in a fixed "market basket" of goods and services purchased either by urban wage earners and clerical workers or by all urban consumers. It is based on a sample of prices for thousands of items, including prices for food, clothing, shelter, fuels, transportation fares, medical care, recreation, education, and other goods and services purchased for day-to-day living. Prices are collected in 87 urban areas across the country and from thousands of housing units and retail establishments. The base year is 1982–1984, written as 1982 – 84 = 100.

This index is used not only to measure inflation, but also as a yardstick for many union contracts, pension plans, and social security and other goverment benefit payments. These are said to be "tied to the Consumer Price Index." When it goes up, these wages, pensions, and government payments go up with it.

The Consumer Price Index, like all composite indexes, must be carefully interpreted. Whatever the index shows, it shows for all the items together and all parts of the country together, and does not necessarily apply to any particular product or place. An index of 166.7 in July of 1999, for example, means that the combined items that cost $100 in 1982–84 would cost $166.70 in July 1999. But it is not necessarily true for $100 worth of clothing or food or automobile repair or any other particular item. Similarly, we cannot be sure that this increase in cost is true for New York City, Chicago, Dallas, Los Angeles, or any other section of the country. To overcome this limitation, separate indexes are computed for different product groups and for different sections of the country.

Other important composite indexes that are also watched carefully as indications of the health of the economy are published every month by agencies of the federal government. Two examples are the *Producer Price Index,* which records changes in commodity prices at the wholesale level, and the *Industrial Production Index,* which measures the output of U.S. factories. ∎

An Index Makes It Easy to Compare Changes in Different Kinds of Data. Since the index numbers show percent change rather than arithmetic change, the size of the data and the units of measurement are not important. We can, for example, compare changes in the inflation rate with changes in the production of goods by using the Consumer Price Index and the Industrial Production Index, even though prices are measured in dollars and production is measured in physical volume (number of cars, tons of steel, etc.).

Now go to Try These Problems (Set IV).

NEW FORMULAS IN THIS CHAPTER

- Mean = $\dfrac{\text{Sum of values}}{\text{Number of values}}$

- Weighted mean = $\dfrac{\text{Sum of (values} \times \text{weights)}}{\text{Sum of weights}}$

- Index = $\dfrac{\text{Given year's value}}{\text{Base year's value}} \times 100$

- Standard deviation = $\sqrt{\dfrac{\text{Sum of the squared distances}}{\text{Number of values}}} = \sqrt{\dfrac{\Sigma d^2}{N}}$

Chapter 12

TRY THESE PROBLEMS (Set IV)

The Fix-It Co. markets a small tool kit. The company statistician decided to construct a price index for this item so that changes in the cost of the tool kit could be compared with changes of cost for other items the company sells. The cost record for the tool kit was:

YEAR	COST
1996	$6.00
1997	6.24
1998	7.08
1999	7.56

1. Using 1996 as the base year, what was the price index for 1997?

2. What percent change in cost does the index for 1997 show? (Use 1996 = 100.)

3. What is the index for 1999 if 1996 = 100?

4. What does the 1999 index show if the base year is 1996?

5. Using 1990 as the base year (1990 = 100), compute a price index for the Litton family's average weekly entertainment expenses shown below.

YEAR	COST	INDEX
1990	$50	100
1992	60	
1994	65	
1996	75	
1998	78	

6. Interpret the indexes in Problem 5 for 1994 and 1996.

Use the following data for Problems 7 through 11. The Consumer Price Index (1982–1984 = 100) was 164.6 in January 1999. For the same month the index for transportation was 140.8 and for medical care 246.3.

7. What was the percent increase for all items from 1982–84 to January 1999?

8. What was the percent change for transportation from the base year?

9. How large was the percent increase in the cost of medical care from 1982–84 to January 1999?

10. Can you determine from the facts given what happened to the price of aspirin between 1982–84 and January 1999?

11. From the values given, can you determine the increase in consumer prices in your area?

12. An index of production was 84 for 1999 (1982 = 100). What percent change does it indicate?

Chapter 12

Name　　　　　Date　　　　　Class

END-OF-CHAPTER PROBLEMS

Problems 1 through 5 are based on Figure 12-18.

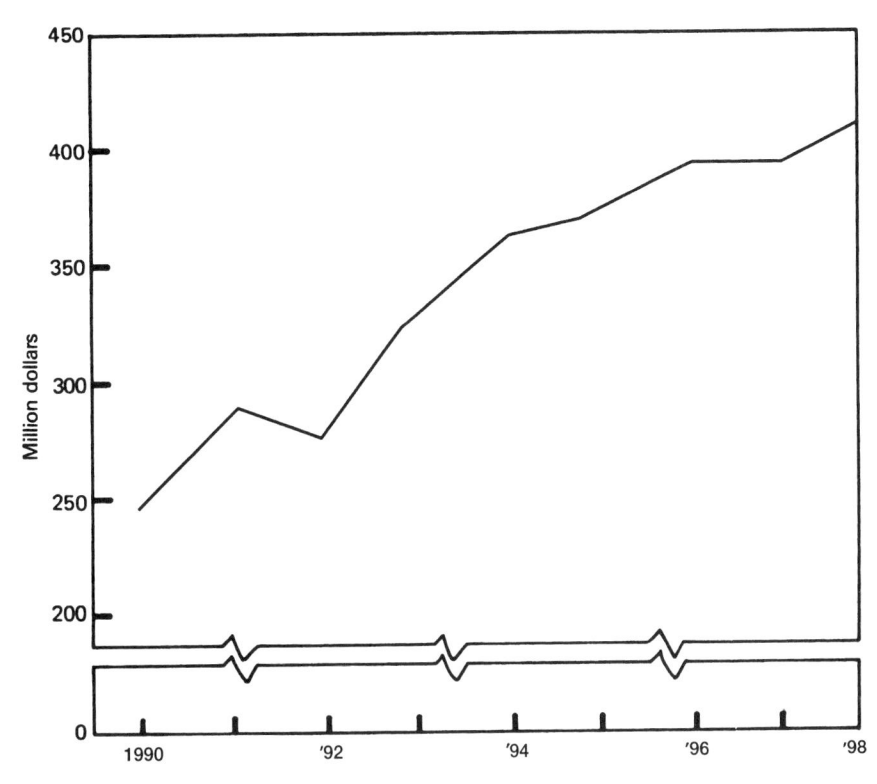

FIGURE 12-18 Sales of Fair Face Cosmetics, Inc., in millions of dollars.

1. How large were sales in 1992?

2. By how much did sales increase between 1992 and 1995?

3. In what year were sales highest?

4. In what year did sales decline?

5. What does ⌇⌇ indicate for this graph?

503

Problems 6 through 8 are based on Figure 12-19.

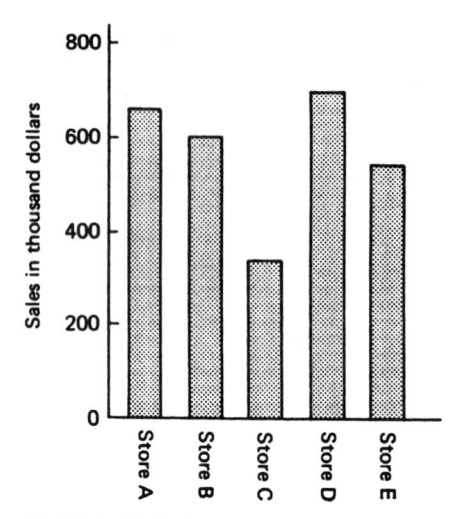

FIGURE 12-19 Sales of stores in the Chicken Tonight Group, 1998, in thousands of dollars.

6. What was the largest amount sold by any store?

7. How much did Store E sell in 1998?

8. What fact shown by the graph should most concern management of the Chicken Tonight Group?

Problems 9 through 11 are based on Figure 12-20.

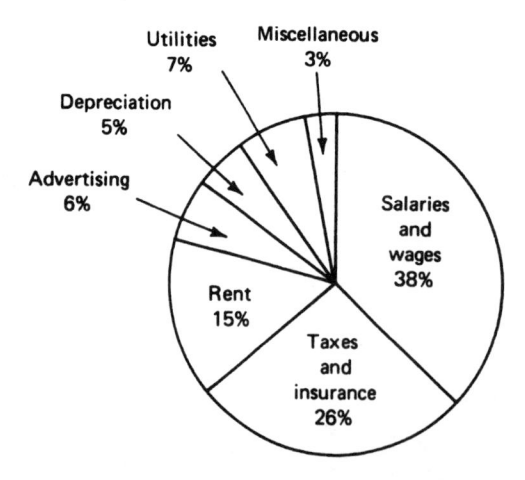

FIGURE 12-20 Operating expenses, Fair Face Cosmetics, Inc., 1999.

504

Chapter 12

END-OF-CHAPTER PROBLEMS

9. What proportion of operating expenses went for rent and utilities?

10. If $39 million was spent on taxes and insurance, how much was spent for advertising?

11. If total operating expenses for the year were $120 million, how much was spent for salaries and wages?

12. The Atlas Home Supply Company had the following monthly sales during the year: January, $25,440; February, $22,320; March, $23,160; April, $23,600; May, $22,200; June, $26,800; July, $24,580; August, $19,100; September, $38,700; October, $37,400; November, $44,100; December, $29,900. Find, to the nearest dollar:

   a. The mean.

   b. The median.

   c. The mode.

13. Find the mean weekly earnings for Gypsum Products' 650 workers if 212 earned $540 a week, 224 earned $585 a week, 145 earned $630 a week, and the rest earned $750 a week.

14. Find the median.

   a. 630, 820, 304, 989, 740, 749, 420

   b. 460, 325, 192, 510, 684, 591

505

15. Joe Hummel's mathematics teacher gave a test every week. Joe's marks were 94, 87, 56, 73, 85, 94, 75, 63, 68, 99, 85, 94, 80, 71, and 74. Find, to the nearest tenth:

   a. The mean.

   b. The median.

   c. The mode.

16. The grades on a civil service promotional exam were as follows. Find the mean grade. (Round to the nearest tenth.)

GRADE	NUMBER OF PEOPLE
95	5
85	39
75	23
65	8

17. A student received the following test grades during a semester: 92, 85, 80, and 80. The final exam grade was 90. If the instructor counts each of the four term tests as 15% of the final grade and the final exam as 40%, what is the student's weighted average for the term?

18. Indicate whether each of the following statements is true or false, and if false, why.

   a. A television reporter interviewing people on the main street of a large city will get a representative sample of the entire population of the city.

   b. A political poll gives exact information on how the whole voting population will vote on election day.

   c. A sample is a part of a whole.

   d. A good statistical sample must be almost as large as the whole population it represents.

   e. The Consumer Price Index is based on a sample.

Chapter 12

END-OF-CHAPTER PROBLEMS

19. There are seven people working in a small office. Their salaries are:

Manager:	$95,000
Bookkeeper:	32,000
Secretary:	24,000
Receptionist:	20,000
Three clerks:	18,000
	18,000
	18,000

Find the mean, median, and mode, and determine which is the best representative of all the salaries. Explain your answer.

20. The Farm Foods Co. packs frozen vegetables, including a combination of carrots, cauliflower, and onions. The company's statistician keeps a price index for each of the items in the mixture so that the company can change the proportions of the three vegetables as the prices change. For the first 3 months of 1999 the price indexes are given in Table 12-7.

TABLE 12-7 INDEX OF PRICES: JANUARY 1995 = 100

1999	CARROTS	CAULIFLOWER	ONIONS
January	100	110	135
February	106	109	130
March	108	115	119

a. What does the March index for cauliflower mean?

b. What does the January index for carrots mean?

c. Which price showed the smallest change between January 1995 and March 1999?

21. Find the range and the standard deviation for the following:

a. 8, 10, 12, 16, 19

b. 1, 2, 3, 4, 4, 28

22. A group of tires has been tested for wear life by the manufacturer with the following results:

Wear life (in thousand miles): 22, 30, 31, 32, 25

   a. Find the mean wear life for the five tires.

   b. Find the standard deviation.

   c. If you had a choice between this brand and another brand that tests showed had the same mean life but a standard deviation of 2.91 thousand miles, which brand would you buy? Why? (Note that the standard deviation is in the same units as the original distribution—in this case, thousand miles.)

23. Using 1995 as the base year, determine the price index for:

   a. A sweater costing $25 in 1995 and $38 today.

   b. A pair of sneakers costing $35 in 1995 and $60 today.

   c. A calculator that sold for $15 in 1995 and costs $8 today.

24. Eddie Chan earned $480 a week in 1994. He earns $600 a week today. If today's price index is 135 with 1994 as the base year, has the increase in his salary kept up with the overall increase in prices?

SUPERPROBLEM

Health House, Inc., packs a 4-ounce mixture of banana chips, pineapple pieces, dates, and carob-coated raisins. Using the following table, calculate the percent change in the cost of the mixture between 1995 and 2000.

	COST PER POUND		PROPORTION IN MIX (%)
	1995 ($)	2000 (1995 = 100)	
Banana chips	2.00	125	35
Pineapple	3.00	120	30
Dates	2.50	130	23
Raisins	4.00	110	12

Chapter 12

Name _____ Date _____ Class _____

TEST YOURSELF

Problems 1 through 6 are based on Figure 12-21.

                                                                                                                     Section

1. What percent of the population was in the 15–24 age group in: 12.1

    a. 1980?

    b. 1997?

2. What percent of the population was 65 and over in 1997?   12.1

3. Which two age groups appear to have shown the greatest change in percent of total population between 1980 and 1997?   12.1

4. If the population totaled 267 million in 1997, approximately how many people were in the 25–34 age group?   12.1

5. If the population totaled 227 million in 1980, approximately how many people were 45–54 years of age?   12.1

6. How might businesses and governments be affected by the percent decrease in the 15–24 age group?   12.1

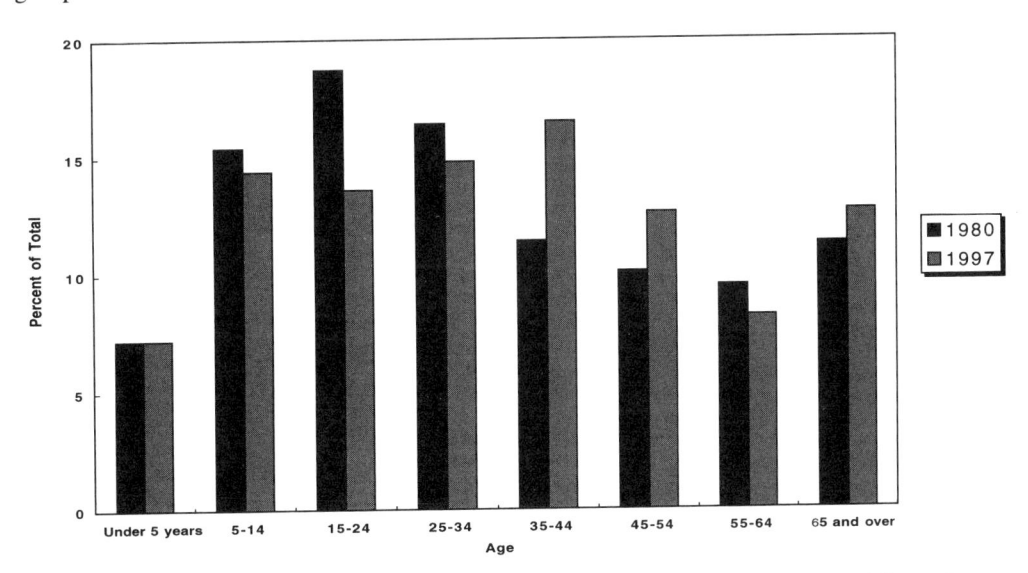

FIGURE 12-21 Population by age, 1980–1997 (percent of total). (*Source:* U.S. Bureau of Census.)

Problems 7 through 10 are based on Figure 12-22.

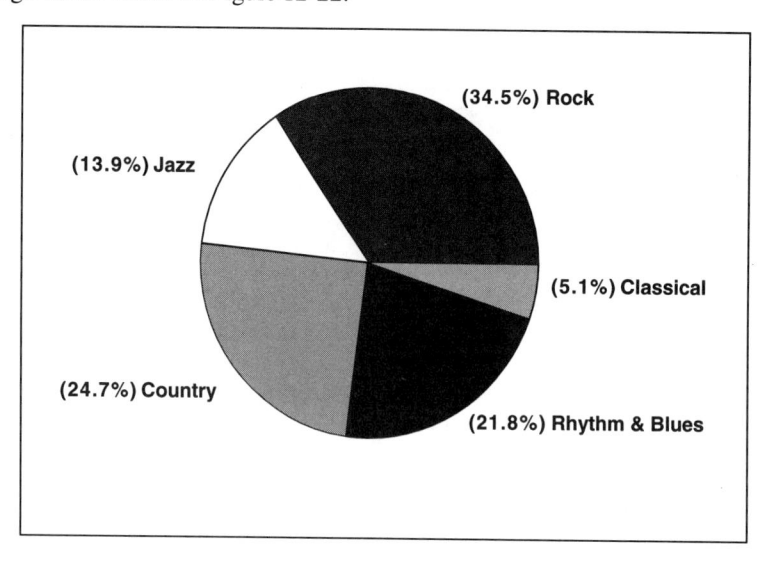

FIGURE 12-22 CD Outlet music sales, by category, 1999.

	Section
7. What percent of CD Outlet's sales comes from country music?	12.1
8. Which category has the smallest percent of music sales?	12.1
9. What part of the store's music sales is not from rock music?	12.1
10. If CD Outlet's total sales were $825,000 for the year, how much revenue was generated from sales of jazz music?	12.1

For Problems 11 through 16, consider the following lengths of service for several employees who are retiring.

EMPLOYEE	LENGTH OF SERVICE (YEARS)
Archer	13
Jones	23
Saam	27
Toral	23
Brown	19

Chapter 12

TEST YOURSELF

	Section
11. What is the arithmetic mean length of service?	12.2
12. What is the median length of service?	12.2
13. What is the mode?	12.2
14. What is the range of length of service?	12.3
15. What is the standard deviation?	12.3
16. If the employees had all been with the company the same length of time, what would standard deviation be?	12.3
17. The following table shows the weekly part-time earnings for a group of students.	12.2

$ EARNINGS	NUMBER OF STUDENTS
75	2
120	3
150	4
270	1

Find the mean amount of earnings.

511

18. Use the following data on the price of crude oil to determine price indexes for 1985, 1990, and 1995. Use 1980 as the base year.

YEAR	PRICE OF CRUDE OIL (PER BARREL)
1980	$21.60
1985	24.10
1990	20.00
1995	14.60

19. What percent change in prices does the 1985 index show if 1980 is the base year?

20. What does the 1995 index show if 1980 = 100?

512

Chapter 12

COMPUTER SPREADSHEET EXERCISES

Numbers can be more easily understood and can make a much stronger impression when presented visually. Microsoft Excel incorporates excellent graphics capabilities. Many different types of charts can be quickly created from data, with just a few keystrokes or clicks of the mouse.

Print only the graphs for each problem. Be sure your graph fits on one page. (If necessary, use the Landscape orientation in File Page Setup to enhance appearance.) Bar and line graphs should have category labels and titles on the x- and y-axes, and a chart title. To begin to construct the graphs for Problems 1–4, type in the data, select the appropriate cells, and click on the Chart Wizard button. The Wizard will automatically guide you through the process of creating a chart. Format adjustments to your graphs can be made by right-clicking on the area that you wish to change.

1. Construct a bar graph with vertical columns and either regular or three-dimensional bars for the following data on the percent of U.S. workers using computers on the job. (To construct a vertical bar graph you must select the Column chart type in the Chart Wizard dialog box.) You will first need to enter the data in a worksheet.

   From the graph, what general conclusion can be stated about educational attainment and computer usage on the job?

   Percent of U.S. Workers Using Computers on the Job: 1998
   by Educational Attainment

Educational Attainment	Percent of Total
Not a high school graduate	10.0
High school graduate	34.2
Some college	50.4
Associate's degree	58.2
Bachelor's degree	68.8
Master's degree	71.2
Doctorate or professional degree	66.9

   Source: U.S. National Center for Education Statistics

2. Construct a pie chart (either regular or three-dimensional) for the following money income of households data. Use the Chart Options command in Chart Wizard to create a legend for the data in the graph and to show the appropriate percentages outside the slices of the pie.

   Percent Distribution of Household Income by
   Income Level: 1996

Household Income	Percent Distribution
Under $25,000	35.8%
$25,000–$49,999	29.9%
$50,000–$74,999	18.0%
$75,000 and over	16.3%

   Source: U.S. Bureau of Census

3. Construct a line graph for the following average annual percent change in earnings data. Use the years as category labels for the x-axis of your graph. Reformat the y-axis scale, with 2% as the lower limit and 4% as the upper limit, with major intervals of 0.5%. To make these changes, right click on the y-axis, choose Format Axis . . . , then select Scale and insert the appropriate values.

   What general trend can be seen in the graph?

Annual Percent Changes in Earnings: 1990 to 1997

Year	Percent
1990	3.6
1991	3.1
1992	2.4
1993	2.5
1994	2.7
1995	2.8
1996	3.3
1997	3.8

4. Create a bar graph for the percent distribution of the labor force data shown in the following chart. This graph will display two bars for each age group—one for 1980 and one for 1997. Use a legend to define which bars belong to which year. Compare the bars over the various age groups. What changes occurred in the labor force between 1980 and 1997?

Percent Distribution of the Civilian Labor Force by Age: 1980 and 1997

Age	1980	1997
15 to 19	8.8%	5.8%
20 to 24	14.9%	9.9%
25 to 34	27.3%	24.5%
35 to 44	19.1%	27.4%
45 to 54	15.8%	20.2%
55 to 64	11.2%	9.3%
65 and over	2.9%	2.9%

Source: U.S. Bureau of Labor Statistics

5. Microsoft Excel has extensive statistical functions. The spreadsheet below gives the results of a math test given to 20 students in a College Algebra class. The professor wants information regarding student performance. Help him compute statistics on the results of the test.

In this exercise, we are going to use the Name command to give a name to a range of cells. There are several advantages to using names in spreadsheet functions. Once defined, names adjust automatically for a movement of the range within the worksheet or for insertions or deletions within the range. For large amounts of data, referring to the data by name is easier than trying to find the first and last cell of the list.

To complete the spreadsheet, first alphabetize the names using the Sort Ascending button on the Microsoft Excel toolbar. Be sure that the scores go with the names when sorting. In Excel, the Name command in the Insert menu can be used to name a range of cells. We are going to name the cells containing the test scores. To use the command, select the range of cells in the Scores column and choose Insert Name Define. Type SCORES in the Names in Workbook text box. The name SCORES can now be used in place of the cell addresses in the Scores column.* For example, the function

$$=\text{AVERAGE(SCORES)}$$

can be used to find the average (mean) test score. Use the following functions and the SCORES name to complete the Statistics portion of the spreadsheet:

---

*If there are no empty rows or columns within the worksheet, Excel will automatically display the selected row or column title in the Names in Workbook text box.

Chapter 12

Name _____ Date _____ Class _____

COMPUTER SPREADSHEET EXERCISES

Calculation	Excel Function
Mean	=AVERAGE
Median	=MEDIAN
Mode	=MODE
Standard Deviation	=STDEVP
Highest Score	=MAX
Lowest Score	=MIN
Range	=MAX-MIN
Number of Scores	=COUNT

The professor has a policy to give a retest if the standard deviation is above 15 or the range is above 40. Construct an IF statement to determine if the retest should be given. Place the word "Yes" in the cell to the right of RETEST if the standard deviation is above 15 or the range is above 50; otherwise insert the word "No". Use the OR logical function in the condition of your IF. The complete formula would take the following general appearance in Microsoft Excel:

=IF(OR(STDEVP>15,RANGE>50),"YES","NO")

where the appropriate cell addresses are substituted for STDEV and RANGE in the formula.

Name	Scores	Statistics	
Shuck, Staci	94	Mean	
Fouts, Larry	76	Median	
Streb, Michael	85	Mode	
Peterson, Chuck	57	Standard Deviation	
Burke, Beth	71	Highest Score	
Hill, Mark	69	Lowest Score	
Jordan, Mary	52	Range	
Smith, Crystal	88	Number of Scores	
Zechman, Aaron	86		
Smith, Anna	80		
Willis, Jamar	79		
Johnston, Phil	63		
Brock, Patricia	76		
Lacey, Lee	48		
Miller, Kerri	99	**RETEST?**	
Verplatse, Allen	92		
Fowler, Brandi	73		
Wiggam, Marilyn	67		
Kocher, Beth	81		
Myers, Lou	90		

6. Compute final grades for the students shown in the spreadsheet that follows. The percentages you will use in calculating the grade are as follows:

Tests	60%
Project	10%
Final	30%

with the lowest test scores being dropped from the grade average.

First, alphabetize the names in the gradebook using the Sort Ascending button. Be sure to move the scores along with the names. Next, compute the test average (to the nearest tenth) using the Excel MIN function in the formula. This function will find the lowest test score (which is dropped) for each student. In Microsoft Excel, for example,

$$=MIN(B5:F5)$$

gives the minimum score recorded for cells B5 through F5. Find the test average by summing the scores, subtracting the lowest score, and dividing by 4. (You can also find the correct number to divide by using the COUNT function to determine the number of tests and then subtracting 1.) Compute the final average (to the nearest tenth) based on the weights given to each category. You must then write an IF statement to determine each student's grade using the following scale:

FINAL AVERAGE	FINAL GRADE
90%	A
80%	B
70%	C
60%	D
below 60%	F

The IF statement

$$=IF(AVG>=90,"A",IF(AVG>=80,"B",IF(AVG>=70,"C",IF(AVG>=60,"D","F"))))$$

will calculate the correct grade when the appropriate cell in the Final Average column is substituted for AVG in the formula.

After the gradebook is completed, find the mean, median, standard deviation, and range for each column except the Grade column (see Exercise 5 for a list of statistical functions). Display these values rounded to the nearest tenth.

	Test 1	Test 2	Test 3	Test 4	Test 5	Test Average	Project	Final Exam	Final Average	Grade
Simopolous, C.	84	79	88	71	67		87	95		
Sites, H.	76	68	49	82	73		71	79		
Cohen, S.	99	92	74	84	97		96	83		
Vu, L.	52	72	87	96	79		68	64		
Domingo, A.	45	67	76	84	61		67	62		
McKenzie, K.	90	83	64	95	73		88	81		
Pratt, D.	63	73	85	80	86		82	97		
Farley, B.	77	98	84	65	70		77	70		
Himic, D.	69	63	59	64	45		62	48		
Croasman, C.	82	87	89	81	92		90	94		

Mean
Median
Standard Dev.
Range

13 ■ Investing

OBJECTIVES

Upon completion of this chapter, the student should be able to:

1. *Read a balance sheet and calculate:*

    working capital;

    current ratio;

    quick assets; and

    quick ratio.

2. *Read an income statement and calculate:*

    return on investment;

    return on total assets; and

    accounts receivable–net sales ratio.

3. *Compute stock dividends and yields.*
4. *Compute bond yields.*
5. *Read and interpret stock and bond quotations.*

Investing should not be thought of as reserved for rich people. Most of us, at some time, have some money available that is not needed for current expenses, and instead of simply putting it into a safe deposit box or checking account, we invest it; that is, we put it into some business activity where, we hope, it will earn money for us. We have already considered the safest kind of investment—a deposit in an interest-bearing bank account in an insured bank. Investment in **securities**—stocks and bonds—is not as safe, but may earn more money.

Many of us, at some time, will be investors in the stock or bond market, either directly or indirectly. More than 75 million people in the United States own stock, with ownership about equally divided between men and women. In addition, millions of other people own securities indirectly through their pension funds, which invest in securities.

Deciding how much to invest and where to put the money is a complex process, but an introduction to the terminology and procedures is essential for any potential investor or anyone who expects to function in the business world.

13.1 READING A FINANCIAL REPORT

There are two principal financial reports issued by corporations that indicate the financial health of the company: the *balance sheet* and the *income statement*. The **balance sheet** shows what the company owes and owns at a particular time, usually the end of either the calendar or fiscal year. The **income statement** summarizes the company's operations during a period of time, usually a year, and shows whether there has been a profit or loss.

TABLE 13-1 HANDWERK AND STILES, INC.
BALANCE SHEET—DECEMBER 31, 1999

ASSETS		LIABILITIES	
Current assets		Current liabilities	
Cash	$ 866,000	Accounts payable	$ 819,000
Accounts receivable	1,706,000	Notes payable	663,000
Notes receivable	371,000	Accrued expenses	158,000
Inventories	1,513,000	Accrued taxes	314,000
Total current assets	4,456,000	*Total current liabilities*	1,954,000
Fixed assets		Long-term liabilities	
Land (at cost)	650,000	7% bonds due in 2017	1,350,000
Building (at cost)	1,300,000	*Total long-term liabilities*	1,350,000
Machinery, equipment, furniture			
and fixtures (at cost)	1,112,000	**STOCKHOLDERS' EQUITY**	
Total fixed assets	3,062,000		
Less accumulated depreciation	394,000	Capital stock	2,450,000
Net fixed assets	2,668,000	Retained earnings	1,370,000
		Total stockholders' equity	3,820,000
Total assets	$7,124,000	Total liabilities & stockholders' equity	$7,124,000

Balance Sheet

Table 13-1 is an illustration of a balance sheet for an apparel manufacturing company. The statement shows a balance (or equality) between two set of figures, total assets on one side, liabilities and stockholders' equity on the other.

The relationship among these quantities can be simply illustrated by the following analogy. Let us represent total assets by a $30,000 car that has a $10,000 bank loan outstanding against it. The balance then is

$$\text{Car value} = \text{Loan outstanding} + \text{Owner's equity}$$

$$\$30,000 = \$10,000 + \$20,000$$

If you own the car and it is sold for $30,000, the bank would get $10,000 (liabilities) and you would get $20,000 (owner's equity).

In accounting, the relationship is stated as:

> Assets = Liabilities + Stockholders' equity

In a corporation the stockholders are the owners.

Many of the terms used in Table 13-1 have already become familiar from work in earlier chapters, and the earlier references will be indicated in parentheses. Only the most commonly found items are discussed here. Other companies may have additional items in the balance sheets.

Assets. Assets are items of value owned by the business, such as cash, machinery and equipment, property, and money owed to the business. The assets portion of the balance sheet is ordinarily broken down into *current assets* and *fixed assets*.

Current assets are those that will be converted to cash within one year (or one operating cycle) of the balance sheet date. For Handwerk and Stiles, in addition to cash, current assets include:

- **Accounts receivable**—the money owed to the company by customers for goods and services purchased on credit (Chapter 4).
- **Inventories** (Chapter 5), which may include raw materials, in-process work, and finished stock. These categories may be shown separately in the balance sheet.
- **Notes receivable**—the short-term promissory notes owed to the company.

There is continual change in current assets, and the figures shown in the balance sheet are only as of that date. When inventories are sold on account, the amount of the sale is added to accounts receivable and subtracted from inventories. When a customer pays for merchandise previously sold, the amount becomes an addition to cash and a subtraction from accounts receivable.

Fixed assets (also referred to as **plant assets** or **plant and equipment**) include land, buildings, machinery, furniture and fixtures, trucks, and the like, which were bought for the company's use and not intended for sale. These assets are used for more than one year and their cost is depreciated over the life of the assets.* In the statement of 13-1 the value of fixed assets is shown at cost, and accumulated depreciation (Chapter 5) is shown separately and deducted.

Liabilities. Liabilities show what the company owes to its employees and creditors. The breakdown here is between **current liabilities**, those that must be paid within one year (or one operating cycle) of the balance sheet date, and **long-term liabilities**, those that need not be paid for more than one year.

Current liabilities for this company include:

- **Accounts payable**—what the company owes to creditors for goods and services it has bought on credit.
- **Notes payable**—what the company owes in short-term debt, like promissory notes to its suppliers and/or bank (Chapters 7 and 8).
- **Accrued expenses**—money owed for salaries and wages, interest on debts, insurance premiums, and so on.
- **Accrued taxes**—the amount of money owed by the company for taxes incurred, but not yet paid.

Long-term liabilities include mortgages (Chapter 11) and bonds, and the company IOUs to investors and banks that have loaned them money.

Stockholders' equity. The difference between assets and liabilities is a measure of how much of the company's assets actually belongs to the owners. In a corporation it is called **stockholders' equity** or net worth and is divided here into two parts, *capital stock* and *retained earnings*.

Capital stock is the value of the stock certificates issued by the company. It represents the value stated on the stock—**par value**—not the market value.

Retained earnings are the accumulated profits of the company that have not been distributed to the shareholders but have been kept (retained) for future use in expanding or improving the business.

What can the potential investor learn from a company's balance sheet? Several comparisons may be used to judge a company's financial condition. It is possible to give some indication of what to look for, but more precise interpretations depend on the particular industry.

---

*Land is not depreciated.

13.2 FINANCIAL RATIOS

The first comparison is between current assets and current liabilities. Both are short-term figures: current assets will be used to pay current liabilities. The comparison can be made by a simple subtraction to find how much money the company has to work with, that is, how large is its **working capital.**

Current Ratio

$$\text{Working capital} = \text{Current assets} - \text{Current liabilities}$$

For Handwerk and Stiles:

$$\text{Working capital} = 4{,}456{,}000 - 1{,}954{,}000 = \$2{,}502{,}000$$

This total can then be compared with totals from previous years. If the company is growing, as the investor would wish, working capital should have increased.

A second method of comparing current assets and current liabilities is in a ratio called the **current ratio** or the **working capital ratio.**

$$\text{Current ratio} = \frac{\text{Current assets}}{\text{Current liabilities}}$$

For Handwerk and Stiles:

$$\text{Current ratio} = \frac{4{,}456{,}000}{1{,}954{,}000} = 2.3 \text{ or } 2.3{:}1$$

In general, a current ratio of 2:1 is considered acceptable, and Handwerk and Stiles passes this test. A current ratio of 2.3:1 means that for every $1 in short-term debts, the company has $2.30 to cover it if the inventory is sold and accounts receivable are collected.

Creditors often perceive a company with a current ratio less than 2:1 as more likely to have financial problems, thus making it difficult for the company to obtain credit.

Quick Ratio (or Acid-Test Ratio)

As the name implies, this ratio is used to indicate how successful the company would be if it had to obtain money quickly. How much of its assets could be converted to cash if a sudden need arose? In addition to cash, **quick assets** may include accounts receivable and notes receivable (promissory notes owed to the company) that can be converted to cash. Inventory is excluded because it may not be possible to convert to cash quickly.

$$\text{Quick assets} = \text{Current assets} - \text{Inventory}$$

$$\text{Quick ratio} = \frac{\text{Quick assets}}{\text{Current liabilities}}$$

For Handwerk and Stiles:

$$\text{Quick assets} = 4{,}456{,}000 - 1{,}513{,}000 = \$2{,}943{,}000$$

$$\text{Quick ratio} = \frac{2{,}943{,}000}{1{,}954{,}000} = 1.5 \text{ or } 1.5{:}1$$

TABLE 13-2 HANDWERK AND STILES, INC.
INCOME STATEMENT—YEAR ENDED DECEMBER 31, 1999

		% OF NET SALES
Net sales	$8,591,000	100.0%
Cost of goods sold	5,780,000	67.3
Gross profit	2,811,000	32.7
Operating expenses	1,400,000	16.3
Operating profit	1,411,000	16.4
Taxes	700,000	8.1
Net income	$711,000	8.3%

For a company in sound financial condition, the quick ratio should be at least 1:1 and, again, Handwerk and Stiles passes the test. The ratio of 1.5:1 means that for every $1 of current liabilities the company has $1.50 quickly available to cover it.

Additional tests of financial soundness depend on data that come from the second financial statement, the income statement.

Income Statement

Table 13-2 shows an income statement for Handwerk and Stiles for 1999 and with it an analysis of each item as a percent of total sales. This is one form of the statement. Often, 2 years' data are shown with or without a percent comparison between the 2 years.

In practice, income statements are generally more detailed than the one shown here. Net sales may be shown as gross sales less returns and allowances. Expenses may be broken down into salaries, utilities, insurance, and so on. But the essential entries in the statement are those that show how the company's gross profit is used and how much is left for net income.

All the percents shown in the second column are calculated in the same way using $R \times B = P$ with net sales as the base. For cost of goods sold:

$$R \times B = P$$

$$R \times 8,591,000 = 5,780,000$$

$$R = \frac{5,780,000}{8,591,000}$$

$$R = 0.6727 = 67.3\%$$

Although the percent for net income is of greatest interest to a potential investor, the operating profit also is very important. It is an indication of the company's efficiency, particularly when compared with values from previous years and with those of other companies in the same industry. Taxes, the only item deducted from operating profit, are fixed by law and do not reflect the quality of management.

Several useful ratios are based on data from both the balance sheet and the income statement.

Return on Investment. The ratio of net income to stockholders' equity shows the percent return on investment. Stockholders would like this ratio to be high and should compare this figure with those from earlier years and with figures from other companies in the same industry.

$$\text{Return on investment} = \frac{\text{Net income}}{\text{Stockholders' equity}}$$

522 CHAPTER 13   INVESTING

For Handwerk and Stiles:

$$\text{Return on investment} = \frac{711{,}000}{3{,}820{,}000} = 0.1861 = 18.6\%$$

Return on total assets. The ratio of net income to total assets is another measure of a firm's profitability. Here, too, comparisons should be made with ratios from earlier years and with competitors' records. Investors would like this ratio to increase.

$$\text{Return on total assets} = \frac{\text{Net income}}{\text{Total assets}}$$

For Handwerk and Stiles:

$$\text{Return on total assets} = \frac{711{,}000}{7{,}124{,}000} = 0.0998 = 10.0\%$$

Accounts receivable–net sales ratio. The accounts receivable–net sales ratio indicates whether the firm is keeping its extension of credit to a safe level and whether this policy is changing.

$$\text{Accounts receivable–net sales ratio} = \frac{\text{Accounts receivable}}{\text{Net sales}}$$

For Handwerk and Stiles:

$$\text{Accounts receivable–net sales ratio} = \frac{1{,}706{,}000}{8{,}591{,}000} = 0.1985 = 19.9\%$$

Additional comparisons are significant in reviewing a company's financial position. For example, inventory levels are compared with sales and with current assets. In industries where inventories show wide seasonal fluctuations, like toys, automobiles, and apparel, only an average inventory figure should be used. Other comparisons involve securities issued by the company, and these will be considered later in the chapter.

Chapter 13

Name	Date	Class

TRY THESE PROBLEMS (Set I)

Answers to the following questions are based on the balance sheet and income statement for Lush Lawns, Inc., shown in Table 13-3. Round all ratios and percents to one decimal place where necessary.

TABLE 13-3   LUSH LAWNS, INC.
BALANCE SHEET—DECEMBER 31, 1999

ASSETS			LIABILITIES	
Current assets			Current liabilities	
Cash	$ 50,000		Notes payable	$ 17,500
Accounts receivable	24,300		Accounts payable	20,400
Notes receivable	10,000		Accrued expenses	5,300
Inventories	33,000		Accrued taxes	7,200
Total current assets	117,300		*Total current liabilities*	50,400
Fixed assets			Long-term liabilities	
Land (at cost)	31,500		Mortgage payable	49,000
Building (at cost)	45,000		*Total long-term liabilities*	49,000
Machinery, equipment (at cost)	34,000		STOCKHOLDERS' EQUITY	
Total fixed assets	110,500			
Less depreciation	30,000		Capital stock	58,000
Net fixed assets	80,500		Retained earnings	40,400
			Total stockholders' equity	98,400
Total assets	$197,800		Total liabilities & stockholders' equity	$197,800

INCOME STATEMENT—YEAR ENDED DECEMBER 31, 1999

Net sales	$159,000
Cost of goods sold	95,000
Gross profit	64,000
Operating expenses	34,600
Operating profit	29,400
Taxes	12,600
Net income	$ 16,800

1. What percent of net sales is each of the items in the income statment?

2. How much working capital does the company have?

3. Find the current ratio.

4. Another name for the current ratio is the _____.

5. What is the quick ratio for the company?

6. Another name for the quick ratio is the _____.

7. Find the return on investment for 1999.

8. Find the return on total assets for 1999.

9. What is the accounts receivable–net sales ratio for the company?

10. Based on the current ratio and the quick ratio, is the company in a good financial position?

Securities

Stocks and bonds are referred to as **securities.** Both are means by which a company raises money. When a corporation issues stock, it is selling part of the business to new owners. Every share of **stock** is a share in the ownership of the company that issued it. A **bond** is an IOU—a promissory note, evidence of a debt of the company to the bondholders. Formerly, buyers of stocks and bonds were issued certificates verifying the number of shares they owned. (See Figure 13-1.) Today, most companies involved in stock and bond trading have gone to a computerized "book-entry" form of ownership and will issue certificates only for a fee.

A stockholder's return on his investment, called **dividends**, varies according to whether the business makes a profit or loses money and whether the management decides to distribute the profit that it has. A bondholder expects a fixed percent return on the face value of the bond.

A stockholder often buys stock with the hope that the value of the investment will increase as the business grows or there is a general advance in stock prices. A bondholder expects the company to pay the face amount of the bond on the maturity date.

If a company goes out of business, its stockholders may collect little or nothing. The company's debts are paid first, and this includes bonds.

13.3 STOCKS

The balance sheet for Handwerk and Stiles, Inc., has the following item listed first under Stockholder's Equity.

Capital Stock $2,450,000

Consider this entry in greater detail.

FIGURE 13-1 Stock certificate.

Capital Stock

9,400 shares of cumulative preferred, 7% par value $100 each	$ 940,000
151,000 shares of common stock, par value $10 each	1,510,000
Total capital stock	$2,450,000

The two main categories of stock, **preferred stock** and **common stock**, are included. Also noted is the **par value** of each.

- **Par value** is the original price set by the company when the stock was issued. Handwerk and Stiles set a par value of $100 per share for its preferred stock and $10 per share for its common stock. Par value may have little relationship to the market value of the stock. The *true* value of a share of stock is how much someone else is willing to pay for it. Many stocks are issued with little or no par value.

- **Common stock** gives the owner the right to share in the profits and to vote on company policy. His share of the profits will vary from year to year and will depend on how much profit the company makes and how much is distributed to the stockholders. Handwerk and Stiles has issued 151,000 shares of common stock.

- **Preferred stock** has a fixed percent of its par value as the dividend to be paid each year. For Handwerk and Stiles, Inc., the preferred stock pays a dividend of 7% on $100, or $7 a year. This dividend must be paid before any dividends can be distributed to owners of common stock. Preferred stocks generally do not give the owner voting rights in the company.

- **Cumulative preferred** stock is stock such that if the company cannot pay the dividend one year, it continues as an obligation for the next year. If Handwerk and Stiles did not pay the preferred dividend one year, they would have to pay $14 a share the next year before distributing any common stock dividends. Cumulative preferreds are more common than noncumulative.

There are other features sometimes given to preferred stock to make it more attractive to buyers. These include:

- **Participating preferred** stock gives the owner the possibility of sharing in the company's profits in addition to the fixed-percent dividend. Most preferred stock is nonparticipating.

- **Convertible preferred** stock gives the owner the right to exchange shares for common stock under specified conditions.

- **Callable preferred** stock gives the company the right to buy back the stock under defined conditions.

Stock trading. A **stock exchange**, sometimes called a **stock market**, is a place where stocks that are registered with the exchange are bought and sold. The two largest exchanges in the United States are the New York Stock Exchange (NYSE), and the National Association of Securities Dealers Automated Quotations (NASDAQ). Generally, the New York Stock Exchange lists the largest and most established companies, while high technology or smaller and less established companies dominate the NASDAQ market.

Stockbrokers act as agents in buying and selling stock. Full-service stockbrokers earn commission based on the value and size of the transactions. Commissions will sometimes vary according to whether the customer is trading in **round lots** (multiples of 100 shares) or **odd lots** (fewer than 100 shares). It may be a much higher percent for a small order than for a large one. There are also discount stockbrokers, operating mainly on the Internet, most of whom charge a flat fee for any trade. The commissions for discount or online trades are much lower than those charged by traditional stockbrokers, but customers generally receive little or no guidance. Internet trading has grown rapidly in the recent past and is now the predominant method of buying and selling stocks.*

Several averages (Chapter 12) are used to record the price changes of stocks on the various exchanges. The most familiar is the **Dow Jones Industrial Average**, which tracks the performance of 30 "blue-chip" corporations listed on the New York Stock Exchange and has a long historical record. The **Standard & Poor's 500 Index** includes a far larger and more varied group of companies and therefore is the more accurate gauge of *overall* stock performance. The **NASDAQ Composite Index** is often used to measure the performance of NASDAQ stocks.

Stock dividends. Assume that Handwerk and Stiles had net income of $600,000 in one year. What would this mean for its stockholders?

First, the preferred stockholders would receive their dividends of 7% of par, or $7 a share. This stock is cumulative preferred and we are assuming that the dividend was paid in earlier years.

Then the directors decide how much of the profit is to be distributed as dividends to owners of common stock and how much will be *retained* for future use. They declare an annual dividend of $2.50 per share of common stock.

Following are the amounts that would be used for each of these purposes.

Preferred stock dividend ($7 × 9,400)	$ 65,800
Common stock dividend ($2.50 × 151,000)	377,500
Addition to retained earnings (600,000 − 65,800 − 377,500)	156,700
Total	$600,000

■ **Example 1**   Soup-to-Nuts, Inc., is a food packing company with the following stockholder's equity:

Capital Stock	
Preferred stock, 6% cumulative, $100 par value, 500 shares outstanding	$ 50,000
Common stock, $5 par value, 40,000 shares outstanding	200,000
Accumulated retained earnings	82,000
Total stockholder's equity	$332,000

In 1998 the company had an unprofitable year and did not pay any dividends, including the 6% on preferred stock. In 1999, business improved and net income of $59,000 was recorded. The directors declared a $1 per share annual dividend for the common stock.

1. How much was distributed to owners of preferred stock?

   Preferred stock dividends. Since no dividends were paid in 1998 and the preferred stock is cumulative, the 1999 payment must include 6% of par for 1998 as well. Each share of preferred stock gets 6% of $100 per year, or $6 for 1998 and $6 for 1999.

   Total preferred dividends = 12 × 500 shares = $6,000

---

*Fees for trading stocks online currently range from $7 to $35 per trade.

2. How much was distributed to owners of common stock?

Common stock dividends. The $1 dividend is paid on 40,000 shares. $40,000 is distributed to owners of common stock.

3. What change occurred in retained earnings?

Retained earnings. The profit remaining after all dividends are paid is added to retained earnings.

$$\begin{array}{r} \$\ 6{,}000 \text{ preferred} \\ \underline{40{,}000} \text{ common} \end{array}$$

Total dividends $46,000

Addition to retained earnings = Net income − Dividends

$$= 59{,}000 - 46{,}000$$

Addition to retained earnings = $13,000

The addition to retained earnings was $13,000. Total retained earnings increased to $95,000 (82,000 + 13,000). ∎

Stock yields. A dividend on common stock may be considered good or bad by a stockholder depending on how much was paid for the stock. In the past, stock prices have shown wide swings.

We will assume that the stock of Handwerk and Stiles followed this pattern. One stockholder bought shares at $85, and a second bought at $60. What percent yield did each have on his investment when he received the $2.50 annual dividend? In business mathematics terms the question is, What percent of the price per share is the annual dividend?

$$R \times B = P$$

% Yield × Price per share = Annual dividend

$$\% \text{ Yield} = \frac{\text{Annual dividend}}{\text{Price per share}}$$

For the first investor:

$$\% \text{ Yield} = \frac{2.50}{85.00} = 0.0294 \text{ or } 2.9\%$$

For the second investor:

$$\% \text{ Yield} = \frac{2.50}{60.00} = 0.0416 \text{ or } 4.2\%$$

For an investor deciding whether or not to buy the stock, the yield would be calculated on the basis of the current price and the dividends the company is paying. This calculation is called the **current yield**.

$$\text{Current yield} = \frac{\text{Annual dividend}}{\text{Current price}}$$

If the stock was selling at $70 and paying a $2.50 annual dividend:

$$\text{Current yield} = \frac{2.50}{70.00} = 0.0357 \text{ or } 3.6\%$$

■ **Example 2** Arthur and Bernard own stock in Soup-to-Nuts, Inc. Arthur bought his stock at $21 per share, and Bernard paid $18.

1. What is the yield for each if the company declares a $1 annual dividend?

$$\% \text{ Yield} = \frac{\text{Annual dividend}}{\text{Price per share}}$$

$$\text{For Arthur: } \% \text{ Yield} = \frac{1}{21} = 0.0476 \text{ or } 4.8\%$$

$$\text{For Bernard: } \% \text{ Yield} = \frac{1}{18} = 0.0555 \text{ or } 5.6\%$$

2. If the current price is $15, what is the current yield?

$$\text{Current yield} = \frac{\text{Annual dividend}}{\text{Current price}} = \frac{1}{15} = 0.0666 \text{ or } 6.7\% \quad ■$$

Price-Earnings Ratio

Another ratio often used by investors is the **price-earnings ratio** (PE ratio), found by dividing the price of the stock by its annual earnings per share.

$$\text{PE ratio} = \frac{\text{Price per share}}{\text{Annual earnings per share}}$$

To find annual earnings per share, use

$$\text{Earnings per share} = \frac{\text{Net income}}{\text{Number of outstanding shares}}$$

Price-earnings ratios can vary widely depending on the company's industry, the outlook for future earnings growth, and economic conditions. In general, a low PE ratio gives better prospects for future price increases. However, the PE ratio may be low because of poor future prospects and not because the stock is undervalued. Conversely, a PE ratio may be high not because the stock is overvalued, but because investors feel that earnings will increase in the future.

■ **Example 3** Carcorp reported net income of $820,000 and has 80,000 outstanding shares of stock. What is the PE ratio if the current price is $131?

$$\text{Earnings per share} = \frac{820{,}000}{80{,}000} = \$10.25$$

$$\text{PE ratio} = \frac{131}{10.25} = 12.8 \quad ■$$

Instead of investing in the shares of a single company, an investor might want to invest in the stocks of several different companies. This can be done by buying shares of a **mutual fund**.

A mutual fund receives money from many investors and uses this money to purchase a variety of stocks and often bonds or other income producing securities—depending upon the investment objectives of the fund. Many mutual funds specialize in a particular type of investment, while others are more general in their investment philosophy.

Mutual funds have become the most popular security investment in the United States. This undoubtedly is due to their professional management and the reduction of risk to the individual investor—since the investment is spread among the many different securities of the fund.

13.4 BONDS

Bonds are direct obligations of the corporation or government agency that issues them, and every effort will be made by the issuer to pay the interest and principal when due. Corporate bonds may be backed by a mortgage on specific property, and government bonds may be backed by specific revenues. A company may establish a sinking fund (Chapter 11) to redeem its bonds periodically or at the maturity date. Investment in bonds, therefore, offers greater security than investment in stock. But there is a trade-off, since bonds offer less chance of capital growth, particularly when compared with common stock. Bond prices do fluctuate, but in a much narrower range than prices of common stocks.

Every bond has a stated rate of interest. The life of a bond begins on the issue date and ends on the maturity date, which is the date on which the principal of the bond will be repaid. Bonds typically have a fairly long term, such as 10, 20, or 30 years, although their life can be as short as 2 years.

Most corporations issue bonds in $1,000 denominations. This amount is called the **face value** or **par value.** Bond prices are quoted as a percent of face value. For example, if a bond is priced at 97 it means that the price is 97% of $1,000, or $970.

Basic types of bonds include:

- **Callable bonds** may be paid off by the issuer before maturity under conditions specified when the bonds are issued. If the investor buys the bond to provide steady income over a period of years, this would be a disadvantage.

- **Convertible bonds** may be exchanged for other securities, usually the common stock of the same company, under specified conditions.

- **Debentures** are bonds that are backed only by the general credit of the borrower and do not have a claim on specific property.

- **Government bonds** are debts of the U.S. government and are the highest-grade security investment. Income is taxable by the federal government.

- **Municipal bonds** are issued by a state, a political subdivision of a state, or a state agency or authority. Income from these bonds is not subject to federal income taxes and may not be subject to taxes within the issuing state. For an investor in a high income bracket this is a main consideration, although not as important as it was when the tax brackets were higher.

Bond trading. Bonds, like stocks, are traded through brokers on exchanges. They are listed by the name of the company, the interest rate, and the redemption date. Brokerage firms can set their own commission charge for bond transactions, the typical fee being $5 to $10 to buy or sell a $1,000 bond.

Bond yields. The dollar income from a bond is set by the terms stated on its face value and remains constant. A $1,000 bond bearing 8% interest will pay

$$\$1{,}000 \times 0.08 = \$80$$

per year, usually in two installments. The percent yield from the bond to the investor, however, is 8% only if the face value is paid. If less than face value is paid, yield will be more than 8%.

If more than face value is paid, yield will be less than 8%. For example, if the $1,000 bond is bought at 96 ($960):

$$\% \text{ yield} = \frac{80}{960} = 0.08\bar{33} \text{ or } 8.3\%$$

If the $1,000 bond is bought at 104:

$$\% \text{ yield} = \frac{80}{1,040} = 0.0769 \text{ or } 7.7\%$$

Current yield is calculated as for stocks.

$$\text{Current yield} = \frac{\text{Annual interest}}{\text{Current price}}$$

If a $1,000, 7% bond is priced at 92 ($920), current yield would be:

$$\text{Current yield} = \frac{70}{920} = 0.0760 \text{ or } 7.6\%$$

If a $1,000, 7% bond is priced at 108 ($1,080), current yield is:

$$\text{Current yield} = \frac{70}{1,080} = 0.0648 \text{ or } 6.5\%$$

For a bond, however, current yield is seldom used alone. The bond has a definite maturity date, and the difference between the price paid and the face value to be paid at maturity must also be considered when calculating yield. If the buyer expects to hold the bond to maturity, the important figure is **yield to maturity.** Accurate calculation of yield to maturity is complex, and tables are available when precise yields are needed. The idea, however, can be illustrated by a $1,000, 9% bond with 10 years left to maturity. The bond is purchased at 110 ($1,100) and pays $90 a year.

When the bond matures, the owner will receive only the par value of $1,000 and will have lost $100 ($1,100 − $1,000) on the transaction. For each of the 10 years before maturity the owner has lost $10, and the return per year is really $80 instead of $90. The yield on basis is:

$$\text{Yield} = \frac{80}{1,000} = 8\%$$

But this change still omits the 10-year loss in interest on the $100 paid above par.

The precise yield to maturity can be closely approximated by the following calculation. Instead of using par value in the denominator of the yield fraction, use the average of par value and cost of the bond. For this bond the average is:

$$\frac{1,000 + 1,100}{2} = \frac{2,100}{2} = \$1,050$$

Approximate yield to maturity: $\frac{80}{1,050} = 0.0761$ or 7.6%

(Accurate table value is 7.53%.)

The two-step calculation can be combined into one formula, which looks worse than it is.

532 CHAPTER 13 INVESTING

$$\text{Approximate yield to maturity} = \frac{I + \left(\frac{P-C}{n}\right)}{\frac{P+C}{2}}$$

where I is the annual interest in dollars,
P is the par value,
C is the cost or current price,
n is the number of years to maturity.

An alternative form of the formula derived by algebra* is easier to use.

$$\text{Approximate yield to maturity} = \frac{2(nI + P - C)}{n(P + C)}$$

For a 9% bond, bought at 110, with 10 years left to maturity:

$$\text{Approximate yield to maturity} = \frac{2(10 \times 90 + 1{,}000 - 1{,}100)}{10(1{,}000 + 1{,}100)}$$

$$= \frac{2(900 - 100)}{10(2{,}100)} = \frac{2 \times 800}{21{,}000}$$

$$\text{Approximate yield to maturity} = \frac{1{,}600}{21{,}000} = 0.0761 \text{ or } 7.6\%$$

■ **Example 1**   Find the approximate yield to maturity when a 9%, $1,000 bond that will mature in 10 years is bought at 90. (This is the same bond used in the preceding illustration, but it is now bought at a discount, that is, below its par value.)

When the bond is paid at maturity, the buyer will receive $100 more than he paid ($1,000 − $900). Divided over 10 years, this gives an additional $10, and the annual return is $100 ($90 in interest and $10 in capital growth).

The average of the par value and the cost is:

$$\frac{1{,}000 + 900}{2} = \frac{1{,}900}{2} = \$950$$

$$\text{Approximate yield to maturity} = \frac{100}{950} = 0.1052 \text{ or } 10.5\%$$

By formula:

$$\text{Approximate yield to maturity} = \frac{2(nI + P - C)}{n(P + C)}$$

$$= \frac{2(10 \times 90 + 1{,}000 - 900)}{10(1{,}000 + 900)}$$

$$= \frac{2 \times 1{,}000}{10 \times 1{,}900} = \frac{2{,}000}{19{,}000}$$

Approximate yield to maturity = 0.1052 or 10.5% ■

* Multiply numerator and denominator by $2n$.

$$\frac{I + \left(\frac{P-C}{n}\right)}{\frac{P+C}{2}} \times \frac{2n}{2n} = \frac{2n\left[I + \left(\frac{P-C}{n}\right)\right]}{2n\left(\frac{P+C}{2}\right)} = \frac{2(nI + P - C)}{n(P + C)}$$

Chapter 13

TRY THESE PROBLEMS (Set II)

Use the following information for Problems 1 through 3:

You own 100 shares of 6% cumulative preferred stock with a par value of $65.

1. How large a dividend will you receive each year?

2. If you paid $67 per share for the stock, what is your yield?

3. If the dividend was not paid in 1998 and 1999, how much is due you in 2000?

Problems 4 through 7 are based on the following:

The U. R. Rich Co. had a net income of $163,000 in 1999 and declared a 70¢ dividend on its common stock. Its balance sheet showed:

Capital Stock

Cumulative preferred, 8%, 65,000 shares outstanding, $10 par value	$650,000
Common stock, $6 par value, 80,000 shares	480,000
Retained earnings	850,000

4. How much must be paid to all owners of preferred stock if the dividend was not paid the previous year?

5. How much would be paid to all owners of common stock?

6. How much (if any) will be added to retained earnings?

7. If the current price of the common stock is $9, what is the current yield?

8. Nasty Chemicals reported net income of $1,600,000 last year. If there are 250,000 shares outstanding and the current price is $80, what is the PE ratio?

Use the following information for Problems 9 and 10:

A 7% bond, with par value $1,000, is bought at 96.

9. What is the current yield?

10. If the bond matures in 10 years, what is the approximate yield to maturity?

11. Find the approximate yield to maturity for a $1,000, 6% bond, bought at 102 that will mature in 7 years.

Reading Stock and Bond Quotations

Stock and bond exchanges are where stock and bond trading takes place. Transactions involving stocks listed on the New York, American, and NASDAQ stock exchanges are reported in daily newspapers and in most financial publications. Some daily newspapers contain bond quotations, but *Barron's* and the *Wall Street Journal* publish complete information on the subject.

13.5 STOCK QUOTATIONS

A portion of the stock market page from the *Wall Street Journal* is shown in Figure 13-2. Stocks are listed alphabetically. To read the table, find the line containing the name of the stock of interest. Company names are given, as well as the corresponding stock symbol. For example, Johnson and Johnson is listed as JohnsJohns and the symbol used for trading is JNJ. Then, to read the stock quotation, read across the table. The two numbers in front of the company's

52 WEEKS					YLD		VOL				NET
HI	LO	STOCK	SYM	DIV	%	PE	100s	HI	LO	CLOSE	CHG
$89\frac{3}{4}$	$63\frac{1}{2}$	JohnsJohns	JNJ	1.00	1.2	32	23902	$83\frac{1}{4}$	81	$83\frac{1}{4}$	$+\frac{1}{16}$

name represent the highest price ($89\frac{3}{4}$) and lowest price ($63\frac{1}{2}$), respectively, that this stock sold for within the last 52 weeks. Parts of a dollar are quoted as fractions rather than cents. Thus $\frac{3}{4}$ means $0.75 and $\frac{1}{2}$ means $0.50. Following the company name and its symbol is the total dividends paid per share during the last 12 months. The table shows that owners of JNJ received $1.00 in dividends during the last year for each share owned. Dividends increase when the company is doing well and decrease when business is doing poorly. The 1.2 in the next column is the current yield, in percent. This means that the dividend of $1.00 per share is 1.2% of the current price of the stock. After the current yield is the price-earnings ratio (32)—the price of a share of stock divided by the last 12 months' earnings per share. Then comes the number of shares traded per day in hundreds—23,902. Next is $83\frac{1}{4}$, the highest price reached by the stock for the day, while 81 was the lowest. Following the low price for the day is $83\frac{1}{4}$, the price paid in the last transaction for the day. The last number, $+\frac{1}{16}$, is the difference between the price paid for the last share today and the last share on the previous day.

Some additional information contained in the table:

a. Price changes (up or down) of 5% or more are in boldface type (see International Paper).

b. A new 52 week high or low is indicated by up or down arrows next to a stock (see Lucent Technology).

c. Stocks having unusual daily volume are underlined (see Madeco).

■ **Example 1**

Use the stock table shown in Figure 13-2 to answer the following:

1. Find the highest price for the last 52 weeks for IBM.

   The highest price in the last 52 weeks was $192\frac{3}{4}$ or $192.75, the first number on the left.

2. What is the dividend for Lockheed Martin?

   The dividend is $0.88 per share, given just after the company symbol, LMT.

3. What is the current yield for International Flavor?

   The current yield is 3.4%, one column to the right of the dividend.

4. Find the sales volume for MagneTek.

   The volume was 144,700 shares.

-I-I-I-

	52 Weeks Hi	52 Weeks Lo	Stock	Sym	Div	Yld %	Pe	Vol 100s	Hi	Lo	Close	Net Chg
	14 7/8	13	InsrdMuniFd	PIF	.77	5.4	...	288	14 7/16	14 5/16	14 3/8	− 1/16
n	24 1/4	12 3/8	IntgtElec	IEE		...	25	589	22 13/16	21 1/2	22 1/8	− 11/16
	39 15/16	9 1/2	IntgtHS	IHS	.02	.2	dd	3337	13 15/16	12 13/16	13 1/4	+ 7/16
	6 3/16	1 3/8	Intellicall	ICL		...	cc	1036	3	2 13/16	2 13/16	...
	34 1/4	13 1/4	InterimSvc	IS		...	19	3755	23 1/8	22 1/8	23 1/8	+ 1
	7 3/16	1 7/8	Interlake	IK		...	11	1126	7	6 15/16	6 15/16	+ 11/16
	8 1/8	2 7/16	IntlCer ADR	ICM		...	...	30	3 1/4	3 3/16	3 3/16	− 1/8
	36	26 11/16	IntAlum	IAL	1.20	4.0	10	7	29 15/16	29 7/8	29 7/8	...
	192 3/4	95 5/8	IBM	IBM	.88	.5	30	45884	192	185 5/8	187 9/16	− 2 5/8
	28 5/16	25 1/8	IBM pf		1.88	6.9	...	62	27 7/16	27 5/16	27 3/8	+ 1/16
	50 5/16	32 1/16	IntFlavor	IFF	1.52f	3.4	23	1480	44 3/8	42 7/8	44 5/16	+ 1 1/16
	28 11/16	16 1/8	IntGameTch	IGT	.12	.5	20	3588	23 7/16	22 7/16	22 7/8	− 1/16
	35	10 1/8	IntHomeFood	IHF		...	cc	2171	20 15/16	20 3/8	20 1/2	+ 1/4
	31 7/16	15 1/8	IntMultfood	IMC	.80	3.1	dd	403	26 1/8	25 3/8	25 7/16	− 7/8
n	25 15/18	25	IntPapCapill	IPP	.37p	...	...	321	25 13/16	25 11/16	25 11/16	− 1/16
	55 1/4	**35 1/2**	**IntPaper**	**IP**	**1.00**	**2.2**	**44**	**51936**	**46**	**42 13/16**	**45 1/2**	**+2 15/16**
	14 3/4	4 1/4	IntHectifr	IRF		...	54	3113	10 13/16	10 1/8	10 3/4	+ 5/8
	17 5/16	14 1/2	IntShiphld	ISH	.25	1.6	21	15	15 7/8	15 7/8	15 7/8	+ 3/16
	20 1/2	9 9/16	IntSpcPdt	ISP		...	14	435	13 3/16	13	13 1/8	...
	18 15/16	9 7/8	Interpool	IPX	.15	1.0	12	184	15 9/16	15	15	− 11/16
	80 5/8	45 1/8	IntpubGp	IPG	.60	.8	37	5551	76 7/8	75 1/8	76 3/4	+ 7/8
	38	23 3/8	IntstBaker	IBC	.28	1.1	15	1592	26 3/8	25 3/8	26 1/4	+ 3/8
	35 3/8	28	IntstEngy	LNT	2.00	6.5	20	747	31	30 5/8	30 13/16	− 1/8
	41 1/4	20 3/16	InstJhnson	IJL	.20	.6	15	30	31 1/8	31 1/8	31 1/8	...
	7 5/8	3 3/8	IntertanInc	ITN		...	dd	955	6 5/16	6 1/4	6 5/16	...
▲	32 5/8	18	IntimtBrnd	IBI	.56	1.7	26	9510	33 3/16	32 3/8	32 7/16	− 1/8
	21 7/8	10 3/16	Intrawest g	IDR	.16	...	...	131	18 5/16	18 1/16	18 1/16	− 1/4
s	20	13 1/16	InvscoGlbl	GHS	3.71e	18.7	...	709	20	19 11/16	19 13/16	+ 3/16
	16 5/16	14 9/16	InvGrdMuni	PPM	.90	5.6	...	277	16	15 7/8	15 15/16	...
	13 3/8	2 15/16	Iomega	IOM		...	dd	150960	10 3/16	9 3/16	9 3/8	− 1/4
	45 7/8	22 1/2	Ionics	ION		...	22	516	30 1/4	29 1/2	30 1/4	+ 1/2

-J-J-J-

	52 Weeks Hi	52 Weeks Lo	Stock	Sym	Div	Yld %	Pe	Vol 100s	Hi	Lo	Close	Net Chg
	5 7/8	2 3/16	JAlexndrs	JAX		...	dd	95	4 1/2	4 5/16	4 1/2	+ 3/16
	26 3/4	25 5/16	JCPL Cap MIPS		2.14	8.3	...	14	25 11/18	25 9/16	25 11/16	+ 1/16
s	23 1/8	18 1/2	JDN Rlty	JDB	1.44	6.5	14	1320	22 7/16	22	22 1/4	+ 5/16
n	25 3/4	24	JDN Rlty pfA		2.34p	...	...	17	24 7/8	24 13/16	24 7/8	+ 1/16
	20 3/4	13	JLG Ind	JLG	.02	.1	14	998	16	15 3/8	15 15/16	+ 9/16
	39 1/2	9 1/4	JLK Direct A	JLK		...	10	127	9 13/16	9 5/8	9 3/4	+ 1/8
	26	18 1/2	JP Realty	JPR	1.861	9.1	14	198	20 11/16	20 3/8	20 3/8	− 3/16
	60	44 3/4	JSB Fnl	JSB	1.80f	3.2	12	113	56 9/16	55 3/4	56 9/16	+ 1/2
▲	77 3/8	23	JabilCircuit	JBL		...	53	4883	79	74	74 7/8	− 2 1/2
	13 15/16	9 1/8	Jackpot	J		...	12	455	9 5/16	9 1/8	9 1/4	+ 1/8
	42 3/8	24 3/4	JacobEngrg	JEC		...	19	1063	40	39 3/16	40	+ 1/4
	4 3/4	**1 7/16**	**Jakarta GrFd**	**JGF**		**...**		**315**	**2 13/16**	**2 5/8**	**2 11/16**	**− 3/16**
	9	5 1/4	JapanEquity	JEQ		...	...	235	7 1/2	7 3/16	7 1/2	− 1/16
	7	3 5/8	JapanOTC	JOF	.01e	.2	...	272	5 13/16	5 9/16	5 9/16	− 3/16
	10 1/2	3 1/2	JF China	JFC		...	...	170	6	5 7/8	5 7/8	− 1/16
	8 3/16	4 1/2	JF India	JFI		...	...	257	5 11/16	5 5/8	5 11/16	+ 1/8
	59 1/2	16 9/16	JeffriesGp	JEF	.20	.4	17	1576	51 1/8	49 7/8	51 1/8	+ 1 1/8
s	78 3/8	48 5/8	JeffPilot	JP	1.18	1.5	21	1783	77 3/8	74 5/8	77 3/8	+2 13/16
	39 1/2	13 1/4	JSmurfit	JS	.68e	3.3	...	301	20 5/8	19 3/4	20 1/2	+ 1/4
	148	**80 3/8**	**JeffPilot ACES**		**5.26**	**4.5**	**...**	**21**	**118**	**112 1/2**	**118**	**+ 6 5/8**
	7 3/8	4 1/4	JennyCraig	JC		...	13	105	6	5 7/8	5 7/8	− 1/16
	14	3 3/8	JilinChem	JCC	.31e	5.9	...	88	5 7/16	5 1/8	5 1/4	− 3/16
	31 7/8	14 15/16	JoAnnStrs A	JASA		...	24	72	16 3/4	16 1/16	16 1/8	− 5/8
	27 1/2	13 1/2	JoAnnStrs B	JASB		...	...	24	14	13 11/16	13 15/16	+ 3/8
	41 3/4	31 9/16	JohnNuveen	JNC	1.04	2.8	18	64	37 3/4	37 1/2	37 3/4	+ 3/8
	8 15/16	3 1/8	JQHamons A	JQH		...	dd	150	3 15/16	3 7/8	3 7/8	+ 1/16
	18 1/8	9 5/8	JohnsManvil	JM	.24	1.5	15	777	15 15/16	15 3/8	15 1/2	− 1/4
	89 3/4	63 1/2	JohnsJohns	JnJ	1.00	1.2	32	23902	83 1/4	81	83 1/4	+ 1/16
	61 7/8	40 1/2	JohnsCtrl	JCI	1.00f	1.7	16	1992	59 15/16	58 3/8	59 7/8	+ 1 3/16
	6 1/8	2 13/16	JohnstnInd	JII	.40	12.5	dd	57	3 3/16	3 3/16	3 3/16	+ 1/16
s	37 3/4	15 7/8	JonesApparel	JNY		...	12	4425	23 7/8	22 1/8	23 5/16	− 5/16
n	25 1/2	25 1/15	JoseSea QUIDS		2.00	7.9	...	1916	25 5/16	25 1/8	25 3/16	...
	27 1/8	18 3/4	Jostens	JOS	.88	3.4	17	692	25 13/16	25 5/16	25 3/4	− 1/8
	23 7/8	12 1/8	JrnlRgstr	JRC		...	15	4570	14 3/8	13	14 3/8	− 1/8

-L-L-L-

	52 Weeks Hi	52 Weeks Lo	Stock	Sym	Div	Yld %	Pe	Vol 100s	Hi	Lo	Close	Net Chg
s	40 5/8	16 5/8	LinenThing	LIN		...	47	3220	38 5/8	36	37 7/8	− 5/8
	68	47 7/16	Litton	LIT		...	16	566	63 3/16	62	62 3/16	− 5/16
	54 7/8	25	LizClabrn	LIZ	.45	1.3	12	3338	35 3/16	34 1/16	34 5/8	− 1/4
s▼	58 15/16	39 1/2	LockhdMartin	LMT	.88	2.2	12	129026	40 7/8	39 3/8	39 15/16	+ 1/16
	22 1/2	**2 1/4**	**Lodglan**	**LOD**		**...**	**dd**	**1508**	**5 7/8**	**5 1/4**	**5 1/2**	**− 5/16**
	28 7/8	7	LoewenGp	LWN	.10j	...	17	6198	7 15/16	7 5/16	7 5/8	− 1/4
	27 5/8	16 1/2	Loewen Gp pfA		2.36	12.0	...	29	19 11/16	19 3/8	19 11/16	...
n	20	7 1/4	LoewsCinEnt	LCP		...	...	749	8 7/8	8 5/8	8 3/4	...
	108 1/4	78	LoewsCp	LTR	1.00	1.0	6	2774	104 1/2	100 1/2	103 3/4	+ 3 5/16
s	42 7/16	25 5/8	LoneStar	LCE	.10	.3	13	865	36 7/8	36 3/8	36 7/16	...
	66	33 1/2	LoneStar wt			...	...	54	54 1/2	54	54 1/4	− 1/4
	33 7/16	**8 5/8**	**LoneStarTech**	**LSS**		**...**	**38**	**1170**	**12**	**11 1/16**	**12**	**+ 11/16**
	44 1/2	26	LongsDrg	LDG	.56	1.5	23	1091	37 3/16	36 5/8	37 1/8	+ 1/4
	17 7/16	9 7/16	LongvwFibr	LFB	.08m	.7	dd	915	12 5/16	11 11/16	12 1/4	+ 1/2
	33 15/16	10 3/4	Loral Space	LOR		...	dd	19606	20 1/16	19 3/16	19 1/4	− 3/8
	20 5/8	10 1/2	L Dreyfus NG	LD		...	dd	253	15	14 3/8	14 7/8	+ 1/2
	23 7/8	25 1/4	LA Cap QUIPS		2.25	8.8	...	4	25 11/16	25 11/16	25 11/16	...
	24 3/16	16 3/8	LA Pac	LPX	.56	2.9	dd	2786	19 3/4	19 1/16	19 5/8	+ 9/16
s▲	54 3/4	21 9/16	Lowes Cos	LOW	.12	.2	43	21874	55 1/8	54 1/16	54 11/16	+ 7/8
	40 3/16	22 3/8	Lubrizol	LZ	1.04	4.0	13	1972	26 1/8	25 5/8	26 1/16	− 1/4
	19 1/2	13 3/8	LubysCafe	LUB	.80	5.2	77	411	15 1/2	15 1/16	15 7/16	− 1/16
	47 7/16	29	LucasVarity	LVA	.99e	2.6	...	2539	37 9/16	36 5/8	37 9/16	− 3/16
s▲	119 7/8	36 23/32	LucentTech	LU	.16	.1	cc	65239	120	114 7/8	115 1/4	− 1 1/4
s	19 1/2	7 1/2	LuxottGp	LUX	.15e	1.4	...	751	11	10 3/4	10 7/8	− 1/8
	20 1/4	9 1/16	Lydall	LDL		...	13	181	12 1/8	11 13/16	12 1/8	− 1/8
	38 1/8	15	Lyondell	LYO	.90	5.1	11	2564	18	17	17 11/16	+ 9/16

-M-M-M-

	52 Weeks Hi	52 Weeks Lo	Stock	Sym	Div	Yld %	Pe	Vol 100s	Hi	Lo	Close	Net Chg
	39 7/8	22 9/16	MGM Grand	mGG		...	23	1774	33 1/8	30 15/16	32 3/16	+ 1 1/16
	26 13/16	16 3/4	MI SchottHm	MHO	.20	.9	7	287	21 1/2	21 1/4	21 3/8	...
	4 3/16	3	ML Macdmia	NUT	.30	9.4	25	168	3 3/16	3 1/8	3 3/16	...
	26	13 3/4	MMI Cos	MMI	.32	1.8	dd	332	17 11/16	17 3/8	17 11/16	+ 3/8
n	32 3/4	27 9/16	MONY Gp	MNY		...	...	1562	31 3/8	29 5/8	30 1/8	− 13/16
	26 1/4	25 1/8	MP&L QUIPS		2.01	7.9	...	4	25 1/2	25 1/2	25 1/2	− 1/8
s	33 1/2	12 1/4	MSC InDirA	MSM		...	31	1517	22 5/8	20 1/2	21 5/16	− 1 1/16
	17 3/16	10 1/8	MYR Gp	MYR	.14	1.2	11	21	12	11 5/8	11 5/8	− 1/4
	20 1/2	7 5/8	MacGray	TUC		...	...	196	11 1/4	11 1/8	11 1/8	...
s	42 3/8	23 85/256	MacDermd	MRD	.08	.2	.32	523	41 5/8	40 5/8	41 1/4	+ 3/16
	30 3/8	22 1/4	Macerich	MAC	1.94f	7.4	28	1216	26 3/8	25 7/8	26 5/16	+ 1/16
	40 15/16	26 1/8	MackCaliRlty	CLI	2.20	7.2	53	882	31	30 1/2	30 11/16	− 1/16
	12 3/8	5 3/16	MacNealSch	MNS		...	9	496	7 1/4	6 7/8	6 7/8	− 5/16
	17 5/8	4	Madeco ADR	MAD	.80e	9.8	...	30233	8 1/8	7 5/8	8 1/8	+ 1/4
	28 7/16	**6 3/8**	**MagelinHlth**	**MGL**		**...**	**22**	**1799**	**9 7/8**	**9 3/16**	**9 7/8**	**+ 1/2**
	79 1/2	53 7/8	MagnaInt g	MGA	1.32	...	...	2119	62 15/16	62	62 3/4	+ 1/2
	20 11/16	8 15/16	MagneTek	MAG		...	9	1447	11 15/16	11 7/16	11 1/2	− 5/16
s	24 15/16	5 3/4	MailWell	MWL		...	13	1314	12 3/16	11 3/4	12	+ 3/16
	18 5/16	14 1/2	MalanRlty	MAL	1.70	11.6	51	52	15 1/16	14 11/16	14 11/16	− 1/2
	9 15/16	2 15/16	MalaysFd	MF	.03e	.6	...	140	5 1/4	5	5 1/8	+ 1/8
	40 1/16	19 3/4	Mallinckrodt	MKG	.66	2.1	15	3500	31 9/16	30 1/2	31 3/16	+ 9/16
	12 1/2	9 9/16	MngdHghInc	MHY	1.01	9.7	...	885	10 3/8	10 5/16	10 3/8	+ 1/16
	14 13/16	11 1/8	MgdHiYld	PHT	1.26	10.7	...	115	11 15/16	11 13/16	11 13/16	...
n	15 1/8	11 3/16	MngdHiYldPl	HYF	1.50f	11.9	...	946	12 3/4	12 9/16	12 9/16	...
	12 7/16	10 11/16	MgdMuni	MMU	.59a	5.2	...	548	11 3/8	11 3/16	11 1/4	− 1/8
	12 1/2	10 5/8	MgdMuni II	MTU	.59a	5.3	...	315	11 5/16	11 1/8	11 1/8	− 3/16
	47	24 1/2	Manitowc	MtW	.45	1.0	16	302	43 13/16	43 1/16	43 7/16	− 3/8
	45 3/4	18 1/4	Manpower	MAN	.18	.7	18	1970	26 7/8	25 3/8	26 1/8	+ 5/8
	27 1/4	22	MfrHome	MHC	1.45	5.8	...	502	25 1/4	24 13/16	25	...
	18 7/16	**12 7/16**	**MarcusCp**	**MCS**	**.22**	**1.5**	**16**	**260**	**14 15/16**	**14 1/8**	**14 15/16**	**+ 7/8**
	26 11/16	7	MarineDrill	MRL		...	7	6118	9 1/2	8 15/16	9 3/16	− 1/16
n	14 3/16	7 5/16	Marinemax	HZO		...	...	19	9 3/4	9 5/8	9 5/8	...
	21 1/2	2 3/4	MarinNtwk	MPN		...	dd	1639	4 5/16	4 1/8	4 5/16	+ 1/16
	11 5/8	6 5/16	Maritrans	TUG	.40f	6.3	14	258	6 7/16	6 3/8	6 3/8	...
	24 1/8	12 1/8	MarldV	IV	.20	1.4	14	7276	14 3/16	13 9/16	14	+ 7/16
	187	132	MarkelCp	MKL		...	18	95	183 1/2	178 1/2	178 3/4	− 3 1/4
n	37 15/16	19 3/8	MarriottInt A	MAR	.20	.6	25	8770	34 7/16	32 5/8	34	+ 1 7/16
s	64 5/16	43 3/8	MarshMcL	MMC	1.60	2.6	29	7336	61 1/8	59	61	+ 1 1/2
	34 13/16	21 1/2	Marshl Ind	MI		...	15	262	25 5/8	23 11/16	24 9/16	− 9/16
	62 3/16	**35 13/16**	**MartnMarMat**	**MLM**	**.52**	**.9**	**25**	**1180**	**59 7/8**	**57 1/4**	**57 7/8**	**− 9/16**

FIGURE 13-2 Stock information from the *Wall Street Journal*. Reprinted by permission of the *Wall Street Journal*, © January 11, 1999, Dow Jones & Company, Inc. All Rights Reserved Worldwide.

5. Assume that 100 shares of Marine Drill were bought at the high price for the day. If the broker received a 3% commission, what was the total amount paid?

$$\text{Total amount paid} = \text{Trading cost} + \text{Commission}$$
$$= (100 \times 9\tfrac{1}{2}) + 0.03\,(100 \times 9\tfrac{1}{2})$$
$$= 950 + 28.50$$
$$= \$978.50 \quad \blacksquare$$

Determining Rate of Return on a Stock Investment

Return on a stock investment is based on the price paid, the current dividend, broker commissions, and selling price.

Assume that the following transactions involving the PQR Corp. have occurred:

Purchased 200 shares at $37.50

Sold 200 shares 1 year later at $52

3% broker commission to buy or sell

Current dividend of $1 per share

The total costs of purchasing and selling are:

PURCHASED		SOLD	
200 shares at $37.50	$7,500	200 shares at $52	$10,400
Broker's commission (0.03 × $7,500)	225	Broker's commission (0.03 × $10,400)	312
Total cost	$7,725	Total received	$10,088

The total gain on the shares of stock can be calculated as follows:

Total received	$10,088
Total cost	7,725
Net gain	2,363
Dividends ($1 share)	200
Total gain	$2,563

To find the rate of return, the total gain (or loss) is divided by the base (total cost).

$$\text{Rate of return} = \frac{\text{Total gain}}{\text{Total cost}}$$

$$= \frac{\$2,563}{\$7,725} = 33.2\%$$

13.6 BOND QUOTATIONS

Bond prices fluctuate, depending on the rate of interest the bond pays, the current market rate, and the financial status of the issuer. If a bond sells at a price below its face value, it is said to be

Bonds	Cur Yld.	Vol.	Close	Net Chg.
AES05	...	6	$105\frac{1}{2}$	$-2\frac{1}{2}$
AMR 9s16	7.6	25	119	$-1\frac{1}{8}$
ATT $4\frac{3}{8}$99	4.4	5	$99\frac{1}{2}$	$+\frac{1}{16}$
ATT $7\frac{1}{8}$02	6.8	65	$105\frac{1}{4}$	$-\frac{1}{8}$
ATT $6\frac{3}{4}$04	6.4	39	$105\frac{5}{8}$	$-\frac{1}{4}$
ATT 7s05	6.5	10	$107\frac{1}{4}$	$-\frac{1}{2}$
ATT 8.2s05	7.9	20	$103\frac{7}{8}$	...
ATT $7\frac{1}{2}$06	6.7	30	$112\frac{1}{2}$	...
ATT $8\frac{1}{8}$22	7.6	45	107	$-\frac{3}{4}$
ATT $8\frac{1}{8}$24	7.6	23	$107\frac{1}{4}$	-1
ATT $8\frac{5}{8}$31	7.7	23	$111\frac{7}{8}$	$+\frac{3}{8}$
Aames $10\frac{1}{2}$02	12.4	82	85	$+2$
AlldC zr01	...	25	$85\frac{5}{8}$	$-\frac{1}{8}$
AlldC zr09	...	110	$50\frac{1}{8}$	$-\frac{1}{2}$
Alza 5s06	cv	25	$133\frac{1}{8}$	$-\frac{7}{8}$
Amresco $8\frac{3}{4}$99	8.8	35	99	...
Amresco 10s03	11.8	194	85	$+\frac{3}{4}$
Amresco 10s04	11.8	415	$84\frac{7}{8}$	$+2\frac{3}{8}$
AnnTaylr $8\frac{3}{4}$00	8.6	182	$101\frac{1}{4}$	$-1\frac{1}{8}$
Argosy 12s01	cv	95	$100\frac{7}{8}$	$+\frac{1}{4}$
Argosy $13\frac{1}{4}$04	11.8	50	112	$-\frac{1}{2}$
ARch $10\frac{7}{8}$05	8.4	14	129	...
BellPa $7\frac{1}{8}$12	7.0	9	$101\frac{7}{8}$	$-\frac{1}{8}$
BellsoT $6\frac{1}{2}$00	6.4	70	$101\frac{3}{8}$	...
BellsoT $6\frac{3}{8}$04	6.1	32	105	$-\frac{1}{8}$
BellsoT $5\frac{7}{8}$09	5.7	30	103	$-\frac{1}{2}$
BellsoT $8\frac{1}{4}$32	7.3	28	$112\frac{1}{4}$	$-\frac{1}{2}$
BellsoT $7\frac{3}{8}$32	7.3	8	$108\frac{1}{4}$	...
BellsoT $7\frac{1}{2}$33	6.9	11	$108\frac{1}{4}$	$\frac{1}{2}$
BellsoT $6\frac{3}{4}$33	6.6	195	$102\frac{1}{4}$	$\frac{1}{4}$
BethSt $8\frac{1}{8}$01	8.3	30	$100\frac{7}{8}$	... $\frac{5}{8}$
Bevrly 9s06	8.5	99	$105\frac{5}{8}$	$\frac{3}{8}$
Bluegrn $8\frac{1}{4}$12	cv	25	103	$+1$
Bordn $8\frac{1}{8}$16	8.4	73	$99\frac{1}{2}$	$-\frac{5}{8}$
BosCelts 6s38	9.3	121	$64\frac{3}{8}$	$+2\frac{1}{8}$
BurN 6.55s20k	6.8	20	$96\frac{1}{4}$	-1
ChaseM $7\frac{3}{4}$99	7.7	3	$100\frac{9}{16}$	...
ChaseM 8s04	7.8	5	$102\frac{1}{4}$	$+\frac{1}{8}$
ChaseM $7\frac{7}{8}$04	7.7	26	102	$+\frac{7}{8}$
ChespkE $9\frac{1}{8}$06	12.3	47	74	$+2$
ChckFul 7s12	cv	22	94	$+\frac{1}{2}$
ChryF $13\frac{1}{4}$99	12.7	34	$104\frac{5}{8}$	$-\frac{3}{32}$
Clardge $11\frac{3}{4}$02f	...	1	84	$+13$
ClrkOll $9\frac{1}{2}$04	9.4	351	$101\frac{1}{4}$	$-\frac{1}{4}$
Coastl $8\frac{1}{8}$02	7.8	25	$104\frac{3}{8}$	$-1\frac{5}{8}$
CoeurDA $7\frac{1}{4}$05	11.7	75	62	$+1$
Coeur $6\frac{3}{8}$04	cv	26	$60\frac{1}{2}$	...
CompUSA $9\frac{1}{2}$00	9.4	15	$101\frac{1}{8}$	$+\frac{1}{8}$
Consec $8\frac{1}{8}$03	7.9	50	$103\frac{1}{2}$	$-\frac{3}{8}$
ConPort 10s06	13.3	20	75	$+4$
Converse 7s04	cv	167	$41\frac{3}{4}$	$+1\frac{1}{4}$
Dr Hrtn 10s06	9.4	10	$106\frac{1}{2}$	$+1$
DVI $9\frac{7}{8}$04	10.0	40	99	...
DataGen 6s04	cv	150	98	-2
DukeEn $6\frac{1}{4}$04	6.1	10	102	$+\frac{7}{8}$
DukeEn $6\frac{7}{8}$23	6.8	10	$101\frac{3}{4}$	$-\frac{1}{4}$
DukeEn $7\frac{7}{8}$24	7.6	65	$103\frac{3}{4}$	...
DukeEn $6\frac{3}{4}$25	6.7	56	$101\frac{1}{4}$	$+\frac{7}{8}$

Bonds	Cur Yld.	Vol.	Close	Net Chg.
GrnTrFn $10\frac{1}{4}$02	9.6	5	107	$-\frac{3}{8}$
Hallwd 10s05	13.0	10	77	...
Hallwd 7s00	7.3	40	96	$+1$
HlthcrR 6.55s02	cv	58	$91\frac{1}{2}$	$+\frac{1}{2}$
Hlthso $9\frac{1}{2}$01	9.3	35	$102\frac{5}{8}$	$-\frac{1}{8}$
Hexcel 7s03	cv	15	90	...
Hills $12\frac{1}{2}$03f	...	1315	$93\frac{3}{8}$	$+1\frac{1}{8}$
Hilton 5s06	cv	98	96	$+1$
InldStl 7.9s07	8.0	5	$99\frac{1}{4}$	...
IBM $6\frac{3}{8}$00	6.3	62	$101\frac{1}{8}$	$-\frac{1}{2}$
IBM $7\frac{1}{4}$02	6.8	54	$106\frac{3}{8}$	$-\frac{1}{4}$
IBM $7\frac{1}{2}$13	6.5	5	116	$-\frac{5}{8}$
IBM $8\frac{3}{8}$19	6.7	5	$124\frac{3}{8}$	$+1$
IBM 7s25	6.6	1	$105\frac{3}{4}$	$-1\frac{7}{8}$
IBM $6\frac{1}{2}$28	6.3	1	$102\frac{5}{8}$	$-\frac{7}{8}$
IPap dc$5\frac{1}{8}$12	5.8	2	$88\frac{3}{8}$	$+1\frac{7}{8}$
JCPL $7\frac{1}{8}$04	7.0	6	$101\frac{1}{2}$	$-\frac{7}{8}$
KaufB $9\frac{3}{8}$03	9.2	94	$101\frac{7}{8}$	...
KaufB $7\frac{3}{4}$04	7.7	115	$100\frac{1}{2}$	$+\frac{1}{2}$
KaufB $9\frac{5}{8}$06	9.0	110	107	$-\frac{1}{8}$
KentE $4\frac{1}{2}$04	cv	40	$77\frac{1}{4}$	$+\frac{1}{2}$
LeasSol $6\frac{7}{8}$03	cv	260	30	$+5\frac{1}{8}$
Leucadia $8\frac{1}{4}$05	8.0	59	$103\frac{1}{2}$	$-\frac{3}{4}$
Leucadia $7\frac{7}{8}$06	7.7	15	102	$+\frac{1}{4}$
Leucadia $7\frac{3}{4}$13	7.7	72	$100\frac{1}{4}$	$+\frac{1}{2}$
LibPrp 8s01	cv	5	120	$3\frac{3}{8}$
Loews $3\frac{1}{8}$07	cv	145	82	$\frac{3}{8}$
LslsLt 7s04	6.8	5	$103\frac{5}{8}$	$-1\frac{7}{8}$
LglsLt 9s22	7.9	5	$114\frac{1}{8}$	$-1\frac{3}{8}$
LglsLt 8.2s23	8.2	7	$109\frac{1}{4}$	...
LouGs $7\frac{1}{2}$02	7.3	43	103	$+1\frac{1}{4}$
Lucent 6.9s01	6.7	5	$102\frac{5}{8}$	$-1\frac{5}{8}$
Lucent $7\frac{1}{4}$06	6.6	1	$110\frac{1}{8}$	$-1\frac{7}{8}$
MDC Hld $8\frac{3}{8}$08	8.3	10	$100\frac{1}{2}$	$+\frac{1}{2}$
Malan $9\frac{1}{2}$04	cv	10	97	$+\frac{7}{8}$
Mascotch 03	cv	28	$80\frac{5}{8}$	...
Medtrst $7\frac{1}{2}$01	cv	10	98	$+\frac{3}{4}$
MPac $4\frac{3}{4}$20f	...	5	72	...
MPac 5s45f	...	15	66	$+1$
Mobil $8\frac{3}{8}$01	7.8	8	$107\frac{1}{8}$	...
Moran $8\frac{3}{4}$08f	cv	10	96	$-3\frac{1}{2}$
Nabis 8.3s99	8.3	27	$100\frac{3}{16}$	$-\frac{9}{32}$
NETelTel $4\frac{1}{2}$02	4.7	43	$96\frac{1}{2}$	$-\frac{1}{2}$
NETelTel $7\frac{3}{8}$07	7.3	90	$101\frac{1}{2}$	$-\frac{1}{4}$
NETel Tel $6\frac{3}{8}$08	6.3	20	$100\frac{1}{2}$	$-\frac{1}{8}$
NETelTel $6\frac{7}{8}$23	6.7	20	$103\frac{1}{4}$	...
NJBTI $7\frac{1}{4}$11	7.1	16	$102\frac{1}{8}$	...
NJBTI $7\frac{3}{8}$12	7.2	50	102	$-\frac{1}{2}$
NYEG $7\frac{5}{8}$01	7.5	5	$101\frac{5}{8}$	$+\frac{1}{8}$
NYTel $4\frac{1}{4}$00	4.3	1	$98\frac{11}{32}$	$-\frac{5}{32}$
NYTel $7\frac{1}{4}$24	6.8	125	106	$-\frac{7}{8}$
NYTel $6\frac{1}{4}$04	6.1	5	$102\frac{1}{4}$	$-1\frac{1}{8}$
NYTel $6\frac{1}{2}$05	6.2	2	$104\frac{1}{2}$	$-\frac{1}{2}$
NYTel 6s07	6.0	10	100	...
NYTel $6\frac{1}{8}$10	6.0	15	103	$-1\frac{3}{4}$
NYTel $7\frac{3}{8}$11	7.2	67	$102\frac{1}{4}$	$-\frac{3}{8}$
NYTel 6.70s23	6.5	11	$102\frac{1}{2}$	...
NYTel 7s25	6.8	16	$103\frac{1}{4}$	$-\frac{1}{4}$

FIGURE 13-3 Bond information from the *Wall Street Journal.* Reprinted by permission of the *Wall Street Journal,* © January 11, 1999. Dow Jones & Company, Inc. All Rights Reserved Worldwide.

sold at a **discount.** This may happen if the rate of interest on the bond is not as high as the current market rate. However, if the bond rate is higher than the current market rate, it could sell for more than its face value, or at a **premium.**

The selling price of corporate and government bonds can be found in the business section of daily newspapers and in many financial publications. A portion of the bond page of the *Wall Street Journal* is shown in Figure 13-3. Bond information is stated differently than the information given on stocks. For example, the table gives the following information for the IBM $7\frac{1}{2}\%$ bonds of 2013:

BONDS	CUR YLD	VOL	CLOSE	NET CHG.
IBM $7\frac{1}{2}$ 13	6.5	5	116	$-\frac{5}{8}$

First comes the name of the issuing corporation, usually abbreviated, the bond's interest rate ($7\frac{1}{2}\%$), and the year the bond matures (2013). The number 6.5 is the current yield. Next is the daily trading volume. Five $1,000 bonds were traded. (Note that two zeros were not added as was the case with the sales volume for stocks.) Following volume is the current price of the bond. Bond prices are not given in dollar amounts, but as a percent of face value (usually $1,000). Thus 116 means that the bond is selling for 116% of its face value. The actual price paid for the bond would be

$$1.16 \times \$1{,}000 = \$1{,}160$$

The last number, $-\frac{5}{8}$, is the percent difference between the price paid for the last bond today and the price paid for the last bond on the previous day. In dollar terms the drop in value is:

$$0.625\% \text{ of } \$1{,}000 = .00625 \times \$1{,}000 = \$6.25$$

from yesterday's close.

■ **Example 1**

Use the bond table shown in Figure 13-3 to answer the following:

1. What is the closing price of the New England Telephone and Telegraph $6\frac{7}{8}\%$ bonds of 2023?

   The closing price is $103\frac{1}{4}\%$ of $1,000. In dollar terms this is $1.0325 \times \$1{,}000 = \$1{,}032.50$

2. Find the interest rate for the Chesapeake Energy bonds of 2006.

   The rate of $9\frac{1}{8}\%$ follows the name.

3. Find the change from the previous day for the price of the AT&T 7% bonds of 2005.

   The last trade was down $\frac{1}{2}\%$ from the previous day's close. In dollar terms this is

   $$0.005 \times \$1{,}000 = \$5.00$$

4. What is the current yield for the Duke Energy $7\frac{7}{8}\%$ bonds?

   The current yield is 7.6%, stated in the second column from the left.

5. Find the total amount paid to purchase 15 Borden $8\frac{3}{8}\%$ bonds of 2016. Assume a sales charge of $10 per bond.

   The selling price for one bond is

   $$99\frac{1}{2}\% \times \$1{,}000 = 0.995 \times \$1{,}000 = \$995$$

   The total amount paid is the cost of the bonds plus the amount of commission.

   $$(15 \times \$995) + (15 \times \$10) = \$14{,}925 + \$150 = \$15{,}075$$

Now go to Try These Problems (Set III).

NEW FORMULAS IN THIS CHAPTER

- Assets = Liabilities + Stockholders' equity

- Working capital = Current assets − Current liabilities

- Current (or working capital) ratio = $\dfrac{\text{Current assets}}{\text{Current liabilities}}$

- Quick (or acid-test) ratio = $\dfrac{\text{Quick assets}}{\text{Current liabilities}}$

- Return on investment = $\dfrac{\text{Net income}}{\text{Stockholders' equity}}$

- Return on total assets = $\dfrac{\text{Net income}}{\text{Total assets}}$

- Accounts receivable–net sales ratio = $\dfrac{\text{Accounts receivable}}{\text{Net sales}}$

- % Yield = $\dfrac{\text{Annual dividend}}{\text{Price per share}}$

CHAPTER 13 INVESTING

- Current yield for a stock = $\dfrac{\text{Annual dividend}}{\text{Current price}}$

- Earnings per share = $\dfrac{\text{Net income}}{\text{Number of outstanding shares}}$

- PE ratio = $\dfrac{\text{Price per share}}{\text{Annual earnings per share}}$

- Current yield for a bond = $\dfrac{\text{Annual interest}}{\text{Current price}}$

- Approximate bond yield to maturity = $\dfrac{2\,(nI + P - C)}{n\,(P + C)}$

    where I is the annual interest in dollars,
    P is the par value,
    C is the cost or current price,
    n is the number of years to maturity

- Rate of return on a stock investment = $\dfrac{\text{Total gain}}{\text{Total cost}}$

Chapter 13

Name	Date	Class

TRY THESE PROBLEMS (Set III)

Use the stock table shown in Figure 13-2 to answer Problems 1 through 7.

1. What is the highest price for the last 52 weeks for International Paper?

2. Find the sales volume for Loral Space and Communication.

3. What is the dividend for Manufactured Home Company?

4. What is the current yield for Johnson Controls?

5. Find the low trading price for the day for Lyondell Petrochemicals.

6. What is the change from the previous day for Jeffries Group?

7. N. Vestor bought 800 shares of Ionics, Inc. at the low price for the day. If the broker received 3% commission, what was the total amount paid for the stock?

8. Calculate the rate of return on Tom and Jerry, Inc., given the following:

   Purchased 400 shares at $12.75
   Sold 400 shares 1 year later at $14.25
   3% broker commission to buy or sell
   Current dividend of $0.30 per share

Use the bond table shown in Figure 13-3 to answer Problems 9 through 13.

9. What is the year that the IBM $8\frac{3}{8}$% bonds will be paid off by the company?

10. Find the change from the previous day for the New York Telephone $6\frac{1}{8}$% bonds of 2010.

11. What is the annual interest for the AT&T $8\frac{1}{8}$% bonds of 2022?

12. Find the closing price for Leucadia National $8\frac{1}{4}$% bonds.

13. Sarah Smart purchased 10 Duke Energy $6\frac{7}{8}$% bonds of 2023. Find the total purchase price if her broker charges a sales charge of $10 per bond.

Chapter 13

END-OF-CHAPTER PROBLEMS

1. Define the following terms:

Common stock	Quick assets
Preferred stock	Current liabilities
Cumulative preferred	Accounts receivable
Callable bond	Debenture
Participating preferred	Fixed assets
Premium	Odd-lot

2. Complete the income statement in Table 13-4 by filling in the missing quantities and the percent of net sales for each entry.

   TABLE 13-4 HOOK, LYNE & SINKER CO.: INCOME STATEMENT—DECEMBER 31, 19—

		% OF NET SALES
Net Sales	$9,123,000	
Cost of goods sold		
Gross Profit	4,856,000	
Operating expenses	1,912,000	
Operating Profit		
Taxes	410,000	
Net income		

3. A company balance sheet shows total current assets of $617,000 and inventory values at $412,000. What is the value of the company's quick assets?

4. If the company in Problem 3 has current liabilities equal to $197,000, what is its quick ratio? What does the ratio indicate?

5. Currents assets of a company are $29 million and current liabilities are $18 million. How large is the company's working capital?

Problems 6 through 12 are based on the financial statements of Goodman and Outlaw, Inc., Table 13-5.

TABLE 13-5 GOODMAN AND OUTLAW, INC.:
INCOME STATEMENT—YEAR ENDING DECEMBER 31, 1999

Net Sales	$1,826,100
Cost of goods sold	1,005,220
Gross Profit	820,880
Operating expenses	723,250
Operating Profit	97,630
Taxes	34,200
Net income	$ 63,420

BALANCE SHEET—DECEMBER 31, 1999

ASSETS		LIABILITIES	
Current assets		Current liabilities	
Cash	$ 30,100	Accounts payable	$ 85,340
Accounts receivable	94,650	Accrued taxes & expenses	56,000
Inventories	182,230	Total current liabilities	141,340
Total current assets	306,980	Long-term liabilities	183,740
		STOCKHOLDERS' EQUITY	172,350
Fixed assets	190,450		
Total assets	$497,430	Total liabilities & stockholders' equity	$497,430

6. What was the company's working capital?

7. What percent of sales was operating profit?

8. What percent of sales was net income?

9. Find the company's quick ratio. Would it be considered adequate?

10. What was the return on total investment for 1999?

11. What was the return on total assets for 1999?

12. Find the accounts receivable–net sales ratio. If this ratio was 4.5% for 1998, what would the change indicate?

Chapter 13

END-OF-CHAPTER PROBLEMS

Use the balance sheet for Handwerk and Stiles, 1998, shown in Table 13-6 and the 1999 balance sheet in Table 13-1 to answer Problems 13 through 16.

TABLE 13-6 HANDWERK AND STILES, INC.
BALANCE SHEET—DECEMBER 31, 1998

ASSETS		LIABILITIES	
Current assets		Current liabilities	
Cash	$ 902,000	Accounts payable	$ 947,000
Accounts receivable	1,654,000	Notes payable	798,000
Notes receivable	559,000	Accrued expenses	293,000
Inventories	1,511,000	Accrued taxes	299,000
Total current assets	4,626,000	*Total current liabilities*	2,337,000
Fixed assets		Long-term liabilities	
Land (at cost)	650,000	7% bonds due in 2017	1,350,000
Building (at cost)	1,300,000	*Total long-term liabilities*	1,350,000
Machinery, equipment, furniture and fixtures (at cost)	1,055,000	**STOCKHOLDERS' EQUITY**	
Total fixed assets	3,005,000		
		Capital stock	2,450,000
Less accumulated depreciation	389,000	Retained earnings	1,105,000
Net fixed assets	2,616,000	*Total stockholders' equity*	3,555,000
Total assets	$7,242,000	Total liabilities & stockholders' equity	$7,242,000

13. Find the working capital for 1998 and the percent change from 1998 to 1999. What does this change indicate about the company's position?

14. Find the current ratio for 1998 and compare it with the 1999 ratio.

15. Find the quick ratio for 1998. Did the company have a better or worse ratio in 1999?

16. The company had a net income of $699,000 in 1998. What was the return on total assets?

17. A company with an annual net income of $550,000 decided to distribute all the profits as dividends. There are 60,000 shares of cumulative preferred, 8%, $50 par value stock, and 100,000 shares of common stock outstanding. What is the maximum dividend it can declare on its common stock?

18. If the company in Problem 17 declared a $2 dividend on its common stock, how much could it add to retained earnings?

19. If a company's return on investment in 1999 was 12.9%, and stockholders' equity was $2,670,000, what was the net income in that year?

20. The No Bugs Software Co. has current assets of $152,000 and inventory valued at $26,250. If the company's current ratio is 1.8:

    a. How large are its current liabilities?

    b. What is the company's quick ratio?

    c. Is the quick ratio adequate?

21. A firm had a net income of $682,000 in 1998 and retained earnings of $895,000 at the beginning of the year. If the directors decided to pay the required dividend on its 70,000 shares of 7% cumulative preferred, $100 par value stock, what would retained earnings be after payment of the dividend?

22. The Jolt Electric Co. had a net income of $72,000 in 1999 and declared a dividend of 72¢ on 70,000 shares of common stock. It also paid 5% on 10,000 shares of cumulative preferred stock with a par value of $20.

    a. How much was added to retained earnings after the dividends were paid?

    b. If Ben Bender bought his stock at $5.75 a share, what was his yield?

23. Elkin Engineering reported annual earnings of $550,000 last year and has 100,000 shares outstanding. What is the price-earnings ratio if the current price is $92\frac{3}{4}$?

24. If a bond is bought at par, what is the relationship between current yield and yield to maturity?

Chapter 13

Name _____ Date _____ Class _____

END-OF-CHAPTER PROBLEMS

25. A $1,000, 6% municipal bond maturing on July 1, 2015, is bought for $90\frac{1}{2}$ on July 1, 2000.

   a. How much interest will the bond pay each year?

   b. Find the current yield.

   c. Find the approximate yield to maturity.

26. An investor has a choice of two bonds, both maturing in 5 years. Bond A pays 8% and is selling at $106\frac{1}{2}$. Bond B pays $6\frac{1}{2}\%$ and is selling at 96.

   a. Which bond has the higher current yield?

   b. Which bond has the higher yield to maturity?

Use the stock table shown in Figure 13-2 to answer Problems 27 through 32.

27. What was the high trading price for the day for Jones Apparel?

28. Find the low for the last 52 weeks for International Game Technology.

29. What is the change from the previous day for Longs Drug Stores?

30. Find the dividend for Jostens.

31. What is the current yield for Maritrans, Inc.?

32. Candy Cash bought 300 shares of Lucent Technology at the closing price for the day. If the broker received 3% commission, what was the total amount that she paid?

33. Calculate the rate of return on Supercorp, given the following information:

   Purchased 500 shares at $8.00
   Sold 500 shares 1 year later at $9.25
   3% broker commission to buy or sell
   Current dividend of $0.20 per share

Use the bond table shown in Figure 13-3 to answer Problems 34 through 38.

34. What is the closing price of the Kaufman and Broad $9\frac{5}{8}$% bonds of 2006?

35. Find the number of Chesapeake Energy $9\frac{1}{8}$% bonds sold during the day.

36. What was the change from the previous day for Amresco 10% bonds of 2004?

37. What is the year that Lucent Technology $7\frac{1}{4}$% bonds will be paid off by the company?

38. Debbie Duright has purchased 20 AT&T $7\frac{1}{2}$% bonds of 2006. Find the total purchase price if her broker charges a sales charge of $10 per bond.

Chapter 13

TEST YOURSELF

Section 13.2

1. Pete, Maureen, and Mitchell are students at State University who have set up a used book operation. Their expenses are low, since they use space in a dormitory and advertise in the school's newspaper. Their income statement and balance sheet just after the start of the spring semester are shown below. Answer the following questions using these statements.

   a. How much working capital does the company have?

   b. Find the current ratio.

   c. What is the quick ratio for the company?

   d. Find the return on investment.

   e. Find the return on total assets.

   f. What is the accounts receivable–net sales ratio for the company?

   g. What percent of net sales is net income?

   h. Based on the current ratio and quick ratio, is the company in a good financial position?

PETE, MAUREEN, AND MITCHELL, INC.
INCOME STATEMENT FOR 3 MONTHS ENDING MARCH 31, 2000

Net sales	$32,000
Cost of goods sold	22,750
Gross profit	9,250
Operating expenses	4,890
Operating profit	4,360
Taxes	2,940
Net income	$ 1,420

PETE, MAUREEN, AND MITCHELL, INC.
BALANCE SHEET, MARCH 31, 1997

ASSETS		LIABILITIES	
Current assets		Current liabilities	
Cash	$2,400	Accounts payable	$ 530
Accounts receivable	1,250	Notes payable	1,300
Inventories	520	Total	1,830
Total	4,170	Long-term liabilities	0
		Stockholders' equity	2,850
Fixed assets	510	Total liabilities & stockholders'	
Total assets	$4,680	equity	$4,680

549

	Section
2. The common stock of Data Source, Inc. is selling for $21. If the company declares an 82¢ annual dividend, what is the current yield?	13.3

3. You own 800 shares of 6% cumulative preferred stock with a par value of $70.   13.3

   a. How large a dividend would you receive each year?

   b. If you paid $76.50 per share for the stock, what is your yield?

   c. If the dividend is not paid in 1998 and 1999, how much is due you in 2000?

4. A broker tells his client that two good bonds are available. The ABC Co. 6% bond, maturing in 8 years, is selling at 95. The XYZ Co. 7% bond maturing in 6 years sells at par.   13.4

   a. On the basis of current yields, which would you buy?

   b. What is the approximate yield to maturity of the bond you chose?

5. Boswell Computers reported annual earnings of $210,000 last year. If there are 120,000 shares outstanding and the current price is $42\frac{7}{8}$, what is the price-earnings ratio?   13.3

6. Use the stock table shown in Figure 13-2 to answer the following: 13.5

   a. What was the low trading price for the day for JSB Financial?

   b. Find the sales volume for Liz Claiborne.

   c. What is the change from the previous day for Linens N Things?

   d. Find the current yield for Marsh and McLennan.

Chapter 13

TEST YOURSELF

	Section
7. Bertha Bucks bought 100 shares of Lucent Technology at the closing price for the day. If her broker received a 3% commission, what was the total amount that she paid?	13.5
8. Calculate the rate of return on Weals Motors, given the following information: Purchased 400 shares at $15.75 Sold 400 shares 1 year later at $14.125 3% broker commission to buy or sell Current dividend of $0.50 per share	13.5

9. Use the bond table shown in Figure 13-3 to answer the following:
   a. What is the closing price of the Chase Manhattan 8% bonds of 2004?
 	
   b. Find the change from the previous day for the Duke Energy $6\frac{3}{4}$% bonds of 2025.
 	
   c. What is the annual interest for the IBM 7% bonds of 2025?
 	
   d. Find the sales volume for the New Jersey Bell Telephone $7\frac{1}{4}$% bonds of 2011.
	13.6

10. Mike O'Keefe purchased 5 AT&T $7\frac{1}{2}$% bonds of 2006. Find the total purchase price if his broker charges a commission of $10 per bond.	13.6

Chapter 13

Name	Date	Class

COMPUTER SPREADSHEET EXERCISES

Find the missing values for the following spreadsheets. Use formulas where possible and $ signs for dollar amounts. Align decimal points in columns where they appear. All labels and numbers in your spreadsheets should appear as shown. If your spreadsheet is too wide to fit across the width of the page, use the Landscape orientation and the Fit Columns to Page scaling option to print lengthwise.

1. Shown below is the asset portion of the balance sheet of a computer retailer for two consecutive years. Fill in the missing values by obtaining sums and differences in the necessary locations. Then compute the percent increase or decrease amounts. Use a minus sign to indicate a percent decrease.

```
                        HACKER'S HEAVEN
                         Balance Sheet
               December 31, 2000 and December 31, 2001
```

			% Increase
Assets	2000	2001	or Decrease
Current assets			
Cash	$ 96,000	$ 102,000	
Accounts receivable	43,400	41,700	
Notes receivable	21,750	24,200	
Inventories	67,200	63,405	
Total current assets			
Fixed assets			
Land (at cost)	63,000	63,000	
Buildings (at cost)	83,800	95,400	
Machinery, equipment furniture and fixtures (at cost)	65,810	67,120	
Total fixed assets			
Less accumulated depreciation	59,000	54,390	
Net fixed assets			
Total assets	$ _____	$ _____	

2. Find the earnings per share, price-earnings ratio, and current yield of the following New York Stock Exchange stocks. Arrange the rows in ascending alphabetical order by using the Sort Ascending button on the standard Microsoft Excel toolbar (be sure to move the entire row). Round the PE ratio and current yield to the nearest tenth. If the PE ratio or current yield is negative, record a series of five dots in that cell. (Try to write an IF statement that will record the correct positive value, but will record ". . . . ." for a negative value.)

Name	Price	Net Income (1,000s)	Number of Shares (1,000s)	Earning per Share	Dividend per Share	PE Ratio	Current Yield %
Echo Bay Mines	$9.25	(59,700)	99,117		$0.00		
Alcan Aluminum	$21.63	543,000	222,677		$1.12		
Unilever N.V.	$84.50	1,980,000	279,200		$2.39		
Texaco, Inc.	$63.00	3,006,000	258,161		$3.05		
Oracle Systems	$13.00	(81,000)	133,688		$0.00		
St. Paul Companies	$65.75	391,000	42,411		$2.40		
Maxus Energy	$8.38	(7,000)	100,223		$0.00		
Cypress Minerals	$19.13	111,000	38,934		$0.80		
Illinois Tool Works	$65.50	182,000	54,805		$0.66		
NYNEX	$72.75	949,000	198,995		$4.56		
Pfizer, Inc.	$57.88	801,000	330,832		$1.20		
E-Systems	$44.13	94,300	31,205		$0.75		

3. Fill in the missing values. Use an IF statement and the tables for round-lot and odd-lot fees to compute the brokerage commission. The round-lot and odd-lot rates apply to the entire purchase. Quotes are displayed as decimal values rounded to the nearest cent. The first 5 transactions are purchases, the last 5 are sales.

To compute the brokerage commission, use the IF and MOD functions. MOD(x,y) calculates the remainder (modulus) of x divided by y. For example,

=MOD(7,4)

returns the value 3. This function can determine whether the transaction is a round-lot or odd-lot. If dividing the number of shares by 100 gives a remainder of 0, the transaction is a round-lot; if the remainder is not equal to 0 (<>0), the transaction is an odd-lot. The logical operator AND will test two conditions occurring together, such as round-lot and over $3,000. If we put all this together, we might end up with the following IF statement to compute the correct commission:

=IF(AND(MOD(NUMB,100)=0,COST<=3000),COST*0.04,IF
(AND(MOD(NUMB,100)=0,COST>3000),COST*0.03,IF
(AND(MOD(NUMB,100)<>0,COST<=2000),COST*0.06,COST*0.05)))

Of course, the cell addresses in the Number of shares and Cost of shares columns must be substituted for NUMB and COST in the formula. (Note that even a fairly simple problem situation can yield a lengthy spreadsheet formula.) Use the Drawing toolbar to reproduce the lines on the spreadsheet.

Round-lot Commissions			Odd-lot Commissions		
Transaction Amount		Rate	Transaction Amount		Rate
up to	$3,000	4%	up to	$2,000	6%
over	$3,000	3%	over	$2,000	5%

Quote	Number of Shares	Cost of Shares	Brokerage Commission	Total Cost	Proceeds of Sale
22.63	100			_____	
67.50	75			_____	
8.13	500			_____	
19.75	2000			_____	
5.88	5000			_____	
11.00	1200				_____
99.25	40				_____
31.38	150				_____
49.50	30				_____
3.25	2200				_____

4. Compute the current yields and the approximate yields to maturity for the following bond investments. Quotes and interest rates are displayed in decimal form and are rounded to the nearest thousandth. Round dollar amounts to the nearest cent, and current yield and yield to maturity values to the nearest tenth of a percent. Assume all bonds have a $1,000 par value. Use one of the formulas in Section 13.4 to compute yield to maturity.

Name	Quote	Current Price	Interest Rate	Annual Interest	Current Yield	Years to Maturity	Yield to Maturity
AlaPw	106.000		9.250%			6.50	
BethSt	98.875		8.450%			13.00	
Bevrly	96.000		5.500%			7.00	
CaterpInc	111.500		9.375%			3.25	
FordCr	94.500		6.375%			9.75	
GrandCas	107.750		10.125%			12.00	
IBM	103.250		7.250%			20.00	
PacBell	93.750		6.875%			18.50	
Revl	102.625		10.875%			3.25	
StoneCn	86.500		6.750%			6.00	
USX	92.250		5.750%			5.75	
UtdAir	117.875		10.670%			4.50	

A ■ Review of Arithmetic

What happened to the extra horse?

A rich man who owned seventeen horses wanted to distribute them among his three sons. He thought he would give half to his eldest son, a third to his second son, and a ninth to his youngest son. But the division was not possible. Half of seventeen is eight and one-half; one-third of seventeen is five and two-thirds; and a ninth of seventeen is one and eight-ninths.

Having found no solution, he consulted a wise friend who offered to give him an extra horse. He would then have eighteen horses and could divide them as he wished. So he accepted the extra horse and then gave half to his eldest son, one-third to the middle son, and one-ninth to the youngest. When the division was completed, each son stood separately with his horses. The oldest had nine, the second had six, and the youngest had two. The rich man was astounded. The division had been correct but the total distribution was only seventeen.

Why?

Answer is on page 562.

Multiples and Factors

Any number that is exactly *divisible* by a given number is a **multiple** of the given number. For example, 24 is a multiple of 2, 3, 4, 6, 8, and 12, since it is divisible by each of these numbers. Saying that 24 is a multiple of 3 is the same as saying that 3 multiplied by some whole number will give 24. Any number is a multiple of itself and 1.

Any number that can be *divided* into a given number without a remainder is a **factor** of the given number. For example, 2, 3, 4, 6, 8, and 12 are factors of 24. A number that has no factor except itself and 1 is a **prime number.** The following series shows all the prime numbers up to 40.

$$1, 2, 3, 5, 7, 11, 13, 17, 19, 23, 29, 31, 37$$

If a number is a multiple of two or more different numbers, it is called a **common multiple.** Thus, 24 is a common multiple of 6 and 2, but so are 36, 48, 54, and so on. The smallest of the common multiples of a set of numbers is the **least common multiple.** The least common multiple of 6 and 2 is 6.

FRACTIONS

The word **fraction** means literally a broken number. If a week is broken into seven equal parts, or 7 days, then 2 of the 7 days make up $\frac{2}{7}$ of the week, and 4 days are $\frac{4}{7}$ of the week. More generally, the fraction $\frac{5}{28}$, for example, implies that 5 of 28 equal parts are being taken.

A fraction has three parts. The figure below the line, the **denominator,** tells into how many equal parts the whole number has been broken. The figure above the line, the **numerator,** tells how many of these parts are taken. The third part of the fraction, *the line itself,* indicates that the numerator is divided by the denominator. In fact, the most general definition of a fraction states that a fraction is an indicated division. By this definition $\frac{2}{3}, \frac{18}{7}, \frac{12}{6}, \frac{3}{4}$ are all fractions, and even integers are special fractions with denominators of 1. For example, $5 = \frac{5}{1}$, $89 = \frac{89}{1}$.

555

The preceding illustrations are **common fractions** because numerator and denominator are integers. Included are:

- **Proper fractions** in which the numerator is smaller than the denominator ($\frac{2}{3}$, $\frac{3}{4}$).
- **Improper fractions** in which the numerator is larger than the denominator ($\frac{18}{7}$, $\frac{12}{6}$).

Improper fractions can be converted to **mixed numbers,** which are whole numbers plus a proper fraction. When 17 is divided by 7, we have 2 and a remainder of 3, written as $2\frac{3}{7}$. If 12 is divided by 6, the result is 2. The remaining proper fraction in this case has a numerator of zero.

The following rule is the most important when dealing with fractions. Practically all operations involving fractions make use of it.

> **Rule A–1.** Multiplying or dividing both terms of a fraction by the same number does not change the value of the fraction.

The following examples show how this rule is applied to change the form of a fraction.

■ **Example 1**

1. Change $\frac{1}{4}$ to twelfths.

$$\frac{1}{4} = \frac{?}{12}$$

$12 \div 4 = 3$ *Divide 12 by 4 to find the multiplier.*

$$\frac{1 \times 3}{4 \times 3} = \frac{3}{12}$$ *Multiply numerator and denominator by the common multiplier, 3, to get the equivalent fraction, $\frac{3}{12}$.*

$$\frac{1}{4} = \frac{3}{12}$$

2. What fraction with a numerator of 6 is equal to $\frac{3}{4}$?

$$\frac{6}{?} = \frac{3}{4}$$

$6 \div 3 = 2$ *Divide 6 by 3 to find the multiplier.*

$$\frac{6}{?} = \frac{3 \times 2}{4 \times 2} = \frac{6}{8}$$ *Multiply numerator and denominator by 2.*

$$\frac{6}{8} = \frac{3}{4}$$

3. Change $\frac{6}{16}$ to eighths.

$$\frac{6}{16} = \frac{?}{8}$$

$16 \div 8 = 2$ *Divide 16 by 8 to find the divisor.*

$$\frac{6 \div 2}{16 \div 2} = \frac{3}{8}$$ *Divide numerator and denominator by 2.*

$$\frac{6}{16} = \frac{3}{8}$$

FRACTIONS 557

Part 3 also illustrates the process of reducing a fraction to lowest terms, that is, to the equivalent fraction with the smallest possible numerator and denominator. A fraction is reduced to lowest terms by finding the **largest common factor** for numerator and denominator and dividing both by it. In mathematical calculations, fractions should usually be reduced to lowest terms to simplify the work. ∎

■ **Example 2**

Reduce $\frac{18}{48}$ to lowest terms.

For simple fractions the largest common factor can be determined by inspection. For this fraction it is 6.

$$\frac{18 \div 6}{48 \div 6} = \frac{3}{8}$$ *Divide both terms by the largest common factor, 6.*

$$\frac{18}{48} = \frac{3}{8}$$

If the largest common factor cannot be readily found, any common factor may be removed and the process repeated until the fraction is in lowest terms. Thus, $\frac{18}{48}$ can first be divided by 2 and then by 3.

$$\frac{18 \div 2}{48 \div 2} = \frac{9}{24}$$

$$\frac{9 \div 3}{24 \div 3} = \frac{3}{8}$$ ∎

Improper Fractions ↔ Mixed Numbers

Although the improper fraction is quite "proper" mathematically, it is customary to change it to a mixed number when it is in a problem statement or answer. In calculations, on the other hand, the mixed number is often changed to an improper fraction.

Rule A-2. To change an improper fraction to a mixed number, divide the numerator by the denominator and write the fractional remainder in lowest terms.

■ **Example 3**

1. Change $\frac{31}{20}$ to a mixed number.

$$31 \div 20 = 1\frac{11}{20}$$ *The remainder is already in lowest terms.*

$$\frac{31}{20} = 1\frac{11}{20}$$

2. Change $\frac{33}{9}$ to a mixed number.

$$33 \div 9 = 3\frac{6}{9}$$

$$3\frac{6}{9} = 3\frac{2}{3}$$ *$\frac{6}{9}$ is reduced to lowest terms by dividing both parts by the largest common factor, 3.*

$$\frac{33}{9} = 3\frac{2}{3}$$

558 APPENDIX A REVIEW OF ARITHMETIC

3. Change $\dfrac{36}{4}$ to a mixed number.

$$36 \div 4 = 9 \qquad \text{\textit{The remainder is zero.}}$$

$$\dfrac{36}{4} = 9 \quad \blacksquare$$

Rule A-3. To change a mixed number to an improper fraction, multiply the whole number by the denominator of the fraction and add the numerator. The result is the numerator of the improper fraction. The denominator remains unchanged.

■ **Example 4** Change the following mixed numbers to improper fractions.

1. $3\dfrac{2}{7}$

$$3\dfrac{2}{7} = \dfrac{3 \times 7 + 2}{7}$$

$$3\dfrac{2}{7} = \dfrac{23}{7}$$

2. $12\dfrac{5}{6}$

$$12\dfrac{5}{6} = \dfrac{12 \times 6 + 5}{6}$$

$$12\dfrac{5}{6} = \dfrac{77}{6}$$

3. $2\dfrac{5}{24}$

$$2\dfrac{5}{24} = \dfrac{2 \times 24 + 5}{24}$$

$$2\dfrac{5}{24} = \dfrac{53}{24} \quad \blacksquare$$

Like and Unlike Fractions

Like fractions have the same denominators. They may be added or subtracted by adding or subtracting the numerators. For example,

$$\dfrac{7}{19} + \dfrac{5}{19} = \dfrac{12}{19}$$

$$\dfrac{7}{19} - \dfrac{5}{19} = \dfrac{2}{19}$$

(Division and multiplication will be considered later.)

When the denominators of fractions are not equal, the fractions are called **unlike fractions.** Addition and subtraction cannot be performed without changing them to fractions with the same denominators. Trying to add $\frac{3}{4}$ and $\frac{2}{5}$ would be as incorrect as adding feet and inches. In both cases there must be a change in form to make the quantities being added similar.

To change unlike fractions to like fractions, it is necessary to find a common denominator, and it is advantageous to find the **least common denominator** (LCD). The LCD is the smallest common multiple of all the denominators.

Consider the example $\frac{3}{4} + \frac{2}{5}$. The number 5 is prime and cannot have any factors in common with 4. Therefore, the LCD is $4 \times 5 = 20$, and both fractions will be changed to twentieths.

FRACTIONS 559

$$\frac{3}{4} = \frac{15}{20}$$

Numerator and denominator are multiplied by 5.

$$+\frac{2}{5} = +\frac{8}{20}$$

Numerator and denominator are multiplied by 4.

$$\frac{23}{20} = 1\frac{3}{20}$$

Mixed numbers are usually preferred to improper fraction for answers.

$$\frac{3}{4} + \frac{2}{5} = 1\frac{3}{20}$$

■ **Example 5**

1. Add $\frac{5}{9}, \frac{2}{3},$ and $\frac{1}{6}$.

   A common denominator can always be found by multiplying the denominators, but this does not usually give the LCD and makes the calculation more difficult than is necessary. To find the LCD, separate each denominator into prime factors. The LCD will contain all the prime factors, each one the maximum number of times it occurs in any one of the denominators. For this problem the denominators are separated as follows.

   $9 = 3 \times 3$ 3 is prime $6 = 3 \times 2$

   LCD = $3 \times 3 \times 2 = 18$

   The factor 3 occurs twice in 9 and only once in the other denominators. The factor 2 occurs only once, in 6.

   The three fractions are now converted to eighteenths and added.

   $$\frac{5}{9} = \frac{10}{18}$$
   $$\frac{2}{3} = \frac{12}{18}$$
   $$\frac{1}{6} = \frac{3}{18}$$
   $$\frac{25}{18} = 1\frac{7}{18}$$

   $$\frac{5}{9} + \frac{2}{3} + \frac{1}{6} = 1\frac{7}{18}$$

2. Subtract $\frac{3}{20}$ from $\frac{7}{15}$ $\left(\frac{7}{15} - \frac{3}{20}\right)$.

   The denominators are:

   $15 = 5 \times 3$ and $20 = 5 \times 4$.

   LCD = $5 \times 3 \times 4 = 60$

   $$\frac{7}{15} = \frac{28}{60}$$
   $$-\frac{3}{20} = -\frac{9}{60}$$
   $$\frac{19}{60}$$

   $$\frac{7}{15} - \frac{3}{20} = \frac{19}{60}$$

560 APPENDIX A   REVIEW OF ARITHMETIC

3. Evaluate $\dfrac{11}{15} + \dfrac{5}{6} - \dfrac{3}{7}$.

$15 = 5 \times 3 \qquad 6 = 3 \times 2 \qquad 7 \text{ is prime}$

$\text{LCD} = 5 \times 3 \times 2 \times 7 = 210$ *None of the prime factors occurs more than once in any denominator.*

$\dfrac{11}{15} = \dfrac{154}{210}$ *The common multiplier is 14, the product of the factors 7 and 2 ($\not{5} \times \not{3} \times 2 \times 7$).*

$\dfrac{5}{6} = \dfrac{175}{210}$ *The common multiplier is 35, the product of the factors 5 and 7 ($5 \times \not{3} \times \not{2} \times 7$).*

$\dfrac{3}{7} = \dfrac{90}{210}$ *The common multiplier is 30, the product of all the factors except 7 ($5 \times 3 \times 2 \times \not{7}$).*

$\dfrac{11}{15} + \dfrac{5}{6} - \dfrac{3}{7} = \dfrac{154}{210} + \dfrac{175}{210} - \dfrac{90}{210}$

$= \dfrac{239}{210} = 1\dfrac{29}{210}$

$\dfrac{11}{15} + \dfrac{5}{6} - \dfrac{3}{7} = 1\dfrac{29}{210}$ ■

Addition of Mixed Numbers

A mixed number indicates a sum. The number $2\frac{1}{3}$ may be written $2 + \frac{1}{3}$. When adding mixed numbers, we are really adding a series of integers and a series of fractions; no new methods are necessary. To add $4\frac{5}{7}$ and $6\frac{3}{7}$, for example, we would add the integers and the fractions.

$$\begin{array}{r} 4\dfrac{5}{7} \\ + \; 6\dfrac{3}{7} \\ \hline 10\dfrac{8}{7} \end{array}$$

The improper fraction $\frac{8}{7}$ is changed to a mixed number, and the addition is completed.

$$10\dfrac{8}{7} = 10 + 1\dfrac{1}{7} = 10 + 1 + \dfrac{1}{7} = 11\dfrac{1}{7}$$

In other problems it may be necessary to change unlike fractions to like fractions.

■ **Example 6**

Add: $4\dfrac{5}{8} + 2\dfrac{1}{2} + \dfrac{1}{3}$.

$8 = 2 \times 2 \times 2 \qquad 2 \text{ is prime} \qquad 3 \text{ is prime}$

$\text{LCD} = 2 \times 2 \times 2 \times 3 = 24$ *Change all fractions to twenty-fourths.*

$4\dfrac{5}{8} = 4\dfrac{15}{24}$

$2\dfrac{1}{2} = 2\dfrac{12}{24}$

$\dfrac{1}{3} = \dfrac{8}{24}$

$\overline{6\dfrac{35}{24}}$ *Add integers and like fractions.*

FRACTIONS 561

$$6\frac{35}{24} = 6 + 1\frac{11}{24}$$ *Change the improper fraction to a mixed number.*

$$6\frac{35}{24} = 6 + 1 + \frac{11}{24} = 7\frac{11}{24}$$

$$4\frac{5}{8} + 2\frac{1}{2} + \frac{1}{3} = 7\frac{11}{24}$$

Check: Approximate the answer to check the results. The answer must be more than 6 because the sum of the integers is 6. Each of the fractions is close to $\frac{1}{2}$, giving an additional sum of approximately $1\frac{1}{2}$. The answer should be near $7\frac{1}{2}$. ∎

Subtraction of Mixed Numbers

As in addition, whole numbers are subtracted from whole numbers and fractions from fractions. The fractions must have the same denominators.

■ **Example 7**

1. Subtract: $12\frac{7}{11} - 9\frac{4}{11}$.

$$12\frac{7}{11}$$
$$-9\frac{4}{11}$$
$$\overline{3\frac{3}{11}}$$

$$12\frac{7}{11} - 9\frac{4}{11} = 3\frac{3}{11}$$

2. Subtract: $8\frac{4}{5} - 5\frac{1}{8}$.

$$8\frac{4}{5} = 8\frac{32}{40}$$
$$-5\frac{1}{8} = -5\frac{5}{40}$$
$$\overline{3\frac{27}{40}}$$

$$8\frac{4}{5} - 5\frac{1}{8} = 3\frac{27}{40}$$

3. Subtract: $3\frac{2}{3} - \frac{11}{12}$.

$$3\frac{2}{3} = 3\frac{8}{12}$$
$$-\frac{11}{12} = -\frac{11}{12}$$

An additional step is necessary when, as is this case, a fraction $\left(\frac{11}{12}\right)$ must be subtracted from a smaller fraction $\left(\frac{8}{12}\right)$. This step, often called "borrowing," consists of regrouping a whole number and fraction. The integer 1 is taken from the whole number and converted to a fraction with the common denominator. For this problem:

$$3\frac{8}{12} = 2 + 1 + \frac{8}{12} = 2 + \frac{12}{12} + \frac{8}{12} = 2\frac{20}{12}$$

The result of this regrouping is always an improper fraction.

Continuing the problem:

$$3\frac{8}{12} = 2\frac{20}{12}$$
$$-\frac{11}{12} = -\frac{11}{12}$$
$$2\frac{9}{12} = 2\frac{3}{4}$$

$$3\frac{2}{3} - \frac{11}{12} = 2\frac{3}{4}$$

4. Subtract: $25\frac{7}{15} - 8\frac{1}{2}$.

$$25\frac{7}{15} = 25\frac{14}{30} = 24\frac{44}{30}$$
$$-8\frac{1}{2} = -8\frac{15}{30}$$
$$16\frac{29}{30}$$

$$25\frac{7}{15} - 8\frac{1}{2} = 16\frac{29}{30} \quad \blacksquare$$

Answer to "What happened to the extra horse?"

When the rich man decided to give away $\frac{1}{2}$, $\frac{1}{3}$, and $\frac{1}{9}$ of his horses, he thought the three fractions would add up to one, representing the total number of horses. But they do not.

$$\frac{1}{2} + \frac{1}{3} + \frac{1}{9} = \frac{9}{18} + \frac{6}{18} + \frac{2}{18} = \frac{17}{18}$$

He was distributing only 17 out of the 18 horses.

Multiplication of Fractions

The general rule for multiplying fractions is as follows:

Rule A-4. The product of two or more fractions is the product of their numerators divided by the product of their denominators. Integers, with denominators equal to 1, are included.

Thus,

$$\frac{1}{2} \times \frac{1}{5} \times \frac{3}{7} = \frac{1 \times 1 \times 3}{2 \times 5 \times 7} = \frac{3}{70}$$

(Common denominators are not needed in multiplication of fractions.)

If a mixed number is involved, it is frequently simpler to change it to an improper fraction before multiplying. Accordingly,

$$\frac{1}{3} \times 1\frac{1}{7} \times 3\frac{1}{3} = \frac{1}{3} \times \frac{8}{7} \times \frac{10}{3} = \frac{1 \times 8 \times 10}{3 \times 7 \times 3} = \frac{80}{63} = 1\frac{17}{63}$$

FRACTIONS 563

Whenever possible, shorten the computation by cancellation; that is, divide out factors common to both numerator and denominator. Cancellation is an application of Rule A-1. Note that:

- The order of cancellations may vary, but the final result is the same.
- Cancellation is permitted only with fractions connected by multiplication signs.

In the first illustration each step of the cancellation is shown separately, but in practice (and in subsequent illustrations) there is no rewriting of the problem.

■ **Example 8**

1. Multiply: $2\frac{1}{4} \times 3\frac{1}{3} \times \frac{1}{5}$.

$$2\frac{1}{4} \times 3\frac{1}{3} \times \frac{1}{5} = \frac{9}{4} \times \frac{10}{3} \times \frac{1}{5}$$

Change the mixed numbers to improper fractions.

The next four steps may be done in any order.

$$\frac{\overset{3}{\cancel{9}}}{4} \times \frac{10}{\cancel{3}} \times \frac{1}{5} = \frac{3}{4} \times \frac{10}{1} \times \frac{1}{5}$$

The common factor 3 is divided out.

$$\frac{3}{4} \times \frac{\overset{2}{\cancel{10}}}{1} \times \frac{1}{\cancel{5}} = \frac{3}{4} \times \frac{2}{1} \times \frac{1}{1}$$

The common factor 5 is divided out.

$$\frac{3}{\underset{2}{\cancel{4}}} \times \frac{\overset{1}{\cancel{2}}}{1} \times \frac{1}{1} = \frac{3}{2} \times \frac{1}{1} \times \frac{1}{1}$$

The common factor 2 is divided out.

$$\frac{3}{2} \times \frac{1}{1} \times \frac{1}{1} = \frac{3 \times 1 \times 1}{2 \times 1 \times 1}$$

$$= \frac{3}{2} = 1\frac{1}{2}$$

$$2\frac{1}{4} \times 3\frac{1}{3} \times \frac{1}{5} = 1\frac{1}{2}$$

2. Evaluate $1\frac{5}{6} \times 3 \times \frac{2}{3}$.

$$1\frac{5}{6} \times 3 \times \frac{2}{3} = \frac{11}{6} \times \frac{3}{1} \times \frac{2}{3}$$

Change the mixed number to an improper fraction and put the denominator 1 under the integer 3.

$$= \frac{11}{\underset{3}{\cancel{6}}} \times \frac{\overset{1}{\cancel{3}}}{1} \times \frac{\overset{1}{\cancel{2}}}{\underset{1}{\cancel{3}}}$$

Common factors are 2 and 3. The order in which they are divided out is not important.

$$= \frac{11}{3} = 3\frac{2}{3}$$

$$1\frac{5}{6} \times 3 \times \frac{2}{3} = 3\frac{2}{3} \quad ■$$

564 APPENDIX A REVIEW OF ARITHMETIC

Division of Fractions The rule for division by a fraction is as follows:

> **Rule A-5.** To divide by a fraction, invert the fraction and multiply. $\left(\frac{A}{B}\text{ inverted is }\frac{B}{A}\right)$. In applying this rule, invert only the divisor and change the division sign to a multiplication sign. The dividend remains the same.

■ **Example 9**

1. Divide 10 by $\frac{3}{4}$.

   The dividend is 10; the divisor is $\frac{3}{4}$.

   $$10 \div \frac{3}{4} = 10 \times \frac{4}{3}$$ *Invert the divisor and multiply.*

   $$= \frac{40}{3}$$

   $$10 \div \frac{3}{4} = 13\frac{1}{3}$$

2. Divide $\frac{3}{10}$ by $\frac{3}{4}$.

   The dividend is $\frac{3}{10}$; the divisor is $\frac{3}{4}$.

   $$\frac{3}{10} \div \frac{3}{4} = \frac{3}{10} \times \frac{4}{3}$$ *Invert the divisor and multiply.*

   $$= \frac{\cancel{3}^{1}}{\cancel{10}_{5}} \times \frac{\cancel{4}^{2}}{\cancel{3}_{1}}$$ *Cancel the common factors 3 and 2.*

   $$\frac{3}{10} \div \frac{3}{4} = \frac{2}{5}$$

3. Evaluate $2\frac{1}{3} \div 1\frac{1}{2}$.

   $$2\frac{1}{3} \div 1\frac{1}{2} = \frac{7}{3} \div \frac{3}{2}$$ *Change the mixed numbers to improper fractions.*

   $$= \frac{7}{3} \times \frac{2}{3} = \frac{14}{9}$$ *Invert the divisor and multiply.*

   $$2\frac{1}{3} \div 1\frac{1}{2} = 1\frac{5}{9}$$

4. Divide 14 by $\frac{7}{10}$.

   $$14 \div \frac{7}{10} = \frac{\cancel{14}^{2}}{1} \times \frac{10}{\cancel{7}_{1}}$$ *Invert the divisor and multiply. Cancel the common factor.*

   $$14 \div \frac{7}{10} = 20$$

5. Divide $\dfrac{3}{20}$ by 5.

$$\dfrac{3}{20} \div 5 = \dfrac{3}{20} \times \dfrac{1}{5}$$

Putting the denominator 1 under the integer 5 makes the divisor a fraction that can be inverted.

$$\dfrac{3}{20} \div 5 = \dfrac{3}{100} \quad \blacksquare$$

Complex Fractions

A **complex fraction** has a fraction in one or both of its terms, thus:

$$\dfrac{\dfrac{3}{5}}{\dfrac{3}{4}}, \quad \dfrac{4}{\dfrac{5}{8}}, \quad \dfrac{3\dfrac{1}{3}}{2}$$

These may be read as three-fifths divided by three-fourths, four divided by five-eighths, and three and one-third divided by two, respectively. Thought of as indicated division, a complex fraction is first simplified by writing as in each of the following:

$$\dfrac{\dfrac{3}{5}}{\dfrac{3}{4}} = \dfrac{3}{5} \div \dfrac{3}{4}$$

$$\dfrac{4}{\dfrac{5}{8}} = 4 \div \dfrac{5}{8}$$

$$\dfrac{3\dfrac{1}{3}}{2} = 3\dfrac{1}{3} \div 2$$

Then proceed as before for the division of fractions.

$$\dfrac{\dfrac{3}{5}}{\dfrac{3}{4}} = \dfrac{3}{5} \div \dfrac{3}{4} = \dfrac{\cancel{3}^{1}}{5} \times \dfrac{4}{\cancel{3}_{1}} = \dfrac{4}{5}$$

$$\dfrac{4}{\dfrac{5}{8}} = \dfrac{4}{1} \div \dfrac{5}{8} = \dfrac{4}{1} \times \dfrac{8}{5} = \dfrac{32}{5} = 6\dfrac{2}{5}$$

$$\dfrac{3\dfrac{1}{3}}{2} = 3\dfrac{1}{3} \div 2 = \dfrac{10}{3} \div \dfrac{2}{1} = \dfrac{\cancel{10}^{5}}{3} \times \dfrac{1}{\cancel{2}_{1}} = \dfrac{5}{3} = 1\dfrac{2}{3}$$

Complex fractions may also contain an indicated operation in the numerator or denominator, or both. To simplify such a fraction, simplify the numerator and denominator and continue as before.

■ **Example 10**

$$\frac{\frac{1}{2}+\frac{1}{3}}{\frac{9}{5}+\frac{1}{5}} = \frac{\frac{3}{6}+\frac{2}{6}}{\frac{10}{5}} = \frac{\frac{5}{6}}{\frac{2}{1}}$$

$$= \frac{5}{6} \div \frac{2}{1} = \frac{5}{6} \times \frac{1}{2}$$

$$= \frac{5}{12}$$

$$\frac{\frac{1}{2}+\frac{1}{3}}{\frac{9}{5}+\frac{1}{5}} = \frac{5}{12} \quad \blacksquare$$

DECIMALS

A decimal, or more accurately a **decimal fraction**, is a fraction whose denominator is 10, 10 × 10, 10 × 10 × 10, or any number of tens multiplied together. The fractions $\frac{7}{10}$, $\frac{12}{100}$, and $\frac{215}{1,000}$ are all decimal fractions. Decimal fractions, along with whole numbers, make up our decimal system of numbers. Table A-1 shows what place value is represented by each place of a number. We can proceed from bottom to top in the table to any larger place value. With decimal fractions we can proceed below the units place to any smaller place value. Note that we can continue indefinitely in either direction.

Note that as we proceed from top to bottom, each place value is one-tenth of the preceding place. Column B shows the numbers written in fraction form, and in Column C the numbers have been written using the *decimal point*. In this form they are usually called **decimals** instead of decimal fractions.

Any decimal fraction may be written in the shortened decimal form by a simple, mechanical process. To write a decimal fraction as a decimal, begin at the right-hand digit of the numerator and count off to the left as many places as there are zeros in the denominator. Place

TABLE A-1 PLACE VALUES, INCLUDING DECIMAL FRACTIONS

A	B	C
Millions	1,000,000	1,000,000
Hundred thousands	100,000	100,000
Ten thousands	10,000	10,000
Thousands	1,000	1,000
Hundreds	100	100
Tens	10	10
Units	1	1
Tenths	$\frac{1}{10}$	0.1
Hundredths	$\frac{1}{100}$	0.01
Thousandths	$\frac{1}{1000}$	0.001
Ten thousandths	$\frac{1}{10,000}$	0.0001
Hundred thousandths	$\frac{1}{100,000}$	0.00001
Millionths	$\frac{1}{1,000,000}$	0.000001

the decimal point to the left of the last digit counted. Drop the denominator. If there are not enough digits, as many place-holding zeros as necessary are added to the left of the numerator.

$$\frac{23}{10,000} = 0.0 \underbrace{0}_{\text{Place-holding zeros added}} 2 \; 3 \quad \overset{④ \leftarrow ③ \leftarrow ② \leftarrow ①}{}$$

In $\frac{23}{10,000}$, beginning with the digit 3, we count off four places to the left, adding two zeros as we count. The decimal point is placed at the extreme left. When a decimal fraction is written in the shortened form, there will always be as many decimal places in the decimal form as there are zeros in the denominator of the fraction form. When the decimal point is not shown in a number, it is always assumed to be to the right of the last digit.

The value of a decimal is not changed by adding zeros at the right-hand end of the number. Since

$$\frac{5}{10} = \frac{50}{100} = \frac{500}{1000}$$

it is also true that the equivalent decimals 0.5, 0.50, and 0.500 are equal. This is not true of whole numbers. The decimals 0.3, 0.30, and 0.3000 are equal, but 3, 30, and 30,000 are not equal. It should also be noted that zeros directly after the decimal point do change values. Thus, 0.3 is not equal to 0.03 or 0.003.

Dividing and Multiplying by 10, 100, 1,000, . . .

Division of any number by 10, 100, 1,000, . . . is just an exercise in placing the decimal point of a decimal fraction. Thus, $5,031 \div 100$ may be thought of as the decimal fraction $\frac{5,031}{100}$. To remove the denominator, move the decimal point two places left for the two zeros in 100.

$$\frac{5,031}{100} = 5031.00 = 50.31$$

■ **Example 11**

1. $401 \div 10 = 40.1$ *There is one zero in* 10; *move the decimal point one place to the left.*

2. $2 \div 1,000 = 0.002$ *There are three zeros in* 1,000; *move the decimal three places to the left.*

If the dividend already contains a decimal point, begin counting with the first number to the left of the decimal point.

3. $243.6 \div 100 = 2.436$ *There are two zeros in* 100; *move the decimal point two places to the left.* ■

Rule A-6. To *divide* by 10, 100, 1,000, . . . , move the decimal point in the dividend to the *left* as many places as there are zeros in the divisor. Add place-holding zeros if necessary.

568 APPENDIX A REVIEW OF ARITHMETIC

> **Rule A-7.** To *multiply* any number by 10, 100, 1,000, . . . , move the decimal point in the number to the *right* as many places as there are zeros in the multiplier. Add zeros if necessary.

Other decimal forms. A **mixed decimal** is a whole number with a fraction in decimal form. Mixed decimals have the same meaning as mixed numbers. The mixed decimal 160.32 is read "one hundred sixty and thirty-two hundredths." The word "and," as with mixed numbers, means plus. When a decimal stands alone, a zero is often used in the units place for clarity. For example, .28 may be written 0.28.

A **complex decimal** contains a common fraction. The number $0.3\frac{1}{3}$ is a complex decimal and is read "three and one-third tenths." The number $2.87\frac{1}{2}$ means 2 and $87\frac{1}{2}$ hundredths. The fraction in each case is part of the last digit. It is not an additional decimal place.

Decimal ↔ Fraction

Any decimal may be changed to an equivalent fraction by the following rule. The fraction should then be reduced to lowest terms.

> **Rule A-8.** To change a decimal to a fraction, write out the number without the decimal point. This is the numerator of the fraction. The denominator is 1 followed by as many zeros as there are decimal places.

■ **Example 12**

Change the following decimals to fractions.

1. $0.12 = \dfrac{12}{100}$ — *There are two decimal places and therefore two zeros in the denominator.*

   $0.12 = \dfrac{3}{25}$ — *Divide both terms by the common factor 4.*

2. $0.0625 = \dfrac{625}{10,000}$ — *There are four decimal places and four zeros in the denominator. The initial zero in the numerator is omitted because it has no function after the decimal point is removed.*

   $0.0625 = \dfrac{25}{400} = \dfrac{1}{16}$ — *The fraction is reduced by dividing twice by 25. Division once by 625 is correct but less obvious.*

3. $0.12\frac{1}{2} = \dfrac{12\frac{1}{2}}{100}$ — *There are two decimal places; $\frac{1}{2}$ is part of the second decimal place.*

   $= \dfrac{\frac{25}{2}}{100}$ — *Change the numerator to an improper fraction.*

$$= \frac{25}{2} \div \frac{100}{1}$$ *Rewrite in division form.*

$$= \frac{\cancel{25}}{2} \times \frac{1}{\cancel{100}_{4}}$$ *Invert the divisor and multiply. Cancel.*

$$= 0.12\frac{1}{2} = \frac{1}{2} \times \frac{1}{4} = \frac{1}{8}$$

4. When changing a mixed decimal to a fraction, change only the part after the decimal point.

$$1.78 = 1\frac{78}{100}$$ *There are two decimal places and two zeros in the denominator.*

$$1.78 = 1\frac{39}{50}$$ *Reduce the fraction to lowest terms.*

Changing a fraction to a decimal is done by performing the indicated division. ■

Rule A-9. To change a fraction to a decimal, divide the numerator by the denominator.

■ **Example 13** Change the following fractions to equivalent decimals.

1. $\frac{1}{4} = 1 \div 4 = 0.25$

$$\begin{array}{r} 0.25 \\ 4\overline{)1.00} \end{array}$$

2. $\frac{7}{25} = 7 \div 25 = 0.28$

$$\begin{array}{r} 0.28 \\ 25\overline{)7.00} \\ \underline{5\ 0} \\ 2\ 00 \\ \underline{2\ 00} \end{array}$$

3. $1\frac{9}{20}$

Place the decimal point after the whole number and convert only the fractional part to a decimal.

$\frac{9}{20} = 9 \div 20 = 0.45$

$$\begin{array}{r} 0.45 \\ 20\overline{)9.00} \\ \underline{8\ 0} \\ 1\ 00 \\ \underline{1\ 00} \end{array}$$

$$1\frac{9}{20} = 1.45$$

4. $\dfrac{5}{13} = 5 \div 13 = 0.3846$

```
        0.3846
    13)5.0000
       3 9
       1 10
       1 04
          60
          52
          80
          78
```

When a fraction generates a decimal that continues indefinitely, it must be stopped at some point. The result may then be shown in either of two ways.

1. We may write the decimal by rounding off* at an appropriate point. In business mathematics many answers are rounded to two decimal places for dollars and cents in the final answer. In that case the division is carried to three decimal places. If the third place is 5 or more, the previous digit is increased by 1. If the third decimal place is less than 5, it is dropped. In the division of 5 by 13 this would give:

$$\dfrac{5}{13} = 0.384 \quad \text{or} \quad 0.38$$

2. We may stop the division and write the remainder, as in any division problem.

$$\dfrac{5}{13} = 0.38 \dfrac{6}{13} \quad \blacksquare$$

Addition and Subtraction of Decimals

When integers are added or subtracted, units are written under units, tens under tens, hundreds under hundreds, and so forth. When decimals are added or subtracted, a similar practice is followed. Tenths are written under tenths, hundredths under hundredths, and so on, with the result that the decimal points fall in a straight line. The decimal point in the answer lines up in the same way—for example,

```
     2.18
    34.35
     0.14
     4.90
    -----
    41.57
```

and

```
    45.76
   -31.87
   ------
    13.89
```

In these illustrations, all the values have the same number of decimal places. In other problems this may not be true. For convenience, zeros may be added to the figure with the smaller number of decimal places—for example,

```
       5    =    5.00
   -3.24    =   -3.24
                -----
                 1.76
```

*See also Chapter 1.

Multiplication of Decimals

Decimals are multiplied like whole numbers; the only additional problem is the location of the decimal point.

> **Rule A-10.** To multiply numbers when one or more contains a decimal, multiply as though the numbers were integers. Mark off as many decimal places in the product as there are decimal places in the numbers together, beginning at the right-hand end.

■ **Example 14**

Do the following multiplications.

1. $0.2 \times 0.3 = 0.06$ — *Each number has one decimal place; therefore, the product has two decimal places.*

2. $3.7 \times 0.02 = 0.074$ — *The numbers have one and two decimal places, respectively. The product has three.*

3. $5.4 \times 0.0073 = 0.03942$ — *There are one and four decimal places, respectively. The product has five.*

$$
\begin{array}{r}
0.0073 \\
\times5.4 \\
\hline
0292 \\
0365 \\
\hline
0.03942
\end{array}
$$ ■

Division of Decimals

In division of decimals two points should be emphasized:

- The decimal point in the quotient is placed directly above the decimal point in the dividend.
- The place values are rigid; that is, tenths in the quotient appear above tenths in the dividend, hundredths over hundredths, and so on.

When the divisor is an integer, the division is carried out the same way as division of an integer by an integer. The following problem illustrates the two points just given.

Divide 5.372 by 32. Carry to three decimal places and round to two.

$$
\begin{array}{r}
.167 \\
32\overline{)5.372} \\
\underline{3\,2} \\
2\,17 \\
\underline{1\,92} \\
252 \\
\underline{224} \\
28
\end{array}
$$

The result of the division is 0.17. The 6 in the second decimal place is increased to 7 because the third decimal place is more than 5.

Rule A-11. To divide by a decimal, first move the decimal point in the divisor enough places to the right to make the divisor an integer. Then move the decimal point in the dividend the same number of places.

When moving the decimal point by Rule A-11, we are, in effect, multiplying both the numerator and denominator of a fraction (an indicated division) by the same number. This simplifies the problem to division by an integer.

■ **Example 15**

Divide 5.648 by 0.45. Carry the answer to four decimal places and round to three.

$$\begin{array}{r} 12.5511 \\ 0.45\overline{)5.64.8000} \\ \underline{4\ 5} \\ 1\ 14 \\ \underline{90} \\ 248 \\ \underline{225} \\ 230 \\ \underline{225} \\ 50 \\ \underline{45} \\ 50 \end{array}$$

$5.648 \div 0.45 = 12.5511$, or 12.551. The 1 in the fourth decimal place is dropped and does not increase the previous digit because it is less than 5. ■

NEW RULES IN THIS APPENDIX

- **Rule A-1.** Multiplying or dividing both terms of a fraction by the same number does not change the value of the fraction.
- **Rule A-2.** To change an improper fraction to a mixed number, divide the numerator by the denominator and write the fractional remainder in lowest terms.
- **Rule A-3.** To change a mixed number to an improper fraction, multiply the whole number by the denominator of the fraction and add the numerator. The result is the numerator of the improper fraction. The denominator remains unchanged.
- **Rule A-4.** The product of two or more fractions is the product of their numerators divided by the product of their denominators. Integers, with denominators equal to 1, are included.
- **Rule A-5.** To divide by a fraction, invert the fraction and multiply.
- **Rule A-6.** To divide by 10, 100, 1,000, . . . , move the decimal point in the dividend to the left as many places as there are zeros in the divisor. Add place-holding zeros if necessary.
- **Rule A-7.** To multiply any number by 10, 100, 1,000, . . . , move the decimal point in the number to the right as many places as there are zeros in the multiplier. Add zeros if necessary.
- **Rule A-8.** To change a decimal to a fraction, write out the number without the decimal point. This is the numerator of the fraction. The denominator is 1 followed by as many zeros as there are decimal places.

- **Rule A-9.** To change a fraction to a decimal, divide the numerator by the denominator.
- **Rule A-10.** To multiply numbers when one or more contain a decimal, multiply as though the numbers were integers. Mark off as many decimal places in the product as there are decimal places in the numbers together, beginning at the right-hand end.
- **Rule A-11.** To divide by a decimal, first move the decimal point in the divisor enough places to the right to make the divisor an integer. Then move the decimal point in the dividend the same number of places.

Appendix A

Name　　　Date　　　Class

EXERCISES

Multiply:

1. $2{,}981 \times 104$
2. $409 \times 6{,}007$
3. $30{,}078 \times 200$

4. $4{,}920 \times 450$
5. $8{,}070 \times 109$

Divide and round to the nearest hundredth:

6. 20,886 by 59
7. 31 into 8,568
8. 4,784 by 23

9. 22,236 by 109
10. 402 into 4,024

Change the following fractions and mixed numbers to decimal equivalents, rounded to two decimal places. If the fractions are not in lowest terms, reduce first to simplify the arithmetic.

11. $\dfrac{5}{8}$
12. $\dfrac{17}{9}$
13. $1\dfrac{8}{9}$
14. $8\dfrac{3}{7}$

15. $\dfrac{122}{6}$
16. $\dfrac{20}{12}$
17. $7\dfrac{3}{7}$
18. $\dfrac{13}{15}$

19. $\dfrac{38}{42}$
20. $\dfrac{9}{32}$

Change the following decimals to equivalent fractions in lowest terms:

21. 0.9
22. 0.78
23. 0.06
24. 0.042

25. 0.0012
26. 1.23
27. 2.05
28. 4.24

29. 0.424
30. $0.42\dfrac{2}{5}$

Do the following multiplications and divisions by moving the decimal point:

31. $3.67 \div 10$
32. $0.629 \times 1{,}000$
33. $1.02 \times 1{,}000$
34. $0.05 \div 100$

575

35. $16.29 \div 10{,}000$ **36.** 75×100 **37.** $85 \div 100$ **38.** $3{,}178 \div 1{,}000$

39. $2.659 \div 100$ **40.** $3.72 \times 1{,}000$

Change to all decimals or all fractions first, and then add:

41. $\dfrac{5}{12} + \dfrac{7}{18}$ **42.** $\dfrac{3}{7} + 1\dfrac{2}{9}$ **43.** $3\dfrac{7}{15} + 7\dfrac{3}{8}$

44. $\dfrac{7}{8} + 1\dfrac{3}{4} + 6\dfrac{1}{2}$ **45.** $0.875 + 1.75 + 6.5$ **46.** $\dfrac{5}{7} + \dfrac{12}{5} + \dfrac{1}{2} + \dfrac{3}{35}$

47. $0.689 + 5.782 + 12.87 + 7.9$ **48.** $0.72 + \dfrac{5}{8}$

49. $0.007 + 11.31 + 3\dfrac{1}{5}$ **50.** $7.43 + 1\dfrac{3}{4} + 99.2 + 4\dfrac{3}{8}$

Evaluate and reduce fractions to lowest terms:

51. $34.784 - 24.247$ **52.** $5.89 - 2.356$ **53.** $1.0076 - 0.29$

54. $29 - 13.004$ **55.** $7\dfrac{3}{5} - 5\dfrac{1}{7}$ **56.** $13\dfrac{2}{9} - 7\dfrac{3}{7}$

57. $3\dfrac{7}{9} - \dfrac{8}{15}$ **58.** $4\dfrac{1}{7} + 2\dfrac{1}{9} + 5\dfrac{11}{63}$

59. $27.328 - 17.011 + 4 - 8.2168$ **60.** $10.007 + 2.16 - 3\dfrac{1}{8}$

Multiply and reduce fractions to lowest terms:

61. $\dfrac{1}{8} \times \dfrac{3}{7}$ **62.** $2\dfrac{7}{8} \times 3\dfrac{1}{2}$ **63.** $\dfrac{11}{25} \times \dfrac{5}{14} \times \dfrac{3}{77} \times \dfrac{2}{27}$ **64.** $5\dfrac{2}{5} \times 10$

Appendix A

Name   Date   Class

EXERCISES

65. $125 \times \dfrac{9}{25}$

66. $10\dfrac{1}{2} \times 13\dfrac{1}{7}$

67. $15\dfrac{6}{7} \times \dfrac{14}{15} \times \dfrac{5}{3}$

68. 1.46×4.89

69. 0.67×0.004

70. 3.005×0.091

Reduce fractions to lowest terms and round decimals to one decimal place. Divide:

71. $\dfrac{3}{8} \div \dfrac{4}{9}$

72. $\dfrac{13}{8} \div \dfrac{39}{5}$

73. $\dfrac{8}{9} \div 2\dfrac{2}{7}$

74. $7\dfrac{1}{2} \div 2\dfrac{1}{4}$

75. $15 \div 3\dfrac{3}{5}$

76. $14.56 \div 14$

77. $179.567 \div 0.04$

78. $15 \div 3.06$

79. $57.2 \div 5.11$

80. $12.7778 \div 2.32$

B ■ The Metric System

METRIC UNITS OF MEASURE

Here are a few questions to get you thinking metric.

1. *If you took your temperature when you were feeling well, you would expect the thermometer to register:*
   a. 41 *degrees* C.
   b. 30 *degrees* C.
   c. 37 *degrees* C.
2. *The speed limit on most highways is:*
   a. 55 *kilometers per hour.*
   b. 70 *kilometers per hour.*
   c. 90 *kilometers per hour.*
3. *If you were serving hamburgers to a family of four, allowing one hamburger weighing one-quarter pound for each, how much meat would you buy?*
   a. 250 *grams.*
   b. 450 *grams.*
   c. 1 *kilogram.*
4. *In buying gasoline by the liter rather than by the gallon, would you:*
   a. *buy more liters than gallons?*
   b. *buy fewer liters than gallons?*
   c. *buy the same number of liters as gallons?*

   Answers: 1-c; 2-c; 3-b; 4-a

In order to more successfully compete in the global marketplace, business and industry in the United States have had to gradually adopt the metric system of measurement. The motivation for the change is the fact that the metric system has been designated as the international standard of measurement, and all of our major trading partners have adopted it. The Metric Conversion Act of 1975 originally committed the United States to the increasing use of and voluntary conversion to metric measurement. At first, the metric changeover was found predominantly among manufacturers with overseas interests; pharmaceutical companies went metric in America in the early 1970s. In 1988, however, a provision of The Omnibus Trade and Competitiveness Act *required* federal agencies to use the metric system for business-related activities. Today, American businesses typically dual-dimension their domestic products and use all metric dimensions on products exported overseas.

Almost without noticing it, Americans have already gone metric in many aspects of daily living. The shutters of 8, 16, and 35 millimeter cameras click busily. Auto mechanics work on foreign cars manufactured in metric countries. Many household products, like packaged and canned goods, carry both English and metric units on their labels. Wine and some soft drinks are sold in liters. Weather reports give the temperatures in Celsius and Fahrenheit readings.

Most of us have developed a sense or feel for the English measurements we use every day. We know, for example, that someone 6 feet 4 inches is very tall, that a pound of steak is a hearty serving for one person, that 90°F means a hot day. A similar feeling for metric units can be readily developed because there are basic units that are almost the same size as the English units they replace.

- A **meter** is a little longer than a *yard* (1.1 yards).
- A **liter** is a little more than a *quart* (1.1 quarts).
- A **gram** is a little more than the *weight of a paper clip*. 500 grams is just over a pound (about 1.1 pounds).

This means that if you need 3 yards of material for a dress, you can buy 3 meters and not have much left over. If the family drinks 4 quarts of milk a week, they will drink 4 liters and not know the difference. A 1-pound serving of steak may have one extra bite if it weighs 500 grams or half a kilogram. But be careful with distances and car speeds. If the speed limit is 80 kilometers, don't put your foot down on the gas. It's only 50 miles an hour.

Metric is a decimal system that uses multiples of 10. Each basic metric unit can be multiplied or divided by factors of 10 to get larger or smaller units. In our system we measure length in inches, feet, yards, and miles. In the metric system, length measurements are decimal fractions or multiples of meters. Our liquid volume may be teaspoons, tablespoons, cups, pints, quarts, or gallons. Metric liquid measurements are decimal fractions or multiples of a liter. We measure weight in ounces, pounds, and tons. Metric weights are based on grams. (The metric ton is part of the decimal system and is equal to 1 million grams or 1,000 kilograms.)

The common multiples, or prefixes, used with the basic metric units are:

- **milli:** one thousandth (0.001)
- **centi:** one hundredth (0.01)
- **deci:** one tenth (0.1)
- **kilo:** one thousand times (1,000)

Thus, a kilometer means 1,000 meters, a centimeter is $\frac{1}{100}$ of a meter, a millimeter is $\frac{1}{1,000}$ of a meter, and deciliter is $\frac{1}{10}$ of a liter.

Table B-1 gives the conversion factors with two significant digits after the decimal point. For ordinary use, even more approximate conversions are often adequate. For example, 2.5 centimeters to an inch may be used instead of 2.54, and 0.6 miles to one kilometer will give a usable conversion instead of 0.62.

■ **Example 1**   Measures of Distance

1. A building lot is 50 × 100 feet. What are its dimensions in meters?

   50 ft × 30.48 = 1,524 cm   *There are 30.48 centimeters per foot.*

   1,524 cm × 0.01 = 15.24 m   *One centimeter equals 0.01 meter.*

   A building lot 50 × 100 feet is about 15 × 30 meters.

2. A road sign says, "100 meters to the next gas station." How far is it in feet?

   100 m × 3.28 = 328 ft   *There are 3.28 feet per meter.*

   It is 328 feet to the gas station.   ■

TABLE B-1 METRIC CONVERSION FACTORS

CONVERTING TO METRIC

ENGLISH UNITS	METRIC UNITS	METRIC SYMBOL
Length		
inch (in)	2.54 centimeters	cm
foot (ft)	30.48 centimeters	cm
yard (yd)	0.91 meter	m
miles (mi)	1.61 kilometers	km
Weight (Dry)		
ounce (oz)	28.35 grams	g
pound (lb)	453.59 grams	g
short ton (2,000 pounds)	0.91 metric ton	t
Volume (Liquid)		
teaspoon (tsp)	4.93 milliliters	mL
tablespoon (tbsp)	14.79 milliliters	mL
ounce (oz)	29.57 milliliters	mL
pint (pt)	473.17 milliliters	mL
quart (qt)	0.95 liter	L
gallon (gal)	3.79 liters	L

CONVERTING FROM METRIC

METRIC SYMBOL	METRIC UNITS	ENGLISH UNITS
Length		
mm	millimeter	0.039 inch (in)
cm	centimeter	0.39 inch (in)
m	meter	39.37 inches (in) or 3.28 feet (ft)
km	kilometer	0.62 mile (mi)
Weight (Dry)		
g	gram	0.035 ounce (oz)
kg	kilogram	2.20 pounds (lb)
t	metric ton (1,000 kg)	1.10 short tons
Volume (Liquid)		
mL	milliliter	0.034 ounce (oz) or 0.068 tablespoon (tbsp) or 0.20 teaspoon (tsp)
L	liter	1.06 quarts (qt) or 2.11 pints (pt) or 0.26 gallon (gal)

■ **Example 2**

Measures of Weight

1. A package of meat is labeled 1.4 pounds. What is the equivalent metric weight?

$$1.4 \text{ lb} \times 453.59 = 635.026 \text{ g}$$

One pound equals 453.59 grams.

The package could be labeled 635 grams.

2. A customer bought 250 grams of cheese. How much is this in ounces?

$$250 \text{ g} \times 0.035 = 8.75 \text{ oz}$$

There is 0.035 ounce in a gram.

The cheese weighed almost 9 ounces. ■

■ **Example 3**

Measures of Volume

1. If the gas tank of a car holds 15 gallons, how many liters would fill it up?

$$15 \text{ gal} \times 3.79 = 56.85 \text{ L}$$

There are 3.79 liters to a gallon.

The 15-gallon tank is also a 57-liter tank.

2. A bottle of apple juice is marked 3 liters. How many quarts does it hold?

$$3 \text{ L} \times 1.06 = 3.18 \text{ qt}$$

There are 1.06 quarts in a liter.

The 3-liter bottle holds $3\frac{1}{5}$ quarts. ■

Temperature

The metric temperature scale must be learned by experience, since it has no simple reference unit in the Fahrenheit scale. The **Celsius** scale (no longer called centigrade) is graduated from the freezing point of water (0°C) to the boiling point of water (100°C). The scale in Figure B-1 will tell you at about what temperature on the Celsius scale you should go to bed and call the doctor and when you need a coat outdoors.

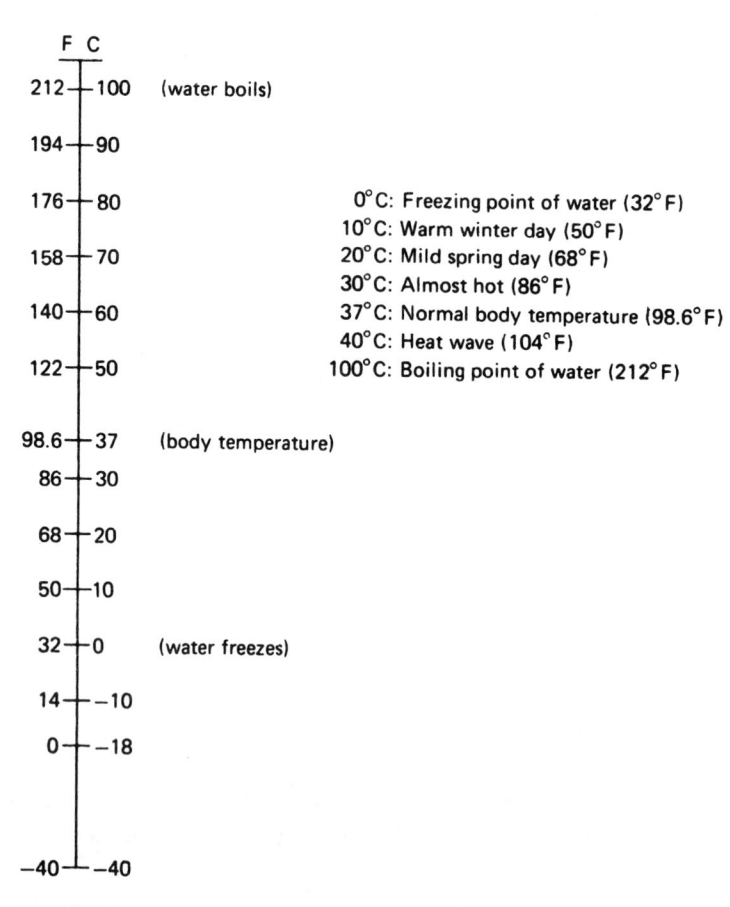

```
  F    C
 212 — 100   (water boils)
 194 —  90
 176 —  80           0°C:  Freezing point of water (32°F)
                    10°C:  Warm winter day (50°F)
 158 —  70          20°C:  Mild spring day (68°F)
                    30°C:  Almost hot (86°F)
 140 —  60          37°C:  Normal body temperature (98.6°F)
                    40°C:  Heat wave (104°F)
 122 —  50         100°C:  Boiling point of water (212°F)
98.6 —  37   (body temperature)
  86 —  30
  68 —  20
  50 —  10
  32 —   0   (water freezes)
  14 — -10
   0 — -18

 -40 — -40
```

FIGURE B-1 Fahrenheit and Celsius scales.

To convert from one scale to the other, use the formula

$$°C = \frac{5}{9}(°F - 32)$$

where °C is the Celsius temperature and °F is the Fahrenheit temperature.

■ **Example 4**

1. What Celsius temperature is equivalent to 90°F?

    $°C = \frac{5}{9}(°F - 32)$

    $= \frac{5}{9}(90 - 32)$ *Substitute 90 for °F.*

    $= \frac{5}{9}(58)$ *Complete arithmetic within parentheses.*

    $= \frac{290}{9}$

    $°C = 32.22$ or $32°$

   Check by reading the scale in Figure B-1.

2. What Fahrenheit temperature is equivalent to 25°C?

    $°C = \frac{5}{9}(°F - 32)$

    $25 = \frac{5}{9}(°F - 32)$ *Substitute 25 for °C.*

    $\frac{9}{5} \times 25 = \frac{9}{5} \times \frac{5}{9}(°F - 32)$ *Multiply both sides of the equation by $\frac{9}{5}$ to simplify the calculation.*

    $45 = °F - 32$

    $45 + 32 = °F$ *Collect similar terms.*

    $°F = 77°$

   Check by reading the scale in Figure B-1. ■

Appendix B

EXERCISES

Correct answers to two decimal places except for temperature conversions. Convert the following measures of length or distance to metric equivalents.

1. 21 ft
2. $2\frac{1}{2}$ yd
3. 5 in

4. 52 mi
5. 2′3″
6. $4\frac{1}{2}$ mi

Convert the following measures of length or distance to the English system.

7. 22 mm
8. $4\frac{1}{4}$ m
9. 18 cm

10. 55 km
11. 100 m
12. 206 cm

Convert the following English measures of weight to metric equivalents.

13. 5 oz
14. $2\frac{1}{2}$ lb
15. 4.2 oz

16. 2.8 short tons
17. $4\frac{1}{4}$ lbs
18. 1 lb 2 oz

Convert the following metric weights to English equivalents.

19. 3 g
20. $2\frac{1}{2}$ kg
21. 3 t
22. 490 g

23. 62 kg
24. 356 g

Convert the following measures of volume to metric equivalents.

25. $1\frac{1}{2}$ qt
26. 2 tsp
27. 4.7 gal
28. 22 oz (liquid)

29. 4 tbsp
30. 3 pt

585

Convert the following metric measures of volume to English equivalents.

31. 3 mL **32.** 3 L **33.** 25 mL **34.** $4\frac{3}{4}$ L

35. 72 L **36.** 4.2 mL

Convert the following Fahrenheit temperatures to Celsius; correct to the nearest degree.

37. 32°F **38.** 212°F **39.** 100°F **40.** 72°F

41. 85°F **42.** 10°F

Convert the following Celsius temperatures to Fahrenheit; correct to the nearest degree.

43. 0°C **44.** 100°C **45.** 35°C **46.** 80°C

47. 50°C **48.** 24°C

Appendix B

PROBLEMS

Without using the conversion tables, complete the statements in Problems 1 to 5.

1. A liter is about equal to a _____.

2. A _____ is less than one-half inch.

3. A kilometer is not much more than half a _____.

4. A pound is about half a _____.

5. A meter is not much more than a _____.

6. If a road sign said "200 meters to gasoline," how many feet would you have to go to reach the gas station?

7. If the weather report said 25°C, what Fahrenheit temperature would you dress for?

8. A house was built on land 1,200 m above sea level. How many feet above sea level was it?

9. A company packing raisins uses a 15-oz package. What metric equivalent should they put on the label?

10. A shopper carried home 400 g of bread, 350 g of cheese, and 600 g of fruit. How many pounds did the purchase weigh?

11. A patient with a fever recorded a temperature of 102°F. What is the equivalent Celsius reading?

12. If the speed limit is 55 miles an hour, how fast can you go in kilometers without risking a ticket?

13. A gas tank holds 17 gallons. How many liters of gasoline would you need to fill the tank?

14. A recipe calls for half a pint of sour cream. How many deciliters of sour cream would the cook need?

15. During a storm, a city received 4.2 in. of rain in a short time. How many millimeters is this?

SUPERPROBLEM

You are in Europe driving at 50 miles an hour in a car that gives you 15 miles per gallon of gas. You expect to travel 255 miles.

1. How many liters of gas would you need to cover the distance?

2. How long would it take you to get there?

3. If you started the trip when the temperature was 84°F and it dropped 10°F while you were traveling, what would the Celsus scale read when you arrived?

4. If you went into a store to buy 5 ounces of cheese and a pint bottle of milk, how much would the cheese weigh in grams and with what volume would the milk container be labeled?

C ∎ Tables

TABLE C-1 INCOME TAX WITHHOLDING
SINGLE PERSON—WEEKLY PAYROLL PERIOD

(For Wages Paid in 1999)

If the wages are—		And the number of withholding allowances claimed is—										
At least	But less than	0	1	2	3	4	5	6	7	8	9	10
		The amount of income tax to be withheld is—										
$0	$55	0	0	0	0	0	0	0	0	0	0	0
55	60	1	0	0	0	0	0	0	0	0	0	0
60	65	2	0	0	0	0	0	0	0	0	0	0
65	70	2	0	0	0	0	0	0	0	0	0	0
70	75	3	0	0	0	0	0	0	0	0	0	0
75	80	4	0	0	0	0	0	0	0	0	0	0
80	85	5	0	0	0	0	0	0	0	0	0	0
85	90	5	0	0	0	0	0	0	0	0	0	0
90	95	6	0	0	0	0	0	0	0	0	0	0
95	100	7	0	0	0	0	0	0	0	0	0	0
100	105	8	0	0	0	0	0	0	0	0	0	0
105	110	8	1	0	0	0	0	0	0	0	0	0
110	115	9	1	0	0	0	0	0	0	0	0	0
115	120	10	2	0	0	0	0	0	0	0	0	0
120	125	11	3	0	0	0	0	0	0	0	0	0
125	130	11	4	0	0	0	0	0	0	0	0	0
130	135	12	4	0	0	0	0	0	0	0	0	0
135	140	13	5	0	0	0	0	0	0	0	0	0
140	145	14	6	0	0	0	0	0	0	0	0	0
145	150	14	7	0	0	0	0	0	0	0	0	0
150	155	15	7	0	0	0	0	0	0	0	0	0
155	160	16	8	0	0	0	0	0	0	0	0	0
160	165	17	9	1	0	0	0	0	0	0	0	0
165	170	17	10	2	0	0	0	0	0	0	0	0
170	175	18	10	2	0	0	0	0	0	0	0	0
175	180	19	11	3	0	0	0	0	0	0	0	0
180	185	20	12	4	0	0	0	0	0	0	0	0
185	190	20	13	5	0	0	0	0	0	0	0	0
190	195	21	13	5	0	0	0	0	0	0	0	0
195	200	22	14	6	0	0	0	0	0	0	0	0
200	210	23	15	7	0	0	0	0	0	0	0	0
210	220	25	17	9	1	0	0	0	0	0	0	0
220	230	26	18	10	2	0	0	0	0	0	0	0
230	240	28	20	12	4	0	0	0	0	0	0	0
240	250	29	21	13	5	0	0	0	0	0	0	0
250	260	31	23	15	7	0	0	0	0	0	0	0
260	270	32	24	16	8	0	0	0	0	0	0	0
270	280	34	26	18	10	2	0	0	0	0	0	0
280	290	35	27	19	11	3	0	0	0	0	0	0
290	300	37	29	21	13	5	0	0	0	0	0	0
300	310	38	30	22	14	6	0	0	0	0	0	0
310	320	40	32	24	16	8	0	0	0	0	0	0
320	330	41	33	25	17	9	1	0	0	0	0	0
330	340	43	35	27	19	11	3	0	0	0	0	0
340	350	44	36	28	20	12	4	0	0	0	0	0
350	360	46	38	30	22	14	6	0	0	0	0	0
360	370	47	39	31	23	15	7	0	0	0	0	0
370	380	49	41	33	25	17	9	1	0	0	0	0
380	390	50	42	34	26	18	10	3	0	0	0	0
390	400	52	44	36	28	20	12	4	0	0	0	0
400	410	53	45	37	29	21	13	6	0	0	0	0
410	420	55	47	39	31	23	15	7	0	0	0	0
420	430	56	48	40	32	24	16	9	1	0	0	0
430	440	58	50	42	34	26	18	10	2	0	0	0
440	450	59	51	43	35	27	19	12	4	0	0	0
450	460	61	53	45	37	29	21	13	5	0	0	0
460	470	62	54	46	38	30	22	15	7	0	0	0
470	480	64	56	48	40	32	24	16	8	0	0	0
480	490	65	57	49	41	33	25	18	10	2	0	0
490	500	67	59	51	43	35	27	19	11	3	0	0
500	510	68	60	52	44	36	28	21	13	5	0	0
510	520	70	62	54	46	38	30	22	14	6	0	0
520	530	71	63	55	47	39	31	24	16	8	0	0
530	540	74	65	57	49	41	33	25	17	9	1	0
540	550	77	66	58	50	42	34	27	19	11	3	0
550	560	80	68	60	52	44	36	28	20	12	4	0
560	570	82	69	61	53	45	37	30	22	14	6	0
570	580	85	71	63	55	47	39	31	23	15	7	0
580	590	88	73	64	56	48	40	33	25	17	9	1
590	600	91	76	66	58	50	42	34	26	18	10	2

From the U.S. Internal Revenue Service, 1999.

TABLE C-1 INCOME TAX WITHHOLDING (cont.)
SINGLE PERSON—WEEKLY PAYROLL PERIOD

(For Wages Paid in 1999)

If the wages are—		And the number of withholding allowances claimed is—										
At least	But less than	0	1	2	3	4	5	6	7	8	9	10
		The amount of income tax to be withheld is—										
$600	$610	94	79	67	59	51	43	36	28	20	12	4
610	620	96	81	69	61	53	45	37	29	21	13	5
620	630	99	84	70	62	54	46	39	31	23	15	7
630	640	102	87	72	64	56	48	40	32	24	16	8
640	650	105	90	75	65	57	49	42	34	26	18	10
650	660	108	93	78	67	59	51	43	35	27	19	11
660	670	110	95	81	68	60	52	45	37	29	21	13
670	680	113	98	83	70	62	54	46	38	30	22	14
680	690	116	101	86	71	63	55	48	40	32	24	16
690	700	119	104	89	74	65	57	49	41	33	25	17
700	710	122	107	92	77	66	58	51	43	35	27	19
710	720	124	109	95	80	68	60	52	44	36	28	20
720	730	127	112	97	83	69	61	54	46	38	30	22
730	740	130	115	100	85	71	63	55	47	39	31	23
740	750	133	118	103	88	73	64	57	49	41	33	25
750	760	136	121	106	91	76	66	58	50	42	34	26
760	770	138	123	109	94	79	67	60	52	44	36	28
770	780	141	126	111	97	82	69	61	53	45	37	29
780	790	144	129	114	99	85	70	63	55	47	39	31
790	800	147	132	117	102	87	73	64	56	48	40	32
800	810	150	135	120	105	90	75	66	58	50	42	34
810	820	152	137	123	108	93	78	67	59	51	43	35
820	830	155	140	125	111	96	81	69	61	53	45	37
830	840	158	143	128	113	99	84	70	62	54	46	38
840	850	161	146	131	116	101	87	72	64	56	48	40
850	860	164	149	134	119	104	89	75	65	57	49	41
860	870	166	151	137	122	107	92	77	67	59	51	43
870	880	169	154	139	125	110	95	80	68	60	52	44
880	890	172	157	142	127	113	98	83	70	62	54	46
890	900	175	160	145	130	115	101	86	71	63	55	47
900	910	178	163	148	133	118	103	89	74	65	57	49
910	920	180	165	151	136	121	106	91	77	66	58	50
920	930	183	168	153	139	124	109	94	79	68	60	52
930	940	186	171	156	141	127	112	97	82	69	61	53
940	950	189	174	159	144	129	115	100	85	71	63	55
950	960	192	177	162	147	132	117	103	88	73	64	56
960	970	194	179	165	150	135	120	105	91	76	66	58
970	980	197	182	167	153	138	123	108	93	79	67	59
980	990	200	185	170	155	141	126	111	96	81	69	61
990	1,000	203	188	173	158	143	129	114	99	84	70	62
1,000	1,010	206	191	176	161	146	131	117	102	87	72	64
1,010	1,020	208	193	179	164	149	134	119	105	90	75	65
1,020	1,030	211	196	181	167	152	137	122	107	93	78	67
1,030	1,040	214	199	184	169	155	140	125	110	95	81	68
1,040	1,050	217	202	187	172	157	143	128	113	98	83	70
1,050	1,060	220	205	190	175	160	145	131	116	101	86	71
1,060	1,070	222	207	193	178	163	148	133	119	104	89	74
1,070	1,080	225	210	195	181	166	151	136	121	107	92	77
1,080	1,090	228	213	198	183	169	154	139	124	109	95	80
1,090	1,100	231	216	201	186	171	157	142	127	112	97	83
1,100	1,110	234	219	204	189	174	159	145	130	115	100	85
1,110	1,120	236	221	207	192	177	162	147	133	118	103	88
1,120	1,130	239	224	209	195	180	165	150	135	121	106	91
1,130	1,140	242	227	212	197	183	168	153	138	123	109	94
1,140	1,150	245	230	215	200	185	171	156	141	126	111	97
1,150	1,160	248	233	218	203	188	173	159	144	129	114	99
1,160	1,170	252	235	221	206	191	176	161	147	132	117	102
1,170	1,180	255	238	223	209	194	179	164	149	135	120	105
1,180	1,190	258	241	226	211	197	182	167	152	137	123	108
1,190	1,200	261	244	229	214	199	185	170	155	140	125	111
1,200	1,210	264	248	232	217	202	187	173	158	143	128	113
1,210	1,220	267	251	235	220	205	190	175	161	146	131	116
1,220	1,230	270	254	237	223	208	193	178	163	149	134	119
1,230	1,240	273	257	240	225	211	196	181	166	151	137	122
1,240	1,250	276	260	244	228	213	199	184	169	154	139	125

$1,250 and over   Use Table 1(a) for a **SINGLE person** on page 34. Also see the instructions on page 32.

From the U.S. Internal Revenue Service, 1999.

TABLE C-1 INCOME TAX WITHHOLDING (cont.)
MARRIED PERSON—WEEKLY PAYROLL PERIOD

(For Wages Paid in 1999)

If the wages are—		And the number of withholding allowances claimed is—										
At least	But less than	0	1	2	3	4	5	6	7	8	9	10
		The amount of income tax to be withheld is—										
$0	$125	0	0	0	0	0	0	0	0	0	0	0
125	130	1	0	0	0	0	0	0	0	0	0	0
130	135	1	0	0	0	0	0	0	0	0	0	0
135	140	2	0	0	0	0	0	0	0	0	0	0
140	145	3	0	0	0	0	0	0	0	0	0	0
145	150	4	0	0	0	0	0	0	0	0	0	0
150	155	4	0	0	0	0	0	0	0	0	0	0
155	160	5	0	0	0	0	0	0	0	0	0	0
160	165	6	0	0	0	0	0	0	0	0	0	0
165	170	7	0	0	0	0	0	0	0	0	0	0
170	175	7	0	0	0	0	0	0	0	0	0	0
175	180	8	0	0	0	0	0	0	0	0	0	0
180	185	9	1	0	0	0	0	0	0	0	0	0
185	190	10	2	0	0	0	0	0	0	0	0	0
190	195	10	2	0	0	0	0	0	0	0	0	0
195	200	11	3	0	0	0	0	0	0	0	0	0
200	210	12	4	0	0	0	0	0	0	0	0	0
210	220	14	6	0	0	0	0	0	0	0	0	0
220	230	15	7	0	0	0	0	0	0	0	0	0
230	240	17	9	1	0	0	0	0	0	0	0	0
240	250	18	10	2	0	0	0	0	0	0	0	0
250	260	20	12	4	0	0	0	0	0	0	0	0
260	270	21	13	5	0	0	0	0	0	0	0	0
270	280	23	15	7	0	0	0	0	0	0	0	0
280	290	24	16	8	0	0	0	0	0	0	0	0
290	300	26	18	10	2	0	0	0	0	0	0	0
300	310	27	19	11	3	0	0	0	0	0	0	0
310	320	29	21	13	5	0	0	0	0	0	0	0
320	330	30	22	14	6	0	0	0	0	0	0	0
330	340	32	24	16	8	0	0	0	0	0	0	0
340	350	33	25	17	9	1	0	0	0	0	0	0
350	360	35	27	19	11	3	0	0	0	0	0	0
360	370	36	28	20	12	4	0	0	0	0	0	0
370	380	38	30	22	14	6	0	0	0	0	0	0
380	390	39	31	23	15	7	0	0	0	0	0	0
390	400	41	33	25	17	9	1	0	0	0	0	0
400	410	42	34	26	18	10	2	0	0	0	0	0
410	420	44	36	28	20	12	4	0	0	0	0	0
420	430	45	37	29	21	13	5	0	0	0	0	0
430	440	47	39	31	23	15	7	0	0	0	0	0
440	450	48	40	32	24	16	8	1	0	0	0	0
450	460	50	42	34	26	18	10	2	0	0	0	0
460	470	51	43	35	27	19	11	4	0	0	0	0
470	480	53	45	37	29	21	13	5	0	0	0	0
480	490	54	46	38	30	22	14	7	0	0	0	0
490	500	56	48	40	32	24	16	8	0	0	0	0
500	510	57	49	41	33	25	17	10	2	0	0	0
510	520	59	51	43	35	27	19	11	3	0	0	0
520	530	60	52	44	36	28	20	13	5	0	0	0
530	540	62	54	46	38	30	22	14	6	0	0	0
540	550	63	55	47	39	31	23	16	8	0	0	0
550	560	65	57	49	41	33	25	17	9	1	0	0
560	570	66	58	50	42	34	26	19	11	3	0	0
570	580	68	60	52	44	36	28	20	12	4	0	0
580	590	69	61	53	45	37	29	22	14	6	0	0
590	600	71	63	55	47	39	31	23	15	7	0	0
600	610	72	64	56	48	40	32	25	17	9	1	0
610	620	74	66	58	50	42	34	26	18	10	2	0
620	630	75	67	59	51	43	35	28	20	12	4	0
630	640	77	69	61	53	45	37	29	21	13	5	0
640	650	78	70	62	54	46	38	31	23	15	7	0
650	660	80	72	64	56	48	40	32	24	16	8	0
660	670	81	73	65	57	49	41	34	26	18	10	2
670	680	83	75	67	59	51	43	35	27	19	11	3
680	690	84	76	68	60	52	44	37	29	21	13	5
690	700	86	78	70	62	54	46	38	30	22	14	6
700	710	87	79	71	63	55	47	40	32	24	16	8
710	720	89	81	73	65	57	49	41	33	25	17	9
720	730	90	82	74	66	58	50	43	35	27	19	11
730	740	92	84	76	68	60	52	44	36	28	20	12

From the U.S. Internal Revenue Service, 1999.

TABLE C-1 INCOME TAX WITHHOLDING (cont.)
MARRIED PERSON—WEEKLY PAYROLL PERIOD

(For Wages Paid in 1999)

If the wages are—		And the number of withholding allowances claimed is—										
At least	But less than	0	1	2	3	4	5	6	7	8	9	10
		The amount of income tax to be withheld is—										
$740	$750	93	85	77	69	61	53	46	38	30	22	14
750	760	95	87	79	71	63	55	47	39	31	23	15
760	770	96	88	80	72	64	56	49	41	33	25	17
770	780	98	90	82	74	66	58	50	42	34	26	18
780	790	99	91	83	75	67	59	52	44	36	28	20
790	800	101	93	85	77	69	61	53	45	37	29	21
800	810	102	94	86	78	70	62	55	47	39	31	23
810	820	104	96	88	80	72	64	56	48	40	32	24
820	830	105	97	89	81	73	65	58	50	42	34	26
830	840	107	99	91	83	75	67	59	51	43	35	27
840	850	108	100	92	84	76	68	61	53	45	37	29
850	860	110	102	94	86	78	70	62	54	46	38	30
860	870	111	103	95	87	79	71	64	56	48	40	32
870	880	113	105	97	89	81	73	65	57	49	41	33
880	890	114	106	98	90	82	74	67	59	51	43	35
890	900	116	108	100	92	84	76	68	60	52	44	36
900	910	117	109	101	93	85	77	70	62	54	46	38
910	920	119	111	103	95	87	79	71	63	55	47	39
920	930	122	112	104	96	88	80	73	65	57	49	41
930	940	124	114	106	98	90	82	74	66	58	50	42
940	950	127	115	107	99	91	83	76	68	60	52	44
950	960	130	117	109	101	93	85	77	69	61	53	45
960	970	133	118	110	102	94	86	79	71	63	55	47
970	980	136	121	112	104	96	88	80	72	64	56	48
980	990	138	124	113	105	97	89	82	74	66	58	50
990	1,000	141	126	115	107	99	91	83	75	67	59	51
1,000	1,010	144	129	116	108	100	92	85	77	69	61	53
1,010	1,020	147	132	118	110	102	94	86	78	70	62	54
1,020	1,030	150	135	120	111	103	95	88	80	72	64	56
1,030	1,040	152	138	123	113	105	97	89	81	73	65	57
1,040	1,050	155	140	126	114	106	98	91	83	75	67	59
1,050	1,060	158	143	128	116	108	100	92	84	76	68	60
1,060	1,070	161	146	131	117	109	101	94	86	78	70	62
1,070	1,080	164	149	134	119	111	103	95	87	79	71	63
1,080	1,090	166	152	137	122	112	104	97	89	81	73	65
1,090	1,100	169	154	140	125	114	106	98	90	82	74	66
1,100	1,110	172	157	142	128	115	107	100	92	84	76	68
1,110	1,120	175	160	145	130	117	109	101	93	85	77	69
1,120	1,130	178	163	148	133	118	110	103	95	87	79	71
1,130	1,140	180	166	151	136	121	112	104	96	88	80	72
1,140	1,150	183	168	154	139	124	113	106	98	90	82	74
1,150	1,160	186	171	156	142	127	115	107	99	91	83	75
1,160	1,170	189	174	159	144	130	116	109	101	93	85	77
1,170	1,180	192	177	162	147	132	118	110	102	94	86	78
1,180	1,190	194	180	165	150	135	120	112	104	96	88	80
1,190	1,200	197	182	168	153	138	123	113	105	97	89	81
1,200	1,210	200	185	170	156	141	126	115	107	99	91	83
1,210	1,220	203	188	173	158	144	129	116	108	100	92	84
1,220	1,230	206	191	176	161	146	132	118	110	102	94	86
1,230	1,240	208	194	179	164	149	134	120	111	103	95	87
1,240	1,250	211	196	182	167	152	137	122	113	105	97	89
1,250	1,260	214	199	184	170	155	140	125	114	106	98	90
1,260	1,270	217	202	187	172	158	143	128	116	108	100	92
1,270	1,280	220	205	190	175	160	146	131	117	109	101	93
1,280	1,290	222	208	193	178	163	148	134	119	111	103	95
1,290	1,300	225	210	196	181	166	151	136	122	112	104	96
1,300	1,310	228	213	198	184	169	154	139	124	114	106	98
1,310	1,320	231	216	201	186	172	157	142	127	115	107	99
1,320	1,330	234	219	204	189	174	160	145	130	117	109	101
1,330	1,340	236	222	207	192	177	162	148	133	118	110	102
1,340	1,350	239	224	210	195	180	165	150	136	121	112	104
1,350	1,360	242	227	212	198	183	168	153	138	124	113	105
1,360	1,370	245	230	215	200	186	171	156	141	126	115	107
1,370	1,380	248	233	218	203	188	174	159	144	129	116	108
1,380	1,390	250	236	221	206	191	176	162	147	132	118	110

From the U.S. Internal Revenue Service, 1999.

TABLE C-2 PERCENTAGE METHOD TABLES

Payroll Period	One Withholding Allowance
Weekly	$ 52.88
Biweekly	105.77
Semimonthly	114.58
Monthly	229.17
Quarterly	687.50
Semiannually	1,375.00
Annually	2,750.00
Daily or miscellaneous (each day of the payroll period)	10.58

From the U.S. Internal Revenue Service, 1999.

TABLE C-2 PERCENTAGE METHOD TABLES (cont.)

(For Wages Paid in 1999)

TABLE 1—WEEKLY Payroll Period

(a) SINGLE person (including head of household)—

If the amount of wages (after subtracting withholding allowances) is:
Not over $51 . . . . . $0

Over—	But not over—	The amount of income tax to withhold is:	of excess over—
$51	—$525	15%	—$51
$525	—$1,125	$71.10 plus 28%	—$525
$1,125	—$2,535	$239.10 plus 31%	—$1,125
$2,535	—$5,475	$676.20 plus 36%	—$2,535
$5,475	. . . . .	$1,734.60 plus 39.6%	—$5,475

(b) MARRIED person—

If the amount of wages (after subtracting withholding allowances) is:
Not over $124 . . . . . $0

Over—	But not over—	The amount of income tax to withhold is:	of excess over—
$124	—$913	15%	—$124
$913	—$1,894	$118.35 plus 28%	—$913
$1,894	—$3,135	$393.03 plus 31%	—$1,894
$3,135	—$5,531	$777.74 plus 36%	—$3,135
$5,531	. . . . .	$1,640.30 plus 39.6%	—$5,531

TABLE 2—BIWEEKLY Payroll Period

(a) SINGLE person (including head of household)—

If the amount of wages (after subtracting withholding allowances) is:
Not over $102 . . . . . $0

Over—	But not over—	The amount of income tax to withhold is:	of excess over—
$102	—$1,050	15%	—$102
$1,050	—$2,250	$142.20 plus 28%	—$1,050
$2,250	—$5,069	$478.20 plus 31%	—$2,250
$5,069	—$10,950	$1,352.09 plus 36%	—$5,069
$10,950	. . . . .	$3,469.25 plus 39.6%	—$10,950

(b) MARRIED person—

If the amount of wages (after subtracting withholding allowances) is:
Not over $248 . . . . . $0

Over—	But not over—	The amount of income tax to withhold is:	of excess over—
$248	—$1,827	15%	—$248
$1,827	—$3,788	$236.85 plus 28%	—$1,827
$3,788	—$6,269	$785.93 plus 31%	—$3,788
$6,269	—$11,062	$1,555.04 plus 36%	—$6,269
$11,062	. . . . .	$3,280.52 plus 39.6%	—$11,062

TABLE 3—SEMIMONTHLY Payroll Period

(a) SINGLE person (including head of household)—

If the amount of wages (after subtracting withholding allowances) is:
Not over $110 . . . . . $0

Over—	But not over—	The amount of income tax to withhold is:	of excess over—
$110	—$1,138	15%	—$110
$1,138	—$2,438	$154.20 plus 28%	—$1,138
$2,438	—$5,492	$518.20 plus 31%	—$2,438
$5,492	—$11,863	$1,464.94 plus 36%	—$5,492
$11,863	. . . . .	$3,758.50 plus 39.6%	—$11,863

(b) MARRIED person—

If the amount of wages (after subtracting withholding allowances) is:
Not over $269 . . . . . $0

Over—	But not over—	The amount of income tax to withhold is:	of excess over—
$269	—$1,979	15%	—$269
$1,979	—$4,104	$256.50 plus 28%	—$1,979
$4,104	—$6,792	$851.50 plus 31%	—$4,104
$6,792	—$11,983	$1,684.78 plus 36%	—$6,792
$11,983	. . . . .	$3,553.54 plus 39.6%	—$11,983

TABLE 4—MONTHLY Payroll Period

(a) SINGLE person (including head of household)—

If the amount of wages (after subtracting withholding allowances) is:
Not over $221 . . . . . $0

Over—	But not over—	The amount of income tax to withhold is:	of excess over—
$221	—$2,275	15%	—$221
$2,275	—$4,875	$308.10 plus 28%	—$2,275
$4,875	—$10,983	$1,036.10 plus 31%	—$4,875
$10,983	—$23,725	$2,929.58 plus 36%	—$10,983
$23,725	. . . . .	$7,516.70 plus 39.6%	—$23,725

(b) MARRIED person—

If the amount of wages (after subtracting withholding allowances) is:
Not over $538 . . . . . $0

Over—	But not over—	The amount of income tax to withhold is:	of excess over—
$538	—$3,958	15%	—$538
$3,958	—$8,208	$513.00 plus 28%	—$3,958
$8,208	—$13,583	$1,703.00 plus 31%	—$8,208
$13,583	—$23,967	$3,369.25 plus 36%	—$13,583
$23,967	. . . . .	$7,107.49 plus 39.6%	—$23,967

From the U.S. Internal Revenue Service, 1999.

TABLE C-3 THE NUMBER OF EACH DAY OF THE YEAR

Day of Month	Jan.	Feb.	Mar.	Apr.	May	June	July	Aug.	Sept.	Oct.	Nov.	Dec.	Day of Month
1	1	32	60	91	121	152	182	213	244	274	305	335	1
2	2	33	61	92	122	153	183	214	245	275	306	336	2
3	3	34	62	93	123	154	184	215	246	276	307	337	3
4	4	35	63	94	124	155	185	216	247	277	308	338	4
5	5	36	64	95	125	156	186	217	248	278	309	339	5
6	6	37	65	96	126	157	187	218	249	279	310	340	6
7	7	38	66	97	127	158	188	219	250	280	311	341	7
8	8	39	67	98	128	159	189	220	251	281	312	342	8
9	9	40	68	99	129	160	190	221	252	282	313	343	9
10	10	41	69	100	130	161	191	222	253	283	314	344	10
11	11	42	70	101	131	162	192	223	254	284	315	345	11
12	12	43	71	102	132	163	193	224	255	285	316	346	12
13	13	44	72	103	133	164	194	225	256	286	317	347	13
14	14	45	73	104	134	165	195	226	257	287	318	348	14
15	15	46	74	105	135	166	196	227	258	288	319	349	15
16	16	47	75	106	136	167	197	228	259	289	320	350	16
17	17	48	76	107	137	168	198	229	260	290	321	351	17
18	18	49	77	108	138	169	199	230	261	291	322	352	18
19	19	50	78	109	139	170	200	231	262	292	323	353	19
20	20	51	79	110	140	171	201	232	263	293	324	354	20
21	21	52	80	111	141	172	202	233	264	294	325	355	21
22	22	53	81	112	142	173	203	234	265	295	326	356	22
23	23	54	82	113	143	174	204	235	266	296	327	357	23
24	24	55	83	114	144	175	205	236	267	297	328	358	24
25	25	56	84	115	145	176	206	237	268	298	329	359	25
26	26	57	85	116	146	177	207	238	269	299	330	360	26
27	27	58	86	117	147	178	208	239	270	300	331	361	27
28	28	59	87	118	148	179	209	240	271	301	332	362	28
29	29		88	119	149	180	210	241	272	302	333	363	29
30	30		89	120	150	181	211	242	273	303	334	364	30
31	31		90		151		212	243		304		365	31

Note: In leap years, after February 28, add 1 to the tabular number.

TABLE C-4 ANNUAL PERCENTAGE RATE TABLE FOR MONTHLY PAYMENT PLANS

NUMBER OF PAYMENTS	10.00%	10.25%	10.50%	10.75%	11.00%	11.25%	11.50%	11.75%	12.00%	12.25%	12.50%	12.75%	13.00%	13.25%	13.50%	13.75%
\multicolumn{17}{c}{(FINANCE CHARGE PER $100 OF AMOUNT FINANCED)}
1	0.83	0.85	0.87	0.90	0.92	0.94	0.96	0.98	1.00	1.02	1.04	1.06	1.08	1.10	1.12	1.15
2	1.25	1.28	1.31	1.35	1.38	1.41	1.44	1.47	1.50	1.53	1.57	1.60	1.63	1.66	1.69	1.72
3	1.67	1.71	1.76	1.80	1.84	1.88	1.92	1.96	2.01	2.05	2.09	2.13	2.17	2.22	2.26	2.30
4	2.09	2.14	2.20	2.25	2.30	2.35	2.41	2.46	2.51	2.57	2.62	2.67	2.72	2.78	2.83	2.88
5	2.51	2.58	2.64	2.70	2.77	2.83	2.89	2.96	3.02	3.08	3.15	3.21	3.27	3.34	3.40	3.46
6	2.94	3.01	3.08	3.16	3.23	3.31	3.38	3.45	3.53	3.60	3.68	3.75	3.83	3.90	3.97	4.05
7	3.36	3.45	3.53	3.62	3.70	3.78	3.87	3.95	4.04	4.12	4.21	4.29	4.38	4.47	4.55	4.64
8	3.79	3.88	3.98	4.07	4.17	4.26	4.36	4.46	4.55	4.65	4.74	4.84	4.94	5.03	5.13	5.22
9	4.21	4.32	4.43	4.53	4.64	4.75	4.85	4.96	5.07	5.17	5.28	5.39	5.49	5.60	5.71	5.82
10	4.64	4.76	4.88	4.99	5.11	5.23	5.35	5.46	5.58	5.70	5.82	5.94	6.05	6.17	6.29	6.41
11	5.07	5.20	5.33	5.45	5.58	5.71	5.84	5.97	6.10	6.23	6.36	6.49	6.62	6.75	6.88	7.01
12	5.50	5.64	5.78	5.92	6.06	6.20	6.34	6.48	6.62	6.76	6.90	7.04	7.18	7.32	7.46	7.60
13	5.93	6.08	6.23	6.38	6.53	6.68	6.84	6.99	7.14	7.29	7.44	7.59	7.75	7.90	8.05	8.20
14	6.36	6.52	6.69	6.85	7.01	7.17	7.34	7.50	7.66	7.82	7.99	8.15	8.31	8.48	8.64	8.81
15	6.80	6.97	7.14	7.32	7.49	7.66	7.84	8.01	8.19	8.36	8.53	8.71	8.88	9.06	9.23	9.41
16	7.23	7.41	7.60	7.78	7.97	8.15	8.34	8.53	8.71	8.90	9.08	9.27	9.46	9.64	9.83	10.02
17	7.67	7.86	8.06	8.25	8.45	8.65	8.84	9.04	9.24	9.44	9.63	9.83	10.03	10.23	10.43	10.63
18	8.10	8.31	8.52	8.73	8.93	9.14	9.35	9.56	9.77	9.98	10.19	10.40	10.61	10.82	11.03	11.24
19	8.54	8.76	8.98	9.20	9.42	9.64	9.86	10.08	10.30	10.52	10.74	10.96	11.18	11.41	11.63	11.85
20	8.98	9.21	9.44	9.67	9.90	10.13	10.37	10.60	10.83	11.06	11.30	11.53	11.76	12.00	12.23	12.46
21	9.42	9.66	9.90	10.15	10.39	10.63	10.88	11.12	11.36	11.61	11.85	12.10	12.34	12.59	12.84	13.08
22	9.86	10.12	10.37	10.62	10.88	11.13	11.39	11.64	11.90	12.16	12.41	12.67	12.93	13.19	13.44	13.70
23	10.30	10.57	10.84	11.10	11.37	11.63	11.90	12.17	12.44	12.71	12.97	13.24	13.51	13.78	14.05	14.32
24	10.75	11.02	11.30	11.58	11.86	12.14	12.42	12.70	12.98	13.26	13.54	13.82	14.10	14.38	14.66	14.95
25	11.19	11.48	11.77	12.06	12.35	12.64	12.93	13.22	13.52	13.81	14.10	14.40	14.69	14.98	15.28	15.57
26	11.64	11.94	12.24	12.54	12.85	13.15	13.45	13.75	14.06	14.36	14.67	14.97	15.28	15.59	15.89	16.20
27	12.09	12.40	12.71	13.03	13.34	13.66	13.97	14.29	14.60	14.92	15.24	15.56	15.87	16.19	16.51	16.83
28	12.53	12.86	13.18	13.51	13.84	14.16	14.49	14.82	15.15	15.48	15.81	16.14	16.47	16.80	17.13	17.46
29	12.98	13.32	13.66	14.00	14.33	14.67	15.01	15.35	15.70	16.04	16.38	16.72	17.07	17.41	17.75	18.10
30	13.43	13.78	14.13	14.48	14.83	15.19	15.54	15.89	16.24	16.60	16.95	17.31	17.66	18.02	18.38	18.74
31	13.89	14.25	14.61	14.97	15.33	15.70	16.06	16.43	16.79	17.16	17.53	17.90	18.27	18.63	19.00	19.38
32	14.34	14.71	15.09	15.46	15.84	16.21	16.59	16.97	17.35	17.73	18.11	18.49	18.87	19.25	19.63	20.02
33	14.79	15.18	15.57	15.95	16.34	16.73	17.12	17.51	17.90	18.29	18.69	19.08	19.47	19.87	20.26	20.66
34	15.25	15.65	16.05	16.44	16.85	17.25	17.65	18.05	18.46	18.86	19.27	19.67	20.08	20.49	20.90	21.31
35	15.70	16.11	16.53	16.94	17.35	17.77	18.18	18.60	19.01	19.43	19.85	20.27	20.69	21.11	21.53	21.95
36	16.16	16.58	17.01	17.43	17.86	18.29	18.71	19.14	19.57	20.00	20.43	20.87	21.30	21.73	22.17	22.60
37	16.62	17.06	17.49	17.93	18.37	18.81	19.25	19.69	20.13	20.58	21.02	21.46	21.91	22.36	22.81	23.25
38	17.08	17.53	17.98	18.43	18.88	19.33	19.78	20.24	20.69	21.15	21.61	22.07	22.52	22.99	23.45	23.91
39	17.54	18.00	18.46	18.93	19.39	19.86	20.32	20.79	21.26	21.73	22.20	22.67	23.14	23.61	24.09	24.56
40	18.00	18.48	18.95	19.43	19.90	20.38	20.86	21.34	21.82	22.30	22.79	23.27	23.76	24.25	24.73	25.22
41	18.47	18.95	19.44	19.93	20.42	20.91	21.40	21.89	22.39	22.88	23.38	23.88	24.38	24.88	25.38	25.88
42	18.93	19.43	19.93	20.43	20.93	21.44	21.94	22.45	22.96	23.47	23.98	24.49	25.00	25.51	26.03	26.55
43	19.40	19.91	20.42	20.94	21.45	21.97	22.49	23.01	23.53	24.05	24.57	25.10	25.62	26.15	26.68	27.21
44	19.86	20.39	20.91	21.44	21.97	22.50	23.03	23.57	24.10	24.64	25.17	25.71	26.25	26.79	27.33	27.88
45	20.33	20.87	21.41	21.95	22.49	23.03	23.58	24.12	24.67	25.22	25.77	26.32	26.88	27.43	27.99	28.55
46	20.80	21.35	21.90	22.46	23.01	23.57	24.13	24.69	25.25	25.81	26.37	26.94	27.51	28.08	28.65	29.22
47	21.27	21.83	22.40	22.97	23.53	24.10	24.68	25.25	25.82	26.40	26.98	27.56	28.14	28.72	29.31	29.89
48	21.74	22.32	22.90	23.48	24.06	24.64	25.23	25.81	26.40	26.99	27.58	28.18	28.77	29.37	29.97	30.57
49	22.21	22.80	23.39	23.99	24.58	25.18	25.78	26.38	26.98	27.59	28.19	28.80	29.41	30.02	30.63	31.24
50	22.69	23.29	23.89	24.50	25.11	25.72	26.33	26.95	27.56	28.18	28.80	29.42	30.04	30.67	31.29	31.92
51	23.16	23.78	24.40	25.02	25.64	26.26	26.89	27.52	28.15	28.78	29.41	30.05	30.68	31.32	31.96	32.60
52	23.64	24.27	24.90	25.53	26.17	26.81	27.45	28.09	28.73	29.38	30.02	30.67	31.32	31.98	32.63	33.29
53	24.11	24.76	25.40	26.05	26.70	27.35	28.00	28.66	29.32	29.98	30.64	31.30	31.97	32.63	33.30	33.97
54	24.59	25.25	25.91	26.57	27.23	27.90	28.56	29.23	29.91	30.58	31.25	31.93	32.61	33.29	33.98	34.66
55	25.07	25.74	26.41	27.09	27.77	28.44	29.13	29.81	30.50	31.18	31.87	32.56	33.26	33.95	34.65	35.35
56	25.55	26.23	26.92	27.61	28.30	28.99	29.69	30.39	31.09	31.79	32.49	33.20	33.91	34.62	35.33	36.04
57	26.03	26.73	27.43	28.13	28.84	29.54	30.25	30.97	31.68	32.39	33.11	33.83	34.56	35.28	36.01	36.74
58	26.51	27.23	27.94	28.66	29.37	30.10	30.82	31.55	32.27	33.00	33.74	34.47	35.21	35.95	36.69	37.43
59	27.00	27.72	28.45	29.18	29.91	30.65	31.39	32.13	32.87	33.61	34.36	35.11	35.86	36.62	37.37	38.13
60	27.48	28.22	28.96	29.71	30.45	31.20	31.96	32.71	33.47	34.23	34.99	35.75	36.52	37.29	38.06	38.83

TABLE C-4 ANNUAL PERCENTAGE RATE TABLE FOR MONTHLY PAYMENT PLANS (*cont.*)

NUMBER OF PAYMENTS	14.00%	14.25%	14.50%	14.75%	15.00%	15.25%	15.50%	15.75%	16.00%	16.25%	16.50%	16.75%	17.00%	17.25%	17.50%	17.75%
	(FINANCE CHARGE PER $100 OF AMOUNT FINANCED)															
1	1.17	1.19	1.21	1.23	1.25	1.27	1.29	1.31	1.33	1.35	1.37	1.40	1.42	1.44	1.46	1.48
2	1.75	1.78	1.82	1.85	1.88	1.91	1.94	1.97	2.00	2.04	2.07	2.10	2.13	2.16	2.19	2.22
3	2.34	2.38	2.43	2.47	2.51	2.55	2.59	2.64	2.68	2.72	2.76	2.80	2.85	2.89	2.93	2.97
4	2.93	2.99	3.04	3.09	3.14	3.20	3.25	3.30	3.36	3.41	3.46	3.51	3.57	3.62	3.67	3.73
5	3.53	3.59	3.65	3.72	3.78	3.84	3.91	3.97	4.04	4.10	4.16	4.23	4.29	4.35	4.42	4.48
6	4.12	4.20	4.27	4.35	4.42	4.49	4.57	4.64	4.72	4.79	4.87	4.94	5.02	5.09	5.17	5.24
7	4.72	4.81	4.89	4.98	5.06	5.15	5.23	5.32	5.40	5.49	5.58	5.66	5.75	5.83	5.92	6.00
8	5.32	5.42	5.51	5.61	5.71	5.80	5.90	6.00	6.09	6.19	6.29	6.38	6.48	6.58	6.67	6.77
9	5.92	6.03	6.14	6.25	6.35	6.46	6.57	6.68	6.78	6.89	7.00	7.11	7.22	7.32	7.43	7.54
10	6.53	6.65	6.77	6.88	7.00	7.12	7.24	7.36	7.48	7.60	7.72	7.84	7.96	8.08	8.19	8.31
11	7.14	7.27	7.40	7.53	7.66	7.79	7.92	8.05	8.18	8.31	8.44	8.57	8.70	8.83	8.96	9.09
12	7.74	7.89	8.03	8.17	8.31	8.45	8.59	8.74	8.88	9.02	9.16	9.30	9.45	9.59	9.73	9.87
13	8.36	8.51	8.66	8.81	8.97	9.12	9.27	9.43	9.58	9.73	9.89	10.04	10.20	10.35	10.50	10.66
14	8.97	9.13	9.30	9.46	9.63	9.79	9.96	10.12	10.29	10.45	10.62	10.78	10.95	11.11	11.28	11.45
15	9.59	9.76	9.94	10.11	10.29	10.47	10.64	10.82	11.00	11.17	11.35	11.53	11.71	11.88	12.06	12.24
16	10.20	10.39	10.58	10.77	10.95	11.14	11.33	11.52	11.71	11.90	12.09	12.28	12.46	12.65	12.84	13.03
17	10.82	11.02	11.22	11.42	11.62	11.82	12.02	12.22	12.42	12.62	12.83	13.03	13.23	13.43	13.63	13.83
18	11.45	11.66	11.87	12.08	12.29	12.50	12.72	12.93	13.14	13.35	13.57	13.78	13.99	14.21	14.42	14.64
19	12.07	12.30	12.52	12.74	12.97	13.19	13.41	13.64	13.86	14.09	14.31	14.54	14.76	14.99	15.22	15.44
20	12.70	12.93	13.17	13.41	13.64	13.88	14.11	14.35	14.59	14.82	15.06	15.30	15.54	15.77	16.01	16.25
21	13.33	13.58	13.82	14.07	14.32	14.57	14.82	15.06	15.31	15.56	15.81	16.06	16.31	16.56	16.81	17.07
22	13.96	14.22	14.48	14.74	15.00	15.26	15.52	15.78	16.04	16.30	16.57	16.83	17.09	17.36	17.62	17.88
23	14.59	14.87	15.14	15.41	15.68	15.96	16.23	16.50	16.78	17.05	17.32	17.60	17.88	18.15	18.43	18.70
24	15.23	15.51	15.80	16.08	16.37	16.65	16.94	17.22	17.51	17.80	18.09	18.37	18.66	18.95	19.24	19.53
25	15.87	16.17	16.46	16.76	17.06	17.35	17.65	17.95	18.25	18.55	18.85	19.15	19.45	19.75	20.05	20.36
26	16.51	16.82	17.13	17.44	17.75	18.06	18.37	18.68	18.99	19.30	19.62	19.93	20.24	20.56	20.87	21.19
27	17.15	17.47	17.80	18.12	18.44	18.76	19.09	19.41	19.74	20.06	20.39	20.71	21.04	21.37	21.69	22.02
28	17.80	18.13	18.47	18.80	19.14	19.47	19.81	20.15	20.48	20.82	21.16	21.50	21.84	22.18	22.52	22.86
29	18.45	18.79	19.14	19.49	19.83	20.18	20.53	20.88	21.23	21.58	21.94	22.29	22.64	22.99	23.35	23.70
30	19.10	19.45	19.81	20.17	20.54	20.90	21.26	21.62	21.99	22.35	22.72	23.08	23.45	23.81	24.18	24.55
31	19.75	20.12	20.49	20.87	21.24	21.61	21.99	22.37	22.74	23.12	23.50	23.88	24.26	24.64	25.02	25.40
32	20.40	20.79	21.17	21.56	21.95	22.33	22.72	23.11	23.50	23.89	24.28	24.68	25.07	25.46	25.86	26.25
33	21.06	21.46	21.85	22.25	22.65	23.06	23.46	23.86	24.26	24.67	25.07	25.48	25.88	26.29	26.70	27.11
34	21.72	22.13	22.54	22.95	23.37	23.78	24.19	24.61	25.03	25.44	25.86	26.28	26.70	27.12	27.54	27.97
35	22.38	22.80	23.23	23.65	24.08	24.51	24.94	25.36	25.79	26.23	26.66	27.09	27.52	27.96	28.39	28.83
36	23.04	23.48	23.92	24.35	24.80	25.24	25.68	26.12	26.57	27.01	27.46	27.90	28.35	28.80	29.25	29.70
37	23.70	24.16	24.61	25.06	25.51	25.97	26.42	26.88	27.34	27.80	28.26	28.72	29.18	29.64	30.10	30.57
38	24.37	24.84	25.30	25.77	26.24	26.70	27.17	27.64	28.11	28.59	29.06	29.53	30.01	30.49	30.96	31.44
39	25.04	25.52	26.00	26.48	26.96	27.44	27.92	28.41	28.89	29.38	29.87	30.36	30.85	31.34	31.83	32.32
40	25.71	26.20	26.70	27.19	27.69	28.18	28.68	29.18	29.68	30.18	30.68	31.18	31.68	32.19	32.69	33.20
41	26.39	26.89	27.40	27.91	28.41	28.92	29.44	29.95	30.46	30.97	31.49	32.01	32.52	33.04	33.56	34.08
42	27.06	27.58	28.10	28.62	29.15	29.67	30.19	30.72	31.25	31.78	32.31	32.84	33.37	33.90	34.44	34.97
43	27.74	28.27	28.81	29.34	29.88	30.42	30.96	31.50	32.04	32.58	33.13	33.67	34.22	34.76	35.31	35.86
44	28.42	28.97	29.52	30.07	30.62	31.17	31.72	32.28	32.83	33.39	33.95	34.51	35.07	35.63	36.19	36.76
45	29.11	29.67	30.23	30.79	31.36	31.92	32.49	33.06	33.63	34.20	34.77	35.35	35.92	36.50	37.08	37.66
46	29.79	30.36	30.94	31.52	32.10	32.68	33.26	33.84	34.43	35.01	35.60	36.19	36.78	37.37	37.96	38.56
47	30.48	31.07	31.66	32.25	32.84	33.44	34.03	34.63	35.23	35.83	36.43	37.04	37.64	38.25	38.86	39.46
48	31.17	31.77	32.37	32.98	33.59	34.20	34.81	35.42	36.03	36.65	37.27	37.88	38.50	39.13	39.75	40.37
49	31.86	32.48	33.09	33.71	34.34	34.96	35.59	36.21	36.84	37.47	38.10	38.74	39.37	40.01	40.65	41.29
50	32.55	33.18	33.82	34.45	35.09	35.73	36.37	37.01	37.65	38.30	38.94	39.59	40.24	40.89	41.55	42.20
51	33.25	33.89	34.54	35.19	35.84	36.49	37.15	37.81	38.46	39.12	39.79	40.45	41.11	41.78	42.45	43.12
52	33.95	34.61	35.27	35.93	36.60	37.27	37.94	38.61	39.28	39.96	40.63	41.31	41.99	42.67	43.36	44.04
53	34.65	35.32	36.00	36.68	37.36	38.04	38.72	39.41	40.10	40.79	41.48	42.17	42.87	43.57	44.27	44.97
54	35.35	36.04	36.73	37.42	38.12	38.82	39.52	40.22	40.92	41.63	42.33	43.04	43.75	44.47	45.18	45.90
55	36.05	36.76	37.46	38.17	38.88	39.60	40.31	41.03	41.74	42.47	43.19	43.91	44.64	45.37	46.10	46.83
56	36.76	37.48	38.20	38.92	39.65	40.38	41.11	41.84	42.57	43.31	44.05	44.79	45.53	46.27	47.02	47.77
57	37.47	38.20	38.94	39.68	40.42	41.16	41.91	42.65	43.40	44.15	44.91	45.66	46.42	47.18	47.94	48.71
58	38.18	38.93	39.68	40.43	41.19	41.95	42.71	43.47	44.23	45.00	45.77	46.54	47.32	48.09	48.87	49.65
59	38.89	39.66	40.42	41.19	41.96	42.74	43.51	44.29	45.07	45.85	46.64	47.42	48.21	49.01	49.80	50.60
60	39.61	40.39	41.17	41.95	42.74	43.53	44.32	45.11	45.91	46.71	47.51	48.31	49.12	49.92	50.73	51.55

TABLE C-4 ANNUAL PERCENTAGE RATE TABLE FOR MONTHLY PAYMENT PLANS (cont.)

ANNUAL PERCENTAGE RATE TABLE FOR MONTHLY PAYMENT PLANS
SEE INSTRUCTIONS FOR USE OF TABLES

FRB-105-M

ANNUAL PERCENTAGE RATE

(FINANCE CHARGE PER $100 OF AMOUNT FINANCED)

NUMBER OF PAYMENTS	18.00%	18.25%	18.50%	18.75%	19.00%	19.25%	19.50%	19.75%	20.00%	20.25%	20.50%	20.75%	21.00%	21.25%	21.50%	21.75%
1	1.50	1.52	1.54	1.56	1.58	1.60	1.62	1.65	1.67	1.69	1.71	1.73	1.75	1.77	1.79	1.81
2	2.26	2.29	2.32	2.35	2.38	2.41	2.44	2.48	2.51	2.54	2.57	2.60	2.63	2.66	2.70	2.73
3	3.01	3.06	3.10	3.14	3.18	3.23	3.27	3.31	3.35	3.39	3.44	3.48	3.52	3.56	3.60	3.65
4	3.78	3.83	3.88	3.94	3.99	4.04	4.10	4.15	4.20	4.25	4.31	4.36	4.41	4.47	4.52	4.57
5	4.54	4.61	4.67	4.74	4.80	4.86	4.93	4.99	5.06	5.12	5.18	5.25	5.31	5.37	5.44	5.50
6	5.32	5.39	5.46	5.54	5.61	5.69	5.76	5.84	5.91	5.99	6.06	6.14	6.21	6.29	6.36	6.44
7	6.09	6.18	6.26	6.35	6.43	6.52	6.60	6.69	6.78	6.86	6.95	7.04	7.12	7.21	7.29	7.38
8	6.87	6.96	7.06	7.16	7.26	7.35	7.45	7.55	7.64	7.74	7.84	7.94	8.03	8.13	8.23	8.33
9	7.65	7.76	7.87	7.97	8.08	8.19	8.30	8.41	8.52	8.63	8.73	8.84	8.95	9.06	9.17	9.28
10	8.43	8.55	8.67	8.79	8.91	9.03	9.15	9.27	9.39	9.51	9.63	9.75	9.88	10.00	10.12	10.24
11	9.22	9.35	9.49	9.62	9.75	9.88	10.01	10.14	10.28	10.41	10.54	10.67	10.80	10.94	11.07	11.20
12	10.02	10.16	10.30	10.44	10.59	10.73	10.87	11.02	11.16	11.31	11.45	11.59	11.74	11.88	12.02	12.17
13	10.81	10.97	11.12	11.28	11.43	11.59	11.74	11.90	12.05	12.21	12.36	12.52	12.67	12.83	12.99	13.14
14	11.61	11.78	11.95	12.11	12.28	12.45	12.61	12.78	12.95	13.11	13.28	13.45	13.62	13.79	13.95	14.12
15	12.42	12.59	12.77	12.95	13.13	13.31	13.49	13.67	13.85	14.03	14.21	14.39	14.57	14.75	14.93	15.11
16	13.22	13.41	13.60	13.80	13.99	14.18	14.37	14.56	14.75	14.94	15.13	15.33	15.52	15.71	15.90	16.10
17	14.04	14.24	14.44	14.64	14.85	15.05	15.25	15.46	15.66	15.86	16.07	16.27	16.48	16.68	16.89	17.09
18	14.85	15.07	15.28	15.49	15.71	15.93	16.14	16.36	16.57	16.79	17.01	17.22	17.44	17.66	17.88	18.09
19	15.67	15.90	16.12	16.35	16.58	16.81	17.03	17.26	17.49	17.72	17.95	18.18	18.41	18.64	18.87	19.10
20	16.49	16.73	16.97	17.21	17.45	17.69	17.93	18.17	18.41	18.66	18.90	19.14	19.38	19.63	19.87	20.11
21	17.32	17.57	17.82	18.07	18.33	18.58	18.83	19.09	19.34	19.60	19.85	20.11	20.36	20.62	20.87	21.13
22	18.15	18.41	18.68	18.94	19.21	19.47	19.74	20.01	20.27	20.54	20.81	21.08	21.34	21.61	21.88	22.15
23	18.98	19.26	19.54	19.81	20.09	20.37	20.65	20.93	21.21	21.49	21.77	22.05	22.33	22.61	22.90	23.18
24	19.82	20.11	20.40	20.69	20.98	21.27	21.56	21.86	22.15	22.44	22.74	23.03	23.33	23.62	23.92	24.21
25	20.66	20.96	21.27	21.57	21.87	22.18	22.48	22.79	23.10	23.40	23.71	24.02	24.32	24.63	24.94	25.25
26	21.50	21.82	22.14	22.45	22.77	23.09	23.41	23.73	24.04	24.36	24.68	25.01	25.33	25.65	25.97	26.29
27	22.35	22.68	23.01	23.34	23.67	24.00	24.33	24.67	25.00	25.33	25.67	26.00	26.34	26.67	27.01	27.34
28	23.20	23.55	23.89	24.23	24.58	24.92	25.27	25.61	25.96	26.30	26.65	27.00	27.35	27.70	28.05	28.40
29	24.06	24.41	24.77	25.13	25.49	25.84	26.20	26.56	26.92	27.28	27.64	28.00	28.37	28.73	29.09	29.46
30	24.92	25.29	25.66	26.03	26.40	26.77	27.14	27.52	27.89	28.26	28.64	29.01	29.39	29.77	30.14	30.52
31	25.78	26.16	26.55	26.93	27.32	27.70	28.09	28.47	28.86	29.25	29.64	30.03	30.42	30.81	31.20	31.59
32	26.65	27.04	27.44	27.84	28.24	28.64	29.04	29.44	29.84	30.24	30.64	31.05	31.45	31.85	32.26	32.67
33	27.52	27.93	28.34	28.75	29.16	29.57	29.99	30.40	30.82	31.23	31.65	32.07	32.49	32.91	33.33	33.75
34	28.39	28.81	29.24	29.66	30.09	30.52	30.95	31.37	31.80	32.23	32.67	33.10	33.53	33.96	34.40	34.83
35	29.27	29.71	30.14	30.58	31.02	31.47	31.91	32.35	32.79	33.24	33.68	34.13	34.58	35.03	35.47	35.92
36	30.15	30.60	31.05	31.51	31.96	32.42	32.87	33.33	33.79	34.25	34.71	35.17	35.63	36.09	36.56	37.02
37	31.03	31.50	31.97	32.43	32.90	33.37	33.84	34.32	34.79	35.26	35.74	36.21	36.69	37.16	37.64	38.12
38	31.92	32.40	32.88	33.37	33.85	34.33	34.82	35.30	35.79	36.28	36.77	37.26	37.75	38.24	38.73	39.23
39	32.81	33.31	33.80	34.30	34.80	35.30	35.80	36.30	36.80	37.30	37.81	38.31	38.82	39.32	39.83	40.34
40	33.71	34.22	34.73	35.24	35.75	36.26	36.78	37.29	37.81	38.33	38.85	39.37	39.89	40.41	40.93	41.46
41	34.61	35.13	35.66	36.18	36.71	37.24	37.77	38.30	38.83	39.36	39.89	40.43	40.96	41.50	42.04	42.58
42	35.51	36.05	36.59	37.13	37.67	38.21	38.76	39.30	39.85	40.40	40.95	41.50	42.05	42.60	43.15	43.71
43	36.42	36.97	37.52	38.08	38.63	39.19	39.75	40.31	40.87	41.44	42.00	42.57	43.13	43.70	44.27	44.84
44	37.33	37.89	38.46	39.03	39.60	40.18	40.75	41.33	41.90	42.48	43.06	43.64	44.22	44.81	45.39	45.98
45	38.24	38.82	39.41	39.99	40.58	41.17	41.75	42.35	42.94	43.53	44.13	44.72	45.32	45.92	46.52	47.12
46	39.16	39.75	40.35	40.95	41.55	42.16	42.76	43.37	43.98	44.58	45.20	45.81	46.42	47.03	47.65	48.27
47	40.08	40.69	41.30	41.92	42.54	43.15	43.77	44.40	45.02	45.64	46.27	46.90	47.53	48.16	48.79	49.42
48	41.00	41.63	42.26	42.89	43.52	44.15	44.79	45.43	46.07	46.71	47.35	47.99	48.64	49.28	49.93	50.58
49	41.93	42.57	43.22	43.86	44.51	45.16	45.81	46.46	47.12	47.77	48.43	49.09	49.75	50.41	51.08	51.74
50	42.86	43.52	44.18	44.84	45.50	46.17	46.83	47.50	48.17	48.84	49.52	50.19	50.87	51.55	52.23	52.91
51	43.79	44.47	45.14	45.82	46.50	47.18	47.86	48.55	49.23	49.92	50.61	51.30	51.99	52.69	53.38	54.08
52	44.73	45.42	46.11	46.80	47.50	48.20	48.89	49.59	50.30	51.00	51.71	52.41	53.12	53.83	54.55	55.26
53	45.67	46.38	47.08	47.79	48.50	49.22	49.93	50.65	51.37	52.09	52.81	53.53	54.26	54.98	55.71	56.44
54	46.62	47.34	48.06	48.79	49.51	50.24	50.97	51.70	52.44	53.17	53.91	54.65	55.39	56.14	56.88	57.63
55	47.57	48.30	49.04	49.78	50.52	51.27	52.02	52.76	53.52	54.27	55.02	55.78	56.54	57.30	58.06	58.82
56	48.52	49.27	50.03	50.78	51.54	52.30	53.06	53.83	54.60	55.37	56.14	56.91	57.68	58.46	59.24	60.02
57	49.47	50.24	51.01	51.79	52.56	53.34	54.12	54.90	55.68	56.47	57.25	58.04	58.84	59.63	60.43	61.22
58	50.43	51.22	52.00	52.79	53.58	54.38	55.17	55.97	56.77	57.57	58.38	59.18	59.99	60.80	61.62	62.43
59	51.39	52.20	53.00	53.80	54.61	55.42	56.23	57.05	57.87	58.68	59.51	60.33	61.15	61.98	62.81	63.64
60	52.36	53.18	54.00	54.82	55.64	56.47	57.30	58.13	58.96	59.80	60.64	61.48	62.32	63.17	64.01	64.86

600 APPENDIX C TABLES

TABLE C-4 ANNUAL PERCENTAGE RATE TABLE FOR MONTHLY PAYMENT PLANS (*cont.*)

ANNUAL PERCENTAGE RATE TABLE FOR MONTHLY PAYMENT PLANS
SEE INSTRUCTIONS FOR USE OF TABLES

FRB-106-M

(FINANCE CHARGE PER $100 OF AMOUNT FINANCED)

NUMBER OF PAYMENTS	22.00%	22.25%	22.50%	22.75%	23.00%	23.25%	23.50%	23.75%	24.00%	24.25%	24.50%	24.75%	25.00%	25.25%	25.50%	25.75%
1	1.83	1.85	1.87	1.90	1.92	1.94	1.96	1.98	2.00	2.02	2.04	2.06	2.08	2.10	2.12	2.15
2	2.76	2.79	2.82	2.85	2.88	2.92	2.95	2.98	3.01	3.04	3.07	3.10	3.14	3.17	3.20	3.23
3	3.69	3.73	3.77	3.82	3.86	3.90	3.94	3.98	4.03	4.07	4.11	4.15	4.20	4.24	4.28	4.32
4	4.62	4.68	4.73	4.78	4.84	4.89	4.94	5.00	5.05	5.10	5.16	5.21	5.26	5.32	5.37	5.42
5	5.57	5.63	5.69	5.76	5.82	5.89	5.95	6.02	6.08	6.14	6.21	6.27	6.34	6.40	6.46	6.53
6	6.51	6.59	6.66	6.74	6.81	6.89	6.96	7.04	7.12	7.19	7.27	7.34	7.42	7.49	7.57	7.64
7	7.47	7.55	7.64	7.73	7.81	7.90	7.99	8.07	8.16	8.24	8.33	8.42	8.51	8.59	8.68	8.77
8	8.42	8.52	8.62	8.72	8.82	8.91	9.01	9.11	9.21	9.31	9.40	9.50	9.60	9.70	9.80	9.90
9	9.39	9.50	9.61	9.72	9.83	9.94	10.04	10.15	10.26	10.37	10.48	10.59	10.70	10.81	10.92	11.03
10	10.36	10.48	10.60	10.72	10.84	10.96	11.08	11.21	11.33	11.45	11.57	11.69	11.81	11.93	12.06	12.18
11	11.33	11.47	11.60	11.73	11.86	12.00	12.13	12.26	12.40	12.53	12.66	12.80	12.93	13.06	13.20	13.33
12	12.31	12.46	12.60	12.75	12.89	13.04	13.18	13.33	13.47	13.62	13.76	13.91	14.05	14.20	14.34	14.49
13	13.30	13.46	13.61	13.77	13.93	14.08	14.24	14.40	14.55	14.71	14.87	15.03	15.18	15.34	15.50	15.66
14	14.29	14.46	14.63	14.80	14.97	15.13	15.30	15.47	15.64	15.81	15.98	16.15	16.32	16.49	16.66	16.83
15	15.29	15.47	15.65	15.83	16.01	16.19	16.37	16.56	16.74	16.92	17.10	17.28	17.47	17.65	17.83	18.02
16	16.29	16.48	16.68	16.87	17.06	17.26	17.45	17.65	17.84	18.03	18.23	18.42	18.62	18.81	19.01	19.21
17	17.30	17.50	17.71	17.92	18.12	18.33	18.53	18.74	18.95	19.16	19.36	19.57	19.78	19.99	20.20	20.40
18	18.31	18.53	18.75	18.97	19.19	19.41	19.62	19.84	20.06	20.28	20.50	20.72	20.95	21.17	21.39	21.61
19	19.33	19.56	19.79	20.02	20.26	20.49	20.72	20.95	21.19	21.42	21.65	21.89	22.12	22.35	22.59	22.82
20	20.35	20.60	20.84	21.09	21.33	21.58	21.82	22.07	22.31	22.56	22.81	23.05	23.30	23.55	23.79	24.04
21	21.38	21.64	21.90	22.16	22.41	22.67	22.93	23.19	23.45	23.71	23.97	24.23	24.49	24.75	25.01	25.27
22	22.42	22.69	22.96	23.23	23.50	23.77	24.04	24.32	24.59	24.86	25.13	25.41	25.68	25.96	26.23	26.50
23	23.46	23.74	24.03	24.31	24.60	24.88	25.17	25.45	25.74	26.02	26.31	26.60	26.88	27.17	27.46	27.75
24	24.51	24.80	25.10	25.40	25.70	25.99	26.29	26.59	26.89	27.19	27.49	27.79	28.09	28.39	28.69	29.00
25	25.56	25.87	26.18	26.49	26.80	27.11	27.43	27.74	28.05	28.36	28.68	28.99	29.31	29.62	29.94	30.25
26	26.62	26.94	27.26	27.59	27.91	28.24	28.56	28.89	29.22	29.55	29.87	30.20	30.53	30.86	31.19	31.52
27	27.68	28.02	28.35	28.69	29.03	29.37	29.71	30.05	30.39	30.73	31.07	31.42	31.76	32.10	32.45	32.79
28	28.75	29.10	29.45	29.80	30.15	30.51	30.86	31.22	31.57	31.93	32.28	32.64	33.00	33.35	33.71	34.07
29	29.82	30.19	30.55	30.92	31.28	31.65	32.02	32.39	32.76	33.13	33.50	33.87	34.24	34.61	34.98	35.36
30	30.90	31.28	31.66	32.04	32.42	32.80	33.18	33.57	33.95	34.33	34.72	35.10	35.49	35.88	36.26	36.65
31	31.98	32.38	32.77	33.17	33.56	33.96	34.35	34.75	35.15	35.55	35.95	36.35	36.75	37.15	37.55	37.95
32	33.07	33.48	33.89	34.30	34.71	35.12	35.53	35.94	36.36	36.77	37.18	37.60	38.01	38.43	38.84	39.26
33	34.17	34.59	35.01	35.44	35.86	36.29	36.71	37.14	37.57	37.99	38.42	38.85	39.28	39.71	40.14	40.58
34	35.27	35.71	36.14	36.58	37.02	37.46	37.90	38.34	38.78	39.23	39.67	40.11	40.56	41.01	41.45	41.90
35	36.37	36.83	37.28	37.73	38.18	38.64	39.09	39.55	40.01	40.47	40.92	41.38	41.84	42.31	42.77	43.23
36	37.49	37.95	38.42	38.89	39.35	39.82	40.29	40.77	41.24	41.71	42.19	42.66	43.14	43.61	44.09	44.57
37	38.60	39.08	39.56	40.05	40.53	41.02	41.50	41.99	42.48	42.96	43.45	43.94	44.43	44.93	45.42	45.91
38	39.72	40.22	40.72	41.21	41.71	42.21	42.71	43.22	43.72	44.22	44.73	45.23	45.74	46.25	46.75	47.26
39	40.85	41.36	41.87	42.39	42.90	43.42	43.93	44.45	44.97	45.49	46.01	46.53	47.05	47.57	48.10	48.62
40	41.98	42.51	43.04	43.56	44.09	44.62	45.16	45.69	46.22	46.76	47.29	47.83	48.37	48.91	49.45	49.99
41	43.12	43.66	44.20	44.75	45.29	45.84	46.39	46.94	47.48	48.04	48.59	49.14	49.69	50.25	50.80	51.36
42	44.26	44.82	45.38	45.94	46.50	47.06	47.62	48.19	48.75	49.32	49.89	50.46	51.03	51.60	52.17	52.74
43	45.41	45.98	46.56	47.13	47.71	48.29	48.87	49.45	50.03	50.61	51.19	51.78	52.36	52.95	53.54	54.13
44	46.56	47.15	47.74	48.33	48.93	49.52	50.11	50.71	51.31	51.91	52.51	53.11	53.71	54.31	54.92	55.52
45	47.72	48.33	48.93	49.54	50.15	50.76	51.37	51.98	52.59	53.21	53.82	54.44	55.06	55.68	56.30	56.92
46	48.89	49.51	50.13	50.75	51.37	52.00	52.63	53.26	53.89	54.52	55.15	55.78	56.42	57.05	57.69	58.33
47	50.06	50.69	51.33	51.97	52.61	53.25	53.89	54.54	55.18	55.83	56.48	57.13	57.78	58.44	59.09	59.75
48	51.23	51.88	52.54	53.19	53.85	54.51	55.16	55.83	56.49	57.15	57.82	58.49	59.15	59.82	60.50	61.17
49	52.41	53.08	53.75	54.42	55.09	55.77	56.44	57.12	57.80	58.48	59.16	59.85	60.53	61.22	61.91	62.60
50	53.59	54.28	54.96	55.65	56.34	57.03	57.73	58.42	59.12	59.81	60.51	61.21	61.92	62.62	63.33	64.03
51	54.78	55.48	56.19	56.89	57.60	58.30	59.01	59.73	60.44	61.15	61.87	62.59	63.31	64.03	64.75	65.47
52	55.98	56.69	57.41	58.13	58.86	59.58	60.31	61.04	61.77	62.50	63.23	63.97	64.70	65.44	66.18	66.92
53	57.18	57.91	58.65	59.38	60.12	60.87	61.61	62.35	63.10	63.85	64.60	65.35	66.11	66.86	67.62	68.38
54	58.38	59.13	59.88	60.64	61.40	62.16	62.92	63.68	64.44	65.21	65.98	66.75	67.52	68.29	69.07	69.84
55	59.59	60.36	61.13	61.90	62.67	63.45	64.23	65.01	65.79	66.57	67.36	68.14	68.93	69.72	70.52	71.31
56	60.80	61.59	62.38	63.17	63.96	64.75	65.54	66.34	67.14	67.94	68.74	69.55	70.36	71.16	71.97	72.79
57	62.02	62.83	63.63	64.44	65.25	66.06	66.87	67.68	68.50	69.32	70.14	70.96	71.78	72.61	73.44	74.27
58	63.25	64.07	64.89	65.71	66.54	67.37	68.20	69.03	69.86	70.70	71.54	72.38	73.22	74.06	74.91	75.76
59	64.48	65.32	66.15	67.00	67.84	68.68	69.53	70.38	71.23	72.09	72.94	73.80	74.66	75.52	76.39	77.25
60	65.71	66.57	67.42	68.28	69.14	70.01	70.87	71.74	72.61	73.48	74.35	75.23	76.11	76.99	77.87	78.76

TABLE C-4 ANNUAL PERCENTAGE RATE TABLE FOR MONTHLY PAYMENT PLANS (cont.)

ANNUAL PERCENTAGE RATE TABLE FOR MONTHLY PAYMENT PLANS
SEE INSTRUCTIONS FOR USE OF TABLES

FRB-107-M

(FINANCE CHARGE PER $100 OF AMOUNT FINANCED)

NUMBER OF PAYMENTS	26.00%	26.25%	26.50%	26.75%	27.00%	27.25%	27.50%	27.75%	28.00%	28.25%	28.50%	28.75%	29.00%	29.25%	29.50%	29.75%
1	2.17	2.19	2.21	2.23	2.25	2.27	2.29	2.31	2.33	2.35	2.37	2.40	2.42	2.44	2.46	2.48
2	3.26	3.29	3.32	3.36	3.39	3.42	3.45	3.48	3.51	3.54	3.58	3.61	3.64	3.67	3.70	3.73
3	4.36	4.41	4.45	4.49	4.53	4.58	4.62	4.66	4.70	4.74	4.79	4.83	4.87	4.91	4.96	5.00
4	5.47	5.53	5.58	5.63	5.69	5.74	5.79	5.85	5.90	5.95	6.01	6.06	6.11	6.17	6.22	6.27
5	6.59	6.66	6.72	6.79	6.85	6.91	6.98	7.04	7.11	7.17	7.24	7.30	7.37	7.43	7.49	7.56
6	7.72	7.79	7.87	7.95	8.02	8.10	8.17	8.25	8.32	8.40	8.48	8.55	8.63	8.70	8.78	8.85
7	8.85	8.94	9.03	9.11	9.20	9.29	9.37	9.46	9.55	9.64	9.72	9.81	9.90	9.98	10.07	10.16
8	9.99	10.09	10.19	10.29	10.39	10.49	10.58	10.68	10.78	10.88	10.98	11.08	11.18	11.28	11.38	11.47
9	11.14	11.25	11.36	11.47	11.58	11.69	11.80	11.91	12.03	12.14	12.25	12.36	12.47	12.58	12.69	12.80
10	12.30	12.42	12.54	12.67	12.79	12.91	13.03	13.15	13.28	13.40	13.52	13.64	13.77	13.89	14.01	14.14
11	13.46	13.60	13.73	13.87	14.00	14.13	14.27	14.40	14.54	14.67	14.81	14.94	15.08	15.21	15.35	15.48
12	14.64	14.78	14.93	15.07	15.22	15.37	15.51	15.66	15.81	15.95	16.10	16.25	16.40	16.54	16.69	16.84
13	15.82	15.97	16.13	16.29	16.45	16.61	16.77	16.93	17.09	17.24	17.40	17.56	17.72	17.88	18.04	18.20
14	17.00	17.17	17.35	17.52	17.69	17.86	18.03	18.20	18.37	18.54	18.72	18.89	19.06	19.23	19.41	19.58
15	18.20	18.38	18.57	18.75	18.93	19.12	19.30	19.48	19.67	19.85	20.04	20.22	20.41	20.59	20.78	20.96
16	19.40	19.60	19.79	19.99	20.19	20.38	20.58	20.78	20.97	21.17	21.37	21.57	21.76	21.96	22.16	22.36
17	20.61	20.82	21.03	21.24	21.45	21.66	21.87	22.08	22.29	22.50	22.71	22.92	23.13	23.34	23.55	23.77
18	21.83	22.05	22.27	22.50	22.72	22.94	23.16	23.39	23.61	23.83	24.06	24.28	24.51	24.73	24.96	25.18
19	23.06	23.29	23.53	23.76	24.00	24.23	24.47	24.71	24.94	25.18	25.42	25.65	25.89	26.13	26.37	26.61
20	24.29	24.54	24.79	25.04	25.28	25.53	25.78	26.03	26.28	26.53	26.78	27.04	27.29	27.54	27.79	28.04
21	25.53	25.79	26.05	26.32	26.58	26.84	27.11	27.37	27.63	27.90	28.16	28.43	28.69	28.96	29.22	29.49
22	26.78	27.05	27.33	27.61	27.88	28.16	28.44	28.71	28.99	29.27	29.55	29.82	30.10	30.38	30.66	30.94
23	28.04	28.32	28.61	28.90	29.19	29.48	29.77	30.07	30.36	30.65	30.94	31.23	31.53	31.82	32.11	32.41
24	29.30	29.60	29.90	30.21	30.51	30.82	31.12	31.43	31.73	32.04	32.34	32.65	32.96	33.27	33.57	33.88
25	30.57	30.89	31.20	31.52	31.84	32.16	32.48	32.80	33.12	33.44	33.76	34.08	34.40	34.72	35.04	35.37
26	31.85	32.18	32.51	32.84	33.18	33.51	33.84	34.18	34.51	34.84	35.18	35.51	35.85	36.19	36.52	36.86
27	33.14	33.48	33.83	34.17	34.52	34.87	35.21	35.56	35.91	36.26	36.61	36.96	37.31	37.66	38.01	38.36
28	34.43	34.79	35.15	35.51	35.87	36.23	36.59	36.96	37.32	37.68	38.05	38.41	38.78	39.15	39.51	39.88
29	35.73	36.10	36.48	36.85	37.23	37.61	37.98	38.36	38.74	39.12	39.50	39.88	40.26	40.64	41.02	41.40
30	37.04	37.43	37.82	38.21	38.60	38.99	39.38	39.77	40.17	40.56	40.95	41.35	41.75	42.14	42.54	42.94
31	38.35	38.76	39.16	39.57	39.97	40.38	40.79	41.19	41.60	42.01	42.42	42.83	43.24	43.65	44.06	44.48
32	39.68	40.10	40.52	40.94	41.36	41.78	42.20	42.62	43.05	43.47	43.90	44.32	44.75	45.17	45.60	46.03
33	41.01	41.44	41.88	42.31	42.75	43.19	43.62	44.06	44.50	44.94	45.38	45.82	46.26	46.70	47.15	47.59
34	42.35	42.80	43.25	43.70	44.15	44.60	45.05	45.51	45.96	46.42	46.87	47.33	47.79	48.24	48.70	49.16
35	43.69	44.16	44.62	45.09	45.56	46.02	46.49	46.96	47.43	47.90	48.37	48.85	49.32	49.79	50.27	50.74
36	45.05	45.53	46.01	46.49	46.97	47.45	47.94	48.42	48.91	49.40	49.88	50.37	50.86	51.35	51.84	52.33
37	46.41	46.91	47.40	47.90	48.39	48.89	49.39	49.89	50.40	50.90	51.40	51.91	52.41	52.92	53.42	53.93
38	47.77	48.29	48.80	49.31	49.82	50.34	50.86	51.37	51.89	52.41	52.93	53.45	53.97	54.49	55.02	55.54
39	49.15	49.68	50.20	50.73	51.26	51.79	52.33	52.86	53.39	53.93	54.46	55.00	55.54	56.08	56.62	57.16
40	50.53	51.07	51.62	52.16	52.71	53.26	53.81	54.35	54.90	55.46	56.01	56.56	57.12	57.67	58.23	58.79
41	51.92	52.48	53.04	53.60	54.16	54.73	55.29	55.86	56.42	56.99	57.56	58.13	58.70	59.28	59.85	60.42
42	53.32	53.89	54.47	55.05	55.63	56.21	56.79	57.37	57.95	58.54	59.12	59.71	60.30	60.89	61.48	62.07
43	54.72	55.31	55.90	56.50	57.09	57.69	58.29	58.89	59.49	60.09	60.69	61.30	61.90	62.51	63.11	63.72
44	56.13	56.74	57.35	57.96	58.57	59.19	59.80	60.42	61.03	61.65	62.27	62.89	63.51	64.14	64.76	65.39
45	57.55	58.17	58.80	59.43	60.06	60.69	61.32	61.95	62.59	63.22	63.86	64.50	65.13	65.77	66.42	67.06
46	58.97	59.61	60.26	60.90	61.55	62.20	62.84	63.49	64.15	64.80	65.45	66.11	66.76	67.42	68.08	68.74
47	60.40	61.06	61.72	62.38	63.05	63.71	64.38	65.05	65.71	66.38	67.06	67.73	68.40	69.08	69.75	70.43
48	61.84	62.52	63.20	63.87	64.56	65.24	65.92	66.60	67.29	67.98	68.67	69.36	70.05	70.74	71.44	72.13
49	63.29	63.98	64.68	65.37	66.07	66.77	67.47	68.17	68.87	69.58	70.29	70.99	71.70	72.41	73.13	73.84
50	64.74	65.45	66.16	66.88	67.59	68.31	69.03	69.75	70.47	71.19	71.91	72.64	73.37	74.10	74.83	75.56
51	66.20	66.93	67.66	68.39	69.12	69.86	70.59	71.33	72.07	72.81	73.55	74.29	75.04	75.78	76.53	77.28
52	67.67	68.41	69.16	69.91	70.66	71.41	72.16	72.92	73.67	74.43	75.19	75.95	76.72	77.48	78.25	79.02
53	69.14	69.90	70.67	71.43	72.20	72.97	73.74	74.52	75.29	76.07	76.85	77.62	78.41	79.19	79.97	80.76
54	70.62	71.40	72.18	72.97	73.75	74.54	75.33	76.12	76.91	77.71	78.50	79.30	80.10	80.90	81.71	82.51
55	72.11	72.91	73.71	74.51	75.31	76.12	76.92	77.73	78.55	79.36	80.17	80.99	81.81	82.63	83.45	84.27
56	73.60	74.42	75.24	76.06	76.88	77.70	78.53	79.35	80.18	81.02	81.85	82.68	83.52	84.36	85.20	86.04
57	75.10	75.94	76.77	77.61	78.45	79.29	80.14	80.98	81.83	82.68	83.53	84.39	85.24	86.10	86.96	87.82
58	76.61	77.46	78.32	79.17	80.03	80.89	81.75	82.62	83.48	84.35	85.22	86.10	86.97	87.85	88.72	89.60
59	78.12	78.99	79.87	80.74	81.62	82.50	83.38	84.26	85.15	86.03	86.92	87.81	88.71	89.60	90.50	91.40
60	79.64	80.53	81.42	82.32	83.21	84.11	85.01	85.91	86.81	87.72	88.63	89.54	90.45	91.37	92.28	93.20

TABLE C-5 AMOUNT OF 1 AT COMPOUND INTEREST

$$s = (1 + i)^n$$

n	1/4%	7/24%	1/3%	5/12%	n
1	1.0025 0000	1.0029 1667	1.0033 3333	1.0041 6667	1
2	1.0050 0625	1.0058 4184	1.0066 7778	1.0083 5069	2
3	1.0075 1877	1.0087 7555	1.0100 3337	1.0125 5216	3
4	1.0100 3756	1.0117 1781	1.0134 0015	1.0167 7112	4
5	1.0125 6266	1.0146 6865	1.0167 7815	1.0210 0767	5
6	1.0150 9406	1.0176 2810	1.0201 6741	1.0252 6187	6
7	1.0176 3180	1.0205 9618	1.0235 6797	1.0295 3379	7
8	1.0201 7588	1.0235 7292	1.0269 7986	1.0338 2352	8
9	1.0227 2632	1.0265 5834	1.0304 0313	1.0381 3111	9
10	1.0252 8313	1.0295 5247	1.0338 3780	1.0424 5666	10
11	1.0278 4634	1.0325 5533	1.0372 8393	1.0468 0023	11
12	1.0304 1596	1.0355 6695	1.0407 4154	1.0511 6190	12
13	1.0329 9200	1.0385 8736	1.0442 1068	1.0555 4174	13
14	1.0355 7448	1.0416 1657	1.0476 9138	1.0599 3983	14
15	1.0381 6341	1.0446 5462	1.0511 8369	1.0643 5625	15
16	1.0407 5882	1.0477 0153	1.0546 8763	1.0687 9106	16
17	1.0433 6072	1.0507 5732	1.0582 0326	1.0732 4436	17
18	1.0459 6912	1.0538 2203	1.0617 3060	1.0777 1621	18
19	1.0485 8404	1.0568 9568	1.0652 6971	1.0822 0670	19
20	1.0512 0550	1.0599 7829	1.0688 2060	1.0867 1589	20
21	1.0538 3352	1.0630 6990	1.0723 8334	1.0912 4387	21
22	1.0564 6810	1.0661 7052	1.0759 5795	1.0957 9072	22
23	1.0591 0927	1.0692 8018	1.0795 4448	1.1003 5652	23
24	1.0617 5704	1.0723 9891	1.0831 4296	1.1049 4134	24
25	1.0644 1144	1.0755 2674	1.0867 5344	1.1095 4526	25
26	1.0670 7247	1.0786 6370	1.0903 7595	1.1141 6836	26
27	1.0697 4015	1.0818 0980	1.0940 1053	1.1188 1073	27
28	1.0724 1450	1.0849 6508	1.0976 5724	1.1234 7244	28
29	1.0750 9553	1.0881 2956	1.1013 1609	1.1281 5358	29
30	1.0777 8327	1.0913 0327	1.1049 8715	1.1328 5422	30
31	1.0804 7773	1.0944 8624	1.1086 7044	1.1375 7444	31
32	1.0831 7892	1.0976 7849	1.1123 6601	1.1423 1434	32
33	1.0858 8687	1.1008 8005	1.1160 7389	1.1470 7398	33
34	1.0886 0159	1.1040 9095	1.1197 9414	1.1518 5346	34
35	1.0913 2309	1.1073 1122	1.1235 2679	1.1566 5284	35
36	1.0940 5140	1.1105 4088	1.1272 7187	1.1614 7223	36
37	1.0967 8653	1.1137 7995	1.1310 2945	1.1663 1170	37
38	1.0995 2850	1.1170 2848	1.1347 9955	1.1711 7133	38
39	1.1022 7732	1.1202 8648	1.1385 8221	1.1760 5121	39
40	1.1050 3301	1.1235 5398	1.1423 7748	1.1809 5142	40
41	1.1077 9559	1.1268 3101	1.1461 8541	1.1858 7206	41
42	1.1105 6508	1.1301 1760	1.1500 0603	1.1908 1319	42
43	1.1133 4149	1.1334 1378	1.1538 3938	1.1957 7491	43
44	1.1161 2485	1.1367 1957	1.1576 8551	1.2007 5731	44
45	1.1189 1516	1.1400 3500	1.1615 4446	1.2057 6046	45
46	1.1217 1245	1.1433 6010	1.1654 1628	1.2107 8446	46
47	1.1245 1673	1.1466 9490	1.1693 0100	1.2158 2940	47
48	1.1273 2802	1.1500 3943	1.1731 9867	1.2208 9536	48
49	1.1301 4634	1.1533 9371	1.1771 0933	1.2259 8242	49
50	1.1329 7171	1.1567 5778	1.1810 3303	1.2310 9068	50

Original source: Simpson, Pirenian, et al. *Mathematics of Finance*, Prentice Hall, 1969.

TABLE C-5 AMOUNT OF 1 AT COMPOUND INTEREST (cont.)

$$s = (1 + i)^n$$

n	1/4%	7/24%	1/3%	5/12%	n
51	1.1358 0414	1.1601 3165	1.1849 6981	1.2362 2002	51
52	1.1386 4365	1.1635 1537	1.1889 1971	1.2413 7114	52
53	1.1414 9026	1.1669 0896	1.1928 8277	1.2465 4352	53
54	1.1443 4398	1.1703 1244	1.1968 5905	1.2517 3745	54
55	1.1472 0484	1.1737 2585	1.2008 4858	1.2569 5302	55
56	1.1500 7285	1.1771 4922	1.2048 5141	1.2621 9033	56
57	1.1529 4804	1.1805 8257	1.2088 6758	1.2674 4946	57
58	1.1558 3041	1.1840 2594	1.2128 9714	1.2727 3050	58
59	1.1587 1998	1.1874 7935	1.2169 4013	1.2780 3354	59
60	1.1616 1678	1.1909 4283	1.2209 9659	1.2833 5868	60
61	1.1645 2082	1.1944 1641	1.2250 6658	1.2887 0601	61
62	1.1674 3213	1.1979 0013	1.2291 5014	1.2940 7561	62
63	1.1703 5071	1.2013 9400	1.2332 4730	1.2994 6760	63
64	1.1732 7658	1.2048 9807	1.2373 5813	1.3048 8204	64
65	1.1762 0977	1.2084 1235	1.2414 8266	1.3103 1905	65
66	1.1791 5030	1.2119 3689	1.2456 2093	1.3157 7872	66
67	1.1820 9817	1.2154 7171	1.2497 7300	1.3212 6113	67
68	1.1850 5342	1.2190 1683	1.2539 3891	1.3267 6638	68
69	1.1880 1605	1.2225 7230	1.2581 1871	1.3322 9458	69
70	1.1909 8609	1.2261 3813	1.2623 1244	1.3378 4580	70
71	1.1939 6356	1.2297 1437	1.2665 2015	1.3434 2016	71
72	1.1969 4847	1.2333 0104	1.2707 4188	1.3490 1774	72
73	1.1999 4084	1.2368 9816	1.2749 7769	1.3546 3865	73
74	1.2029 4069	1.2405 0578	1.2792 2761	1.3602 8298	74
75	1.2059 4804	1.2441 2393	1.2834 9170	1.3659 5082	75
76	1.2089 6291	1.2477 5262	1.2877 7001	1.3716 4229	76
77	1.2119 8532	1.2513 9190	1.2920 6258	1.3773 5746	77
78	1.2150 1528	1.2550 4179	1.2963 6945	1.3830 9645	78
79	1.2180 5282	1.2587 0233	1.3006 9068	1.3888 5935	79
80	1.2210 9795	1.2623 7355	1.3050 2632	1.3946 4627	80
81	1.2241 5070	1.2660 5547	1.3093 7641	1.4004 5729	81
82	1.2272 1108	1.2697 4813	1.3137 4099	1.4062 9253	82
83	1.2302 7910	1.2734 5156	1.3181 2013	1.4121 5209	83
84	1.2333 5480	1.2771 6580	1.3225 1386	1.4180 3605	84
85	1.2364 3819	1.2808 9086	1.3269 2224	1.4239 4454	85
86	1.2395 2928	1.2846 2680	1.3313 4532	1.4298 7764	86
87	1.2426 2811	1.2883 7362	1.3357 8314	1.4358 3546	87
88	1.2457 3468	1.2921 3138	1.3402 3575	1.4418 1811	88
89	1.2488 4901	1.2959 0010	1.3447 0320	1.4478 2568	89
90	1.2519 7114	1.2996 7980	1.3491 8554	1.4538 5829	90
91	1.2551 0106	1.3034 7054	1.3536 8283	1.4599 1603	91
92	1.2582 3882	1.3072 7233	1.3581 9510	1.4659 9902	92
93	1.2613 8441	1.3110 8520	1.3627 2242	1.4721 0735	93
94	1.2645 3787	1.3149 0920	1.3672 6483	1.4782 4113	94
95	1.2676 9922	1.3187 4435	1.3718 2238	1.4844 0047	95
96	1.2708 6847	1.3225 9069	1.3763 9512	1.4905 8547	96
97	1.2740 4564	1.3264 4825	1.3809 8310	1.4967 9624	97
98	1.2772 3075	1.3303 1706	1.3855 8638	1.5030 3289	98
99	1.2804 2383	1.3341 9715	1.3902 0500	1.5092 9553	99
100	1.2836 2489	1.3380 8856	1.3948 3902	1.5155 8426	100

TABLE C-5 AMOUNT OF 1 AT COMPOUND INTEREST (cont.)

$$s = (1 + i)^n$$

n	½%	7/12%	5/8%	2/3%	n
1	1.0050 0000	1.0058 3333	1.0062 5000	1.0066 6667	1
2	1.0100 2500	1.0117 0069	1.0125 3906	1.0133 7778	2
3	1.0150 7513	1.0176 0228	1.0188 6743	1.0201 3363	3
4	1.0201 5050	1.0235 3830	1.0252 3535	1.0269 3452	4
5	1.0252 5125	1.0295 0894	1.0316 4307	1.0337 8075	5
6	1.0303 7751	1.0355 1440	1.0380 9084	1.0406 7262	6
7	1.0355 2940	1.0415 5490	1.0445 7891	1.0476 1044	7
8	1.0407 0704	1.0476 3064	1.0511 0753	1.0545 9451	8
9	1.0459 1058	1.0537 4182	1.0576 7695	1.0616 2514	9
10	1.0511 4013	1.0598 8865	1.0642 8743	1.0687 0264	10
11	1.0563 9583	1.0660 7133	1.0709 3923	1.0758 2732	11
12	1.0616 7781	1.0722 9008	1.0776 3260	1.0829 9951	12
13	1.0669 8620	1.0785 4511	1.0843 6780	1.0902 1950	13
14	1.0723 2113	1.0848 3662	1.0911 4510	1.0974 8763	14
15	1.0776 8274	1.0911 6483	1.0979 6476	1.1048 0422	15
16	1.0830 7115	1.0975 2996	1.1048 2704	1.1121 6958	16
17	1.0884 8651	1.1039 3222	1.1117 3221	1.1195 8404	17
18	1.0939 2894	1.1103 7182	1.1186 8053	1.1270 4794	18
19	1.0993 9858	1.1168 4899	1.1256 7229	1.1345 6159	19
20	1.1048 9558	1.1233 6395	1.1327 0774	1.1421 2533	20
21	1.1104 2006	1.1299 1690	1.1397 8716	1.1497 3950	21
22	1.1159 7216	1.1365 0808	1.1469 1083	1.1574 0443	22
23	1.1215 5202	1.1431 3771	1.1540 7902	1.1651 2046	23
24	1.1271 5978	1.1498 0602	1.1612 9202	1.1728 8793	24
25	1.1327 9558	1.1565 1322	1.1685 5009	1.1807 0718	25
26	1.1384 5955	1.1632 5955	1.1758 5353	1.1885 7857	26
27	1.1441 5185	1.1700 4523	1.1832 0262	1.1965 0242	27
28	1.1498 7261	1.1768 7049	1.1905 9763	1.2044 7911	28
29	1.1556 2197	1.1837 3557	1.1980 3887	1.2125 0897	29
30	1.1614 0008	1.1906 4069	1.2055 2661	1.2205 9236	30
31	1.1672 0708	1.1975 8610	1.2130 6115	1.2287 2964	31
32	1.1730 4312	1.2045 7202	1.2206 4278	1.2369 2117	32
33	1.1789 0833	1.2115 9869	1.2282 7180	1.2451 6731	33
34	1.1848 0288	1.2186 6634	1.2359 4850	1.2534 6843	34
35	1.1907 2689	1.2257 7523	1.2436 7318	1.2618 2489	35
36	1.1966 8052	1.2329 2559	1.2514 4614	1.2702 3705	36
37	1.2026 6393	1.2401 1765	1.2592 6767	1.2787 0530	37
38	1.2086 7725	1.2473 5167	1.2671 3810	1.2872 3000	38
39	1.2147 2063	1.2546 2789	1.2750 5771	1.2958 1153	39
40	1.2207 9424	1.2619 4655	1.2830 2682	1.3044 5028	40
41	1.2268 9821	1.2693 0791	1.2910 4574	1.3131 4661	41
42	1.2330 3270	1.2767 1220	1.2991 1477	1.3219 0092	42
43	1.2391 9786	1.2841 5969	1.3072 3424	1.3307 1360	43
44	1.2453 9385	1.2916 5062	1.3154 0446	1.3395 8502	44
45	1.2516 2082	1.2991 8525	1.3236 2573	1.3485 1559	45
46	1.2578 7892	1.3067 6383	1.3318 9839	1.3575 0569	46
47	1.2641 6832	1.3143 8662	1.3402 2276	1.3665 5573	47
48	1.2704 8916	1.3220 5388	1.3485 9915	1.3756 6610	48
49	1.2768 4161	1.3297 6586	1.3570 2790	1.3848 3721	49
50	1.2832 2581	1.3375 2283	1.3655 0932	1.3940 6946	50

TABLE C-5 AMOUNT OF 1 AT COMPOUND INTEREST (cont.)

$$s = (1 + i)^n$$

n	½%	7/12%	5/8%	2/3%	n
51	1.2896 4194	1.3453 2504	1.3740 4375	1.4033 6325	51
52	1.2960 9015	1.3531 7277	1.3826 3153	1.4127 1901	52
53	1.3025 7060	1.3610 6628	1.3912 7297	1.4221 3713	53
54	1.3090 8346	1.3690 0583	1.3999 6843	1.4316 1805	54
55	1.3156 2887	1.3769 9170	1.4087 1823	1.4411 6217	55
56	1.3222 0702	1.3850 2415	1.4175 2272	1.4507 6992	56
57	1.3288 1805	1.3931 0346	1.4263 8224	1.4604 4172	57
58	1.3354 6214	1.4012 2990	1.4352 9713	1.4701 7799	58
59	1.3421 3946	1.4094 0374	1.4442 6773	1.4799 7918	59
60	1.3488 5015	1.4176 2526	1.4532 9441	1.4898 4571	60
61	1.3555 9440	1.4258 9474	1.4623 7750	1.4997 7801	61
62	1.3623 7238	1.4342 1246	1.4715 1736	1.5097 7653	62
63	1.3691 8424	1.4425 7870	1.4807 1434	1.5198 4171	63
64	1.3760 3016	1.4509 9374	1.4899 6881	1.5299 7399	64
65	1.3829 1031	1.4594 5787	1.4992 8111	1.5401 7381	65
66	1.3898 2486	1.4679 7138	1.5086 5162	1.5504 4164	66
67	1.3967 7399	1.4765 3454	1.5180 8069	1.5607 7792	67
68	1.4037 5785	1.4851 4766	1.5275 6869	1.5711 8310	68
69	1.4107 7664	1.4938 1102	1.5371 1600	1.5816 5766	69
70	1.4178 3053	1.5025 2492	1.5467 2297	1.5922 0204	70
71	1.4249 1968	1.5112 8965	1.5563 8999	1.6028 1672	71
72	1.4320 4428	1.5201 0550	1.5661 1743	1.6135 0217	72
73	1.4392 0450	1.5289 7279	1.5759 0566	1.6242 5885	73
74	1.4464 0052	1.5378 9179	1.5857 5507	1.6350 8724	74
75	1.4536 3252	1.5468 6283	1.5956 6604	1.6459 8782	75
76	1.4609 0069	1.5558 8620	1.6056 3896	1.6569 6107	76
77	1.4682 0519	1.5649 6220	1.6156 7420	1.6680 0748	77
78	1.4755 4622	1.5740 9115	1.6257 7216	1.6791 2753	78
79	1.4829 2395	1.5832 7334	1.6359 3324	1.6903 2172	79
80	1.4903 3857	1.5925 0910	1.6461 5782	1.7015 9053	80
81	1.4977 9026	1.6017 9874	1.6564 4631	1.7129 3446	81
82	1.5052 7921	1.6111 4257	1.6667 9910	1.7243 5403	82
83	1.5128 0561	1.6205 4090	1.6772 1659	1.7358 4972	83
84	1.5203 6964	1.6299 9405	1.6876 9920	1.7474 2205	84
85	1.5279 7148	1.6395 0235	1.6982 4732	1.7590 7153	85
86	1.5356 1134	1.6490 6612	1.7088 6136	1.7707 9868	86
87	1.5432 8940	1.6586 8567	1.7195 4175	1.7826 0400	87
88	1.5510 0585	1.6683 6134	1.7302 8888	1.7944 8803	88
89	1.5587 6087	1.6780 9344	1.7411 0319	1.8064 5128	89
90	1.5665 5468	1.6878 8232	1.7519 8508	1.8184 9429	90
91	1.5743 8745	1.6977 2830	1.7629 3499	1.8306 1758	91
92	1.5822 5939	1.7076 3172	1.7739 5333	1.8428 2170	92
93	1.5901 7069	1.7175 9290	1.7850 4054	1.8551 0718	93
94	1.5981 2154	1.7276 1219	1.7961 9704	1.8674 7456	94
95	1.6061 1215	1.7376 8993	1.8074 2328	1.8799 2439	95
96	1.6141 4271	1.7478 2646	1.8187 1967	1.8924 5722	96
97	1.6222 1342	1.7580 2211	1.8300 8667	1.9050 7360	97
98	1.6303 2449	1.7682 7724	1.8415 2471	1.9177 7409	98
99	1.6384 7611	1.7785 9219	1.8530 3424	1.9305 5925	99
100	1.6466 6849	1.7889 6731	1.8646 1570	1.9434 2965	100

TABLE C-5 AMOUNT OF 1 AT COMPOUND INTEREST (cont.)

$$s = (1 + i)^n$$

n	3/4%	7/8%	1%	1 1/8%	n
1	1.0075 0000	1.0087 5000	1.0100 0000	1.0112 5000	1
2	1.0150 5625	1.0175 7656	1.0201 0000	1.0226 2656	2
3	1.0226 6917	1.0264 8036	1.0303 0100	1.0341 3111	3
4	1.0303 3919	1.0354 6206	1.0406 0401	1.0457 6509	4
5	1.0380 6673	1.0445 2235	1.0510 1005	1.0575 2994	5
6	1.0458 5224	1.0536 6192	1.0615 2015	1.0694 2716	6
7	1.0536 9613	1.0628 8147	1.0721 3535	1.0814 5821	7
8	1.0615 9885	1.0721 8168	1.0828 5671	1.0936 2462	8
9	1.0695 6084	1.0815 6327	1.0936 8527	1.1059 2789	9
10	1.0775 8255	1.0910 2695	1.1046 2213	1.1183 6958	10
11	1.0856 6441	1.1005 7343	1.1156 6835	1.1309 5124	11
12	1.0938 0690	1.1102 0345	1.1268 2503	1.1436 7444	12
13	1.1020 1045	1.1199 1773	1.1380 9328	1.1565 4078	13
14	1.1102 7553	1.1297 1701	1.1494 7421	1.1695 5186	14
15	1.1186 0259	1.1396 0203	1.1609 6896	1.1827 0932	15
16	1.1269 9211	1.1495 7355	1.1725 7864	1.1960 1480	16
17	1.1354 4455	1.1596 3232	1.1843 0443	1.2094 6997	17
18	1.1439 6039	1.1697 7910	1.1961 4748	1.2230 7650	18
19	1.1525 4009	1.1800 1467	1.2081 0895	1.2368 3611	19
20	1.1611 8414	1.1903 3980	1.2201 9004	1.2507 5052	20
21	1.1698 9302	1.2007 5527	1.2323 9194	1.2648 2146	21
22	1.1786 6722	1.2112 6188	1.2447 1586	1.2790 5071	22
23	1.1875 0723	1.2218 6042	1.2571 6302	1.2934 4003	23
24	1.1964 1353	1.2325 5170	1.2697 3465	1.3079 9123	24
25	1.2053 8663	1.2433 3653	1.2824 3200	1.3227 0613	25
26	1.2144 2703	1.2542 1572	1.2952 5631	1.3375 8657	26
27	1.2235 3523	1.2651 9011	1.3082 0888	1.3526 3442	27
28	1.2327 1175	1.2762 6052	1.3212 9097	1.3678 5156	28
29	1.2419 5709	1.2874 2780	1.3345 0388	1.3832 3989	29
30	1.2512 7176	1.2986 9280	1.3478 4892	1.3988 0134	30
31	1.2606 5630	1.3100 5636	1.3613 2740	1.4145 3785	31
32	1.2701 1122	1.3215 1935	1.3749 4068	1.4304 5140	32
33	1.2796 3706	1.3330 8265	1.3886 9009	1.4465 4398	33
34	1.2892 3434	1.3447 4712	1.4025 7699	1.4628 1760	34
35	1.2989 0359	1.3565 1366	1.4166 0276	1.4792 7430	35
36	1.3086 4537	1.3683 8315	1.4307 6878	1.4959 1613	36
37	1.3184 6021	1.3803 5650	1.4450 7647	1.5127 4519	37
38	1.3283 4866	1.3924 3462	1.4595 2724	1.5297 6357	38
39	1.3383 1128	1.4046 1843	1.4741 2251	1.5469 7341	39
40	1.3483 4861	1.4169 0884	1.4888 6373	1.5643 7687	40
41	1.3584 6123	1.4293 0679	1.5037 5237	1.5819 7611	41
42	1.3686 4969	1.4418 1322	1.5187 8989	1.5997 7334	42
43	1.3789 1456	1.4544 2909	1.5339 7779	1.6177 7079	43
44	1.3892 5642	1.4671 5534	1.5493 1757	1.6359 7071	44
45	1.3996 7584	1.4799 9295	1.5648 1075	1.6543 7538	45
46	1.4101 7341	1.4929 4289	1.5804 5885	1.6729 8710	46
47	1.4207 4971	1.5060 0614	1.5962 6344	1.6918 0821	47
48	1.4314 0533	1.5191 8370	1.6122 2608	1.7108 4105	48
49	1.4421 4087	1.5324 7655	1.6283 4834	1.7300 8801	49
50	1.4529 5693	1.5458 8572	1.6446 3182	1.7495 5150	50

TABLE C-5 AMOUNT OF 1 AT COMPOUND INTEREST (cont.)

$$s = (1 + i)^n$$

n	¾%	⅞%	1%	1⅛%	n
51	1.4638 5411	1.5594 1222	1.6610 7814	1.7692 3395	51
52	1.4748 3301	1.5730 5708	1.6776 8892	1.7891 3784	52
53	1.4858 9426	1.5868 2133	1.6944 6581	1.8092 6564	53
54	1.4970 3847	1.6007 0602	1.7114 1047	1.8296 1988	54
55	1.5082 6626	1.6147 1219	1.7285 2457	1.8502 0310	55
56	1.5195 7825	1.6288 4093	1.7458 0982	1.8710 1788	56
57	1.5309 7509	1.6430 9328	1.7632 6792	1.8920 6684	57
58	1.5424 5740	1.6574 7035	1.7809 0060	1.9133 5259	58
59	1.5540 2583	1.6719 7322	1.7987 0960	1.9348 7780	59
60	1.5656 8103	1.6866 0298	1.8166 9670	1.9566 4518	60
61	1.5774 2363	1.7013 6076	1.8348 6367	1.9786 5744	61
62	1.5892 5431	1.7162 4766	1.8532 1230	2.0009 1733	62
63	1.6011 7372	1.7312 6483	1.8717 4443	2.0234 2765	63
64	1.6131 8252	1.7464 1340	1.8904 6187	2.0461 9121	64
65	1.6252 8139	1.7616 9452	1.9093 6649	2.0692 1087	65
66	1.6374 7100	1.7771 0934	1.9284 6015	2.0924 8949	66
67	1.6497 5203	1.7926 5905	1.9477 4475	2.1160 2999	67
68	1.6621 2517	1.8083 4482	1.9672 2220	2.1398 3533	68
69	1.6745 9111	1.8241 6783	1.9868 9442	2.1639 0848	69
70	1.6871 5055	1.8401 2930	2.0067 6337	2.1882 5245	70
71	1.6998 0418	1.8562 3043	2.0268 3100	2.2128 7029	71
72	1.7125 5271	1.8724 7245	2.0470 9931	2.2377 6508	72
73	1.7253 9685	1.8888 5658	2.0675 7031	2.2629 3994	73
74	1.7383 3733	1.9053 8408	2.0882 4601	2.2883 9801	74
75	1.7513 7486	1.9220 5619	2.1091 2847	2.3141 4249	75
76	1.7645 1017	1.9388 7418	2.1302 1975	2.3401 7659	76
77	1.7777 4400	1.9558 3933	2.1515 2195	2.3665 0358	77
78	1.7910 7708	1.9729 5292	2.1730 3717	2.3931 2675	78
79	1.8045 1015	1.9902 1626	2.1947 6754	2.4200 4942	79
80	1.8180 4398	2.0076 3066	2.2167 1522	2.4472 7498	80
81	1.8316 7931	2.0251 9742	2.2388 8237	2.4748 0682	81
82	1.8454 1691	2.0429 1790	2.2612 7119	2.5026 4840	82
83	1.8592 5753	2.0607 9343	2.2838 8390	2.5308 0319	83
84	1.8732 0196	2.0788 2537	2.3067 2274	2.5592 7473	84
85	1.8872 5098	2.0970 1510	2.3297 8997	2.5880 6657	85
86	1.9014 0536	2.1153 6398	2.3530 8787	2.6171 8232	86
87	1.9156 6590	2.1338 7341	2.3766 1875	2.6466 2562	87
88	1.9300 3339	2.1525 4481	2.4003 8494	2.6764 0016	88
89	1.9445 0865	2.1713 7957	2.4243 8879	2.7065 0966	89
90	1.9590 9246	2.1903 7914	2.4486 3267	2.7369 5789	90
91	1.9737 8565	2.2095 4496	2.4731 1900	2.7677 4867	91
92	1.9885 8905	2.2288 7849	2.4978 5019	2.7988 8584	92
93	2.0035 0346	2.2483 8117	2.5228 2869	2.8303 7331	93
94	2.0185 2974	2.2680 5450	2.5480 5698	2.8622 1501	94
95	2.0336 6871	2.2878 9998	2.5735 3755	2.8944 1492	95
96	2.0489 2123	2.3079 1910	2.5992 7293	2.9269 7709	96
97	2.0642 8814	2.3281 1340	2.6252 6565	2.9599 0559	97
98	2.0797 7030	2.3484 8439	2.6515 1831	2.9932 0452	98
99	2.0953 6858	2.3690 3363	2.6780 3349	3.0268 7807	99
100	2.1110 8384	2.3897 6267	2.7048 1383	3.0609 3045	100

TABLE C-5 AMOUNT OF 1 AT COMPOUND INTEREST (cont.)

$$s = (1 + i)^n$$

n	1¼%	1⅜%	1½%	1¾%	n
1	1.0125 0000	1.0137 5000	1.0150 0000	1.0175 0000	1
2	1.0251 5625	1.0276 8906	1.0302 2500	1.0353 0625	2
3	1.0379 7070	1.0418 1979	1.0456 7838	1.0534 2411	3
4	1.0509 4534	1.0561 4481	1.0613 6355	1.0718 5903	4
5	1.0640 8215	1.0706 6680	1.0772 8400	1.0906 1656	5
6	1.0773 8318	1.0853 8847	1.0934 4326	1.1097 0235	6
7	1.0908 5047	1.1003 1256	1.1098 4491	1.1291 2215	7
8	1.1044 8610	1.1154 4196	1.1264 9259	1.1488 8178	8
9	1.1182 9218	1.1307 7918	1.1433 8998	1.1689 8721	9
10	1.1322 7083	1.1463 2740	1.1605 4083	1.1894 4449	10
11	1.1464 2422	1.1620 8940	1.1779 4894	1.2102 5977	11
12	1.1607 5452	1.1780 6813	1.1956 1817	1.2314 3931	12
13	1.1752 6395	1.1942 6656	1.2135 5244	1.2529 8950	13
14	1.1899 5475	1.2106 8773	1.2317 5573	1.2749 1682	14
15	1.2048 2918	1.2273 3469	1.2502 3207	1.2972 2786	15
16	1.2198 8955	1.2442 1054	1.2689 8555	1.3199 2935	16
17	1.2351 3817	1.2613 1843	1.2880 2033	1.3430 2811	17
18	1.2505 7739	1.2786 6156	1.3073 4064	1.3665 3111	18
19	1.2662 0961	1.2962 4316	1.3269 5075	1.3904 4540	19
20	1.2820 3723	1.3140 6650	1.3468 5501	1.4147 7820	20
21	1.2980 6270	1.3321 3492	1.3670 5783	1.4395 3681	21
22	1.3142 8848	1.3504 5177	1.3875 6370	1.4647 2871	22
23	1.3307 1709	1.3690 2048	1.4083 7715	1.4903 6146	23
24	1.3473 5105	1.3878 4451	1.4295 0281	1.5164 4279	24
25	1.3641 9294	1.4069 2738	1.4509 4535	1.5429 8054	25
26	1.3812 4535	1.4262 7263	1.4727 0953	1.5699 8269	26
27	1.3985 1092	1.4458 8388	1.4948 0018	1.5974 5739	27
28	1.4159 9230	1.4657 6478	1.5172 2218	1.6254 1290	28
29	1.4336 9221	1.4859 1905	1.5399 8051	1.6538 5762	29
30	1.4516 1336	1.5063 5043	1.5630 8022	1.6828 0013	30
31	1.4697 5853	1.5270 6275	1.5865 2642	1.7122 4913	31
32	1.4881 3051	1.5480 5986	1.6103 2432	1.7422 1349	32
33	1.5067 3214	1.5693 4569	1.6344 7918	1.7727 0223	33
34	1.5255 6629	1.5909 2415	1.6589 9637	1.8037 2452	34
35	1.5446 3587	1.6127 9940	1.6838 8132	1.8352 8970	35
36	1.5639 4382	1.6349 7539	1.7091 3954	1.8674 0727	36
37	1.5834 9312	1.6574 5630	1.7347 7663	1.9000 8689	37
38	1.6032 8678	1.6802 4633	1.7607 9828	1.9333 3841	38
39	1.6233 2787	1.7033 4971	1.7872 1025	1.9671 7184	39
40	1.6436 1946	1.7267 7077	1.8140 1841	2.0015 9734	40
41	1.6641 6471	1.7505 1387	1.8412 2868	2.0366 2530	41
42	1.6849 6677	1.7745 8343	1.8688 4712	2.0722 6624	42
43	1.7060 2885	1.7989 8396	1.8968 7982	2.1085 3090	43
44	1.7273 5421	1.8237 1999	1.9253 3302	2.1454 3019	44
45	1.7489 4614	1.8487 9614	1.9542 1301	2.1829 7522	45
46	1.7708 0797	1.8742 1708	1.9835 2621	2.2211 7728	46
47	1.7929 4306	1.8999 8757	2.0132 7910	2.2600 4789	47
48	1.8153 5485	1.9261 1240	2.0434 7829	2.2995 9872	48
49	1.8380 4679	1.9525 9644	2.0741 3046	2.3398 4170	49
50	1.8610 2237	1.9794 4464	2.1052 4242	2.3807 8893	50

TABLE C-5 AMOUNT OF 1 AT COMPOUND INTEREST (cont.)

$$s = (1 + i)^n$$

n	1¼%	1⅜%	1½%	1¾%	n
51	1.8842 8515	2.0066 6201	2.1368 2106	2.4224 5274	51
52	1.9078 3872	2.0342 5361	2.1688 7337	2.4648 4566	52
53	1.9316 8670	2.0622 2460	2.2014 0647	2.5079 8046	53
54	1.9558 3279	2.0905 8019	2.2344 2757	2.5518 7012	54
55	1.9802 8070	2.1193 2566	2.2679 4398	2.5965 2785	55
56	2.0050 3420	2.1484 6639	2.3019 6314	2.6419 6708	56
57	2.0300 9713	2.1780 0780	2.3364 9259	2.6882 0151	57
58	2.0554 7335	2.2079 5541	2.3715 3998	2.7352 4503	58
59	2.0811 6676	2.2383 1480	2.4071 1308	2.7831 1182	59
60	2.1071 8135	2.2690 9163	2.4432 1978	2.8318 1628	60
61	2.1335 2111	2.3002 9164	2.4798 6807	2.8813 7306	61
62	2.1601 9013	2.3319 2065	2.5170 6609	2.9317 9709	62
63	2.1871 9250	2.3639 8456	2.5548 2208	2.9831 0354	63
64	2.2145 3241	2.3964 8934	2.5931 4442	3.0343 0785	64
65	2.2422 1407	2.4294 4107	2.6320 4158	3.0884 2574	65
66	2.2702 4174	2.4628 4589	2.6715 2221	3.1424 7319	66
67	2.2986 1976	2.4967 1002	2.7115 9504	3.1974 6647	67
68	2.3273 5251	2.5310 3978	2.7522 6896	3.2534 2213	68
69	2.3564 4442	2.5658 4158	2.7935 5300	3.3103 5702	69
70	2.3858 9997	2.6011 2190	2.8354 5629	3.3682 8827	70
71	2.4157 2372	2.6368 8732	2.8779 8814	3.4272 3331	71
72	2.4459 2027	2.6731 4453	2.9211 5796	3.4872 0990	72
73	2.4764 9427	2.7099 0026	2.9649 7533	3.5482 3607	73
74	2.5074 5045	2.7471 6139	3.0094 4996	3.6103 3020	74
75	2.5387 9358	2.7849 3486	3.0545 9171	3.6735 1098	75
76	2.5705 2850	2.8232 2771	3.1004 1059	3.7377 9742	76
77	2.6026 6011	2.8620 4710	3.1469 1674	3.8032 0888	77
78	2.6351 9336	2.9014 0024	3.1941 2050	3.8697 6503	78
79	2.6681 3327	2.9412 9450	3.2420 3230	3.9374 8592	79
80	2.7014 8494	2.9817 3730	3.2906 6279	4.0063 9192	80
81	2.7352 5350	3.0227 3618	3.3400 2273	4.0765 0378	81
82	2.7694 4417	3.0642 9881	3.3901 2307	4.1478 4260	82
83	2.8040 6222	3.1064 3291	3.4409 7492	4.2204 2984	83
84	2.8391 1300	3.1491 4637	3.4925 8954	4.2942 8737	84
85	2.8746 0191	3.1924 4713	3.5449 7838	4.3694 3740	85
86	2.9105 3444	3.2363 4328	3.5981 5306	4.4459 0255	86
87	2.9469 1612	3.2808 4300	3.6521 2535	4.5237 0584	87
88	2.9837 5257	3.3259 5459	3.7069 0723	4.6028 7070	88
89	3.0210 4948	3.3716 8646	3.7625 1084	4.6834 2093	89
90	3.0588 1260	3.4180 4715	3.8189 4851	4.7653 8080	90
91	3.0970 4775	3.4650 4530	3.8762 3273	4.8487 7496	91
92	3.1357 6085	3.5126 8967	3.9343 7622	4.9336 2853	92
93	3.1749 5786	3.5609 8916	3.9933 9187	5.0199 6703	93
94	3.2146 4483	3.6099 5276	4.0532 9275	5.1078 1645	94
95	3.2548 2789	3.6595 8961	4.1140 9214	5.1972 0324	95
96	3.2955 1324	3.7099 0897	4.1758 0352	5.2881 5429	96
97	3.3367 0716	3.7609 2021	4.2384 4057	5.3806 9699	97
98	3.3784 1600	3.8126 3287	4.3020 1718	5.4748 5919	98
99	3.4206 4620	3.8650 5657	4.3665 4744	5.5706 6923	99
100	3.4634 0427	3.9182 0110	4.4320 4565	5.6681 5594	100

TABLE C-5 AMOUNT OF 1 AT COMPOUND INTEREST (cont.)

$$s = (1 + i)^n$$

n	2%	2¼%	2½%	2¾%	n
1	1.0200 0000	1.0225 0000	1.0250 0000	1.0275 0000	1
2	1.0404 0000	1.0455 0625	1.0506 2500	1.0557 5625	2
3	1.0612 0800	1.0690 3014	1.0768 9063	1.0847 8955	3
4	1.0824 3216	1.0930 8332	1.1038 1289	1.1146 2126	4
5	1.1040 8080	1.1176 7769	1.1314 0821	1.1452 7334	5
6	1.1261 6242	1.1428 2544	1.1596 9342	1.1767 6836	6
7	1.1486 8567	1.1685 3901	1.1886 8575	1.2091 2949	7
8	1.1716 5938	1.1948 3114	1.2184 0290	1.2423 8055	8
9	1.1950 9257	1.2217 1484	1.2488 6297	1.2765 4602	9
10	1.2189 9442	1.2492 0343	1.2800 8454	1.3116 5103	10
11	1.2433 7431	1.2773 1050	1.3120 8666	1.3477 2144	11
12	1.2682 4179	1.3060 4999	1.3448 8882	1.3847 8378	12
13	1.2936 0663	1.3354 3611	1.3785 1104	1.4228 6533	13
14	1.3194 7876	1.3654 8343	1.4129 7382	1.4619 9413	14
15	1.3458 6834	1.3962 0680	1.4482 9817	1.5021 9896	15
16	1.3727 8571	1.4276 2146	1.4845 0562	1.5435 0944	16
17	1.4002 4142	1.4597 4294	1.5216 1826	1.5859 5595	17
18	1.4282 4625	1.4925 8716	1.5596 5872	1.6295 6973	18
19	1.4568 1117	1.5261 7037	1.5986 5019	1.6743 8290	19
20	1.4859 4740	1.5605 0920	1.6386 1644	1.7204 2843	20
21	1.5156 6634	1.5956 2066	1.6795 8185	1.7677 4021	21
22	1.5459 7967	1.6315 2212	1.7215 7140	1.8163 5307	22
23	1.5768 9926	1.6682 3137	1.7646 1068	1.8663 0278	23
24	1.6084 3725	1.7057 6658	1.8087 2595	1.9176 2610	24
25	1.6406 0599	1.7441 4632	1.8539 4410	1.9703 6082	25
26	1.6734 1811	1.7833 8962	1.9002 9270	2.0245 4575	26
27	1.7068 8648	1.8235 1588	1.9478 0002	2.0802 2075	27
28	1.7410 2421	1.8645 4499	1.9964 9502	2.1374 2682	28
29	1.7758 4469	1.9064 9725	2.0464 0739	2.1962 0606	29
30	1.8113 6158	1.9493 9344	2.0975 6758	2.2566 0173	30
31	1.8475 8882	1.9932 5479	2.1500 0677	2.3186 5828	31
32	1.8845 4059	2.0381 0303	2.2037 5694	2.3824 2138	32
33	1.9222 3140	2.0839 6034	2.2588 5086	2.4479 3797	33
34	1.9606 7603	2.1308 4945	2.3153 2213	2.5152 5626	34
35	1.9998 8955	2.1787 9356	2.3732 0519	2.5844 2581	35
36	2.0398 8734	2.2278 1642	2.4325 3532	2.6554 9752	36
37	2.0806 8509	2.2779 4229	2.4933 4870	2.7285 2370	37
38	2.1222 9879	2.3291 9599	2.5556 8242	2.8035 5810	38
39	2.1647 4477	2.3816 0290	2.6195 7448	2.8806 5595	39
40	2.2080 3966	2.4351 8897	2.6850 6384	2.9598 7399	40
41	2.2522 0046	2.4899 8072	2.7521 9043	3.0412 7052	41
42	2.2972 4447	2.5460 0528	2.8209 9520	3.1249 0546	42
43	2.3431 8936	2.6032 9040	2.8915 2008	3.2108 4036	43
44	2.3900 5314	2.6618 6444	2.9638 0808	3.2991 3847	44
45	2.4378 5421	2.7217 5639	3.0379 0328	3.3898 6478	45
46	2.4866 1129	2.7829 9590	3.1138 5086	3.4830 8606	46
47	2.5363 4351	2.8456 1331	3.1916 9713	3.5788 7093	47
48	2.5870 7039	2.9096 3961	3.2714 8956	3.6772 8988	48
49	2.6388 1179	2.9751 0650	3.3532 7680	3.7784 1535	49
50	2.6915 8803	3.0420 4640	3.4371 0872	3.8823 2177	50

TABLE C-5 AMOUNT OF 1 AT COMPOUND INTEREST (cont.)

$$s = (1 + i)^n$$

n	2%	2¼%	2½%	2¾%	n
51	2.7454 1979	3.1104 9244	3.5230 3644	3.9890 8562	51
52	2.8003 2819	3.1804 7852	3.6111 1235	4.0987 8547	52
53	2.8563 3475	3.2520 3929	3.7013 9016	4.2115 0208	53
54	2.9134 6144	3.3252 1017	3.7939 2491	4.3273 1838	54
55	2.9717 3067	3.4000 2740	3.8887 7303	4.4463 1964	55
56	3.0311 6529	3.4765 2802	3.9859 9236	4.5685 9343	56
57	3.0917 8859	3.5547 4990	4.0856 4217	4.6942 2975	57
58	3.1536 2436	3.6347 3177	4.1877 8322	4.8233 2107	58
59	3.2166 9685	3.7165 1324	4.2924 7780	4.9559 6239	59
60	3.2810 3079	3.8001 3479	4.3997 8975	5.0922 5136	60
61	3.3466 5140	3.8856 3782	4.5097 8449	5.2322 8827	61
62	3.4135 8443	3.9730 6467	4.6225 2910	5.3761 7620	62
63	3.4818 5612	4.0624 5862	4.7380 9233	5.5240 2105	63
64	3.5514 9324	4.1538 6394	4.8565 4464	5.6759 3162	64
65	3.6225 2311	4.2473 2588	4.9779 5826	5.8320 1974	65
66	3.6949 7357	4.3428 9071	5.1024 0721	5.9924 0029	66
67	3.7688 7304	4.4406 0576	5.2299 6739	6.1571 9130	67
68	3.8442 5050	4.5405 1939	5.3607 1658	6.3265 1406	68
69	3.9211 3551	4.6426 8107	5.4947 3449	6.5004 9319	69
70	3.9995 5822	4.7471 4140	5.6321 0286	6.6792 5676	70
71	4.0795 4939	4.8539 5208	5.7729 0543	6.8629 3632	71
72	4.1611 4038	4.9631 6600	5.9172 2806	7.0516 6706	72
73	4.2443 6318	5.0748 3723	6.0651 5876	7.2455 8791	73
74	4.3292 5045	5.1890 2107	6.2167 8773	7.4448 4158	74
75	4.4158 3546	5.3057 7405	6.3722 0743	7.6495 7472	75
76	4.5041 5216	5.4251 5396	6.5315 1261	7.8599 3802	76
77	4.5942 3521	5.5472 1993	6.6948 0043	8.0760 8632	77
78	4.6861 1991	5.6720 3237	6.8621 7044	8.2981 7869	78
79	4.7798 4231	5.7996 5310	7.0337 2470	8.5263 7861	79
80	4.8754 3916	5.9301 4530	7.2095 6782	8.7608 5402	80
81	4.9729 4794	6.0635 7357	7.3898 0701	9.0017 7751	81
82	5.0724 0690	6.2000 0397	7.5745 5219	9.2493 2639	82
83	5.1738 5504	6.3395 0406	7.7639 1599	9.5036 8286	83
84	5.2773 3214	6.4821 4290	7.9580 1389	9.7650 3414	84
85	5.3828 7878	6.6279 9112	8.1569 6424	10.0335 7258	85
86	5.4905 3636	6.7771 2092	8.3608 8834	10.3094 9583	86
87	5.6003 4708	6.9296 0614	8.5699 1055	10.5930 0696	87
88	5.7123 5402	7.0855 2228	8.7841 5832	10.8843 1465	88
89	5.8266 0110	7.2449 4653	9.0037 6228	11.1836 3331	89
90	5.9431 3313	7.4079 5782	9.2288 5633	11.4911 8322	90
91	6.0619 9579	7.5746 3688	9.4595 7774	11.8071 9076	91
92	6.1832 3570	7.7450 6621	9.6960 6718	12.1318 8851	92
93	6.3069 0042	7.9193 3020	9.9384 6886	12.4655 1544	93
94	6.4330 3843	8.0975 1512	10.1869 3058	12.8083 1711	94
95	6.5616 9920	8.2797 0921	10.4416 0385	13.1605 4584	95
96	6.6929 3318	8.4660 0267	10.7026 4395	13.5224 6085	96
97	6.8267 9184	8.6564 8773	10.9702 1004	13.8943 2852	97
98	6.9633 2768	8.8512 5871	11.2444 6530	14.2764 2255	98
99	7.1025 9423	9.0504 1203	11.5255 7693	14.6690 2417	99
100	7.2446 4612	9.2540 4630	11.8127 1635	15.0724 2234	100

TABLE C-5 AMOUNT OF 1 AT COMPOUND INTEREST (cont.)

$$s = (1 + i)^n$$

n	3%	3½%	4%	4½%	n
1	1.0300 0000	1.0350 0000	1.0400 0000	1.0450 0000	1
2	1.0609 0000	1.0712 2500	1.0816 0000	1.0920 2500	2
3	1.0927 2700	1.1087 1788	1.1248 6400	1.1411 6613	3
4	1.1255 0881	1.1475 2300	1.1698 5856	1.1925 1860	4
5	1.1592 7407	1.1876 8631	1.2166 5290	1.2461 8194	5
6	1.1940 5230	1.2292 5533	1.2653 1902	1.3022 6012	6
7	1.2298 7387	1.2722 7926	1.3159 3178	1.3608 6183	7
8	1.2667 7008	1.3168 0904	1.3685 6905	1.4221 0061	8
9	1.3047 7318	1.3628 9735	1.4233 1181	1.4860 9514	9
10	1.3439 1638	1.4105 9876	1.4802 4428	1.5529 6942	10
11	1.3842 3387	1.4599 6972	1.5394 5406	1.6228 5305	11
12	1.4257 6089	1.5110 6866	1.6010 3222	1.6958 8143	12
13	1.4685 3371	1.5639 5606	1.6650 7351	1.7721 9610	13
14	1.5125 8972	1.6186 9452	1.7316 7645	1.8519 4492	14
15	1.5579 6742	1.6753 4883	1.8009 4351	1.9352 8244	15
16	1.6047 0644	1.7339 8604	1.8729 8125	2.0223 7015	16
17	1.6528 4763	1.7946 7555	1.9479 0050	2.1133 7681	17
18	1.7024 3306	1.8574 8920	2.0258 1652	2.2084 7877	18
19	1.7535 0605	1.9225 0132	2.1068 4918	2.3078 6031	19
20	1.8061 1123	1.9897 8886	2.1911 2314	2.4117 1402	20
21	1.8602 9457	2.0594 3147	2.2787 6807	2.5202 4116	21
22	1.9161 0341	2.1315 1158	2.3699 1879	2.6336 5201	22
23	1.9735 8651	2.2061 1448	2.4647 1554	2.7521 6635	23
24	2.0327 9411	2.2833 2849	2.5633 0416	2.8760 1383	24
25	2.0937 7793	2.3632 4498	2.6658 3633	3.0054 3446	25
26	2.1565 9127	2.4459 5856	2.7724 6978	3.1406 7901	26
27	2.2212 8901	2.5315 6711	2.8833 6858	3.2820 0956	27
28	2.2879 2768	2.6201 7196	2.9987 0332	3.4296 9999	28
29	2.3565 6551	2.7118 7798	3.1186 5145	3.5840 3649	29
30	2.4272 6247	2.8067 9370	3.2433 9751	3.7453 1813	30
31	2.5000 8035	2.9050 3148	3.3731 3341	3.9138 5745	31
32	2.5750 8276	3.0067 0759	3.5080 5875	4.0899 8104	32
33	2.6523 3524	3.1119 4235	3.6483 8110	4.2740 3018	33
34	2.7319 0530	3.2208 6033	3.7943 1634	4.4663 6154	34
35	2.8138 6245	3.3335 9045	3.9460 8899	4.6673 4781	35
36	2.8982 7833	3.4502 6611	4.1039 3255	4.8773 7846	36
37	2.9852 2668	3.5710 2543	4.2680 8986	5.0968 6049	37
38	3.0747 8348	3.6960 1132	4.4388 1345	5.3262 1921	38
39	3.1670 2698	3.8253 7171	4.6163 6599	5.5658 9908	39
40	3.2620 3779	3.9592 5972	4.8010 2063	5.8163 6454	40
41	3.3598 9893	4.0978 3381	4.9930 6145	6.0781 0094	41
42	3.4606 9589	4.2412 5799	5.1927 8391	6.3516 1548	42
43	3.5645 1677	4.3897 0202	5.4004 9527	6.6374 3818	43
44	3.6714 5227	4.5433 4160	5.6165 1508	6.9361 2290	44
45	3.7815 9584	4.7023 5855	5.8411 7568	7.2482 4843	45
46	3.8950 4372	4.8669 4110	6.0748 2271	7.5744 1961	46
47	4.0118 9503	5.0372 8404	6.3178 1562	7.9152 6849	47
48	4.1322 5188	5.2135 8858	6.5705 2824	8.2714 5557	48
49	4.2562 1944	5.3960 6459	6.8333 4937	8.6436 7107	49
50	4.3839 0602	5.5849 2686	7.1066 8335	9.0326 3627	50

TABLE C-5 AMOUNT OF 1 AT COMPOUND INTEREST (cont.)

$$s = (1 + i)^n$$

n	3%	3½%	4%	4½%	n
51	4.5154 2320	5.7803 9930	7.3909 5068	9.4391 0490	51
52	4.6508 8590	5.9827 1327	7.6865 8871	9.8638 6463	52
53	4.7904 1247	6.1921 0824	7.9940 5226	10.3077 3853	53
54	4.9341 2485	6.4088 3202	8.3138 1435	10.7715 8677	54
55	5.0821 4859	6.6331 4114	8.6463 6692	11.2563 0817	55
56	5.2346 1305	6.8653 0108	8.9922 2160	11.7628 4204	56
57	5.3916 5144	7.1055 8662	9.3519 1046	12.2921 6993	57
58	5.5534 0098	7.3542 8215	9.7259 8688	12.8453 1758	58
59	5.7200 0301	7.6116 8203	10.1150 2635	13.4233 5687	59
60	5.8916 0310	7.8780 9090	10.5196 2741	14.0274 0793	60
61	6.0683 5120	8.1538 2408	10.9404 1250	14.6586 4129	61
62	6.2504 0173	8.4392 0793	11.3780 2900	15.3182 8014	62
63	6.4379 1379	8.7345 8020	11.8331 5016	16.0076 0275	63
64	6.6310 5120	9.0402 9051	12.3064 7617	16.7279 4487	64
65	6.8299 8273	9.3567 0068	12.7987 3522	17.4807 0239	65
66	7.0348 8222	9.6841 8520	13.3106 8463	18.2673 3400	66
67	7.2459 2868	10.0231 3168	13.8431 1201	19.0893 6403	67
68	7.4633 0654	10.3739 4129	14.3968 3649	19.9483 8541	68
69	7.6872 0574	10.7370 2924	14.9727 0995	20.8460 6276	69
70	7.9178 2191	11.1128 2526	15.5716 1835	21.7841 3558	70
71	8.1553 5657	11.5017 7414	16.1944 8308	22.7644 2168	71
72	8.4000 1727	11.9043 3624	16.8422 6241	23.7888 2066	72
73	8.6520 1778	12.3209 8801	17.5159 5290	24.8593 1759	73
74	8.9115 7832	12.7522 2259	18.2165 9102	25.9779 8688	74
75	9.1789 2567	13.1985 5038	18.9452 5466	27.1469 9629	75
76	9.4542 9344	13.6604 9964	19.7030 6485	28.3686 1112	76
77	9.7379 2224	14.1386 1713	20.4911 8744	29.6451 9862	77
78	10.0300 5991	14.6334 6873	21.3108 3494	30.9792 3256	78
79	10.3309 6171	15.1456 4013	22.1632 6834	32.3732 9802	79
80	10.6408 9056	15.6757 3754	23.0497 9907	33.8300 9643	80
81	10.9601 1727	16.2243 8035	23.9717 9103	35.3524 5077	81
82	11.2889 2079	16.7922 4195	24.9306 6267	36.9433 1106	82
83	11.6275 8842	17.3799 7041	25.9278 8918	38.6057 6006	83
84	11.9764 1607	17.9882 6938	26.9650 0475	40.3430 1926	84
85	12.3357 0855	18.6178 5881	28.0436 0494	42.1584 5513	85
86	12.7057 7981	19.2694 8387	29.1653 4914	44.0555 8561	86
87	13.0869 5320	19.9439 1580	30.3319 6310	46.0380 8696	87
88	13.4795 6180	20.6419 5285	31.5452 4163	48.1098 0087	88
89	13.8839 4865	21.3644 2120	32.8070 5129	50.2747 4191	89
90	14.3004 6711	22.1121 7595	34.1193 3334	52.5371 0530	90
91	14.7294 8112	22.8861 0210	35.4841 0668	54.9012 7503	91
92	15.1713 6556	23.6871 1568	36.9034 7094	57.3718 3241	92
93	15.6265 0652	24.5161 6473	38.3796 0978	59.9535 6487	93
94	16.0953 0172	25.3742 3049	39.9147 9417	62.6514 7529	94
95	16.5781 6077	26.2623 2856	41.5113 8594	65.4707 9168	95
96	17.0755 0559	27.1815 1006	43.1718 4138	68.4169 7730	96
97	17.5877 7076	28.1328 6291	44.8987 1503	71.4957 4128	97
98	18.1154 0388	29.1175 1311	46.6946 6363	74.7130 4964	98
99	18.6588 6600	30.1366 2607	48.5624 5018	78.0751 3687	99
100	19.2186 3198	31.1914 0798	50.5049 4818	81.5885 1803	100

TABLE C-5 AMOUNT OF 1 AT COMPOUND INTEREST (cont.)

$$s = (1 + i)^n$$

n	5%	5½%	6%	6½%	n
1	1.0500 0000	1.0550 0000	1.0600 0000	1.0650 0000	1
2	1.1025 0000	1.1130 2500	1.1236 0000	1.1342 2500	2
3	1.1576 2500	1.1742 4138	1.1910 1600	1.2079 4963	3
4	1.2155 0625	1.2388 2465	1.2624 7696	1.2864 6635	4
5	1.2762 8156	1.3069 6001	1.3382 2558	1.3700 8666	5
6	1.3400 9564	1.3788 4281	1.4185 1911	1.4591 4230	6
7	1.4071 0042	1.4546 7916	1.5036 3026	1.5539 8655	7
8	1.4774 5544	1.5346 8651	1.5938 4807	1.6549 9567	8
9	1.5513 2822	1.6190 9427	1.6894 7896	1.7625 7039	9
10	1.6288 9463	1.7081 4446	1.7908 4770	1.8771 3747	10
11	1.7103 3936	1.8020 9240	1.8982 9856	1.9991 5140	11
12	1.7958 5633	1.9012 0749	2.0121 9647	2.1290 9624	12
13	1.8856 4914	2.0057 7390	2.1329 2826	2.2674 8750	13
14	1.9799 3160	2.1160 9146	2.2609 0396	2.4148 7418	14
15	2.0789 2818	2.2324 7649	2.3965 5819	2.5718 4101	15
16	2.1828 7459	2.3552 6270	2.5403 5168	2.7390 1067	16
17	2.2920 1832	2.4848 0215	2.6927 7279	2.9170 4637	17
18	2.4066 1923	2.6214 6627	2.8543 3915	3.1066 5438	18
19	2.5269 5020	2.7656 4691	3.0255 9950	3.3085 8691	19
20	2.6532 9771	2.9177 5749	3.2071 3547	3.5236 4506	20
21	2.7859 6259	3.0782 3415	3.3995 6360	3.7526 8199	21
22	2.9252 6072	3.2475 3703	3.6035 3742	3.9966 0632	22
23	3.0715 2376	3.4261 5157	3.8197 4966	4.2563 8573	23
24	3.2250 9994	3.6145 8990	4.0489 3464	4.5330 5081	24
25	3.3863 5494	3.8133 9235	4.2918 7072	4.8276 9911	25
26	3.5556 7269	4.0231 2893	4.5493 8296	5.1414 9955	26
27	3.7334 5632	4.2444 0102	4.8223 4594	5.4756 9702	27
28	3.9201 2914	4.4778 4307	5.1116 8670	5.8316 1733	28
29	4.1161 3560	4.7241 2444	5.4183 8790	6.2106 7245	29
30	4.3219 4238	4.9839 5129	5.7434 9117	6.6143 6616	30
31	4.5380 3949	5.2580 6861	6.0881 0064	7.0442 9996	31
32	4.7649 4147	5.5472 6238	6.4533 8668	7.5021 7946	32
33	5.0031 8854	5.8523 6181	6.8405 8988	7.9898 2113	33
34	5.2533 4797	6.1742 4171	7.2510 2528	8.5091 5950	34
35	5.5160 1537	6.5138 2501	7.6860 8679	9.0622 5487	35
36	5.7918 1614	6.8720 8538	8.1472 5200	9.6513 0143	36
37	6.0814 0694	7.2500 5008	8.6360 8712	10.2786 3603	37
38	6.3854 7729	7.6488 0283	9.1542 5235	10.9467 4737	38
39	6.7047 5115	8.0694 8699	9.7035 0749	11.6582 8595	39
40	7.0399 8871	8.5133 0877	10.2857 1794	12.4160 7453	40
41	7.3919 8815	8.9815 4076	10.9028 6101	13.2231 1938	41
42	7.7615 8756	9.4755 2550	11.5570 3267	14.0826 2214	42
43	8.1496 6693	9.9966 7940	12.2504 5463	14.9979 9258	43
44	8.5571 5028	10.5464 9677	12.9854 8191	15.9728 6209	44
45	8.9850 0779	11.1265 5409	13.7646 1083	17.0110 9813	45
46	9.4342 5818	11.7385 1456	14.5904 8748	18.1168 1951	46
47	9.9059 7109	12.3841 3287	15.4659 1673	19.2944 1278	47
48	10.4012 6965	13.0652 6017	16.3938 7173	20.5485 4961	48
49	10.9213 3313	13.7838 4948	17.3775 0403	21.8842 0533	49
50	11.4673 9979	14.5419 6120	18.4201 5427	23.3066 7868	50

TABLE C-5 AMOUNT OF 1 AT COMPOUND INTEREST (cont.)

$$s = (1 + i)^n$$

n	5%	5½%	6%	6½%	n
51	12.0407 6978	15.3417 6907	19.5253 6353	24.8216 1279	51
52	12.6428 0826	16.1855 6637	20.6968 8534	26.4350 1762	52
53	13.2749 4868	17.0757 7252	21.9386 9846	28.1532 9377	53
54	13.9386 9611	18.0149 4001	23.2550 2037	29.9832 5786	54
55	14.6356 3092	19.0057 6171	24.6503 2159	31.9321 6963	55
56	15.3674 1246	20.0510 7860	26.1293 4089	34.0077 6065	56
57	16.1357 8309	21.1538 8793	27.6971 0134	36.2182 6509	57
58	16.9425 7224	22.3173 5176	29.3589 2742	38.5724 5233	58
59	17.7897 0085	23.5448 0611	31.1204 6307	41.0796 6173	59
60	18.6791 8589	24.8397 7045	32.9876 9085	43.7498 3974	60
61	19.6131 4519	26.2059 5782	34.9669 5230	46.5935 7932	61
62	20.5938 0245	27.6472 8550	37.0649 6944	49.6221 6198	62
63	21.6234 9257	29.1678 8620	39.2888 6761	52.8476 0251	63
64	22.7046 6720	30.7721 1994	41.6461 9967	56.2826 9667	64
65	23.8399 0056	32.4645 8654	44.1149 7165	59.9410 7195	65
66	25.0318 9559	34.2501 3880	46.7936 6994	63.8372 4163	66
67	26.2834 9037	36.1338 9643	49.6012 9014	67.9866 6234	67
68	27.5976 6488	38.1212 6074	52.5773 6755	72.4057 9539	68
69	28.9775 4813	40.2179 3008	55.7320 0960	77.1121 7209	69
70	30.4264 2554	42.4299 1623	59.0759 3018	82.1244 6327	70
71	31.9477 4681	44.7635 6163	62.6204 8599	87.4625 5339	71
72	33.5451 3415	47.2255 5751	66.3777 1515	93.1476 1936	72
73	35.2223 9086	49.8229 6318	70.3603 7806	99.2022 1461	73
74	36.9835 1040	52.5632 2615	74.5820 0074	105.6503 5856	74
75	38.8326 8592	55.4542 0359	79.0569 2079	112.5176 3187	75
76	40.7743 2022	58.5041 8479	83.8003 3603	119.8312 7794	76
77	42.8130 3623	61.7219 1495	88.8283 5620	127.6203 1101	77
78	44.9536 8804	65.1166 2027	94.1580 5757	135.9156 3122	78
79	47.2013 7244	68.6980 3439	99.8075 4102	144.7501 4725	79
80	49.5614 4107	72.4764 2628	105.7959 9348	154.1589 0683	80
81	52.0395 1312	76.4626 2973	112.1437 5309	164.1792 3577	81
82	54.6414 8878	80.6680 7436	118.8723 7828	174.8508 8609	82
83	57.3735 6322	85.1048 1845	126.0047 2097	186.2161 9369	83
84	60.2422 4138	89.7855 8347	133.5650 0423	198.3202 4628	84
85	63.2543 5344	94.7237 9056	141.5789 0449	211.2110 6229	85
86	66.4170 7112	99.9335 9904	150.0736 3875	224.9397 8134	86
87	69.7379 2467	105.4299 4698	159.0780 5708	239.5608 6712	87
88	73.2248 2091	111.2285 9407	168.6227 4050	255.1323 2349	88
89	76.8860 6195	117.3461 6674	178.7401 0493	271.7159 2451	89
90	80.7303 6505	123.8002 0591	189.4645 1123	289.3774 5961	90
91	84.7668 8330	130.6092 1724	200.8323 8190	308.1869 9448	91
92	89.0052 2747	137.7927 2419	212.8823 2482	328.2191 4912	92
93	93.4554 8884	145.3713 2402	225.6552 6431	349.5533 9382	93
94	98.1282 6328	153.3667 4684	239.1945 8017	372.2743 6441	94
95	103.0346 7645	161.8019 1791	253.5462 5498	396.4721 9810	95
96	108.1864 1027	170.7010 2340	268.7590 3028	422.2428 9098	96
97	113.5957 3078	180.0895 7969	284.8845 7209	449.6886 7889	97
98	119.2755 1732	189.9945 0657	301.9776 4642	478.9184 4302	98
99	125.2392 9319	200.4442 0443	320.0963 0520	510.0481 4181	99
100	131.5012 5785	211.4686 3567	339.3020 8351	543.2012 7103	100

TABLE C-5 AMOUNT OF 1 AT COMPOUND INTEREST (cont.)

$$s = (1 + i)^n$$

n	7%	7½%	8%	8½%	n
1	1.0700 0000	1.0750 0000	1.0800 0000	1.0850 0000	1
2	1.1449 0000	1.1556 2500	1.1664 0000	1.1772 2500	2
3	1.2250 4300	1.2422 9688	1.2597 1200	1.2772 8913	3
4	1.3107 9601	1.3354 6914	1.3604 8896	1.3858 5870	4
5	1.4025 5173	1.4356 2933	1.4693 2808	1.5036 5669	5
6	1.5007 3035	1.5433 0153	1.5868 7432	1.6314 6751	6
7	1.6057 8148	1.6590 4914	1.7138 2427	1.7701 4225	7
8	1.7181 8618	1.7834 7783	1.8509 3021	1.9206 0434	8
9	1.8384 5921	1.9172 3866	1.9990 0463	2.0838 5571	9
10	1.9671 5136	2.0610 3156	2.1589 2500	2.2609 8344	10
11	2.1048 5195	2.2156 0893	2.3316 3900	2.4531 6703	11
12	2.2521 9159	2.3817 7960	2.5181 7012	2.6616 8623	12
13	2.4098 4500	2.5604 1307	2.7196 2373	2.8879 2956	13
14	2.5785 3415	2.7524 4405	2.9371 9362	3.1334 0357	14
15	2.7590 3154	2.9588 7735	3.1721 6911	3.3997 4288	15
16	2.9521 6375	3.1807 9315	3.4259 4264	3.6887 2102	16
17	3.1588 1521	3.4193 5264	3.7000 1805	4.0022 6231	17
18	3.3799 3228	3.6758 0409	3.9960 1950	4.3424 5461	18
19	3.6165 2754	3.9514 8940	4.3157 0106	4.7115 6325	19
20	3.8696 8446	4.2478 5110	4.6609 5714	5.1120 4612	20
21	4.1405 6237	4.5664 3993	5.0338 3372	5.5465 7005	21
22	4.4304 0174	4.9089 2293	5.4365 4041	6.0180 2850	22
23	4.7405 2986	5.2770 9215	5.8714 6365	6.5295 6092	23
24	5.0723 6695	5.6728 7406	6.3411 8074	7.0845 7360	24
25	5.4274 3264	6.0983 3961	6.8484 7520	7.6867 6236	25
26	5.8073 5292	6.5557 1508	7.3963 5321	8.3401 3716	26
27	6.2138 6763	7.0473 9371	7.9880 6147	9.0490 4881	27
28	6.6488 3836	7.5759 4824	8.6271 0639	9.8182 1796	28
29	7.1142 5705	8.1441 4436	9.3172 7490	10.6527 6649	29
30	7.6122 5504	8.7549 5519	10.0626 5689	11.5582 5164	30
31	8.1451 1290	9.4115 7683	10.8676 6944	12.5407 0303	31
32	8.7152 7080	10.1174 4509	11.7370 8300	13.6066 6279	32
33	9.3253 3975	10.8762 5347	12.6760 4964	14.7632 2913	33
34	9.9781 1354	11.6919 7248	13.6901 3361	16.0181 0360	34
35	10.6765 8148	12.5688 7042	14.7853 4429	17.3796 4241	35
36	11.4239 4219	13.5115 3570	15.9681 7184	18.8569 1201	36
37	12.2236 1814	14.5249 0088	17.2456 2558	20.4597 4953	37
38	13.0792 7141	15.6142 6844	18.6252 7563	22.1988 2824	38
39	13.9948 2041	16.7853 3858	20.1152 9768	24.0857 2865	39
40	14.9744 5784	18.0442 3897	21.7245 2150	26.1330 1558	40
41	16.0226 6989	19.3975 5689	23.4624 8322	28.3543 2190	41
42	17.1442 5678	20.8523 7366	25.3394 8187	30.7644 3927	42
43	18.3443 5475	22.4163 0168	27.3666 4042	33.3794 1660	43
44	19.6284 5959	24.0975 2431	29.5559 7166	36.2166 6702	44
45	21.0024 5176	25.9048 3863	31.9204 4939	39.2950 8371	45
46	22.4726 2338	27.8477 0153	34.4740 8534	42.6351 6583	46
47	24.0457 0702	29.9362 7915	37.2320 1217	46.2591 5492	47
48	25.7289 0651	32.1815 0008	40.2105 7314	50.1911 8309	48
49	27.5299 2997	34.5951 1259	43.4274 1899	54.4574 3365	49
50	29.4570 2506	37.1897 4603	46.9016 1251	59.0863 1551	50

TABLE C-5 AMOUNT OF 1 AT COMPOUND INTEREST (*cont.*)

$$s = (1 + i)^n$$

n	7%	7½%	8%	8½%	n
51	31.5190 1682	39.9789 7698	50.6537 4151	64.1086 5233	51
52	33.7253 4799	42.9774 0026	54.7060 4084	69.5578 8778	52
53	36.0861 2235	46.2007 0528	59.0825 2410	75.4703 0824	53
54	38.6121 5092	49.6657 5817	63.8091 2603	81.8852 8444	54
55	41.3150 0148	53.3906 9004	68.9138 5611	88.8455 3362	55
56	44.2070 5159	57.3949 9179	74.4269 6460	96.3974 0398	56
57	47.3015 4520	61.6996 1617	80.3811 2177	104.5911 8332	57
58	50.6126 5336	66.3270 8739	86.8116 1151	113.4814 3390	58
59	54.1555 3910	71.3016 1894	93.7565 4043	123.1273 5578	59
60	57.9464 2683	76.6492 4036	01.2570 6367	133.5931 8102	60
61	62.0026 7671	82.3979 3339	109.3576 2876	144.9486 0141	61
62	66.3428 6408	88.5777 7839	118.1062 3906	157.2692 3253	62
63	70.9868 6457	95.2211 1177	127.5547 3819	170.6371 1729	63
64	75.9559 4509	102.3626 9515	137.7591 1724	185.1412 7226	64
65	81.2728 6124	110.0398 9729	148.7798 4662	200.8782 8041	65
66	86.9619 6153	118.2928 8959	160.6822 3435	217.9529 3424	66
67	93.0492 9884	127.1648 5631	173.5368 1310	236.4789 3365	67
68	99.5627 4976	136.7022 2053	187.4197 5815	256.5796 4301	68
69	106.5321 4224	146.9548 8707	202.4133 3880	278.3889 1267	69
70	113.9893 9220	157.9765 0360	218.6064 0590	302.0519 7024	70
71	121.9686 4965	169.8247 4137	236.0949 1837	327.7263 8771	71
72	130.5064 5513	182.5615 9697	254.9825 1184	355.5831 3067	72
73	139.6419 0699	196.2537 1675	275.3811 1279	385.8076 9678	73
74	149.4168 4047	210.9727 4550	297.4116 0181	418.6013 5100	74
75	159.8760 1931	226.7957 0141	321.2045 2996	454.1824 6584	75
76	171.0673 4066	243.8053 7902	346.9008 9236	492.7879 7543	76
77	183.0420 5451	262.0907 8245	374.6529 6374	534.6749 5335	77
78	195.8549 9832	281.7475 9113	404.6252 0084	580.1223 2438	78
79	209.5648 4820	302.8786 6046	436.9952 1691	629.4327 2195	79
80	224.2343 8758	325.5945 6000	471.9548 3426	682.9345 0332	80
81	239.9307 9471	350.0141 5200	509.7112 2101	740.9839 3610	81
82	256.7259 5034	376.2652 1340	550.4881 1869	803.9675 7067	82
83	274.6967 6686	404.4851 0440	594.5271 6818	872.3048 1418	83
84	293.9255 4054	434.8214 8723	642.0893 4164	946.4507 2338	84
85	314.5003 2838	467.4330 9878	693.4564 8897	1026.8990 3487	85
86	336.5153 5137	502.4905 8119	748.9330 0808	1114.1854 5283	86
87	360.0714 2596	540.1773 7477	808.8476 4873	1208.8912 1633	87
88	385.2764 2578	580.6906 7788	873.5554 6063	1311.6469 6971	88
89	412.2457 7558	624.2424 7872	943.4398 9748	1423.1369 6214	89
90	441.1029 7988	671.0606 6463	1018.9150 8928	1544.1036 0392	90
91	471.9801 8847	721.3902 1447	1100.4282 9642	1675.3524 1025	91
92	505.0188 0166	775.4944 8056	1188.4625 6013	1817.7573 6512	92
93	540.3701 1778	833.6565 6660	1283.5395 6494	1972.2667 4116	93
94	578.1960 2602	896.1808 0910	1386.2227 3014	2139.9094 1416	94
95	618.6697 4784	963.3943 6978	1497.1205 4855	2321.8017 1436	95
96	661.9766 3019	1035.6489 4751	1616.8901 9244	2519.1548 6008	96
97	708.3149 9430	1113.3226 1858	1746.2414 0783	2733.2830 2319	97
98	757.8970 4390	1196.8218 1497	1885.9407 2046	2965.6120 8016	98
99	810.9498 3698	1286.5834 5109	2036.8159 7809	3217.6891 0698	99
100	867.7163 2557	1383.0772 0993	2199.7612 5634	3491.1926 8107	100

TABLE C-6 PRESENT VALUE OF 1 AT COMPOUND INTEREST

$$v^n = (1 + i)^{-n}$$

n	1/4%	7/24%	1/3%	5/12%	n
1	0.9975 0623	0.9970 9182	0.9966 7774	0.9958 5062	1
2	0.9950 1869	0.9941 9209	0.9933 6652	0.9917 1846	2
3	0.9925 3734	0.9913 0079	0.9900 6630	0.9876 0345	3
4	0.9900 6219	0.9884 1791	0.9867 7704	0.9835 0551	4
5	0.9875 9321	0.9855 4341	0.9834 9871	0.9794 2457	5
6	0.9851 3038	0.9826 7726	0.9802 3127	0.9753 6057	6
7	0.9826 7370	0.9798 1946	0.9769 7469	0.9713 1343	7
8	0.9802 2314	0.9769 6996	0.9737 2893	0.9672 8308	8
9	0.9777 7869	0.9741 2875	0.9704 9395	0.9632 6946	9
10	0.9753 4034	0.9712 9580	0.9672 6972	0.9592 7249	10
11	0.9729 0807	0.9684 7110	0.9640 5620	0.9552 9211	11
12	0.9704 8187	0.9656 5460	0.9608 5335	0.9513 2824	12
13	0.9680 6171	0.9628 4630	0.9576 6115	0.9473 8082	13
14	0.9656 4759	0.9600 4617	0.9544 7955	0.9434 4978	14
15	0.9632 3949	0.9572 5418	0.9513 0852	0.9395 3505	15
16	0.9608 3740	0.9544 7030	0.9481 4803	0.9356 3656	16
17	0.9584 4130	0.9516 9453	0.9449 9803	0.9317 5425	17
18	0.9560 5117	0.9489 2682	0.9418 5851	0.9278 8805	18
19	0.9536 6700	0.9461 6717	0.9387 2941	0.9240 3789	19
20	0.9512 8878	0.9434 1554	0.9356 1071	0.9202 0371	20
21	0.9489 1649	0.9406 7191	0.9325 0236	0.9163 8544	21
22	0.9465 5011	0.9379 3627	0.9294 0435	0.9125 8301	22
23	0.9441 8964	0.9352 0857	0.9263 1663	0.9087 9636	23
24	0.9418 3505	0.9324 8881	0.9232 3916	0.9050 2542	24
25	0.9394 8634	0.9297 7696	0.9201 7192	0.9012 7012	25
26	0.9371 4348	0.9270 7300	0.9171 1487	0.8975 3041	26
27	0.9348 0646	0.9243 7690	0.9140 6798	0.8938 0622	27
28	0.9324 7527	0.9216 8864	0.9110 3121	0.8900 9748	28
29	0.9301 4990	0.9190 0820	0.9080 0453	0.8864 0413	29
30	0.9278 3032	0.9163 3556	0.9049 8790	0.8827 2610	30
31	0.9255 1653	0.9136 7068	0.9019 8130	0.8790 6334	31
32	0.9232 0851	0.9110 1356	0.8989 8468	0.8754 1577	32
33	0.9209 0624	0.9083 6416	0.8959 9802	0.8717 8334	33
34	0.9186 0972	0.9057 2247	0.8930 2128	0.8681 6599	34
35	0.9163 1892	0.9030 8847	0.8900 5444	0.8645 6364	35
36	0.9140 3384	0.9004 6212	0.8870 9745	0.8609 7624	36
37	0.9117 5445	0.8978 4341	0.8841 5028	0.8574 0372	37
38	0.9094 8075	0.8952 3231	0.8812 1290	0.8538 4603	38
39	0.9072 1272	0.8926 2881	0.8782 8528	0.8503 0310	39
40	0.9049 5034	0.8900 3288	0.8753 6739	0.8467 7487	40
41	0.9026 9361	0.8874 4450	0.8724 5920	0.8432 6128	41
42	0.9004 4250	0.8848 6365	0.8695 6066	0.8397 6227	42
43	0.8981 9701	0.8822 9030	0.8666 7175	0.8362 7778	43
44	0.8959 5712	0.8797 2444	0.8637 9245	0.8328 0775	44
45	0.8937 2281	0.8771 6604	0.8609 2270	0.8293 5211	45
46	0.8914 9407	0.8746 1508	0.8580 6249	0.8259 1082	46
47	0.8892 7090	0.8720 7153	0.8552 1179	0.8224 8380	47
48	0.8870 5326	0.8695 3539	0.8523 7055	0.8190 7100	48
49	0.8848 4116	0.8670 0662	0.8495 3876	0.8156 7237	49
50	0.8826 3457	0.8644 8520	0.8467 1637	0.8122 8784	50

TABLE C-6 PRESENT VALUE OF 1 AT COMPOUND INTEREST (cont.)

$$v^n = (1+i)^{-n}$$

n	1/4%	7/24%	1/3%	5/12%	n
51	0.8804 3349	0.8619 7112	0.8439 0336	0.8089 1735	51
52	0.8782 3790	0.8594 6435	0.8410 9969	0.8055 6084	52
53	0.8760 4778	0.8569 6487	0.8383 0534	0.8022 1827	53
54	0.8738 6312	0.8544 7266	0.8355 2027	0.7988 8956	54
55	0.8716 8391	0.8519 8769	0.8327 4446	0.7955 7467	55
56	0.8695 1013	0.8495 0995	0.8299 7787	0.7922 7353	56
57	0.8673 4178	0.8470 3942	0.8272 2047	0.7889 8608	57
58	0.8651 7883	0.8445 7608	0.8244 7222	0.7857 1228	58
59	0.8630 2128	0.8421 1989	0.8217 3311	0.7824 5207	59
60	0.8608 6911	0.8396 7085	0.8190 0310	0.7792 0538	60
61	0.8587 2230	0.8372 2893	0.8162 8216	0.7759 7216	61
62	0.8565 8085	0.8347 9412	0.8135 7026	0.7727 5236	62
63	0.8544 4474	0.8323 6638	0.8108 6737	0.7695 4591	63
64	0.8523 1395	0.8299 4571	0.8081 7346	0.7663 5278	64
65	0.8501 8848	0.8275 3207	0.8054 8850	0.7631 7289	65
66	0.8480 6831	0.8251 2545	0.8028 1246	0.7600 0620	66
67	0.8459 5343	0.8227 2584	0.8001 4531	0.7568 5265	67
68	0.8438 4382	0.8203 3320	0.7974 8702	0.7537 1218	68
69	0.8417 3947	0.8179 4752	0.7948 3756	0.7505 8474	69
70	0.8396 4037	0.8155 6878	0.7921 9690	0.7474 7028	70
71	0.8375 4650	0.8131 9695	0.7895 6502	0.7443 6874	71
72	0.8354 5786	0.8108 3202	0.7869 4188	0.7412 8008	72
73	0.8333 7442	0.8084 7397	0.7843 2745	0.7382 0423	73
74	0.8312 9618	0.8061 2278	0.7817 2171	0.7351 4114	74
75	0.8292 2312	0.8037 7843	0.7791 2463	0.7320 9076	75
76	0.8271 5523	0.8014 4089	0.7765 3618	0.7290 5304	76
77	0.8250 9250	0.7991 1015	0.7739 5632	0.7260 2792	77
78	0.8230 3491	0.7967 8619	0.7713 8504	0.7230 1536	78
79	0.8209 8246	0.7944 6899	0.7688 2230	0.7200 1529	79
80	0.8189 3512	0.7921 5853	0.7662 6807	0.7170 2768	80
81	0.8168 9289	0.7898 5479	0.7637 2233	0.7140 5246	81
82	0.8148 5575	0.7875 5774	0.7611 8505	0.7110 8959	82
83	0.8128 2369	0.7852 6738	0.7586 5619	0.7081 3901	83
84	0.8107 9670	0.7829 8368	0.7561 3574	0.7052 0067	84
85	0.8087 7476	0.7807 0662	0.7536 2366	0.7022 7453	85
86	0.8067 5787	0.7784 3618	0.7511 1993	0.6993 6052	86
87	0.8047 4600	0.7761 7234	0.7486 2451	0.6964 5861	87
88	0.8027 3915	0.7739 1509	0.7461 3739	0.6935 6874	88
89	0.8007 3731	0.7716 6440	0.7436 5853	0.6906 9086	89
90	0.7987 4046	0.7694 2026	0.7411 8790	0.6878 2493	90
91	0.7967 4859	0.7671 8264	0.7387 2548	0.6849 7088	91
92	0.7947 6168	0.7649 5153	0.7362 7125	0.6821 2868	92
93	0.7927 7973	0.7627 2691	0.7338 2516	0.6792 9827	93
94	0.7908 0273	0.7605 0876	0.7313 8720	0.6764 7960	94
95	0.7888 3065	0.7582 9706	0.7289 5735	0.6736 7263	95
96	0.7868 6349	0.7560 9179	0.7265 3556	0.6708 7731	96
97	0.7849 0124	0.7538 9294	0.7241 2182	0.6680 9359	97
98	0.7829 4388	0.7517 0048	0.7217 1610	0.6653 2141	98
99	0.7809 9140	0.7495 1439	0.7193 1837	0.6625 6074	99
100	0.7790 4379	0.7473 3467	0.7169 2861	0.6598 1153	100

TABLE C-6 PRESENT VALUE OF 1 AT COMPOUND INTEREST (cont.)

$$v^n = (1 + i)^{-n}$$

n	1/2%	7/12%	5/8%	2/3%	n
1	0.9950 2488	0.9942 0050	0.9937 8882	0.9933 7748	1
2	0.9900 7450	0.9884 3463	0.9876 1622	0.9867 9882	2
3	0.9851 4876	0.9827 0220	0.9814 8196	0.9802 6373	3
4	0.9802 4752	0.9770 0302	0.9753 8580	0.9737 7192	4
5	0.9753 7067	0.9713 3688	0.9693 2750	0.9673 2310	5
6	0.9705 1808	0.9657 0361	0.9633 0683	0.9609 1699	6
7	0.9656 8963	0.9601 0301	0.9573 2356	0.9545 5330	7
8	0.9608 8520	0.9545 3489	0.9513 7745	0.9482 3175	8
9	0.9561 0468	0.9489 9907	0.9454 6827	0.9419 5207	9
10	0.9513 4794	0.9434 9534	0.9395 9580	0.9357 1398	10
11	0.9466 1489	0.9380 2354	0.9337 5980	0.9295 1720	11
12	0.9419 0534	0.9325 8347	0.9279 6005	0.9233 6145	12
13	0.9372 1924	0.9271 7495	0.9221 9632	0.9172 4648	13
14	0.9325 5646	0.9217 9780	0.9164 6840	0.9111 7200	14
15	0.9279 1688	0.9164 5183	0.9107 7604	0.9051 3775	15
16	0.9233 0037	0.9111 3686	0.9051 1905	0.8991 4346	16
17	0.9187 0684	0.9058 5272	0.8994 9719	0.8931 8886	17
18	0.9141 3616	0.9005 9923	0.8939 1025	0.8872 7371	18
19	0.9095 8822	0.8953 7620	0.8883 5802	0.8813 9772	19
20	0.9050 6290	0.8901 8346	0.8828 4027	0.8755 6065	20
21	0.9005 6010	0.8850 2084	0.8773 5679	0.8697 6224	21
22	0.8960 7971	0.8798 8816	0.8719 0736	0.8640 0222	22
23	0.8916 2160	0.8747 8525	0.8664 9179	0.8582 8035	23
24	0.8871 8567	0.8697 1193	0.8611 0985	0.8525 9638	24
25	0.8827 7181	0.8646 6803	0.8557 6135	0.8469 5004	25
26	0.8783 7991	0.8596 5339	0.8504 4606	0.8413 4110	26
27	0.8740 0986	0.8546 6782	0.8451 6378	0.8357 6931	27
28	0.8696 6155	0.8497 1118	0.8399 1432	0.8302 3441	28
29	0.8653 3488	0.8447 8327	0.8346 9746	0.8247 3617	29
30	0.8610 2973	0.8398 8395	0.8295 1300	0.8192 7434	30
31	0.8567 4600	0.8350 1304	0.8243 6075	0.8138 4868	31
32	0.8524 8358	0.8301 7038	0.8192 4050	0.8084 5896	32
33	0.8482 4237	0.8253 5581	0.8141 5205	0.8031 0492	33
34	0.8440 2226	0.8205 6915	0.8090 9520	0.7977 8635	34
35	0.8398 2314	0.8158 1026	0.8040 6976	0.7925 0299	35
36	0.8356 4492	0.8110 7897	0.7990 7554	0.7872 5463	36
37	0.8314 8748	0.8063 7511	0.7941 1234	0.7820 4102	37
38	0.8273 5073	0.8016 9854	0.7891 7997	0.7768 6194	38
39	0.8232 3455	0.7970 4908	0.7842 7823	0.7717 1716	39
40	0.8191 3886	0.7924 2660	0.7794 0693	0.7666 0645	40
41	0.8150 6354	0.7878 3092	0.7745 6590	0.7615 2959	41
42	0.8110 0850	0.7832 6189	0.7697 5493	0.7564 8635	42
43	0.8069 7363	0.7787 1936	0.7649 7384	0.7514 7650	43
44	0.8029 5884	0.7742 0317	0.7602 2245	0.7464 9984	44
45	0.7989 6402	0.7697 1318	0.7555 0057	0.7415 5613	45
46	0.7949 8907	0.7652 4923	0.7508 0802	0.7366 4516	46
47	0.7910 3390	0.7608 1116	0.7461 4462	0.7317 6672	47
48	0.7870 9841	0.7563 9884	0.7415 1018	0.7269 2058	48
49	0.7831 8250	0.7520 1210	0.7369 0453	0.7221 0654	49
50	0.7792 8607	0.7476 5080	0.7323 2748	0.7173 2437	50

TABLE C-6 PRESENT VALUE OF 1 AT COMPOUND INTEREST (cont.)

$$v^n = (1+i)^{-n}$$

n	1/2%	7/12%	5/8%	2/3%	n
51	0.7754 0902	0.7433 1480	0.7277 7886	0.7125 7388	51
52	0.7715 5127	0.7390 0394	0.7232 5849	0.7078 5485	52
53	0.7677 1270	0.7347 1809	0.7187 6620	0.7031 6707	53
54	0.7638 9324	0.7304 5709	0.7143 0182	0.6985 1033	54
55	0.7600 9277	0.7262 2080	0.7098 6516	0.6938 8444	55
56	0.7563 1122	0.7220 0908	0.7054 5606	0.6892 8918	56
57	0.7525 4847	0.7178 2179	0.7010 7434	0.6847 2435	57
58	0.7488 0445	0.7136 5878	0.6967 1985	0.6801 8975	58
59	0.7450 7906	0.7095 1991	0.6923 9239	0.6756 8518	59
60	0.7413 7220	0.7054 0505	0.6880 9182	0.6712 1044	60
61	0.7376 8378	0.7013 1405	0.6838 1796	0.6667 6534	61
62	0.7340 1371	0.6972 4678	0.6795 7064	0.6623 4968	62
63	0.7303 6190	0.6932 0310	0.6753 4970	0.6579 6326	63
64	0.7267 2826	0.6891 8286	0.6711 5499	0.6536 0588	64
65	0.7231 1269	0.6851 8594	0.6669 8632	0.6492 7737	65
66	0.7195 1512	0.6812 1221	0.6628 4355	0.6449 7752	66
67	0.7159 3544	0.6772 6151	0.6587 2651	0.6407 0614	67
68	0.7123 7357	0.6733 3373	0.6546 3504	0.6364 6306	68
69	0.7088 2943	0.6694 2873	0.6505 6898	0.6322 4807	69
70	0.7053 0291	0.6655 4638	0.6465 2818	0.6280 6100	70
71	0.7017 9394	0.6616 8654	0.6425 1248	0.6239 0165	71
72	0.6983 0243	0.6578 4909	0.6385 2172	0.6197 6985	72
73	0.6948 2829	0.6540 3389	0.6345 5574	0.6156 6541	73
74	0.6913 7143	0.6502 4082	0.6306 1440	0.6115 8816	74
75	0.6879 3177	0.6464 6975	0.6266 9754	0.6075 3791	75
76	0.6845 0923	0.6427 2054	0.6228 0501	0.6035 1448	76
77	0.6811 0371	0.6389 9308	0.6189 3666	0.5995 1769	77
78	0.6777 1513	0.6352 8724	0.6150 9233	0.5955 4738	78
79	0.6743 4342	0.6316 0289	0.6112 7188	0.5916 0336	79
80	0.6709 8847	0.6279 3991	0.6074 7516	0.5876 8545	80
81	0.6676 5022	0.6242 9817	0.6037 0203	0.5837 9350	81
82	0.6643 2858	0.6206 7755	0.5999 5232	0.5799 2732	82
83	0.6610 2346	0.6170 7793	0.5962 2591	0.5760 8674	83
84	0.6577 3479	0.6134 9919	0.5925 2264	0.5722 7159	84
85	0.6544 6248	0.6099 4120	0.5888 4238	0.5684 8171	85
86	0.6512 0644	0.6064 0384	0.5851 8497	0.5647 1693	86
87	0.6479 6661	0.6028 8700	0.5815 5028	0.5609 7709	87
88	0.6447 4290	0.5993 9056	0.5779 3817	0.5572 6201	88
89	0.6415 3522	0.5959 1439	0.5743 4849	0.5535 7153	89
90	0.6383 4350	0.5924 5838	0.5707 8111	0.5499 0549	90
91	0.6351 6766	0.5890 2242	0.5672 3589	0.5462 6374	91
92	0.6320 0763	0.5856 0638	0.5637 1268	0.5426 4610	92
93	0.6288 6331	0.5822 1015	0.5602 1136	0.5390 5241	93
94	0.6257 3464	0.5788 3363	0.5567 3179	0.5354 8253	94
95	0.6226 2153	0.5754 7668	0.5532 7383	0.5319 3629	95
96	0.6195 2391	0.5721 3920	0.5498 3734	0.5284 1353	96
97	0.6164 4170	0.5688 2108	0.5464 2220	0.5249 1410	97
98	0.6133 7483	0.5655 2220	0.5430 2828	0.5214 3785	98
99	0.6103 2321	0.5622 4245	0.5396 5543	0.5179 8462	99
100	0.6072 8678	0.5589 8172	0.5363 0353	0.5145 5426	100

TABLE C-6 PRESENT VALUE OF 1 AT COMPOUND INTEREST (cont.)

$$v^n = (1+i)^{-n}$$

n	¾%	⅞%	1%	1⅛%	n
1	0.9925 5583	0.9913 2590	0.9900 9901	0.9888 7515	1
2	0.9851 6708	0.9827 2704	0.9802 9605	0.9778 7407	2
3	0.9778 3333	0.9742 0276	0.9705 9015	0.9669 9537	3
4	0.9705 5417	0.9657 5243	0.9609 8034	0.9562 3770	4
5	0.9633 2920	0.9573 7539	0.9514 6569	0.9455 9970	5
6	0.9561 5802	0.9490 7102	0.9420 4524	0.9350 8005	6
7	0.9490 4022	0.9408 3868	0.9327 1805	0.9246 7743	7
8	0.9419 7540	0.9326 7775	0.9234 8322	0.9143 9054	8
9	0.9349 6318	0.9245 8761	0.9143 3982	0.9042 1808	9
10	0.9280 0315	0.9165 6765	0.9052 8695	0.8941 5881	10
11	0.9210 9494	0.9086 1724	0.8963 2372	0.8842 1142	11
12	0.9142 3815	0.9007 3581	0.8874 4923	0.8743 7470	12
13	0.9074 3241	0.8929 2273	0.8786 6260	0.8646 4742	13
14	0.9006 7733	0.8851 7743	0.8699 6297	0.8550 2835	14
15	0.8939 7254	0.8774 9931	0.8613 4947	0.8455 1629	15
16	0.8873 1766	0.8698 8779	0.8528 2126	0.8361 1005	16
17	0.8807 1231	0.8623 4230	0.8443 7749	0.8268 0846	17
18	0.8741 5614	0.8548 6225	0.8360 1731	0.8176 1034	18
19	0.8676 4878	0.8474 4709	0.8277 3992	0.8085 1455	19
20	0.8611 8985	0.8400 9624	0.8195 4447	0.7995 1995	20
21	0.8547 7901	0.8328 0917	0.8114 3017	0.7906 2542	21
22	0.8484 1589	0.8255 8530	0.8033 9621	0.7818 2983	22
23	0.8421 0014	0.8184 2409	0.7954 4179	0.7731 3210	23
24	0.8358 3140	0.8113 2499	0.7875 6613	0.7645 3112	24
25	0.8296 0933	0.8042 8748	0.7797 6844	0.7560 2583	25
26	0.8234 3358	0.7973 1101	0.7720 4796	0.7476 1516	26
27	0.8173 0380	0.7903 9505	0.7644 0392	0.7392 9806	27
28	0.8112 1966	0.7835 3908	0.7568 3557	0.7310 7348	28
29	0.8051 8080	0.7767 4258	0.7493 4215	0.7229 4040	29
30	0.7991 8690	0.7700 0504	0.7419 2292	0.7148 9780	30
31	0.7932 3762	0.7633 2594	0.7345 7715	0.7069 4467	31
32	0.7873 3262	0.7567 0477	0.7273 0411	0.6990 8002	32
33	0.7814 7158	0.7501 4104	0.7201 0307	0.6913 0287	33
34	0.7756 5418	0.7436 3424	0.7129 7334	0.6836 1223	34
35	0.7698 8008	0.7371 8388	0.7059 1420	0.6760 0715	35
36	0.7641 4896	0.7307 8947	0.6989 2495	0.6684 8667	36
37	0.7584 6051	0.7244 5053	0.6920 0490	0.6610 4986	37
38	0.7528 1440	0.7181 6657	0.6851 5337	0.6536 9578	38
39	0.7472 1032	0.7119 3712	0.6783 6967	0.6464 2352	39
40	0.7416 4796	0.7057 6171	0.6716 5314	0.6392 3216	40
41	0.7361 2701	0.6996 3986	0.6650 0311	0.6321 2080	41
42	0.7306 4716	0.6935 7111	0.6584 1892	0.6250 8855	42
43	0.7252 0809	0.6875 5500	0.6518 9992	0.6181 3454	43
44	0.7198 0952	0.6815 9108	0.6454 4546	0.6112 5789	44
45	0.7144 5114	0.6756 7889	0.6390 5492	0.6044 5774	45
46	0.7091 3264	0.6698 1798	0.6327 2764	0.5977 3324	46
47	0.7038 5374	0.6640 0792	0.6264 6301	0.5910 8355	47
48	0.6986 1414	0.6582 4824	0.6202 6041	0.5845 0784	48
49	0.6934 1353	0.6525 3853	0.6141 1921	0.5780 0528	49
50	0.6882 5165	0.6468 7835	0.6080 3882	0.5715 7506	50

TABLE C-6 PRESENT VALUE OF 1 AT COMPOUND INTEREST (cont.)

$$v^n = (1+i)^{-n}$$

n	3/4%	7/8%	1%	1 1/8%	n
51	0.6831 2819	0.6412 6726	0.6020 1864	0.5652 1637	51
52	0.6780 4286	0.6357 0484	0.5960 5806	0.5589 2843	52
53	0.6729 9540	0.6301 9067	0.5901 5649	0.5527 1044	53
54	0.6679 8551	0.6247 2433	0.5843 1336	0.5465 6162	54
55	0.6630 1291	0.6193 0541	0.5785 2808	0.5404 8120	55
56	0.6580 7733	0.6139 3349	0.5728 0008	0.5344 6843	56
57	0.6531 7849	0.6086 0817	0.5671 2879	0.5285 2256	57
58	0.6483 1612	0.6033 2904	0.5615 1365	0.5226 4282	58
59	0.6434 8995	0.5980 9571	0.5559 5411	0.5168 2850	59
60	0.6386 9970	0.5929 0776	0.5504 4962	0.5110 7887	60
61	0.6339 4511	0.5877 6482	0.5449 9962	0.5053 9319	61
62	0.6292 2592	0.5826 6649	0.5396 0358	0.4997 7077	62
63	0.6245 4185	0.5776 1238	0.5342 6097	0.4942 1090	63
64	0.6198 9266	0.5726 0211	0.5289 7126	0.4887 1288	64
65	0.6152 7807	0.5676 3530	0.5237 3392	0.4832 7602	65
66	0.6106 9784	0.5627 1158	0.5185 4844	0.4778 9965	66
67	0.6061 5170	0.5578 3056	0.5134 1429	0.4725 8309	67
68	0.6016 3940	0.5529 9188	0.5083 3099	0.4673 2568	68
69	0.5971 6070	0.5481 9517	0.5032 9801	0.4621 2675	69
70	0.5927 1533	0.5434 4007	0.4983 1486	0.4569 8566	70
71	0.5883 0306	0.5387 2622	0.4933 8105	0.4519 0177	71
72	0.5839 2363	0.5340 5325	0.4884 9609	0.4468 7443	72
73	0.5795 7681	0.5294 2082	0.4836 5949	0.4419 0302	73
74	0.5752 6234	0.5248 2857	0.4788 7078	0.4369 8692	74
75	0.5709 7998	0.5202 7615	0.4741 2949	0.4321 2551	75
76	0.5667 2952	0.5157 6322	0.4694 3514	0.4273 1818	76
77	0.5625 1069	0.5112 8944	0.4647 8726	0.4225 6433	77
78	0.5583 2326	0.5068 5447	0.4601 8541	0.4178 6337	78
79	0.5541 6701	0.5024 5796	0.4556 2912	0.4132 1470	79
80	0.5500 4170	0.4980 9959	0.4511 1794	0.4086 1775	80
81	0.5459 4710	0.4937 7902	0.4466 5142	0.4040 7194	81
82	0.5418 8297	0.4894 9593	0.4422 2913	0.3995 7670	82
83	0.5378 4911	0.4852 4999	0.4378 5063	0.3951 3148	83
84	0.5338 4527	0.4810 4089	0.4335 1547	0.3907 3570	84
85	0.5298 7123	0.4768 6829	0.4292 2324	0.3863 8882	85
86	0.5259 2678	0.4727 3188	0.4249 7350	0.3820 9031	86
87	0.5220 1169	0.4686 3136	0.4207 6585	0.3778 3961	87
88	0.5181 2575	0.4645 6640	0.4165 9985	0.3736 3621	88
89	0.5142 6873	0.4605 3671	0.4124 7510	0.3694 7956	89
90	0.5104 4043	0.4565 4197	0.4083 9119	0.3653 6916	90
91	0.5066 4063	0.4525 8187	0.4043 4771	0.3613 0448	91
92	0.5028 6911	0.4486 5613	0.4003 4427	0.3572 8503	92
93	0.4991 2567	0.4447 6444	0.3963 8046	0.3533 1029	93
94	0.4954 1009	0.4409 0651	0.3924 5590	0.3493 7976	94
95	0.4917 2217	0.4370 8204	0.3885 7020	0.3454 9297	95
96	0.4880 6171	0.4332 9075	0.3847 2297	0.3416 4941	96
97	0.4844 2850	0.4295 3234	0.3809 1383	0.3378 4861	97
98	0.4808 2233	0.4258 0654	0.3771 4241	0.3340 9010	98
99	0.4772 4301	0.4221 1305	0.3734 0832	0.3303 7340	99
100	0.4736 9033	0.4184 5159	0.3697 1121	0.3266 9805	100

TABLE C-6 PRESENT VALUE OF 1 AT COMPOUND INTEREST (*cont.*)

$$v^n = (1 + i)^{-n}$$

n	1¼%	1⅜%	1½%	1¾%	n
1	0.9876 5432	0.9864 3650	0.9852 2167	0.9828 0098	1
2	0.9754 6106	0.9730 5696	0.9706 6175	0.9658 9777	2
3	0.9634 1833	0.9598 5890	0.9563 1699	0.9492 8528	3
4	0.9515 2428	0.9468 3986	0.9421 8423	0.9329 5851	4
5	0.9397 7706	0.9339 9739	0.9282 6033	0.9169 1254	5
6	0.9281 7488	0.9213 2912	0.9145 4219	0.9011 4254	6
7	0.9167 1593	0.9088 3267	0.9010 2679	0.8856 4378	7
8	0.9053 9845	0.8965 0571	0.8877 1112	0.8704 1157	8
9	0.8942 2069	0.8843 4596	0.8745 9224	0.8554 4135	9
10	0.8831 8093	0.8723 5113	0.8616 6723	0.8407 2860	10
11	0.8722 7746	0.8605 1899	0.8489 3323	0.8262 6889	11
12	0.8615 0860	0.8488 4734	0.8363 8742	0.8120 5788	12
13	0.8508 7269	0.8373 3400	0.8240 2702	0.7980 9128	13
14	0.8403 6809	0.8259 7682	0.8118 4928	0.7843 6490	14
15	0.8299 9318	0.8147 7368	0.7998 5150	0.7708 7459	15
16	0.8197 4635	0.8037 2250	0.7880 3104	0.7576 1631	16
17	0.8096 2602	0.7928 2120	0.7763 8526	0.7445 8605	17
18	0.7996 3064	0.7820 6777	0.7649 1159	0.7317 7990	18
19	0.7897 5866	0.7714 6020	0.7536 0747	0.7191 9401	19
20	0.7800 0855	0.7609 9649	0.7424 7042	0.7068 2458	20
21	0.7703 7881	0.7506 7472	0.7314 9795	0.6946 6789	21
22	0.7608 6796	0.7404 9294	0.7206 8763	0.6827 2028	22
23	0.7514 7453	0.7304 4926	0.7100 3708	0.6709 7817	23
24	0.7421 9707	0.7205 4181	0.6995 4392	0.6594 3800	24
25	0.7330 3414	0.7107 6874	0.6892 0583	0.6480 9632	25
26	0.7239 8434	0.7011 2823	0.6790 2052	0.6369 4970	26
27	0.7150 4626	0.6916 1847	0.6689 8574	0.6259 9479	27
28	0.7062 1853	0.6822 3771	0.6590 9925	0.6152 2829	28
29	0.6974 9978	0.6729 8417	0.6493 5887	0.6046 4697	29
30	0.6888 8867	0.6638 5615	0.6397 6243	0.5942 4764	30
31	0.6803 8387	0.6548 5194	0.6303 0781	0.5840 2716	31
32	0.6719 8407	0.6459 6985	0.6209 9292	0.5739 8247	32
33	0.6636 8797	0.6372 0824	0.6118 1568	0.5641 1053	33
34	0.6554 9429	0.6285 6546	0.6027 7407	0.5544 0839	34
35	0.6474 0177	0.6200 3991	0.5938 6608	0.5448 7311	35
36	0.6394 0916	0.6116 3000	0.5850 8974	0.5355 0183	36
37	0.6315 1522	0.6033 3416	0.5764 4309	0.5262 9172	37
38	0.6237 1873	0.5951 5083	0.5679 2423	0.5172 4002	38
39	0.6160 1850	0.5870 7850	0.5595 3126	0.5083 4400	39
40	0.6084 1334	0.5791 1566	0.5512 6232	0.4996 0098	40
41	0.6009 0206	0.5712 6083	0.5431 1559	0.4910 0834	41
42	0.5934 8352	0.5635 1253	0.5350 8925	0.4825 6348	42
43	0.5861 5656	0.5558 6933	0.5271 8153	0.4742 6386	43
44	0.5789 2006	0.5483 2979	0.5193 9067	0.4661 0699	44
45	0.5717 7290	0.5408 9252	0.5117 1494	0.4580 9040	45
46	0.5647 1397	0.5335 5612	0.5041 5265	0.4502 1170	46
47	0.5577 4219	0.5263 1923	0.4967 0212	0.4424 6850	47
48	0.5508 5649	0.5191 8050	0.4893 6170	0.4348 5848	48
49	0.5440 5579	0.5121 3860	0.4821 2975	0.4273 7934	49
50	0.5373 3905	0.5051 9220	0.4750 0468	0.4200 2883	50

TABLE C-6 PRESENT VALUE OF 1 AT COMPOUND INTEREST (cont.)

$$v^n = (1+i)^{-n}$$

n	$1\frac{1}{4}\%$	$1\frac{3}{8}\%$	$1\frac{1}{2}\%$	$1\frac{3}{4}\%$	n
51	0.5307 0524	0.4983 4003	0.4679 8491	0.4128 0475	51
52	0.5241 5332	0.4915 8079	0.4610 6887	0.4057 0492	52
53	0.5176 8229	0.4849 1323	0.4542 5505	0.3987 2719	53
54	0.5112 9115	0.4783 3611	0.4475 4192	0.3918 6947	54
55	0.5049 7892	0.4718 4820	0.4409 2800	0.3851 2970	55
56	0.4987 4461	0.4654 4829	0.4344 1182	0.3785 0585	56
57	0.4925 8727	0.4591 3518	0.4279 9194	0.3719 9592	57
58	0.4865 0594	0.4529 0770	0.4216 6694	0.3655 9796	58
59	0.4804 9970	0.4467 6468	0.4154 3541	0.3593 1003	59
60	0.4745 6760	0.4407 0499	0.4092 9597	0.3531 3025	60
61	0.4687 0874	0.4347 2749	0.4032 4726	0.3470 5676	61
62	0.4629 2222	0.4288 3106	0.3972 8794	0.3410 8772	62
63	0.4572 0713	0.4230 1461	0.3914 1669	0.3352 2135	63
64	0.4515 6259	0.4172 7705	0.3856 3221	0.3294 5587	64
65	0.4459 8775	0.4116 1731	0.3799 3321	0.3237 8956	65
66	0.4404 8173	0.4060 3434	0.3743 1843	0.3182 2069	66
67	0.4350 4368	0.4005 2709	0.3687 8663	0.3127 4761	67
68	0.4296 7277	0.3950 9454	0.3633 3658	0.3073 6866	68
69	0.4243 6817	0.3897 3568	0.3579 6708	0.3020 8222	69
70	0.4191 2905	0.3844 4949	0.3526 7692	0.2968 8670	70
71	0.4139 5462	0.3792 3501	0.3474 6495	0.2917 8054	71
72	0.4088 4407	0.3740 9126	0.3423 3000	0.2867 6221	72
73	0.4037 9661	0.3690 1727	0.3372 7093	0.2818 3018	73
74	0.3988 1147	0.3640 1210	0.3322 8663	0.2769 8298	74
75	0.3938 8787	0.3590 7483	0.3273 7599	0.2722 1914	75
76	0.3890 2506	0.3542 0451	0.3225 3793	0.2675 3724	76
77	0.3842 2228	0.3494 0026	0.3177 7136	0.2629 3586	77
78	0.3794 7879	0.3446 6117	0.3130 7523	0.2584 1362	78
79	0.3747 9387	0.3399 8636	0.3084 4850	0.2539 6916	79
80	0.3701 6679	0.3353 7495	0.3038 9015	0.2496 0114	80
81	0.3655 9683	0.3308 2609	0.2993 9916	0.2453 0825	81
82	0.3610 8329	0.3263 3893	0.2949 7454	0.2410 8919	82
83	0.3566 2547	0.3219 1263	0.2906 1531	0.2369 4269	83
84	0.3522 2268	0.3175 4637	0.2863 2050	0.2328 6751	84
85	0.3478 7426	0.3132 3933	0.2820 8917	0.2288 6242	85
86	0.3435 7951	0.3089 9071	0.2779 2036	0.2249 2621	86
87	0.3393 3779	0.3047 9971	0.2738 1316	0.2210 5770	87
88	0.3351 4843	0.3006 6556	0.2697 6666	0.2172 5572	88
89	0.3310 1080	0.2965 8748	0.2657 7997	0.2135 1914	89
90	0.3269 2425	0.2925 6472	0.2618 5218	0.2098 4682	90
91	0.3228 8814	0.2885 9652	0.2579 8245	0.2062 3766	91
92	0.3189 0187	0.2846 8214	0.2541 6990	0.2026 9057	92
93	0.3149 6481	0.2808 2085	0.2504 1369	0.1992 0450	93
94	0.3110 7636	0.2770 1194	0.2467 1300	0.1957 7837	94
95	0.3072 3591	0.2732 5468	0.2430 6699	0.1924 1118	95
96	0.3034 4287	0.2695 4839	0.2394 7487	0.1891 0190	96
97	0.2996 9666	0.2658 9237	0.2359 3583	0.1858 4953	97
98	0.2959 9670	0.2622 8594	0.2324 4909	0.1826 5310	98
99	0.2923 4242	0.2587 2843	0.2290 1389	0.1795 1165	99
100	0.2887 3326	0.2552 1916	0.2256 2944	0.1764 2422	100

TABLE C-6 PRESENT VALUE OF 1 AT COMPOUND INTEREST (*cont.*)

$$v^n = (1+i)^{-n}$$

n	2%	2¼%	2½%	2¾%	n
1	0.9803 9216	0.9779 9511	0.9756 0976	0.9732 3601	1
2	0.9611 6878	0.9564 7444	0.9518 1440	0.9471 8833	2
3	0.9423 2233	0.9354 2732	0.9285 9941	0.9218 3779	3
4	0.9238 4543	0.9148 4335	0.9059 5064	0.8971 6573	4
5	0.9057 3081	0.8947 1232	0.8838 5429	0.8731 5400	5
6	0.8879 7138	0.8750 2427	0.8622 9687	0.8497 8491	6
7	0.8705 6018	0.8557 6946	0.8412 6524	0.8270 4128	7
8	0.8534 9037	0.8369 3835	0.8207 4657	0.8049 0635	8
9	0.8367 5527	0.8185 2161	0.8007 2836	0.7833 6385	9
10	0.8203 4830	0.8005 1013	0.7811 9840	0.7623 9791	10
11	0.8042 6304	0.7828 9499	0.7621 4478	0.7419 9310	11
12	0.7884 9318	0.7656 6748	0.7435 5589	0.7221 3440	12
13	0.7730 3253	0.7488 1905	0.7254 2038	0.7028 0720	13
14	0.7578 7502	0.7323 4137	0.7077 2720	0.6839 9728	14
15	0.7430 1473	0.7162 2628	0.6904 6556	0.6656 9078	15
16	0.7284 4581	0.7004 6580	0.6736 2493	0.6478 7424	16
17	0.7141 6256	0.6850 5212	0.6571 9506	0.6305 3454	17
18	0.7001 5937	0.6699 7763	0.6411 6591	0.6136 5892	18
19	0.6864 3076	0.6552 3484	0.6255 2772	0.5972 3496	19
20	0.6729 7133	0.6408 1647	0.6102 7094	0.5812 5057	20
21	0.6597 7582	0.6267 1538	0.5953 8629	0.5656 9398	21
22	0.6468 3904	0.6129 2457	0.5808 6467	0.5505 5375	22
23	0.6341 5592	0.5994 3724	0.5666 9724	0.5358 1874	23
24	0.6217 2149	0.5862 4668	0.5528 7535	0.5214 7809	24
25	0.6095 3087	0.5733 4639	0.5393 9059	0.5075 2126	25
26	0.5975 7928	0.5607 2997	0.5262 3472	0.4939 3796	26
27	0.5858 6204	0.5483 9117	0.5133 9973	0.4807 1821	27
28	0.5743 7455	0.5363 2388	0.5008 7778	0.4678 5227	28
29	0.5631 1231	0.5245 2213	0.4886 6125	0.4553 3068	29
30	0.5520 7089	0.5129 8008	0.4767 4269	0.4431 4421	30
31	0.5412 4597	0.5016 9201	0.4651 1481	0.4312 8391	31
32	0.5306 3330	0.4906 5233	0.4537 7055	0.4197 4103	32
33	0.5202 2873	0.4798 5558	0.4427 0298	0.4085 0708	33
34	0.5100 2817	0.4692 9641	0.4319 0534	0.3975 7380	34
35	0.5000 2761	0.4589 6960	0.4213 7107	0.3869 3314	35
36	0.4902 2315	0.4488 7002	0.4110 9372	0.3765 7727	36
37	0.4806 1093	0.4389 9268	0.4010 6705	0.3664 9856	37
38	0.4711 8719	0.4293 3270	0.3912 8492	0.3566 8959	38
39	0.4619 4822	0.4198 8528	0.3817 4139	0.3471 4316	39
40	0.4528 9042	0.4106 4575	0.3724 3062	0.3378 5222	40
41	0.4440 1021	0.4016 0954	0.3633 4695	0.3288 0995	41
42	0.4353 0413	0.3927 7216	0.3544 8483	0.3200 0968	42
43	0.4267 6875	0.3841 2925	0.3458 3886	0.3114 4495	43
44	0.4184 0074	0.3756 7653	0.3374 0376	0.3031 0944	44
45	0.4101 9680	0.3674 0981	0.3291 7440	0.2949 9702	45
46	0.4021 5373	0.3593 2500	0.3211 4576	0.2871 0172	46
47	0.3942 6836	0.3514 1809	0.3133 1294	0.2794 1773	47
48	0.3865 3761	0.3436 8518	0.3056 7116	0.2719 3940	48
49	0.3789 5844	0.3361 2242	0.2982 1576	0.2646 6122	49
50	0.3715 2788	0.3287 2608	0.2909 4221	0.2575 7783	50

TABLE C-6 PRESENT VALUE OF 1 AT COMPOUND INTEREST (*cont.*)

$$v^n = (1+i)^{-n}$$

n	2%	2¼%	2½%	2¾%	n
51	0.3642 4302	0.3214 9250	0.2838 4606	0.2506 8402	51
52	0.3571 0100	0.3144 1810	0.2769 2298	0.2439 7471	52
53	0.3500 9902	0.3074 9936	0.2701 6876	0.2374 4497	53
54	0.3432 3433	0.3007 3287	0.2635 7928	0.2310 9000	54
55	0.3365 0425	0.2941 1528	0.2571 5052	0.2249 0511	55
56	0.3299 0613	0.2876 4330	0.2508 7855	0.2188 8575	56
57	0.3234 3738	0.2813 1374	0.2447 5956	0.2130 2749	57
58	0.3170 9547	0.2751 2347	0.2387 8982	0.2073 2603	58
59	0.3108 7791	0.2690 6940	0.2329 6568	0.2017 7716	59
60	0.3047 8227	0.2631 4856	0.2272 8359	0.1963 7679	60
61	0.2988 0614	0.2573 5801	0.2217 4009	0.1911 2097	61
62	0.2929 4720	0.2516 9487	0.2163 3179	0.1860 0581	62
63	0.2872 0314	0.2461 5635	0.2110 5541	0.1810 2755	63
64	0.2815 7170	0.2407 3971	0.2059 0771	0.1761 8253	64
65	0.2760 5069	0.2354 4226	0.2008 8557	0.1714 6718	65
66	0.2706 3793	0.2302 6138	0.1959 8593	0.1668 7804	66
67	0.2653 3130	0.2251 9450	0.1912 0578	0.1624 1172	67
68	0.2601 2873	0.2202 3912	0.1865 4223	0.1580 6493	68
69	0.2550 2817	0.2153 9278	0.1819 9241	0.1538 3448	69
70	0.2500 2761	0.2106 5309	0.1775 5358	0.1497 1726	70
71	0.2451 2511	0.2060 1769	0.1732 2300	0.1457 1023	71
72	0.2403 1874	0.2014 8429	0.1689 9805	0.1418 1044	72
73	0.2356 0661	0.1970 5065	0.1648 7615	0.1380 1503	73
74	0.2309 8687	0.1927 1458	0.1608 5478	0.1343 2119	74
75	0.2264 5771	0.1884 7391	0.1569 3149	0.1307 2622	75
76	0.2220 1737	0.1843 2657	0.1531 0389	0.1272 2747	76
77	0.2176 6408	0.1802 7048	0.1493 6965	0.1238 2235	77
78	0.2133 9616	0.1763 0365	0.1457 2649	0.1205 0837	78
79	0.2092 1192	0.1724 2411	0.1421 7218	0.1172 8309	79
80	0.2051 0973	0.1686 2993	0.1387 0457	0.1141 4412	80
81	0.2010 8797	0.1649 1925	0.1353 2153	0.1110 8917	81
82	0.1971 4507	0.1612 9022	0.1320 2101	0.1081 1598	82
83	0.1932 7948	0.1577 4105	0.1288 0098	0.1052 2237	83
84	0.1894 8968	0.1542 6997	0.1256 5949	0.1024 0620	84
85	0.1857 7420	0.1508 7528	0.1225 9463	0.0996 6540	85
86	0.1821 3157	0.1475 5528	0.1196 0452	0.0969 9795	86
87	0.1785 6036	0.1443 0835	0.1166 8733	0.0944 0190	87
88	0.1750 5918	0.1411 3286	0.1138 4130	0.0918 7533	88
89	0.1716 2665	0.1380 2724	0.1110 6468	0.0894 1638	89
90	0.1682 6142	0.1349 8997	0.1083 5579	0.0870 2324	90
91	0.1649 6217	0.1320 1953	0.1057 1296	0.0846 9415	91
92	0.1617 2762	0.1291 1445	0.1031 3460	0.0824 2740	92
93	0.1585 5649	0.1262 7331	0.1006 1912	0.0802 2131	93
94	0.1554 4754	0.1234 9468	0.0981 6500	0.0780 7427	94
95	0.1523 9955	0.1207 7719	0.0957 7073	0.0759 8469	95
96	0.1494 1132	0.1181 1950	0.0934 3486	0.0739 5104	96
97	0.1464 8169	0.1155 2029	0.0911 5598	0.0719 7181	97
98	0.1436 0950	0.1129 7828	0.0889 3264	0.0700 4556	98
99	0.1407 9363	0.1104 9221	0.0867 6355	0.0681 7086	99
100	0.1380 3297	0.1080 6084	0.0846 4737	0.0663 4634	100

TABLE C-6 PRESENT VALUE OF 1 AT COMPOUND INTEREST (cont.)

$$v^n = (1+i)^{-n}$$

n	3%	3½%	4%	4½%	n
1	0.9708 7379	0.9661 8357	0.9615 3846	0.9569 3780	1
2	0.9425 9591	0.9335 1070	0.9245 5621	0.9157 2995	2
3	0.9151 4166	0.9019 4271	0.8889 9636	0.8762 9660	3
4	0.8884 8705	0.8714 4223	0.8548 0419	0.8385 6134	4
5	0.8626 0878	0.8419 7317	0.8219 2711	0.8024 5105	5
6	0.8374 8426	0.8135 0064	0.7903 1453	0.7678 9574	6
7	0.8130 9151	0.7859 9096	0.7599 1781	0.7348 2846	7
8	0.7894 0923	0.7594 1156	0.7306 9021	0.7031 8513	8
9	0.7664 1673	0.7337 3097	0.7025 8674	0.6729 0443	9
10	0.7440 9391	0.7089 1881	0.6755 6417	0.6439 2768	10
11	0.7224 2128	0.6849 4571	0.6495 8093	0.6161 9874	11
12	0.7013 7988	0.6617 8330	0.6245 9705	0.5896 6386	12
13	0.6809 5134	0.6394 0415	0.6005 7409	0.5642 7164	13
14	0.6611 1781	0.6177 8179	0.5774 7508	0.5399 7286	14
15	0.6418 6195	0.5968 9062	0.5552 6450	0.5167 2044	15
16	0.6231 6694	0.5767 0591	0.5339 0818	0.4944 6932	16
17	0.6050 1645	0.5572 0378	0.5133 7325	0.4731 7639	17
18	0.5873 9461	0.5383 6114	0.4936 2812	0.4528 0037	18
19	0.5702 8603	0.5201 5569	0.4746 4242	0.4333 0179	19
20	0.5536 7575	0.5025 6588	0.4563 8695	0.4146 4286	20
21	0.5375 4928	0.4855 7090	0.4388 3360	0.3967 8743	21
22	0.5218 9250	0.4691 5063	0.4219 5539	0.3797 0089	22
23	0.5066 9175	0.4532 8563	0.4057 2633	0.3633 5013	23
24	0.4919 3374	0.4379 5713	0.3901 2147	0.3477 0347	24
25	0.4776 0557	0.4231 4699	0.3751 1680	0.3327 3060	25
26	0.4636 9473	0.4088 3767	0.3606 8923	0.3184 0248	26
27	0.4501 8906	0.3950 1224	0.3468 1657	0.3046 9137	27
28	0.4370 7675	0.3816 5434	0.3334 7747	0.2915 7069	28
29	0.4243 4636	0.3687 4815	0.3206 5141	0.2790 1502	29
30	0.4119 8676	0.3562 7841	0.3083 1867	0.2670 0002	30
31	0.3999 8715	0.3442 3035	0.2964 6026	0.2555 0241	31
32	0.3883 3703	0.3325 8971	0.2850 5794	0.2444 9991	32
33	0.3770 2625	0.3213 4271	0.2740 9417	0.2339 7121	33
34	0.3660 4490	0.3104 7605	0.2635 5209	0.2238 9589	34
35	0.3553 8340	0.2999 7686	0.2534 1547	0.2142 5444	35
36	0.3450 3243	0.2898 3272	0.2436 6872	0.2050 2817	36
37	0.3349 8294	0.2800 3161	0.2342 9685	0.1961 9921	37
38	0.3252 2615	0.2705 6194	0.2252 8543	0.1877 5044	38
39	0.3157 5355	0.2614 1250	0.2166 2061	0.1796 6549	39
40	0.3065 5684	0.2525 7247	0.2082 8904	0.1719 2870	40
41	0.2976 2800	0.2440 3137	0.2002 7793	0.1645 2507	41
42	0.2889 5922	0.2357 7910	0.1925 7493	0.1574 4026	42
43	0.2805 4294	0.2278 0590	0.1851 6820	0.1506 6054	43
44	0.2723 7178	0.2201 0231	0.1780 4635	0.1441 7276	44
45	0.2644 3862	0.2126 5924	0.1711 9841	0.1379 6437	45
46	0.2567 3653	0.2054 6787	0.1646 1386	0.1320 2332	46
47	0.2492 5876	0.1985 1968	0.1582 8256	0.1263 3810	47
48	0.2419 9880	0.1918 0645	0.1521 9476	0.1208 9771	48
49	0.2349 5029	0.1853 2024	0.1463 4112	0.1156 9158	49
50	0.2281 0708	0.1790 5337	0.1407 1262	0.1107 0965	50

TABLE C-6 PRESENT VALUE OF 1 AT COMPOUND INTEREST (*cont.*)

$$v^n = (1 + i)^{-n}$$

n	3%	3½%	4%	4½%	n
51	0.2214 6318	0.1729 9843	0.1353 0059	0.1059 4225	51
52	0.2150 1280	0.1671 4824	0.1300 9672	0.1013 8014	52
53	0.2087 5029	0.1614 9589	0.1250 9300	0.0970 1449	53
54	0.2026 7019	0.1560 3467	0.1202 8173	0.0928 3683	54
55	0.1967 6717	0.1507 5814	0.1156 5551	0.0888 3907	55
56	0.1910 3609	0.1456 6004	0.1112 0722	0.0850 1347	56
57	0.1854 7193	0.1407 3433	0.1069 3002	0.0813 5260	57
58	0.1800 6984	0.1359 7520	0.1028 1733	0.0778 4938	58
59	0.1748 2508	0.1313 7701	0.0988 6282	0.0744 9701	59
60	0.1697 3309	0.1269 3431	0.0950 6040	0.0712 8901	60
61	0.1647 8941	0.1226 4184	0.0914 0423	0.0682 1915	61
62	0.1599 8972	0.1184 9453	0.0878 8868	0.0652 8148	62
63	0.1553 2982	0.1144 8747	0.0845 0835	0.0624 7032	63
64	0.1508 0565	0.1106 1591	0.0812 5803	0.0597 8021	64
65	0.1464 1325	0.1068 7528	0.0781 3272	0.0572 0594	65
66	0.1421 4879	0.1032 6114	0.0751 2762	0.0547 4253	66
67	0.1380 0853	0.0997 6922	0.0722 3809	0.0523 8519	67
68	0.1339 8887	0.0963 9538	0.0694 5970	0.0501 2937	68
69	0.1300 8628	0.0931 3563	0.0667 8818	0.0479 7069	69
70	0.1262 9736	0.0899 8612	0.0642 1940	0.0459 0497	70
71	0.1226 1880	0.0869 4311	0.0617 4942	0.0439 2820	71
72	0.1190 4737	0.0840 0300	0.0593 7445	0.0420 3655	72
73	0.1155 7998	0.0811 6232	0.0570 9081	0.0402 2637	73
74	0.1122 1357	0.0784 1770	0.0548 9501	0.0384 9413	74
75	0.1089 4521	0.0757 6590	0.0527 8367	0.0368 3649	75
76	0.1057 7205	0.0732 0376	0.0507 5353	0.0352 5023	76
77	0.1026 9131	0.0707 2827	0.0488 0147	0.0337 3228	77
78	0.0997 0030	0.0683 3650	0.0469 2449	0.0322 7969	78
79	0.0967 9641	0.0660 2560	0.0451 1970	0.0308 8965	79
80	0.0939 7710	0.0637 9285	0.0433 8433	0.0295 5948	80
81	0.0912 3990	0.0616 3561	0.0417 1570	0.0282 8658	81
82	0.0885 8243	0.0595 5131	0.0401 1125	0.0270 6850	82
83	0.0860 0236	0.0575 3750	0.0385 6851	0.0259 0287	83
84	0.0834 9743	0.0555 9178	0.0370 8510	0.0247 8744	84
85	0.0810 6547	0.0537 1187	0.0356 5875	0.0237 2003	85
86	0.0787 0434	0.0518 9553	0.0342 8726	0.0226 9860	86
87	0.0764 1198	0.0501 4060	0.0329 6852	0.0217 2115	87
88	0.0741 8639	0.0484 4503	0.0317 0050	0.0207 8579	88
89	0.0720 2562	0.0468 0679	0.0304 8125	0.0198 9070	89
90	0.0699 2779	0.0452 2395	0.0293 0890	0.0190 3417	90
91	0.0678 9105	0.0436 9464	0.0281 8163	0.0182 1451	91
92	0.0659 1364	0.0422 1704	0.0270 9772	0.0174 3016	92
93	0.0639 9383	0.0407 8941	0.0260 5550	0.0166 7958	93
94	0.0621 2993	0.0394 1006	0.0250 5337	0.0159 6132	94
95	0.0603 2032	0.0380 7735	0.0240 8978	0.0152 7399	95
96	0.0585 6342	0.0367 8971	0.0231 6325	0.0146 1626	96
97	0.0568 5769	0.0355 4562	0.0222 7235	0.0139 8685	97
98	0.0552 0164	0.0343 4359	0.0214 1572	0.0133 8454	98
99	0.0535 9383	0.0331 8221	0.0205 9204	0.0128 0817	99
100	0.0520 3284	0.0320 6011	0.0198 0004	0.0122 5663	100

TABLE C-6 PRESENT VALUE OF 1 AT COMPOUND INTEREST (*cont.*)

$$v^n = (1+i)^{-n}$$

n	5%	5½%	6%	6½%	n
1	0.9523 8095	0.9478 6730	0.9433 9623	0.9389 6714	1
2	0.9070 2948	0.8984 5242	0.8899 9644	0.8816 5928	2
3	0.8638 3760	0.8516 1366	0.8396 1928	0.8278 4909	3
4	0.8227 0247	0.8072 1674	0.7920 9366	0.7773 2309	4
5	0.7835 2617	0.7651 3435	0.7472 5817	0.7298 8084	5
6	0.7462 1540	0.7252 4583	0.7049 6054	0.6853 3412	6
7	0.7106 8133	0.6874 3681	0.6650 5711	0.6435 0521	7
8	0.6768 3936	0.6515 9887	0.6274 1237	0.6042 3119	8
9	0.6446 0892	0.6176 2926	0.5918 9846	0.5673 5323	9
10	0.6139 1325	0.5854 3058	0.5583 9478	0.5327 2604	10
11	0.5846 7929	0.5549 1050	0.5267 8753	0.5002 1224	11
12	0.5568 3742	0.5259 8152	0.4969 6936	0.4696 8285	12
13	0.5303 2135	0.4985 6068	0.4688 3902	0.4410 1676	13
14	0.5050 6795	0.4725 6937	0.4423 0096	0.4141 0025	14
15	0.4810 1710	0.4479 3305	0.4172 6506	0.3888 2652	15
16	0.4581 1152	0.4245 8109	0.3936 4628	0.3650 9533	16
17	0.4362 9669	0.4024 4653	0.3713 6442	0.3428 1251	17
18	0.4155 2065	0.3814 6590	0.3503 4379	0.3218 8969	18
19	0.3957 3396	0.3615 7906	0.3305 1301	0.3022 4384	19
20	0.3768 8948	0.3427 2896	0.3118 0473	0.2837 9703	20
21	0.3589 4236	0.3248 6158	0.2941 5540	0.2664 7608	21
22	0.3418 4987	0.3079 2567	0.2775 0510	0.2502 1228	22
23	0.3255 7131	0.2918 7267	0.2617 9726	0.2349 4111	23
24	0.3100 6791	0.2766 5656	0.2469 7855	0.2206 0198	24
25	0.2953 0277	0.2622 3370	0.2329 9863	0.2071 3801	25
26	0.2812 4073	0.2485 6275	0.2198 1003	0.1944 9579	26
27	0.2678 4832	0.2356 0450	0.2073 6795	0.1826 2515	27
28	0.2550 9364	0.2233 2181	0.1956 3014	0.1714 7902	28
29	0.2429 4632	0.2116 7944	0.1845 5674	0.1610 1316	29
30	0.2313 7745	0.2005 4402	0.1741 1013	0.1511 8607	30
31	0.2203 5947	0.1901 8390	0.1642 5484	0.1419 5875	31
32	0.2098 6617	0.1802 6910	0.1549 5740	0.1332 9460	32
33	0.1998 7254	0.1708 7119	0.1461 8622	0.1251 5925	33
34	0.1903 5480	0.1619 6321	0.1379 1153	0.1175 2042	34
35	0.1812 9029	0.1535 1963	0.1301 0522	0.1103 4781	35
36	0.1726 5741	0.1455 1624	0.1227 4077	0.1036 1297	36
37	0.1644 3563	0.1379 3008	0.1157 9318	0.0972 8917	37
38	0.1566 0536	0.1307 3941	0.1092 3885	0.0913 5134	38
39	0.1491 4797	0.1239 2362	0.1030 5552	0.0857 7590	39
40	0.1420 4568	0.1174 6314	0.0972 2219	0.0805 4075	40
41	0.1352 8160	0.1113 3947	0.0917 1905	0.0756 2512	41
42	0.1288 3962	0.1055 3504	0.0865 2740	0.0710 0950	42
43	0.1227 0440	0.1000 3322	0.0816 2962	0.0666 7559	43
44	0.1168 6133	0.0948 1822	0.0770 0908	0.0626 0619	44
45	0.1112 9651	0.0898 7509	0.0726 5007	0.0587 8515	45
46	0.1059 9668	0.0851 8965	0.0685 3781	0.0551 9733	46
47	0.1009 4921	0.0807 4849	0.0646 5831	0.0518 2848	47
48	0.0961 4211	0.0765 3885	0.0609 9840	0.0486 6524	48
49	0.0915 6391	0.0725 4867	0.0575 4566	0.0456 9506	49
50	0.0872 0373	0.0687 6652	0.0542 8836	0.0429 0616	50

TABLE C-6 PRESENT VALUE OF 1 AT COMPOUND INTEREST (cont.)

$$v^n = (1 + i)^{-n}$$

n	5%	5½%	6%	6½%	n
51	0.0830 5117	0.0651 8153	0.0512 1544	0.0402 8747	51
52	0.0790 9635	0.0617 8344	0.0483 1645	0.0378 2861	52
53	0.0753 2986	0.0585 6250	0.0455 8156	0.0355 1982	53
54	0.0717 4272	0.0555 0948	0.0430 0147	0.0333 5195	54
55	0.0683 2640	0.0526 1562	0.0405 6742	0.0313 1638	55
56	0.0650 7276	0.0498 7263	0.0382 7115	0.0294 0505	56
57	0.0619 7406	0.0472 7263	0.0361 0486	0.0276 1038	57
58	0.0590 2291	0.0448 0818	0.0340 6119	0.0259 2524	58
59	0.0562 1230	0.0424 7221	0.0321 3320	0.0243 4295	59
60	0.0535 3552	0.0402 5802	0.0303 1434	0.0228 5723	60
61	0.0509 8621	0.0381 5926	0.0285 9843	0.0214 6218	61
62	0.0485 5830	0.0361 6992	0.0269 7965	0.0201 5229	62
63	0.0462 4600	0.0342 8428	0.0254 5250	0.0189 2233	63
64	0.0440 4381	0.0324 9695	0.0240 1179	0.0177 6745	64
65	0.0419 4648	0.0308 0279	0.0226 5264	0.0166 8305	65
66	0.0399 4903	0.0291 9696	0.0213 7041	0.0156 6484	66
67	0.0380 4670	0.0276 7485	0.0201 6077	0.0147 0877	67
68	0.0362 3495	0.0262 3208	0.0190 1959	0.0138 1105	68
69	0.0345 0948	0.0248 6453	0.0179 4301	0.0129 6812	69
70	0.0328 6617	0.0235 6828	0.0169 2737	0.0121 7664	70
71	0.0313 0111	0.0223 3960	0.0159 6921	0.0114 3346	71
72	0.0298 1058	0.0211 7498	0.0150 6530	0.0107 3565	72
73	0.0283 9103	0.0200 7107	0.0142 1254	0.0100 8042	73
74	0.0270 3908	0.0190 2471	0.0134 0806	0.0094 6518	74
75	0.0257 5150	0.0180 3290	0.0126 4911	0.0088 8750	75
76	0.0245 2524	0.0170 9279	0.0119 3313	0.0083 4507	76
77	0.0233 5737	0.0162 0170	0.0112 5767	0.0078 3574	77
78	0.0222 4512	0.0153 5706	0.0106 2044	0.0073 5751	78
79	0.0211 8582	0.0145 5646	0.0100 1928	0.0069 0846	79
80	0.0201 7698	0.0137 9759	0.0094 5215	0.0064 8681	80
81	0.0192 1617	0.0130 7828	0.0089 1713	0.0060 9090	81
82	0.0183 0111	0.0123 9648	0.0084 1238	0.0057 1916	82
83	0.0174 2963	0.0117 5022	0.0079 3621	0.0053 7010	83
84	0.0165 9965	0.0111 3765	0.0074 8699	0.0050 4235	84
85	0.0158 0919	0.0105 5701	0.0070 6320	0.0047 3460	85
86	0.0150 5637	0.0100 0664	0.0066 6340	0.0044 4563	86
87	0.0143 3940	0.0094 8497	0.0062 8622	0.0041 7430	87
88	0.0136 5657	0.0089 9049	0.0059 3040	0.0039 1953	88
89	0.0130 0626	0.0085 2180	0.0055 9472	0.0036 8031	89
90	0.0123 8691	0.0080 7753	0.0052 7803	0.0034 5569	90
91	0.0117 9706	0.0076 5643	0.0049 7928	0.0032 4478	91
92	0.0112 3530	0.0072 5728	0.0046 9743	0.0030 4674	92
93	0.0107 0028	0.0068 7894	0.0044 3154	0.0028 6079	93
94	0.0101 9074	0.0065 2032	0.0041 8070	0.0026 8619	94
95	0.0097 0547	0.0061 8040	0.0039 4405	0.0025 2224	95
96	0.0092 4331	0.0058 5820	0.0037 2081	0.0023 6831	96
97	0.0088 0315	0.0055 5279	0.0035 1019	0.0022 2376	97
98	0.0083 8395	0.0052 6331	0.0033 1150	0.0020 8804	98
99	0.0079 8471	0.0049 8892	0.0031 2406	0.0019 6060	99
100	0.0076 0449	0.0047 2883	0.0029 4723	0.0018 4094	100

TABLE C-6 PRESENT VALUE OF 1 AT COMPOUND INTEREST (*cont.*)

$$v^n = (1+i)^{-n}$$

n	7%	7½%	8%	8½%	n
1	0.9345 7944	0.9302 3256	0.9259 2593	0.9216 5899	1
2	0.8734 3873	0.8653 3261	0.8573 3882	0.8494 5529	2
3	0.8162 9788	0.8049 6057	0.7938 3224	0.7829 0810	3
4	0.7628 9521	0.7488 0053	0.7350 2985	0.7215 7428	4
5	0.7129 8618	0.6965 5863	0.6805 8320	0.6650 4542	5
6	0.6663 4222	0.6479 6152	0.6301 6963	0.6129 4509	6
7	0.6227 4974	0.6027 5490	0.5834 9040	0.5649 2635	7
8	0.5820 0910	0.5607 0223	0.5402 6888	0.5206 6945	8
9	0.5439 3374	0.5215 8347	0.5002 4897	0.4798 7968	9
10	0.5083 4929	0.4851 9393	0.4631 9349	0.4422 8542	10
11	0.4750 9280	0.4513 4319	0.4288 8286	0.4076 3633	11
12	0.4440 1196	0.4198 5413	0.3971 1376	0.3757 0168	12
13	0.4149 6445	0.3905 6198	0.3676 9792	0.3462 6883	13
14	0.3878 1724	0.3633 1347	0.3404 6104	0.3191 4178	14
15	0.3624 4602	0.3379 6602	0.3152 4170	0.2941 3989	15
16	0.3387 3460	0.3143 8699	0.2918 9047	0.2710 9667	16
17	0.3165 7439	0.2924 5302	0.2702 6895	0.2498 5869	17
18	0.2958 6392	0.2720 4932	0.2502 4903	0.2302 8450	18
19	0.2765 0832	0.2530 6913	0.2317 1206	0.2122 4378	19
20	0.2584 1900	0.2354 1315	0.2145 4821	0.1956 1639	20
21	0.2415 1309	0.2189 8897	0.1986 5575	0.1802 9160	21
22	0.2257 1317	0.2037 1067	0.1839 4051	0.1661 6738	22
23	0.2109 4688	0.1894 9830	0.1703 1528	0.1531 4965	23
24	0.1971 4662	0.1762 7749	0.1576 9934	0.1411 5176	24
25	0.1842 4918	0.1639 7906	0.1460 1790	0.1300 9378	25
26	0.1721 9549	0.1525 3866	0.1352 0176	0.1199 0210	26
27	0.1609 3037	0.1418 9643	0.1251 8682	0.1105 0885	27
28	0.1504 0221	0.1319 9668	0.1159 1372	0.1018 5148	28
29	0.1405 6282	0.1227 8761	0.1073 2752	0.0938 7233	29
30	0.1313 6712	0.1142 2103	0.0993 7733	0.0865 1828	30
31	0.1227 7301	0.1062 5212	0.0920 1605	0.0797 4035	31
32	0.1147 4113	0.0988 3918	0.0852 0005	0.0734 9341	32
33	0.1072 3470	0.0919 4343	0.0788 8893	0.0677 3586	33
34	0.1002 1934	0.0855 2877	0.0730 4531	0.0624 2936	34
35	0.0936 6294	0.0795 6164	0.0676 3454	0.0575 3858	35
36	0.0875 3546	0.0740 1083	0.0626 2458	0.0530 3095	36
37	0.0818 0884	0.0688 4729	0.0579 8572	0.0488 7645	37
38	0.0764 5686	0.0640 4399	0.0536 9048	0.0450 4742	38
39	0.0714 5501	0.0595 7580	0.0497 1341	0.0415 1836	39
40	0.0667 8038	0.0554 1935	0.0460 3093	0.0382 6577	40
41	0.0624 1157	0.0515 5288	0.0426 2123	0.0352 6799	41
42	0.0583 2857	0.0479 5617	0.0394 6411	0.0325 0506	42
43	0.0545 1268	0.0446 1039	0.0365 4084	0.0299 5858	43
44	0.0509 4643	0.0414 9804	0.0338 3411	0.0276 1160	44
45	0.0476 1349	0.0386 0283	0.0313 2788	0.0254 4848	45
46	0.0444 9859	0.0359 0961	0.0290 0730	0.0234 5482	46
47	0.0415 8747	0.0334 0428	0.0268 5861	0.0216 1734	47
48	0.0388 6679	0.0310 7375	0.0248 6908	0.0199 2382	48
49	0.0363 2410	0.0289 0582	0.0230 2693	0.0183 6297	49
50	0.0339 4776	0.0268 8913	0.0213 2123	0.0169 2439	50

TABLE C-6 PRESENT VALUE OF 1 AT COMPOUND INTEREST (cont.)

$$v^n = (1+i)^{-n}$$

n	7%	7½%	8%	8½%	n
51	0.0317 2688	0.0250 1315	0.0197 4188	0.0155 9852	51
52	0.0296 5129	0.0232 6804	0.0182 7952	0.0143 7651	52
53	0.0277 1148	0.0216 4469	0.0169 2548	0.0132 5024	53
54	0.0258 9858	0.0201 3460	0.0156 7174	0.0122 1221	54
55	0.0242 0428	0.0187 2986	0.0145 1087	0.0112 5549	55
56	0.0226 2083	0.0174 2312	0.0134 3599	0.0103 7372	56
57	0.0211 4096	0.0162 0756	0.0124 4073	0.0095 6104	57
58	0.0197 5791	0.0150 7680	0.0115 1920	0.0088 1201	58
59	0.0184 6533	0.0140 2493	0.0106 6592	0.0081 2167	59
60	0.0172 5732	0.0130 4644	0.0098 7585	0.0074 8541	60
61	0.0161 2834	0.0121 3623	0.0091 4431	0.0068 9900	61
62	0.0150 7321	0.0112 8951	0.0084 6695	0.0063 5852	62
63	0.0140 8711	0.0105 0187	0.0078 3977	0.0058 6039	63
64	0.0131 6553	0.0097 6918	0.0072 5905	0.0054 0128	64
65	0.0123 0423	0.0090 8761	0.0067 2134	0.0049 7814	65
66	0.0114 9928	0.0084 5359	0.0062 2346	0.0045 8815	66
67	0.0107 4699	0.0078 6381	0.0057 6247	0.0042 2871	67
68	0.0100 4392	0.0073 1517	0.0053 3562	0.0038 9743	68
69	0.0093 8684	0.0068 0481	0.0049 4039	0.0035 9210	69
70	0.0087 7275	0.0063 3006	0.0045 7443	0.0033 1069	70
71	0.0081 9883	0.0058 8842	0.0042 3558	0.0030 5133	71
72	0.0076 6246	0.0054 7760	0.0039 2184	0.0028 1228	72
73	0.0071 6117	0.0050 9544	0.0036 3133	0.0025 9196	73
74	0.0066 9269	0.0047 3995	0.0033 6234	0.0023 8891	74
75	0.0062 5485	0.0044 0925	0.0031 1328	0.0022 0176	75
76	0.0058 4565	0.0041 0163	0.0028 8267	0.0020 2927	76
77	0.0054 6323	0.0038 1547	0.0026 6914	0.0018 7030	77
78	0.0051 0582	0.0035 4928	0.0024 7142	0.0017 2377	78
79	0.0047 7179	0.0033 0165	0.0022 8835	0.0015 8873	79
80	0.0044 5962	0.0030 7130	0.0021 1885	0.0014 6427	80
81	0.0041 6787	0.0028 5703	0.0019 6190	0.0013 4956	81
82	0.0038 9520	0.0026 5770	0.0018 1657	0.0012 4383	82
83	0.0036 4038	0.0024 7228	0.0016 8201	0.0011 4639	83
84	0.0034 0222	0.0022 9979	0.0015 5742	0.0010 5658	84
85	0.0031 7965	0.0021 3934	0.0014 4205	0.0009 7381	85
86	0.0029 7163	0.0019 9009	0.0013 3523	0.0008 9752	86
87	0.0027 7723	0.0018 5124	0.0012 3633	0.0008 2720	87
88	0.0025 9554	0.0017 2209	0.0011 4475	0.0007 6240	88
89	0.0024 2574	0.0016 0194	0.0010 5995	0.0007 0267	89
90	0.0022 6704	0.0014 9018	0.0009 8144	0.0006 4762	90
91	0.0021 1873	0.0013 8621	0.0009 0874	0.0005 9689	91
92	0.0019 8012	0.0012 8950	0.0008 4142	0.0005 5013	92
93	0.0018 5058	0.0011 9953	0.0007 7910	0.0005 0703	93
94	0.0017 2952	0.0011 1585	0.0007 2138	0.0004 6731	94
95	0.0016 1637	0.0010 3800	0.0006 6795	0.0004 3070	95
96	0.0015 1063	0.0009 6558	0.0006 1847	0.0003 9696	96
97	0.0014 1180	0.0008 9821	0.0005 7266	0.0003 6586	97
98	0.0013 1944	0.0008 3555	0.0005 3024	0.0003 3720	98
99	0.0012 3312	0.0007 7725	0.0004 9096	0.0003 1078	99
100	0.0011 5245	0.0007 2303	0.0004 5459	0.0002 8644	100

TABLE C-7 AMOUNT OF AN ANNUITY OF 1 PER PERIOD

$$s_{\overline{n}|i} = \frac{(1+i)^n - 1}{i}$$

n	1/4%	7/24%	1/3%	5/12%	n
1	1.0000 0000	1.0000 0000	1.0000 0000	1.0000 0000	1
2	2.0025 0000	2.0029 1667	2.0033 3333	2.0041 6667	2
3	3.0075 0625	3.0087 5851	3.0100 1111	3.0125 1736	3
4	4.0150 2502	4.0175 3405	4.0200 4448	4.0250 6952	4
5	5.0250 6258	5.0292 5186	5.0334 4463	5.0418 4064	5
6	6.0376 2523	6.0439 2051	6.0502 2278	6.0628 4831	6
7	7.0527 1930	7.0615 4861	7.0703 9019	7.0881 1018	7
8	8.0703 5110	8.0821 4480	8.0939 5816	8.1176 4397	8
9	9.0905 2697	9.1057 1772	9.1209 3802	9.1514 6749	9
10	10.1132 5329	10.1322 7606	10.1513 4114	10.1895 9860	10
11	11.1385 3642	11.1618 2853	11.1851 7895	11.2320 5526	11
12	12.1663 8277	12.1943 8387	12.2224 6288	12.2788 5549	12
13	13.1967 9872	13.2299 5082	13.2632 0442	13.3300 1739	13
14	14.2297 9072	14.2685 3818	14.3074 1510	14.3855 5913	14
15	15.2653 6520	15.3101 5475	15.3551 0648	15.4454 9896	15
16	16.3035 2861	16.3548 0936	16.4062 9017	16.5098 5520	16
17	17.3442 8743	17.4025 1089	17.4609 7781	17.5786 4627	17
18	18.3876 4815	18.4532 6822	18.5191 8107	18.6518 9063	18
19	19.4336 1727	19.5070 9025	19.5809 1167	19.7296 0684	19
20	20.4822 0131	20.5639 8593	20.6461 8137	20.8118 1353	20
21	21.5334 0682	21.6239 6422	21.7150 0198	21.8985 2942	21
22	22.5872 4033	22.6870 3412	22.7873 8532	22.9897 7330	22
23	23.6437 0843	23.7532 0463	23.8633 4327	24.0855 6402	23
24	24.7028 1770	24.8224 8481	24.9428 8775	25.1859 2054	24
25	25.7645 7475	25.8948 8373	26.0260 3071	26.2908 6187	25
26	26.8289 8619	26.9704 1047	27.1127 8414	27.4004 0713	26
27	27.8960 5865	28.0490 7417	28.2031 6009	28.5145 7549	27
28	28.9657 9880	29.1308 8397	29.2971 7062	29.6333 8622	28
29	30.0382 1330	30.2158 4904	30.3948 2786	30.7568 5867	29
30	31.1133 0883	31.3039 7860	31.4961 4395	31.8850 1224	30
31	32.1910 9210	32.3952 8188	32.6011 3110	33.0178 6646	31
32	33.2715 6983	33.4897 6811	33.7098 0154	34.1554 4090	32
33	34.3547 4876	34.5874 4660	34.8221 6754	35.2977 5524	33
34	35.4406 3563	35.6883 2666	35.9382 4143	36.4448 2922	34
35	36.5292 3722	36.7924 1761	37.0580 3557	37.5966 8268	35
36	37.6205 6031	37.8997 2883	38.1815 6236	38.7533 3552	36
37	38.7146 1171	39.0102 6970	39.3088 3423	39.9148 0775	37
38	39.8113 9824	40.1240 4966	40.4398 6368	41.0811 1945	38
39	40.9109 2673	41.2410 7813	41.5746 6322	42.2522 9078	39
40	42.0132 0405	42.3613 6461	42.7132 4543	43.4283 4199	40
41	43.1182 3706	43.4849 1859	43.8556 2292	44.6092 9342	41
42	44.2260 3265	44.6117 4961	45.0018 0833	45.7951 6548	42
43	45.3365 9774	45.7418 6721	46.1518 1436	46.9859 7866	43
44	46.4499 3923	46.8752 8099	47.3056 5374	48.1817 5358	44
45	47.5660 6408	48.0120 0056	48.4633 3925	49.3825 1088	45
46	48.6849 7924	49.1520 3556	49.6248 8371	50.5882 7134	46
47	49.8066 9169	50.2953 9566	50.7902 9999	51.7990 5581	47
48	50.9312 0842	51.4420 9057	51.9596 0099	53.0148 8521	48
49	52.0585 3644	52.5921 3000	53.1327 9966	54.2357 8056	49
50	53.1886 8278	53.7455 2371	54.3099 0899	55.4617 6298	50

TABLE C-7 AMOUNT OF AN ANNUITY OF 1 PER PERIOD (cont.)

$$s_{\overline{n}|i} = \frac{(1+i)^n - 1}{i}$$

n	1/2%	7/12%	5/8%	2/3%	n
1	1.0000 0000	1.0000 0000	1.0000 0000	1.0000 0000	1
2	2.0050 0000	2.0058 3333	2.0062 5000	2.0066 6667	2
3	3.0150 2500	3.0175 3403	3.1087 8906	3.0200 4444	3
4	4.0301 0013	4.0351 3631	4.0376 5649	4.0401 7807	4
5	5.0502 5063	5.0586 7460	5.0628 9185	5.0671 1259	5
6	6.0755 0188	6.0881 8354	6.0945 3492	6.1008 9335	6
7	7.1058 7939	7.1236 9794	7.1326 2576	7.1415 6597	7
8	8.1414 0879	8.1652 5284	8.1772 0468	8.1891 7641	8
9	9.1821 1583	9.2128 8343	9.2283 1220	9.2437 7092	9
10	10.2280 2641	10.2666 2531	10.2859 8916	10.3053 9606	10
11	11.2791 6654	11.3265 1396	11.3502 7659	11.3740 9870	11
12	12.3355 6237	12.3925 8529	12.4212 1582	12.4499 2602	12
13	13.3972 4018	13.4648 7537	13.4988 4842	13.5329 2553	13
14	14.4642 2639	14.5434 2048	14.5832 1622	14.6231 4503	14
15	15.5365 4752	15.6282 5710	15.6743 6132	15.7206 3266	15
16	16.6142 3026	16.7194 2193	16.7723 2608	16.8254 3688	16
17	17.6973 0141	17.8169 5189	17.8771 5312	17.9376 0646	17
18	18.7857 8791	18.9208 8411	18.9888 8532	19.0571 9051	18
19	19.8797 1685	20.0312 5593	20.1075 6586	20.1842 3844	19
20	20.9791 1544	21.1481 0493	21.2332 3814	21.3188 0003	20
21	22.0840 1101	22.2714 6887	22.3659 4588	22.4609 2536	21
22	23.1944 3107	23.4013 8577	23.5057 3304	23.6106 6487	22
23	24.3104 0322	24.5378 9386	24.6526 4387	24.7680 6930	23
24	25.4319 5524	25.6810 3157	25.8067 2290	25.9331 8976	24
25	26.5591 1502	26.8308 3759	26.9680 1492	27.1060 7769	25
26	27.6919 1059	27.9873 5081	28.1365 6501	28.2867 8488	26
27	28.8303 7015	29.1506 1035	29.3124 1854	29.4753 6344	27
28	29.9745 2200	30.3206 5558	30.4956 2116	30.6718 6586	28
29	31.1243 9461	31.4975 2607	31.6862 1879	31.8763 4497	29
30	32.2800 1658	32.6812 6164	32.8842 5766	33.0888 5394	30
31	33.4414 1666	33.8719 0233	34.0897 8427	34.3094 4630	31
32	34.6086 2375	35.0694 8843	35.3028 4542	35.5381 7594	32
33	35.7816 6686	36.2740 6045	36.5234 8820	36.7750 9711	33
34	36.9605 7520	37.4856 5913	37.7517 6000	38.0202 6443	34
35	38.1453 7807	38.7043 2548	38.9877 0850	39.2737 3286	35
36	39.3361 0496	39.9301 0071	40.2313 8168	40.5355 5774	36
37	40.5327 8549	41.1630 2630	41.4828 2782	41.8057 9479	37
38	41.7354 4942	42.4031 4395	42.7420 9549	43.0845 0009	38
39	42.9441 2666	43.6504 9562	44.0092 3359	44.3717 3009	39
40	44.1588 4730	44.9051 2352	45.2842 9130	45.6675 4163	40
41	45.3796 4153	46.1670 7007	46.5673 1812	46.9719 9191	41
42	46.6065 3974	47.4363 7798	47.8583 6386	48.2851 3852	42
43	47.8395 7244	48.7130 9018	49.1574 7863	49.6070 3944	43
44	49.0787 7030	49.9972 4988	50.4647 1287	50.9377 5304	44
45	50.3241 6415	51.2889 0050	51.7801 1733	52.2773 3806	45
46	51.5757 8497	52.5880 8575	53.1037 4306	53.6258 5365	46
47	52.8336 6390	53.8948 4959	54.4356 4146	54.9833 5934	47
48	54.0978 3222	55.2092 3621	55.7758 6421	56.3499 1507	48
49	55.3683 2138	56.5312 9009	57.1244 6337	57.7255 8117	49
50	56.6451 6299	57.8610 5595	58.4814 9126	59.1104 1837	50

TABLE C-7 AMOUNT OF AN ANNUITY OF 1 PER PERIOD (cont.)

$$s_{\overline{n}|i} = \frac{(1+i)^n - 1}{i}$$

n	¾%	⅞%	1%	1⅛%	n
1	1.0000 0000	1.0000 0000	1.0000 0000	1.0000 0000	1
2	2.0075 0000	2.0087 5000	2.0100 0000	2.0112 5000	2
3	3.0225 5625	3.0263 2656	3.0301 0000	3.0338 7656	3
4	4.0452 2542	4.0528 0692	4.0604 0100	4.0680 0767	4
5	5.0755 6461	5.0882 6898	5.1010 0501	5.1137 7276	5
6	6.1136 3135	6.1327 9133	6.1520 1506	6.1713 0270	6
7	7.1594 8358	7.1864 5326	7.2135 3521	7.2407 2986	7
8	8.2131 7971	8.2493 3472	8.2856 7056	8.3221 8807	8
9	9.2747 7856	9.3215 1640	9.3685 2727	9.4158 1269	9
10	10.3443 3940	10.4030 7967	10.4622 1254	10.5217 4058	10
11	11.4219 2194	11.4941 0662	11.5668 3467	11.6401 1016	11
12	12.5075 8636	12.5946 8005	12.6825 0301	12.7710 6140	12
13	13.6013 9325	13.7048 8350	13.8093 2804	13.9147 3584	13
14	14.7034 0370	14.8248 0123	14.9474 2132	15.0712 7662	14
15	15.8136 7923	15.9545 1824	16.0968 9554	16.2408 2848	15
16	16.9322 8183	17.0941 2028	17.2578 6449	17.4235 3780	16
17	18.0592 7394	18.2436 9383	18.4304 4314	18.6195 5260	17
18	19.1947 1849	19.4033 2615	19.6147 4757	19.8290 2257	18
19	20.3386 7888	20.5731 0526	20.8108 9504	21.0520 9907	19
20	21.4912 1897	21.7531 1993	22.0190 0399	22.2889 3519	20
21	22.6524 0312	22.9434 5973	23.2391 9403	23.5396 8571	21
22	23.8222 9614	24.1442 1500	24.4715 8598	24.8045 0717	22
23	25.0009 6336	25.3554 7688	25.7163 0183	26.0835 5788	23
24	26.1884 7059	26.5773 3730	26.9734 6485	27.3769 9790	24
25	27.3848 8412	27.8098 8900	28.2431 9950	28.6849 8913	25
26	28.5902 7075	29.0532 2553	29.5256 3150	30.0076 9526	26
27	29.8046 9778	30.3074 4126	30.8208 8781	31.3452 8183	27
28	31.0282 3301	31.5726 3137	32.1290 9669	32.6979 1625	28
29	32.2609 4476	32.8488 9189	33.4503 8766	34.0657 6781	29
30	33.5029 0184	34.1363 1970	34.7848 9153	35.4490 0769	30
31	34.7541 7361	35.4350 1249	36.1327 4045	36.8478 0903	31
32	36.0148 2991	36.7450 6885	37.4940 6785	38.2623 4688	32
33	37.2849 4113	38.0665 8820	38.8690 0853	39.6927 9829	33
34	38.5645 7819	39.3996 7085	40.2576 9862	41.1393 4227	34
35	39.8538 1253	40.7444 1797	41.6602 7560	42.6021 5987	35
36	41.1527 1612	42.1009 3163	43.0768 7836	44.0814 3417	36
37	42.4613 6149	43.4693 1478	44.5076 4714	45.5773 5030	37
38	43.7798 2170	44.8496 7128	45.9527 2361	47.0900 9549	38
39	45.1081 7037	46.2421 0591	47.4122 5085	48.6198 5906	39
40	46.4464 8164	47.6467 2433	48.8863 7336	50.1668 3248	40
41	47.7948 3026	49.0636 3317	50.3752 3709	51.7312 0934	41
42	49.1532 9148	50.4929 3996	51.8789 8946	53.3131 8545	42
43	50.5219 4117	51.9347 5319	53.3977 7936	54.9129 5879	43
44	51.9008 5573	53.3891 8228	54.9317 5715	56.5307 2957	44
45	53.2901 1215	54.8563 3762	56.4810 7472	58.1667 0028	45
46	54.6897 8799	56.3363 3058	58.0458 8547	59.8210 7566	46
47	56.0999 6140	57.8292 7347	59.6263 4432	61.4940 6276	47
48	57.5207 1111	59.3352 7961	61.2226 0777	63.1858 7097	48
49	58.9521 1644	60.8544 6331	62.8348 3385	64.8967 1201	49
50	60.3942 5732	62.3869 3986	64.4631 8218	66.6268 0002	50

TABLE C-7 AMOUNT OF AN ANNUITY OF 1 PER PERIOD (cont.)

$$s_{\overline{n}|i} = \frac{(1+i)^n - 1}{i}$$

n	1¼%	1⅜%	1½%	1¾%	n
1	1.0000 0000	1.0000 0000	1.0000 0000	1.0000 0000	1
2	2.0125 0000	2.0137 5000	2.0150 0000	2.0175 0000	2
3	3.0376 5625	3.0414 3906	3.0452 2500	3.0528 0625	3
4	4.0756 2695	4.0832 5885	4.0909 0338	4.1062 3036	4
5	5.1265 7229	5.1394 0366	5.1522 6693	5.1780 8938	5
6	6.1906 5444	6.2100 7046	6.2295 5093	6.2687 0596	6
7	7.2680 3762	7.2954 5893	7.3229 9419	7.3784 0831	7
8	8.3588 8809	8.3957 7149	8.4328 3911	8.5075 3045	8
9	9.4633 7420	9.5112 1335	9.5593 3169	9.6564 1224	9
10	10.5816 6637	10.6419 9253	10.7027 2167	10.8253 9945	10
11	11.7139 3720	11.7883 1993	11.8632 6249	12.0148 4394	11
12	12.8603 6142	12.9504 0933	13.0412 1143	13.2251 0371	12
13	14.0211 1594	14.1284 7745	14.2368 2960	14.4565 4303	13
14	15.1963 7988	15.3227 4402	15.4503 8205	15.7095 3253	14
15	16.3863 3463	16.5334 3175	16.6821 3778	16.9844 4935	15
16	17.5911 6382	17.7607 6644	17.9323 6984	18.2816 7721	16
17	18.8110 5336	19.0049 7697	19.2013 5539	19.6016 0656	17
18	20.0461 9153	20.2662 9541	20.4893 7572	20.9446 3468	18
19	21.2967 6893	21.5449 5697	21.7967 1636	22.3111 6578	19
20	22.5629 7854	22.8412 0013	23.1236 6710	23.7016 1119	20
21	23.8450 1577	24.1552 6663	24.4705 2211	25.1163 8938	21
22	25.1430 7847	25.4874 0155	25.8375 7994	26.5559 2620	22
23	26.4573 6695	26.8378 5332	27.2251 4364	28.0206 5490	23
24	27.7880 8403	28.2068 7380	28.6335 2080	29.5110 1637	24
25	29.1354 3508	29.5947 1832	30.0630 2361	31.0274 5915	25
26	30.4996 2802	31.0016 4569	31.5139 6896	32.5704 3969	26
27	31.8808 7337	32.4279 1832	32.9866 7850	34.1404 2238	27
28	33.2793 8429	33.8738 0220	34.4814 7867	35.7378 7977	28
29	34.6953 7659	35.3395 6698	35.9987 0085	37.3632 9267	29
30	36.1290 6880	36.8254 8602	37.5386 8137	39.0171 5029	30
31	37.5806 8216	38.3318 3646	39.1017 6159	40.6999 5042	31
32	39.0504 4069	39.8588 9921	40.6882 8801	42.4121 9955	32
33	40.5385 7120	41.4069 5907	42.2986 1233	44.1544 1305	33
34	42.0453 0334	42.9763 0476	43.9330 9152	45.9271 1527	34
35	43.5708 6963	44.5672 2895	45.5920 8789	47.7308 3979	35
36	45.1155 0550	46.1800 2835	47.2759 6921	49.5661 2949	36
37	46.6794 4932	47.8150 0374	48.9851 0874	51.4335 3675	37
38	48.2926 4243	49.4724 6004	50.7198 8538	53.3336 2365	38
39	49.8862 2921	51.1527 0636	52.4806 8366	55.2669 6206	39
40	51.4895 5708	52.8560 5608	54.2678 9391	57.2341 3390	40
41	53.1331 7654	54.5828 2685	56.0819 1232	59.2357 3124	41
42	54.7973 4125	56.3333 4072	57.9231 4100	61.2723 5654	42
43	56.4823 0801	58.1079 2415	59.7919 8812	63.3446 2278	43
44	58.1883 3687	59.9069 0811	61.6888 6794	65.4531 5367	44
45	59.9156 9108	61.7306 2810	63.6142 0096	67.5985 8386	45
46	61.6646 3721	63.5794 2423	65.5684 1398	69.7815 5908	46
47	63.4354 4518	65.4536 4131	67.5519 4018	72.0027 3637	47
48	65.2283 8824	67.3536 2888	69.5652 1929	74.2627 8425	48
49	67.0437 4310	69.2797 4128	71.6086 9758	76.5623 8298	49
50	68.8817 8989	71.2323 3772	73.6828 2804	78.9022 2468	50

TABLE C-7 AMOUNT OF AN ANNUITY OF 1 PER PERIOD (cont.)

$$s_{\overline{n}|i} = \frac{(1+i)^n - 1}{i}$$

n	2%	2¼%	2½%	2¾%	n
1	1.0000 0000	1.0000 0000	1.0000 0000	1.0000 0000	1
2	2.0200 0000	2.0225 0000	2.0250 0000	2.0275 0000	2
3	3.0604 0000	3.0680 0625	3.0756 2500	3.0832 5625	3
4	4.1216 0800	4.1370 3639	4.1525 1563	4.1680 4580	4
5	5.2040 4016	5.2301 1971	5.2563 2852	5.2826 6706	5
6	6.3081 2096	6.3477 9740	6.3877 3673	6.4279 4040	6
7	7.4342 8338	7.4906 2284	7.5474 3015	7.6047 0876	7
8	8.5829 6905	8.6591 6186	8.7361 1590	8.8138 3825	8
9	9.7546 2843	9.8539 9300	9.9545 1880	10.0562 1880	9
10	10.9497 2100	11.0757 0784	11.2033 8177	11.3327 6482	10
11	12.1687 1542	12.3249 1127	12.4834 6631	12.6444 1585	11
12	13.4120 8973	13.6022 2177	13.7955 5297	13.9921 3729	12
13	14.6803 3152	14.9082 7176	15.1404 4179	15.3769 2107	13
14	15.9739 3815	16.2437 0788	16.5189 5284	16.7997 8639	14
15	17.2934 1692	17.6091 9130	17.9319 2666	18.2617 8052	15
16	18.6392 8525	19.0053 9811	19.3802 2483	19.7639 7948	16
17	20.0120 7096	20.4330 1957	20.8647 3045	21.3074 8892	17
18	21.4123 1238	21.8927 6251	22.3863 4871	22.8934 4487	18
19	22.8405 5863	23.3853 4966	23.9460 0743	24.5230 1460	19
20	24.2973 6980	24.9115 2003	25.5446 5761	26.1973 9750	20
21	25.7833 1719	26.4720 2923	27.1832 7405	27.9178 2593	21
22	27.2989 8354	28.0676 4989	28.8628 5590	29.6855 6615	22
23	28.8449 6321	29.6991 7201	30.5844 2730	31.5019 1921	23
24	30.4218 6247	31.3674 0338	32.3490 3798	33.3682 2199	24
25	32.0302 9972	33.0731 6996	34.1577 6393	35.2858 4810	25
26	33.6709 0572	34.8173 1628	36.0117 0803	37.2562 0892	26
27	35.3443 2383	36.6007 0590	37.9120 0073	39.2807 5467	27
28	37.0512 1031	38.4242 2178	39.8598 0075	41.3609 7542	28
29	38.7922 3451	40.2887 6677	41.8562 9577	43.4984 0224	29
30	40.5680 7921	42.1952 6402	43.9027 0316	45.6946 0830	30
31	42.3794 4079	44.1446 5746	46.0002 7074	47.9512 1003	31
32	44.2270 2961	46.1379 1226	48.1502 7751	50.2698 6831	32
33	46.1115 7020	48.1760 1528	50.3540 3445	52.6522 8969	33
34	48.0338 0160	50.2599 7563	52.6128 8531	55.1002 2765	34
35	49.9944 7763	52.3908 2508	54.9282 0744	57.6154 8391	35
36	51.9943 6719	54.5696 1864	57.3014 1263	60.1999 0972	36
37	54.0342 5453	56.7974 3506	59.7339 4794	62.8554 0724	37
38	56.1149 3962	59.0753 7735	62.2272 9664	65.5839 3094	38
39	58.2372 3841	61.4045 7334	64.7829 7906	68.3874 8904	39
40	60.4019 8318	63.7861 7624	67.4025 5354	71.2681 4499	40
41	62.6100 2284	66.2213 6521	70.0876 1737	74.2280 1898	41
42	64.8622 2330	68.7113 4592	72.8398 0781	77.2692 8950	42
43	67.1594 6777	71.2573 5121	75.6608 0300	80.3941 9496	43
44	69.5026 5712	73.8606 4161	78.5523 2308	83.6050 3532	44
45	71.8927 1027	76.5225 0605	81.5161 3116	86.9041 7379	45
46	74.3305 6447	79.2442 6243	84.5540 3443	90.2940 3857	46
47	76.8171 7576	82.0272 5834	87.6678 8530	93.7771 2463	47
48	79.3535 1927	84.8728 7165	90.8595 8243	97.3559 9556	48
49	81.9405 8966	87.7825 1126	94.1310 7199	101.0332 8544	49
50	84.5794 0145	90.7576 1776	97.4843 4879	104.8117 0079	50

TABLE C-7 AMOUNT OF AN ANNUITY OF 1 PER PERIOD (cont.)

$$s_{\overline{n}|i} = \frac{(1+i)^n - 1}{i}$$

n	3%	3½%	4%	4½%	n
1	1.0000 0000	1.0000 0000	1.0000 0000	1.0000 0000	1
2	2.0300 0000	2.0350 0000	2.0400 0000	2.0450 0000	2
3	3.0909 0000	3.1062 2500	3.1216 0000	3.1370 2500	3
4	4.1836 2700	4.2149 4288	4.2464 6400	4.2781 9113	4
5	5.3091 3581	5.3624 6588	5.4163 2256	5.4707 0973	5
6	6.4684 0988	6.5501 5218	6.6329 7546	6.7168 9166	6
7	7.6624 6218	7.7794 0751	7.8982 9448	8.0191 5179	7
8	8.8923 3605	9.0516 8677	9.2142 2626	9.3800 1362	8
9	10.1591 0613	10.3684 9581	10.5827 9531	10.8021 1423	9
10	11.4638 7931	11.7313 9316	12.0061 0712	12.2882 0937	10
11	12.8077 9569	13.1419 9192	13.4863 5141	13.8411 7879	11
12	14.1920 2956	14.6019 6164	15.0258 0546	15.4640 3184	12
13	15.6177 9045	16.1130 3030	16.6268 3768	17.1599 1327	13
14	17.0863 2416	17.6769 8636	18.2919 1119	18.9321 0937	14
15	18.5989 1389	19.2956 8088	20.0235 8764	20.7840 5429	15
16	20.1568 8130	20.9710 2971	21.8245 3114	22.7193 3673	16
17	21.7615 8774	22.7050 1575	23.6975 1239	24.7417 0689	17
18	23.4144 3537	24.4996 9130	25.6454 1288	26.8550 8370	18
19	25.1168 6844	26.3571 8050	27.6712 2940	29.0635 6246	19
20	26.8703 7449	28.2796 8181	29.7780 7858	31.3714 2277	20
21	28.6764 8572	30.2694 7068	31.9692 0172	33.7831 3680	21
22	30.5367 8030	32.3289 0215	34.2479 6979	36.3033 7795	22
23	32.4528 8370	34.4604 1373	36.6178 8858	38.9370 2996	23
24	34.4264 7022	36.6665 2821	39.0826 0412	41.6891 9631	24
25	36.4592 6432	38.9498 5669	41.6459 0829	44.5652 1015	25
26	38.5530 4225	41.3131 0168	44.3117 4462	47.5706 4460	26
27	40.7096 3352	43.7590 6024	47.0842 1440	50.7113 2361	27
28	42.9309 2252	46.2906 2734	49.9675 8298	53.9933 3317	28
29	45.2188 5020	48.9107 9930	52.9662 8630	57.4230 3316	29
30	47.5754 1571	51.6226 7728	56.0849 3775	61.0070 6966	30
31	50.0026 7818	54.4294 7098	59.3283 3526	64.7523 8779	31
32	52.5027 5852	57.3345 0247	62.7014 6867	68.6662 4524	32
33	55.0778 4128	60.3412 1005	66.2095 2742	72.7562 2628	33
34	57.7301 7652	63.4531 5240	69.8579 0851	77.0302 5646	34
35	60.4620 8181	66.6740 1274	73.6522 2486	81.4966 1800	35
36	63.2759 4427	70.0076 0318	77.5983 1385	86.1639 6581	36
37	66.1742 2259	73.4578 6930	81.7022 4640	91.0413 4427	37
38	69.1594 4927	77.0288 9472	85.9703 3626	96.1382 0476	38
39	72.2342 3275	80.7249 0604	90.4091 4971	101.4644 2398	39
40	75.4012 5973	84.5502 7775	95.0255 1570	107.0303 2306	40
41	78.6632 9753	88.5095 3747	99.8265 3633	112.8466 8760	41
42	82.0231 9645	92.6073 7128	104.8195 9778	118.9247 8854	42
43	85.4838 9234	96.8486 2928	110.0123 8169	125.2764 0402	43
44	89.0484 0911	101.2383 3130	115.4128 7696	131.9138 4220	44
45	92.7198 6139	105.7816 7290	121.0293 9204	138.8499 6510	45
46	96.5014 5723	110.4840 3145	126.8705 6772	146.0982 1353	46
47	100.3965 0095	115.3509 7255	132.9453 9043	153.6726 3314	47
48	104.4083 9598	120.3882 5659	139.2632 0604	161.5879 0163	48
49	108.5406 4785	125.6018 4557	145.8337 3429	169.8593 5720	49
50	112.7968 6729	130.9979 1016	152.6670 8366	178.5030 2828	50

TABLE C-7 AMOUNT OF AN ANNUITY OF 1 PER PERIOD (cont.)

$$s_{\overline{n}|i} = \frac{(1+i)^n - 1}{i}$$

n	5%	5½%	6%	6½%	n
1	1.0000 0000	1.0000 0000	1.0000 0000	1.0000 0000	1
2	2.0500 0000	2.0550 0000	2.0600 0000	2.0650 0000	2
3	3.1525 0000	3.1680 2500	3.1836 0000	3.1992 2500	3
4	4.3101 2500	4.3422 6638	4.3746 1600	4.4071 7463	4
5	5.5256 3125	5.5810 9103	5.6370 9296	5.6936 4098	5
6	6.8019 1281	6.8880 5103	6.9753 1854	7.0637 2764	6
7	8.1420 0845	8.2668 9384	8.3938 3765	8.5228 6994	7
8	9.5491 0888	9.7215 7300	9.8974 6791	10.0768 5648	8
9	11.0265 6432	11.2562 5951	11.4913 1598	11.7318 5215	9
10	12.5778 9254	12.8753 5379	13.1807 9494	13.4944 2254	10
11	14.2067 8716	14.5834 9825	14.9716 4264	15.3715 6001	11
12	15.9171 2652	16.3855 9065	16.8699 4120	17.3707 1141	12
13	17.7129 8285	18.2867 9814	18.8821 3767	19.4998 0765	13
14	19.5986 3199	20.2925 7203	21.0150 6593	21.7672 9515	14
15	21.5785 6359	22.4086 6350	23.2759 6988	24.1821 6933	15
16	23.6574 9177	24.6411 3999	25.6725 2808	26.7540 1034	16
17	25.8403 6636	26.9964 0269	28.2128 7976	29.4930 2101	17
18	28.1323 8467	29.4812 0483	30.9056 5255	32.4100 6738	18
19	30.5390 0391	32.1026 7110	33.7599 9170	35.5167 2176	19
20	33.0659 5410	34.8683 1801	36.7855 9120	38.8253 0867	20
21	35.7192 5181	37.7860 7550	39.9927 2668	42.3489 5373	21
22	38.5052 1440	40.8643 0965	43.3922 9028	46.1016 3573	22
23	41.4304 7512	44.1118 4669	46.9958 2769	50.0982 4205	23
24	44.5019 9887	47.5379 9825	50.8155 7735	54.3546 2778	24
25	47.7270 9882	51.1525 8816	54.8645 1200	58.8876 7859	25
26	51.1134 5376	54.9659 8051	59.1563 8272	63.7153 7769	26
27	54.6691 2645	58.9891 0943	63.7057 6568	68.8568 7725	27
28	58.4025 8277	63.2335 1045	68.5281 1162	74.3325 7427	28
29	62.3227 1191	67.7113 5353	73.6397 9832	80.1641 9159	29
30	66.4388 4750	72.4354 7797	79.0581 8622	86.3748 6405	30
31	70.7607 8988	77.4194 2926	84.8016 7739	92.9892 3021	31
32	75.2988 2937	82.6774 9787	90.8897 7803	100.0335 3017	32
33	80.0637 7084	88.2247 6025	97.3431 6471	107.5357 0963	33
34	85.0659 5938	94.0771 2207	104.1837 5460	115.5255 3076	34
35	90.3203 0735	100.2513 6378	111.4347 7987	124.0346 9026	35
36	95.8363 2272	106.7651 8879	119.1208 6666	133.0969 4513	36
37	101.6281 3886	113.6372 7417	127.2681 1866	142.7482 4656	37
38	107.7095 4580	120.8873 2425	135.9042 0578	153.0268 8259	38
39	114.0950 2309	128.5361 2708	145.0584 5813	163.9736 2995	39
40	120.7997 7424	136.6056 1407	154.7619 6562	175.6319 1590	40
41	127.8397 6295	145.1189 2285	165.0476 8356	188.0479 9044	41
42	135.2317 5110	154.1004 6360	175.9505 4457	201.2711 0981	42
43	142.9933 3866	163.5759 8910	187.5075 7724	215.3537 3195	43
44	151.1430 0559	173.5726 6850	199.7580 3188	230.3517 2453	44
45	159.7001 5587	184.1191 6527	212.7435 1379	246.3245 8662	45
46	168.6851 6366	195.2457 1936	226.5081 2462	263.3356 8475	46
47	178.1194 2185	206.9842 3392	241.0986 1210	281.4525 0426	47
48	188.0253 9294	219.3683 6679	256.5645 2882	300.7469 1704	48
49	198.4266 6259	232.4336 2696	272.9584 0055	321.2954 6665	49
50	209.3479 9572	246.2174 7645	290.3359 0458	343.1796 7198	50

TABLE C-7 AMOUNT OF AN ANNUITY OF 1 PER PERIOD (cont.)

$$s_{\overline{n}|i} = \frac{(1+i)^n - 1}{i}$$

n	7%	7½%	8%	8½%	n
1	1.0000 0000	1.0000 0000	1.0000 0000	1.0000 0000	1
2	2.0700 0000	2.0750 0000	2.0800 0000	2.0850 0000	2
3	3.2149 0000	3.2306 2500	3.2464 0000	3.2622 2500	3
4	4.4399 4300	4.4729 2188	4.5061 1200	4.5395 1413	4
5	5.7507 3901	5.8083 9102	5.8666 0096	5.9253 7283	5
6	7.1532 9074	7.2440 2034	7.3359 2904	7.4290 2952	6
7	8.6540 2109	8.7873 2187	8.9228 0336	9.0604 9702	7
8	10.2598 0257	10.4463 7101	10.6366 2763	10.8306 3927	8
9	11.9779 8875	12.2298 4883	12.4875 5784	12.7512 4361	9
10	13.8164 4796	14.1470 8750	14.4865 6247	14.8350 9932	10
11	15.7835 9932	16.2081 1906	16.6454 8746	17.0960 8276	11
12	17.8884 5127	18.4237 2799	18.9771 2646	19.5492 4979	12
13	20.1406 4286	20.8055 0759	21.4952 9658	22.2109 3603	13
14	22.5504 8786	23.3659 2066	24.2149 2030	25.0988 6559	14
15	25.1290 2201	26.1183 6470	27.1521 1393	28.2322 6916	15
16	27.8880 5355	29.0772 4206	30.3242 8304	31.6320 1204	16
17	30.8402 1730	32.2580 3521	33.7502 2569	35.3207 3306	17
18	33.9990 3251	35.6773 8785	37.4502 4374	39.3229 9538	18
19	37.3789 6479	39.3531 9194	41.4462 6324	43.6654 4998	19
20	40.9954 9232	43.3046 8134	45.7619 6430	48.3770 1323	20
21	44.8651 7678	47.5525 3244	50.4229 2144	53.4890 5936	21
22	49.0057 3916	52.1189 7237	55.4567 5516	59.0356 2940	22
23	53.4361 4090	57.0278 9530	60.8932 9557	65.0536 5790	23
24	58.1766 7076	62.3049 8744	66.7647 5922	71.5832 1882	24
25	63.2490 3772	67.9778 6150	73.1059 3995	78.6677 9242	25
26	68.6764 7036	74.0762 0112	79.9544 1515	86.3545 5478	26
27	74.4838 2328	80.6319 1620	87.3507 6836	94.6946 9193	27
28	80.6976 9091	87.6793 0991	95.3388 2983	103.7437 4075	28
29	87.3465 2927	95.2552 5816	103.9659 3622	113.5619 5871	29
30	94.4607 8632	103.3994 0252	113.2832 1111	124.2147 2520	30
31	102.0730 4137	112.1543 5771	123.3458 6800	135.7729 7684	31
32	110.2181 5426	121.5659 3454	134.2135 3744	148.3136 7987	32
33	118.9334 2506	131.6833 7963	145.9506 2044	161.9203 4266	33
34	128.2587 6481	142.5596 3310	158.6266 7007	176.6835 7179	34
35	138.2368 7835	154.2516 0558	172.3168 0368	192.7016 7539	35
36	148.9134 5984	166.8204 7600	187.1021 4797	210.0813 1780	36
37	160.3374 0202	180.3320 1170	203.0703 1981	228.9382 2981	37
38	172.5610 2017	194.8569 1258	220.3159 4540	249.3979 7935	38
39	185.6402 9158	210.4711 8102	238.9412 2103	271.5968 0759	39
40	199.6351 1199	227.2565 1960	259.0565 1871	295.6825 3624	40
41	214.6095 6983	245.3007 5857	280.7810 4021	321.8155 5182	41
42	230.6322 3972	264.6983 1546	304.2435 2342	350.1698 7372	42
43	247.7764 9650	285.5506 8912	329.5830 0530	380.9343 1299	43
44	266.1208 5125	307.9669 9080	356.9496 4572	414.3137 2959	44
45	285.7493 1084	332.0645 1511	386.5056 1738	450.5303 9661	45
46	306.7517 6260	357.9693 5375	418.4260 6677	489.8254 8032	46
47	329.2243 8598	385.8170 5528	452.9001 5211	532.4606 4615	47
48	353.2700 9300	415.7533 3442	490.1321 6428	578.7198 0107	48
49	378.9989 9951	447.9348 3451	530.3427 3742	628.9109 8416	49
50	406.5289 2947	482.5299 4709	573.7701 5642	683.3684 1782	50

TABLE C-8 PRESENT VALUE OF ANNUITY OF 1 PER POUND

$$a_{\overline{n}|i} = \frac{1-(1+i)^{-n}}{i}$$

n	1/4%	7/24%	1/3%	5/12%	n
1	0.9975 0623	0.9970 9182	0.9966 7774	0.9958 5062	1
2	1.9925 2492	1.9912 8390	1.9900 4426	1.9875 6908	2
3	2.9850 6227	2.9825 8470	2.9801 1056	2.9751 7253	3
4	3.9751 2446	3.9710 0260	3.9668 8760	3.9586 7804	4
5	4.9627 1766	4.9565 4601	4.9503 8631	4.9381 0261	5
6	5.9478 4804	5.9392 2327	5.9306 1759	5.9134 6318	6
7	6.9305 2174	6.9190 4273	6.9075 9228	6.8847 7661	7
8	7.9107 4487	7.8960 1269	7.8813 2121	7.8520 5969	8
9	8.8885 2357	8.8701 4144	8.8518 1516	8.8153 2915	9
10	9.8638 6391	9.8414 3725	9.8190 8487	9.7746 0164	10
11	10.8367 7198	10.8099 0834	10.7831 4107	10.7298 9374	11
12	11.8072 5384	11.7755 6295	11.7439 9442	11.6812 2198	12
13	12.7753 1555	12.7384 0915	12.7016 5557	12.6286 0280	13
14	13.7409 6314	13.6984 5542	13.6561 3512	13.5720 5257	14
15	14.7042 0264	14.6557 0959	14.6074 4364	14.5115 8762	15
16	15.6650 4004	15.6101 7990	15.5555 9167	15.4472 2418	16
17	16.6234 8133	16.5618 7442	16.5005 8970	16.3789 7843	17
18	17.5795 3250	17.5108 0125	17.4424 4821	17.3068 6648	18
19	18.5331 9950	18.4569 6842	18.3811 7762	18.2309 0438	19
20	19.4844 8828	19.4003 8396	19.3167 8832	19.1511 0809	20
21	20.4334 0477	20.3410 5587	20.2492 9069	20.0674 9352	21
22	21.3799 5488	21.2789 9213	21.1786 9504	20.9800 7653	22
23	22.3241 4452	22.2142 0071	22.1050 1167	21.8888 7289	23
24	23.2659 7957	23.1466 8952	23.0282 5083	22.7938 9831	24
25	24.2054 6591	24.0764 6648	23.9484 2275	23.6951 6843	25
26	25.1426 0939	25.0035 3949	24.8655 3763	24.5926 9884	26
27	26.0774 1585	25.9279 1639	25.7796 0561	25.4865 0506	27
28	27.0098 9112	26.8496 0503	26.6906 3682	26.3766 0254	28
29	27.9400 4102	27.7686 1324	27.5986 4135	27.2630 0668	29
30	28.8678 7134	28.6849 4879	28.5036 2925	28.1457 3278	30
31	29.7933 8787	29.5986 1947	29.4056 1055	29.0247 9612	31
32	30.7165 9638	30.5096 3303	30.3045 9523	29.9002 1189	32
33	31.6375 0262	31.4179 9720	31.2005 9325	30.7719 9524	33
34	32.5561 1234	32.3237 1967	32.0936 1454	31.6401 6122	34
35	33.4724 3126	33.2268 0814	32.9836 6898	32.5047 2486	35
36	34.3864 6510	34.1272 7025	33.8707 6642	33.3657 0109	36
37	35.2982 1955	35.0251 1366	34.7549 1670	34.2231 0481	37
38	36.2077 0030	35.9203 4597	35.6361 2960	35.0769 5084	38
39	37.1149 1302	36.8129 7478	36.5144 1488	35.9272 5394	39
40	38.0198 6336	37.7030 0767	37.3897 8228	36.7740 2881	40
41	38.9225 5697	38.5904 5217	38.2622 4147	37.6172 9009	41
42	39.8229 9947	39.4753 1582	39.1318 0213	38.4570 5236	42
43	40.7211 9648	40.3576 0612	39.9984 7388	39.2933 3013	43
44	41.6171 5359	41.2373 3056	40.8622 6633	40.1261 3788	44
45	42.5108 7640	42.1144 9659	41.7231 8903	40.9554 8999	45
46	43.4023 7047	42.9891 1167	42.5812 5153	41.7814 0081	46
47	44.2916 4137	43.8611 8320	43.4364 6332	42.6038 8461	47
48	45.1786 9463	44.7307 1859	44.2888 3387	43.4229 5562	48
49	46.0635 3580	45.5977 2521	45.1383 7263	44.2386 2799	49
50	46.9461 7037	46.4622 1042	45.9850 8900	45.0509 1582	50

TABLE C-8 PRESENT VALUE OF ANNUITY OF 1 PER POUND (*cont.*)

$$a_{\overline{n}|i} = \frac{1 - (1 + i)^{-n}}{i}$$

n	½%	7/12%	5/8%	2/3%	n
1	0.9950 2488	0.9942 0050	0.9937 8882	0.9933 7748	1
2	1.9850 9938	1.9826 3513	1.9814 0504	1.9801 7631	2
3	2.9702 4814	2.9653 3733	2.9628 8699	2.9604 4004	3
4	3.9504 9566	3.9423 4034	3.9382 7279	3.9342 1196	4
5	4.9258 6633	4.9136 7723	4.9076 0029	4.9015 3506	5
6	5.8963 8441	5.8793 8084	5.8709 0712	5.8624 5205	6
7	6.8620 7404	6.8394 8385	6.8282 3068	6.8170 0535	7
8	7.8229 5924	7.7940 1875	7.7796 0813	7.7652 3710	8
9	8.7790 6392	8.7430 1781	8.7250 7640	8.7071 8917	9
10	9.7304 1186	9.6865 1315	9.6646 7220	9.6429 0315	10
11	10.6770 2673	10.6245 3669	10.5984 3200	10.5724 2035	11
12	11.6189 3207	11.5571 2016	11.5263 9205	11.4957 8180	12
13	12.5561 5131	12.4842 9511	12.4485 8837	12.4130 2828	13
14	13.4887 0777	13.4060 9291	13.3650 5676	13.3242 0028	14
15	14.4166 2465	14.3225 4473	14.2758 3281	14.2293 3802	15
16	15.3399 2502	15.2336 8160	15.1809 5186	15.1284 8148	16
17	16.2586 3186	16.1395 3432	16.0804 4905	16.0216 7035	17
18	17.1727 6802	17.0401 3354	16.9743 5931	16.9089 4405	18
19	18.0823 5624	17.9355 0974	17.8627 1733	17.7903 4177	19
20	18.9874 1915	18.8256 9320	18.7455 5759	18.6659 0242	20
21	19.8879 7925	19.7107 1404	19.6229 1438	19.5356 6466	21
22	20.7840 5896	20.5906 0220	20.4948 2174	20.3996 6688	22
23	21.6756 8055	21.4653 8745	21.3613 1353	21.2579 4723	23
24	22.5628 6622	22.3350 9938	22.2224 2338	22.1105 4361	24
25	23.4456 3803	23.1997 6741	23.0781 8473	22.9574 9365	25
26	24.3240 1794	24.0594 2079	23.9286 3079	23.7988 3475	26
27	25.1980 2780	24.9140 8862	24.7737 9457	24.6346 0406	27
28	26.0676 8936	25.7637 9979	25.6137 0889	25.4648 3847	28
29	26.9330 2423	26.6085 8307	26.4484 0635	26.2895 7464	29
30	27.7940 5397	27.4484 6702	27.2779 1935	27.1088 4898	30
31	28.6507 9997	28.2834 8006	28.1022 8010	27.9226 9766	31
32	29.5032 8355	29.1136 5044	28.9215 2060	28.7311 5662	32
33	30.3515 2592	29.9390 0625	29.7356 7265	29.5342 6154	33
34	31.1955 4818	30.7595 7540	30.5447 6785	30.3320 4789	34
35	32.0353 7132	31.5753 8566	31.3488 3761	31.1245 5088	35
36	32.8710 1624	32.3864 6463	32.1479 1315	31.9118 0551	36
37	33.7025 0372	33.1928 3974	32.9420 2550	32.6938 4653	37
38	34.5298 5445	33.9945 3828	33.7312 0546	33.4707 0848	38
39	35.3530 8900	34.7915 8736	34.5154 8369	34.2424 2564	39
40	36.1722 2786	35.5840 1396	35.2948 9062	35.0090 3209	40
41	36.9872 9141	36.3718 4487	36.0694 5652	35.7705 6168	41
42	37.7982 9991	37.1551 0676	36.8392 1145	36.5270 4803	42
43	38.6052 7354	37.9338 2612	37.6041 8529	37.2785 2453	43
44	39.4082 3238	38.7080 2929	38.3644 0774	38.0250 2437	44
45	40.2071 9640	39.4777 4248	39.1199 0831	38.7665 8050	45
46	41.0021 8547	40.2429 9170	39.8707 1634	39.5032 2566	46
47	41.7932 1937	41.0038 0287	40.6168 6096	40.2349 9238	47
48	42.5803 1778	41.7602 0170	41.3583 7114	40.9619 1296	48
49	43.3635 0028	42.5122 1380	42.0952 7566	41.6840 1949	49
50	44.1427 8635	43.2598 6460	42.8276 0314	42.4013 4387	50

TABLE C-8 PRESENT VALUE OF ANNUITY OF 1 PER POUND (cont.)

$$a_{\overline{n}|i} = \frac{1-(1+i)^{-n}}{i}$$

n	¾%	⅞%	1%	1⅛%	n
1	0.9925 5583	0.9913 2590	0.9900 9901	0.9888 7515	1
2	1.9777 2291	1.9740 5294	1.9703 9506	1.9667 4923	2
3	2.9555 5624	2.9482 5570	2.9409 8521	2.9337 4460	3
4	3.9261 1041	3.9140 0813	3.9019 6555	3.8899 8230	4
5	4.8894 3961	4.8713 8352	4.8534 3124	4.8355 8200	5
6	5.8455 9763	5.8204 5454	5.7954 7647	5.7706 6205	6
7	6.7946 3785	6.7612 9323	6.7281 9453	6.6953 3948	7
8	7.7366 1325	7.6939 7098	7.6516 7775	7.6097 3002	8
9	8.6715 7642	8.6185 5859	8.5660 1758	8.5139 4810	9
10	9.5995 7958	9.5351 2624	9.4713 0453	9.4081 0690	10
11	10.5206 7452	10.4437 4348	10.3676 2825	10.2923 1832	11
12	11.4349 1267	11.3444 7929	11.2550 7747	11.1666 9302	12
13	12.3423 4508	12.2374 0202	12.1337 4007	12.0313 4044	13
14	13.2430 2242	13.1225 7945	13.0037 0304	12.8863 6880	14
15	14.1369 9495	14.0000 7876	13.8650 5252	13.7318 8509	15
16	15.0243 1261	14.8699 6656	14.7178 7378	14.5679 9514	16
17	15.9050 2492	15.7323 0885	15.5622 5127	15.3948 0360	17
18	16.7791 8107	16.5871 7111	16.3982 6858	16.2124 1395	18
19	17.6468 2984	17.4346 1820	17.2260 0850	17.0209 2850	19
20	18.5080 1969	18.2747 1445	18.0455 5297	17.8204 4845	20
21	19.3627 9870	19.1075 2361	18.8569 8313	18.6110 7387	21
22	20.2112 1459	19.9331 0891	19.6603 7934	19.3929 0371	22
23	21.0533 1473	20.7515 3300	20.4558 2113	20.1660 3580	23
24	21.8891 4614	21.5628 5799	21.2433 8726	20.9305 6693	24
25	22.7187 5547	22.3671 4547	22.0231 5570	21.6865 9276	25
26	23.5421 8905	23.1644 5647	22.7952 0366	22.4342 0792	26
27	24.3594 9286	23.9548 5152	23.5596 0759	23.1735 0598	27
28	25.1707 1251	24.7383 9060	24.3164 4316	23.9045 7946	28
29	25.9758 9331	25.5151 3319	25.0657 8530	24.6275 1986	29
30	26.7750 8021	26.2851 3823	25.8077 0822	25.3424 1766	30
31	27.5683 1783	27.0484 6417	26.5422 8537	26.0493 6233	31
32	28.3556 5045	27.8051 6894	27.2695 8947	26.7484 4236	32
33	29.1371 2203	28.5553 0998	27.9896 9255	27.4397 4522	33
34	29.9127 7621	29.2989 4422	28.7026 6589	28.1233 5745	34
35	30.6826 5629	30.0361 2809	29.4085 8009	28.7993 6460	35
36	31.4468 0525	30.7669 1757	30.1075 0504	29.4678 5127	36
37	32.2052 6576	31.4913 6810	30.7995 0994	30.1289 0114	37
38	32.9580 8016	32.2095 3467	31.4846 6330	30.7825 9692	38
39	33.7052 9048	32.9214 7179	32.1630 3298	31.4290 2044	39
40	34.4469 3844	33.6272 3350	32.8346 8611	32.0682 5260	40
41	35.1830 6545	34.3268 7335	33.4996 8922	32.7903 7340	41
42	35.9137 1260	35.0204 4446	34.1581 0814	33.3254 6195	42
43	36.6389 2070	35.7079 9947	34.8100 0806	33.9435 9649	43
44	37.3587 3022	36.3895 9055	35.4554 5352	34.5548 5438	44
45	38.0731 8136	37.0652 6944	36.0945 0844	35.1593 1212	45
46	38.7823 1401	37.7350 8743	36.7272 3608	35.7570 4536	46
47	39.4861 6774	38.3990 9535	37.3536 9909	36.3481 2891	47
48	40.1847 8189	39.0573 4359	37.9739 5949	36.9326 3674	48
49	40.8781 9542	39.7098 8212	38.5880 7871	37.5106 4202	49
50	41.5664 4707	40.3567 6047	39.1961 1753	38.0822 1708	50

TABLE C-8 PRESENT VALUE OF ANNUITY OF 1 PER POUND (*cont.*)

$$a_{\overline{n}|i} = \frac{1-(1+i)^{-n}}{i}$$

n	1¼%	1⅜%	1½%	1¾%	n
1	0.9876 5432	0.9864 3650	0.9852 2167	0.9828 0098	1
2	1.9631 1538	1.9594 9346	1.9558 8342	1.9486 9875	2
3	2.9265 3371	2.9193 5237	2.9122 0042	2.8979 8403	3
4	3.8780 5798	3.8661 9222	3.8543 8465	3.8309 4254	4
5	4.8178 3504	4.8001 8962	4.7826 4497	4.7478 5508	5
6	5.7460 0992	5.7215 1874	5.6971 8717	5.6489 9762	6
7	6.6627 2585	6.6303 5140	6.5982 1396	6.5346 4139	7
8	7.5681 2429	7.5268 5712	7.4859 2508	7.4050 5297	8
9	8.4623 4498	8.4112 0308	8.3605 1732	8.2604 9432	9
10	9.3455 2591	9.2835 5421	9.2221 8455	9.1012 2291	10
11	10.2178 0337	10.1440 7320	10.0711 1779	9.9274 9181	11
12	11.0793 1197	10.9929 2054	10.9075 0521	10.7395 4969	12
13	11.9301 8466	11.8302 5454	11.7315 3222	11.5376 4097	13
14	12.7705 5275	12.6562 3136	12.5433 8150	12.3220 0587	14
15	13.6005 4592	13.4710 0504	13.3432 3301	13.0928 8046	15
16	14.4202 9227	14.2747 2754	14.1312 6405	13.8504 9677	16
17	15.2299 1829	15.0675 4874	14.9076 4931	14.5950 8282	17
18	16.0295 4893	15.8496 1651	15.6725 6089	15.3268 6272	18
19	16.8193 0759	16.6210 7671	16.4261 6837	16.0460 5673	19
20	17.5993 1613	17.3820 7320	17.1686 3879	16.7528 8130	20
21	18.3696 9495	18.1327 4792	17.9001 3673	17.4475 4919	21
22	19.1305 6291	18.8732 4086	18.6208 2437	18.1302 6948	22
23	19.8820 3744	19.6036 9012	19.3308 6145	18.8012 4764	23
24	20.6242 3451	20.3242 3193	20.0304 0537	19.4606 8565	24
25	21.3572 6865	21.0350 0067	20.7196 1120	20.1087 8196	25
26	22.0812 5299	21.7361 2890	21.3986 3172	20.7457 3166	26
27	22.7962 9925	22.4277 4737	22.0676 1746	21.3717 2644	27
28	23.5025 1778	23.1099 8508	22.7267 1671	21.9869 5474	28
29	24.2000 1756	23.7829 6925	23.3760 7558	22.5916 0171	29
30	24.8889 0623	24.4468 2540	24.0158 3801	23.1858 4934	30
31	25.5692 9010	25.1016 7734	24.6461 4582	23.7698 7650	31
32	26.2412 7418	25.7476 4719	25.2671 3874	24.3438 5897	32
33	26.9049 6215	26.3848 5543	25.8789 5442	24.9079 6951	33
34	27.5604 5644	27.0134 2089	26.4817 2849	25.4623 7789	34
35	28.2078 5822	27.6334 6080	27.0755 9458	26.0072 5100	35
36	28.8472 6737	28.2450 9080	27.6606 8431	26.5427 5283	36
37	29.4787 8259	28.8484 2496	28.2371 2740	27.0690 4455	37
38	30.1025 0133	29.4435 7579	28.8050 5163	27.5862 8457	38
39	30.7185 1983	30.0306 5430	29.3645 8288	28.0946 2857	39
40	31.3269 3316	30.6097 6996	29.9158 4520	28.5942 2955	40
41	31.9278 3522	31.1810 3079	30.4589 6079	29.0852 3789	41
42	32.5213 1874	31.7445 4332	30.9940 5004	29.5678 0135	42
43	33.1074 7530	32.3004 1264	31.5212 3157	30.0420 6522	43
44	33.6863 9536	32.8487 4243	32.0406 2223	30.5081 7221	44
45	34.2581 6825	33.3896 3495	32.5523 3718	30.9662 6261	45
46	34.8228 8222	33.9231 9108	33.0564 8983	31.4164 7431	46
47	35.3806 2442	34.4495 1031	33.5531 9195	31.8589 4281	47
48	35.9314 8091	34.9686 9081	34.0425 5365	32.2938 0129	48
49	36.4755 3670	35.4808 2941	34.5246 8339	32.7211 8063	49
50	37.0128 7574	35.9860 2161	34.9996 8807	33.1412 0946	50

TABLE C-8 PRESENT VALUE OF ANNUITY OF 1 PER POUND (cont.)

$$a_{\overline{n}|i} = \frac{1-(1+i)^{-n}}{i}$$

n	2%	2¼%	2½%	2¾%	n
1	0.9803 9216	0.9779 9511	0.9756 0976	0.9732 3601	1
2	1.9415 6094	1.9344 6955	1.9274 2415	1.9204 2434	2
3	2.8838 8327	2.8698 9687	2.8560 2356	2.8422 6213	3
4	3.8077 2870	3.7847 4021	3.7619 7421	3.7394 2787	4
5	4.7134 5951	4.6794 5253	4.6458 2850	4.6125 8186	5
6	5.6014 3089	5.5544 7660	5.5081 2536	5.4623 6678	6
7	6.4719 9107	6.4102 4626	6.3493 9060	6.2894 0806	7
8	7.3254 8144	7.2471 8461	7.1701 3717	7.0943 1441	8
9	8.1622 3671	8.0657 0622	7.9708 6553	7.8776 7826	9
10	8.9825 8501	8.8662 1635	8.7520 6393	8.6400 7616	10
11	9.7868 4805	9.6491 1134	9.5142 0871	9.3820 6926	11
12	10.5753 4122	10.4147 7882	10.2577 6460	10.1042 0366	12
13	11.3483 7375	11.1635 9787	10.9831 8497	10.8070 1086	13
14	12.1062 4877	11.8959 3924	11.6909 1217	11.4910 0814	14
15	12.8492 6350	12.6121 6551	12.3813 7773	12.1566 9892	15
16	13.5777 0931	13.3126 3131	13.0550 0266	12.8045 7315	16
17	14.2918 7188	13.9976 8343	13.7121 9772	13.4351 0769	17
18	14.9920 3125	14.6676 6106	14.3533 6363	14.0487 6661	18
19	15.6784 6201	15.3228 9590	14.9788 9134	14.6460 0157	19
20	16.3514 3334	15.9637 1237	15.5891 6229	15.2272 5213	20
21	17.0112 0916	16.5904 2775	16.1845 4857	15.7929 4612	21
22	17.6580 4820	17.2033 5232	16.7654 1324	16.3434 9987	22
23	18.2922 0412	17.8027 8955	17.3321 1048	16.8793 1861	23
24	18.9139 2560	18.3890 3624	17.8849 8583	17.4007 9670	24
25	19.5234 5647	18.9623 8263	18.4243 7642	17.9083 1795	25
26	20.1210 3576	19.5231 1260	18.9506 1114	18.4022 5592	26
27	20.7068 9780	20.0715 0376	19.4640 1087	18.8829 7413	27
28	21.2812 7236	20.6078 2764	19.9648 8866	19.3508 2640	28
29	21.8443 8466	21.1323 4977	20.4535 4991	19.8061 5708	29
30	22.3964 5555	21.6453 2985	20.9302 9259	20.2493 0130	30
31	22.9377 0152	22.1470 2186	21.3954 0741	20.6805 8520	31
32	23.4683 3482	22.6376 7419	21.8491 7796	21.1003 2623	32
33	23.9885 6355	23.1175 2977	22.2918 8094	21.5088 3332	33
34	24.4985 9172	23.5868 2618	22.7237 8628	21.9064 0712	34
35	24.9986 1933	24.0457 9577	23.1451 5734	22.2933 4026	35
36	25.4888 4248	24.4946 6579	23.5562 5107	22.6699 1753	36
37	25.9694 5341	24.9336 5848	23.9573 1812	23.0364 1609	37
38	26.4406 4060	25.3629 9118	24.3486 0304	23.3931 0568	38
39	26.9025 8883	25.7828 7646	24.7303 4443	23.7402 4884	39
40	27.3554 7924	26.1935 2221	25.1027 7505	24.0781 0106	40
41	27.7994 8945	26.5951 3174	25.4661 2200	24.4069 1101	41
42	28.2347 9358	26.9879 0390	25.8206 0683	24.7269 2069	42
43	28.6615 6233	27.3720 3316	26.1664 4569	25.0383 6563	43
44	29.0799 6307	27.7477 0969	26.5038 4945	25.3414 7507	44
45	29.4901 5987	28.1151 1950	26.8330 2386	25.6364 7209	45
46	29.8923 1360	28.4744 4450	27.1541 6962	25.9235 7381	46
47	30.2865 8196	28.8258 6259	27.4674 8255	26.2029 9154	47
48	30.6731 1957	29.1695 4777	27.7731 5371	26.4749 3094	48
49	31.0520 7801	29.5056 7019	28.0713 6947	26.7395 9215	49
50	31.4236 0589	29.8343 9627	28.3623 1168	26.9971 6998	50

TABLE C-8 PRESENT VALUE OF ANNUITY OF 1 PER POUND (cont.)

$$a_{\overline{n}|i} = \frac{1 - (1+i)^{-n}}{i}$$

n	3%	3½%	4%	4½%	n
1	0.9708 7379	0.9661 8357	0.9615 3846	0.9569 3780	1
2	1.9134 6970	1.8996 9428	1.8860 9467	1.8726 6775	2
3	2.8286 1135	2.8016 3698	2.7750 9103	2.7489 6435	3
4	3.7170 9840	3.6730 7921	3.6298 9522	3.5875 2570	4
5	4.5797 0719	4.5150 5238	4.4518 2233	4.3899 7674	5
6	5.4171 9144	5.3285 5302	5.2421 3686	5.1578 7248	6
7	6.2302 8296	6.1145 4398	6.0020 5467	5.8927 0094	7
8	7.0196 9219	6.8739 5554	6.7327 4487	6.5958 8607	8
9	7.7861 0892	7.6076 8651	7.4353 3161	7.2687 9050	9
10	8.5302 0284	8.3166 0532	8.1108 9578	7.9127 1818	10
11	9.2526 2411	9.0015 5104	8.7604 7671	8.5289 1692	11
12	9.9540 0399	9.6633 3433	9.3850 7376	9.1185 8078	12
13	10.6349 5533	10.3027 3849	9.9856 4785	9.6828 5242	13
14	11.2960 7314	10.9205 2028	10.5631 2293	10.2228 2528	14
15	11.9379 3509	11.5174 1090	11.1183 8743	10.7395 4573	15
16	12.5611 0203	12.0941 1681	11.6522 9561	11.2340 1505	16
17	13.1661 1847	12.6513 2059	12.1656 6885	11.7071 9143	17
18	13.7535 1308	13.1896 8173	12.6592 9697	12.1599 9180	18
19	14.3237 9911	13.7098 3742	13.1339 3940	12.5932 9359	19
20	14.8774 7486	14.2124 0330	13.5903 2634	13.0079 3645	20
21	15.4150 2414	14.6979 7420	14.0291 5995	13.4047 2388	21
22	15.9369 1664	15.1671 2484	14.4511 1533	13.7844 2476	22
23	16.4436 0839	15.6204 1047	14.8568 4167	14.1477 7489	23
24	16.9355 4212	16.0583 6760	15.2469 6314	14.4954 7837	24
25	17.4131 4769	16.4815 1459	15.6220 7994	14.8282 0896	25
26	17.8768 4242	16.8903 5226	15.9827 6918	15.1466 1145	26
27	18.3270 3147	17.2853 6451	16.3295 8575	15.4513 0282	27
28	18.7641 0823	17.6670 1885	16.6630 6322	15.7428 7351	28
29	19.1884 5459	18.0357 6700	16.9837 1463	16.0218 8853	29
30	19.6004 4135	18.3920 4541	17.2920 3330	16.2888 8854	30
31	20.0004 2849	18.7362 7576	17.5884 9356	16.5443 9095	31
32	20.3887 6553	19.0688 6547	17.8735 5150	16.7888 9086	32
33	20.7657 9178	19.3902 0818	18.1476 4567	17.0228 6207	33
34	21.1318 3668	19.7006 8423	18.4111 9776	17.2467 5796	34
35	21.4872 2007	20.0006 6110	18.6646 1323	17.4610 1240	35
36	21.8322 5250	20.2904 9381	18.9082 8195	17.6660 4058	36
37	22.1672 3544	20.5705 2542	19.1425 7880	17.8622 3979	37
38	22.4924 6159	20.8410 8736	19.3678 6423	18.0499 9023	38
39	22.8082 1513	21.1024 9987	19.5844 8484	18.2296 5572	39
40	23.1147 7197	21.3550 7234	19.7927 7388	18.4015 8442	40
41	23.4123 9997	21.5991 0371	19.9930 5181	18.5661 0949	41
42	23.7013 5920	21.8348 8281	20.1856 2674	18.7235 4975	42
43	23.9819 0213	22.0626 8870	20.3707 9494	18.8742 1029	43
44	24.2542 7392	22.2827 9102	20.5488 4129	19.0183 8305	44
45	24.5187 1254	22.4954 5026	20.7200 3970	19.1563 4742	45
46	24.7754 4907	22.7009 1813	20.8846 5356	19.2883 7074	46
47	25.0247 0783	22.8994 3780	21.0429 3612	19.4147 0884	47
48	25.2667 0664	23.0912 4425	21.1951 3088	19.5356 0654	48
49	25.5016 5693	23.2765 6450	21.3414 7200	19.6512 9813	49
50	25.7297 6401	23.4556 1787	21.4821 8462	19.7620 0778	50

TABLE C-8 PRESENT VALUE OF ANNUITY OF 1 PER POUND (cont.)

$$a_{\overline{n}|i} = \frac{1-(1+i)^{-n}}{i}$$

n	5%	5½%	6%	6½%	n
1	0.9523 8095	0.9478 6730	0.9433 9623	0.9389 6714	1
2	1.8594 1043	1.8463 1971	1.8333 9267	1.8206 2642	2
3	2.7232 4803	2.6979 3338	2.6730 1195	2.6484 7551	3
4	3.5459 5050	3.5051 5012	3.4651 0561	3.4257 9860	4
5	4.3294 7667	4.2702 8448	4.2123 6379	4.1556 7944	5
6	5.0756 9206	4.9955 3031	4.9173 2433	4.8410 1356	6
7	5.7863 7340	5.6829 6712	5.5823 8144	5.4845 1977	7
8	6.4632 1276	6.3345 6599	6.2097 9381	6.0887 5096	8
9	7.1078 2168	6.9521 9525	6.8016 9227	6.6561 0419	9
10	7.7217 3493	7.5376 2583	7.3600 8705	7.1888 3022	10
11	8.3064 1422	8.0925 3633	7.8868 7458	7.6890 4246	11
12	8.8632 5164	8.6185 1785	8.3838 4394	8.1587 2532	12
13	9.3935 7299	9.1170 7853	8.8526 8296	8.5997 4208	13
14	9.8986 4094	9.5896 4790	9.2949 8393	9.0138 4233	14
15	10.3796 5804	10.0375 8094	9.7122 4899	9.4026 6885	15
16	10.8377 6956	10.4621 6203	10.1058 9527	9.7677 6418	16
17	11.2740 6625	10.8646 0856	10.4772 5969	10.1105 7670	17
18	11.6895 8690	11.2460 7447	10.8276 0348	10.4324 6638	18
19	12.0853 2086	11.6076 5352	11.1581 1649	10.7347 1022	19
20	12.4622 1034	11.9503 8249	11.4699 2122	11.0185 0725	20
21	12.8211 5271	12.2752 4406	11.7640 7662	11.2849 8333	21
22	13.1630 0258	12.5831 6973	12.0415 8172	11.5351 9562	22
23	13.4885 7388	12.8750 4240	12.3033 7898	11.7701 3673	23
24	13.7986 4179	13.1516 9895	12.5503 5753	11.9907 3871	24
25	14.0939 4457	13.4139 3266	12.7833 5616	12.1978 7672	25
26	14.3751 8530	13.6624 9541	13.0031 6619	12.3923 7251	26
27	14.6430 3362	13.8980 9991	13.2105 3414	12.5749 9766	27
28	14.8981 2726	14.1214 2172	13.4061 6428	12.7464 7668	28
29	15.1410 7358	14.3331 0116	13.5907 2102	12.9074 8984	29
30	15.3724 5103	14.5337 4517	13.7648 3115	13.0586 7591	30
31	15.5928 1050	14.7239 2907	13.9290 8599	13.2006 3465	31
32	15.8026 7667	14.9041 9817	14.0840 4339	13.3339 2925	32
33	16.0025 4921	15.0750 6936	14.2302 2961	13.4590 8850	33
34	16.1929 0401	15.2370 3257	14.3681 4114	13.5766 0892	34
35	16.3741 9429	15.3905 5220	14.4982 4636	13.6869 5673	35
36	16.5468 5171	15.5360 6843	14.6209 8713	13.7905 6970	36
37	16.7112 8734	15.6739 9851	14.7367 8031	13.8878 5887	37
38	16.8678 9271	15.8047 3793	14.8460 1916	13.9792 1021	38
39	17.0170 4067	15.9286 6154	14.9490 7468	14.0649 8611	39
40	17.1590 8635	16.0461 2469	15.0462 9687	14.1455 2687	40
41	17.2943 6796	16.1574 6416	15.1380 1592	14.2211 5199	41
42	17.4232 0758	16.2629 9920	15.2245 4332	14.2921 6149	42
43	17.5459 1198	16.3630 3242	15.3061 7294	14.3588 3708	43
44	17.6627 7331	16.4578 5063	15.3831 8202	14.4214 4327	44
45	17.7740 6982	16.5477 2572	15.4558 3209	14.4802 2842	45
46	17.8800 6650	16.6329 1537	15.5243 6990	14.5354 2575	46
47	17.9810 1571	16.7136 6386	15.5890 2821	14.5872 5422	47
48	18.0771 5782	16.7902 0271	15.6500 2661	14.6359 1946	48
49	18.1687 2173	16.8627 5139	15.7075 7227	14.6816 1451	49
50	18.2559 2546	16.9315 1790	15.7618 6064	14.7245 2067	50

TABLE C-8 PRESENT VALUE OF ANNUITY OF 1 PER POUND (cont.)

$$a_{\overline{n}|i} = \frac{1 - (1+i)^{-n}}{i}$$

n	7%	7½%	8%	8½%	n
1	0.9345 7944	0.9302 3256	0.9259 2593	0.9216 5899	1
2	1.8080 1817	1.7955 6517	1.7832 6475	1.7711 1427	2
3	2.6243 1604	2.6005 2574	2.5770 9699	2.5540 2237	3
4	3.3872 1126	3.3493 2627	3.3121 2684	3.2755 9666	4
5	4.1001 9744	4.0458 8490	3.9927 1004	3.9406 4208	5
6	4.7665 3966	4.6938 4642	4.6228 7966	4.5535 8717	6
7	5.3892 8940	5.2966 0132	5.2063 7006	5.1185 1352	7
8	5.9712 9851	5.8573 0355	5.7466 3894	5.6391 8297	8
9	6.5152 3225	6.3788 8703	6.2468 8791	6.1190 6264	9
10	7.0235 8154	6.8640 8096	6.7100 8140	6.5613 4806	10
11	7.4986 7434	7.3154 2415	7.1389 6426	6.9689 8439	11
12	7.9426 8630	7.7352 7827	7.5360 7802	7.3446 8607	12
13	8.3576 5074	8.1258 4026	7.9037 7594	7.6909 5490	13
14	8.7454 6799	8.4891 5373	8.2442 3698	8.0100 9668	14
15	9.1079 1401	8.8271 1974	8.5594 7869	8.3042 3658	15
16	9.4466 4860	9.1415 0674	8.8513 6916	8.5753 3325	16
17	9.7632 2299	9.4339 5976	9.1216 3811	8.8251 9194	17
18	10.0590 8691	9.7060 0908	9.3718 8714	9.0554 7644	18
19	10.3355 9524	9.9590 7821	9.6035 9920	9.2677 2022	19
20	10.5940 1425	10.1944 9136	9.8181 4741	9.4633 3661	20
21	10.8355 2733	10.4134 8033	10.0168 0316	9.6436 2821	21
22	11.0612 4050	10.6171 9101	10.2007 4366	9.8097 9559	22
23	11.2721 8738	10.8066 8931	10.3710 5895	9.9629 4524	23
24	11.4693 3400	10.9829 6680	10.5287 5828	10.1040 9700	24
25	11.6535 8318	11.1469 4586	10.6747 7619	10.2341 9078	25
26	11.8257 7867	11.2994 8452	10.8099 7795	10.3540 9288	26
27	11.9867 0904	11.4413 8095	10.9351 6477	10.4646 0174	27
28	12.1371 1125	11.5733 7763	11.0510 7849	10.5664 5321	28
29	12.2776 7407	11.6961 6524	11.1584 0601	10.6603 2554	29
30	12.4090 4118	11.8103 8627	11.2577 8334	10.7468 4382	30
31	12.5318 1419	11.9166 3839	11.3497 9939	10.8265 8416	31
32	12.6465 5532	12.0154 7757	11.4349 9944	10.9000 7757	32
33	12.7537 9002	12.1074 2099	11.5138 8837	10.9678 1343	33
34	12.8540 0936	12.1929 4976	11.5869 3367	11.0302 4279	34
35	12.9476 7230	12.2725 1141	11.6545 6822	11.0877 8137	35
36	13.0352 0776	12.3465 2224	11.7171 9279	11.1408 1233	36
37	13.1170 1660	12.4153 6953	11.7751 7851	11.1896 8878	37
38	13.1934 7345	12.4794 1351	11.8288 6899	11.2347 3620	38
39	13.2649 2846	12.5389 8931	11.8785 8240	11.2762 5457	39
40	13.3317 0884	12.5944 0866	11.9246 1333	11.3145 2034	40
41	13.3941 2041	12.6459 6155	11.9672 3457	11.3497 8833	41
42	13.4524 4898	12.6939 1772	12.0066 9867	11.3822 9339	42
43	13.5069 6167	12.7385 2811	12.0432 3951	11.4122 5197	43
44	13.5579 0810	12.7800 2615	12.0770 7362	11.4398 6357	44
45	13.6055 2159	12.8186 2898	12.1084 0150	11.4653 1205	45
46	13.6500 2018	12.8545 3858	12.1374 0880	11.4887 6686	46
47	13.6916 0764	12.8879 4287	12.1642 6741	11.5103 8420	47
48	13.7304 7443	12.9190 1662	12.1891 3649	11.5303 0802	48
49	13.7667 9853	12.9479 2244	12.2121 6341	11.5486 7099	49
50	13.8007 4629	12.9748 1157	12.2334 8464	11.5655 9538	50

TABLE C-9 PERIODIC RENT OF ANNUITY WHOSE PRESENT VALUE IS 1

$$\frac{1}{a_{\overline{n}|i}} = \frac{i}{1-(1+i)^{-n}} \qquad \left[\frac{1}{s_{\overline{n}|i}} = \frac{1}{a_{\overline{n}|i}} - i\right]$$

n	1/4%	7/24%	1/3%	5/12%	n
1	1.0025 0000	1.0029 1667	1.0033 3333	1.0041 6667	1
2	0.5018 7578	0.5021 8856	0.5025 0139	0.5031 2717	2
3	0.3350 0139	0.3352 7967	0.3355 5802	0.3361 1496	3
4	0.2515 6445	0.2518 2557	0.2520 8680	0.2526 0958	4
5	0.2015 0250	0.2017 5340	0.2020 0444	0.2025 0693	5
6	0.1681 2803	0.1683 7219	0.1686 1650	0.1691 0564	6
7	0.1442 8928	0.1445 2866	0.1447 6824	0.1452 4800	7
8	0.1264 1035	0.1266 4620	0.1268 8228	0.1273 5512	8
9	0.1125 0462	0.1127 3777	0.1129 7118	0.1134 3876	9
10	0.1013 8015	0.1016 1117	0.1018 4248	0.1023 0596	10
11	0.0922 7840	0.0925 0772	0.0927 3736	0.0931 9757	11
12	0.0846 9370	0.0849 2163	0.0851 4990	0.0856 0748	12
13	0.0782 7595	0.0785 0274	0.0787 2989	0.0791 8532	13
14	0.0727 7510	0.0730 0093	0.0732 2716	0.0736 8082	14
15	0.0680 0777	0.0682 3279	0.0684 5825	0.0689 1045	15
16	0.0638 3642	0.0640 6076	0.0642 8557	0.0647 3655	16
17	0.0601 5587	0.0603 7964	0.0606 0389	0.0610 5387	17
18	0.0568 8433	0.0571 0761	0.0573 3140	0.0577 8053	18
19	0.0539 5722	0.0541 8008	0.0544 0348	0.0548 5191	19
20	0.0513 2288	0.0515 4537	0.0517 6844	0.0522 1630	20
21	0.0489 3947	0.0491 6166	0.0493 8445	0.0498 3183	21
22	0.0467 7278	0.0469 9471	0.0472 1726	0.0476 6427	22
23	0.0447 9455	0.0450 1625	0.0452 3861	0.0456 8531	23
24	0.0429 8121	0.0432 0272	0.0434 2492	0.0438 7139	24
25	0.0413 1298	0.0415 3433	0.0417 5640	0.0422 0270	25
26	0.0397 7312	0.0399 9434	0.0402 1630	0.0406 6247	26
27	0.0383 4736	0.0385 6847	0.0387 9035	0.0392 3645	27
28	0.0370 2347	0.0372 4450	0.0374 6632	0.0379 1239	28
29	0.0357 9093	0.0360 1188	0.0362 3367	0.0366 7974	29
30	0.0346 4059	0.0348 6149	0.0350 8325	0.0355 2936	30
31	0.0335 6449	0.0337 8536	0.0340 0712	0.0344 5330	31
32	0.0325 5569	0.0327 7653	0.0329 9830	0.0334 4458	32
33	0.0316 0806	0.0318 2889	0.0320 5067	0.0324 9708	33
34	0.0307 1620	0.0309 3703	0.0311 5885	0.0316 0540	34
35	0.0298 7533	0.0300 9618	0.0303 1803	0.0307 6476	35
36	0.0290 8121	0.0293 0208	0.0295 2399	0.0299 7090	36
37	0.0283 3004	0.0285 5094	0.0287 7291	0.0292 2003	37
38	0.0276 1843	0.0278 3938	0.0280 6141	0.0285 0875	38
39	0.0269 4335	0.0271 6434	0.0273 8644	0.0278 3402	39
40	0.0263 0204	0.0265 2308	0.0267 4527	0.0271 9310	40
41	0.0256 9204	0.0259 1315	0.0261 3543	0.0263 8352	41
42	0.0251 1112	0.0253 3229	0.0255 5466	0.0260 0303	42
43	0.0245 5724	0.0247 7848	0.0250 0095	0.0254 4961	43
44	0.0240 2855	0.0242 4987	0.0244 7246	0.0249 2141	44
45	0.0235 2339	0.0237 4479	0.0239 6749	0.0244 1675	45
46	0.0230 4022	0.0232 6170	0.0234 8451	0.0239 3409	46
47	0.0225 7762	0.0227 9920	0.0230 2213	0.0234 7204	47
48	0.0221 3433	0.0223 5600	0.0225 7905	0.0230 2929	48
49	0.0217 0915	0.0219 3092	0.0221 5410	0.0226 0468	49
50	0.0213 0099	0.0215 2287	0.0217 4618	0.0221 9711	50

TABLE C-9 PERIODIC RENT OF ANNUITY WHOSE PRESENT VALUE IS 1 (cont.)

$$\frac{1}{a_{\overline{n}|i}} = \frac{i}{1-(1+i)^{-n}} \qquad \left[\frac{1}{s_{\overline{n}|i}} = \frac{1}{a_{\overline{n}|i}} - i\right]$$

n	½%	7/12%	5/8%	2/3%	n
1	1.0050 0000	1.0058 3333	1.0062 5000	1.0066 6667	1
2	0.5037 5312	0.5043 7924	0.5046 9237	0.5050 0554	2
3	0.3366 7221	0.3372 2976	0.3375 0865	0.3377 8762	3
4	0.2531 3279	0.2536 5644	0.2539 1842	0.2541 8051	4
5	0.2030 0997	0.2035 1357	0.2037 6558	0.2040 1772	5
6	0.1695 9546	0.1700 8594	0.1703 3143	0.1705 7709	6
7	0.1457 2854	0.1462 0986	0.1464 5082	0.1466 9198	7
8	0.1278 2886	0.1283 0351	0.1285 4118	0.1287 7907	8
9	0.1139 0736	0.1143 7698	0.1146 1218	0.1148 4763	9
10	0.1027 7057	0.1032 3632	0.1034 6963	0.1037 0321	10
11	0.0936 5903	0.0941 2175	0.0943 5358	0.0945 8572	11
12	0.0860 6643	0.0865 2675	0.0867 5742	0.0869 8843	12
13	0.0796 4224	0.0801 0064	0.0803 3039	0.0805 6052	13
14	0.0741 3609	0.0745 9295	0.0748 2198	0.0750 5141	14
15	0.0693 6436	0.0698 1999	0.0700 4845	0.0702 7734	15
16	0.0651 8937	0.0656 4401	0.0658 7202	0.0661 0049	16
17	0.0615 0579	0.0619 5966	0.0621 8732	0.0624 1546	17
18	0.0582 3173	0.0586 8499	0.0589 1239	0.0591 4030	18
19	0.0553 0253	0.0557 5532	0.0559 8252	0.0562 1027	19
20	0.0526 6645	0.0531 1889	0.0533 4597	0.0535 7362	20
21	0.0502 8163	0.0507 3383	0.0509 6083	0.0511 8843	21
22	0.0481 1380	0.0485 6585	0.0487 9281	0.0490 2041	22
23	0.0461 3465	0.0465 8663	0.0468 1360	0.0470 4123	23
24	0.0443 2061	0.0447 7258	0.0449 9959	0.0452 2729	24
25	0.0426 5186	0.0431 0388	0.0433 3096	0.0435 5876	25
26	0.0411 1163	0.0415 6376	0.0417 9094	0.0420 1886	26
27	0.0396 8565	0.0401 3793	0.0403 6523	0.0405 9331	27
28	0.0383 6167	0.0388 1415	0.0390 4159	0.0392 6983	28
29	0.0371 2914	0.0375 8186	0.0378 0946	0.0380 3789	29
30	0.0359 7892	0.0364 3191	0.0366 5969	0.0368 8832	30
31	0.0349 0304	0.0353 5633	0.0355 8430	0.0358 1316	31
32	0.0338 9453	0.0343 4815	0.0345 7633	0.0348 0542	32
33	0.0329 4727	0.0334 0124	0.0336 2964	0.0338 5898	33
34	0.0320 5586	0.0325 1020	0.0327 3883	0.0329 6843	34
35	0.0312 1550	0.0316 7024	0.0318 9911	0.0321 2898	35
36	0.0304 2194	0.0308 7710	0.0311 0622	0.0313 3637	36
37	0.0296 7139	0.0301 2698	0.0303 5636	0.0305 8680	37
38	0.0289 6045	0.0294 1649	0.0296 4614	0.0298 7687	38
39	0.0282 8607	0.0287 4258	0.0289 7250	0.0292 0354	39
40	0.0276 4552	0.0281 0251	0.0283 3271	0.0285 6406	40
41	0.0270 3631	0.0274 9379	0.0277 2429	0.0279 5595	41
42	0.0264 5622	0.0269 1420	0.0271 4499	0.0273 7697	42
43	0.0259 0320	0.0263 6170	0.0265 9278	0.0268 2509	43
44	0.0253 7541	0.0258 3443	0.0260 6583	0.0262 9847	44
45	0.0248 7117	0.0253 3073	0.0255 6243	0.0257 9541	45
46	0.0243 8894	0.0248 4905	0.0250 8106	0.0253 1439	46
47	0.0239 2733	0.0243 8798	0.0246 2032	0.0248 5399	47
48	0.0234 8503	0.0239 4624	0.0241 7890	0.0244 1292	48
49	0.0230 6087	0.0235 2265	0.0237 5563	0.0239 9001	49
50	0.0226 5376	0.0231 1611	0.0233 4943	0.0235 8416	50

TABLE C-9 PERIODIC RENT OF ANNUITY WHOSE PRESENT VALUE IS 1 (cont.)

$$\frac{1}{a_{\overline{n}|i}} = \frac{i}{1-(1+i)^{-n}} \qquad \left[\frac{1}{s_{\overline{n}|i}} = \frac{1}{a_{\overline{n}|i}} - i\right]$$

n	¾%	⅞%	1%	1⅛%	n
1	1.0075 0000	1.0087 5000	1.0100 0000	1.0112 5000	1
2	0.5056 3200	0.5065 7203	0.5075 1244	0.5084 5323	2
3	0.3383 4579	0.3391 8361	0.3400 2211	0.3408 6130	3
4	0.2547 0501	0.2554 9257	0.2562 8109	0.2570 7058	4
5	0.2045 2242	0.2052 8049	0.2060 3980	0.2068 0034	5
6	0.1710 6891	0.1718 0789	0.1725 4837	0.1732 9034	6
7	0.1471 7488	0.1479 0070	0.1486 2828	0.1493 5762	7
8	0.1292 5552	0.1299 7190	0.1306 9029	0.1314 1071	8
9	0.1153 1929	0.1160 2868	0.1167 4037	0.1174 5432	9
10	0.1041 7123	0.1048 7538	0.1055 8208	0.1062 9131	10
11	0.0950 5094	0.0957 5111	0.0964 5408	0.0971 5984	11
12	0.0874 5148	0.0881 4860	0.0888 4879	0.0895 5203	12
13	0.0810 2188	0.0817 1669	0.0824 1482	0.0831 1626	13
14	0.0755 1146	0.0762 0453	0.0769 0117	0.0776 0138	14
15	0.0707 3639	0.0714 2817	0.0721 2378	0.0728 2321	15
16	0.0665 5879	0.0672 4965	0.0679 4460	0.0686 4363	16
17	0.0628 7321	0.0635 6346	0.0642 5806	0.0649 5698	17
18	0.0595 9766	0.0602 8756	0.0609 8205	0.0616 8113	18
19	0.0566 6740	0.0573 5715	0.0580 5175	0.0587 5120	19
20	0.0540 3063	0.0547 2042	0.0554 1532	0.0561 1531	20
21	0.0516 4543	0.0523 3541	0.0530 3075	0.0537 3145	21
22	0.0494 7748	0.0501 6779	0.0508 6371	0.0515 6525	22
23	0.0474 9846	0.0481 8921	0.0488 8584	0.0495 8833	23
24	0.0456 8474	0.0463 7604	0.0470 7347	0.0477 7701	24
25	0.0440 1650	0.0447 0843	0.0454 0675	0.0461 1144	25
26	0.0424 7693	0.0431 6959	0.0438 6888	0.0445 7479	26
27	0.0410 5176	0.0417 4520	0.0424 4553	0.0431 5273	27
28	0.0397 2871	0.0404 2300	0.0411 2444	0.0418 3299	28
29	0.0384 9723	0.0391 9243	0.0398 9502	0.0406 0498	29
30	0.0373 4816	0.0380 4431	0.0387 4811	0.0394 5953	30
31	0.0362 7352	0.0369 7068	0.0376 7573	0.0383 8866	31
32	0.0352 6634	0.0359 6454	0.0366 7089	0.0373 8535	32
33	0.0343 2048	0.0350 1976	0.0357 2744	0.0364 4349	33
34	0.0334 3053	0.0341 3092	0.0348 3997	0.0355 5763	34
35	0.0325 9170	0.0332 9324	0.0340 0368	0.0347 2299	35
36	0.0317 9973	0.0325 0244	0.0332 1431	0.0339 3529	36
37	0.0310 5082	0.0317 5473	0.0324 6805	0.0331 9072	37
38	0.0303 4157	0.0310 4671	0.0317 6150	0.0324 8589	38
39	0.0296 6893	0.0303 7531	0.0310 9160	0.0318 1773	39
40	0.0290 3016	0.0297 3780	0.0304 5560	0.0311 8349	40
41	0.0284 2276	0.0291 3169	0.0298 5102	0.0305 8069	41
42	0.0278 4452	0.0285 5475	0.0292 7563	0.0300 0709	42
43	0.0272 9338	0.0280 0493	0.0287 2737	0.0294 6064	43
44	0.0267 6751	0.0274 8039	0.0282 0441	0.0289 3949	44
45	0.0262 6521	0.0269 7943	0.0277 0505	0.0284 4197	45
46	0.0257 8495	0.0265 0053	0.0272 2775	0.0279 6652	46
47	0.0253 2532	0.0260 4228	0.0267 7111	0.0275 1173	47
48	0.0248 8504	0.0256 0338	0.0263 3384	0.0270 7632	48
49	0.0244 6292	0.0251 8265	0.0259 1474	0.0266 5910	49
50	0.0240 5787	0.0247 7900	0.0255 1273	0.0262 5898	50

TABLE C-9 PERIODIC RENT OF ANNUITY WHOSE PRESENT VALUE IS 1 (cont.)

$$\frac{1}{a_{\overline{n}|i}} = \frac{i}{1-(1+i)^{-n}} \qquad \left[\frac{1}{s_{\overline{n}|i}} = \frac{1}{a_{\overline{n}|i}} - i\right]$$

n	1¼%	1⅜%	1½%	1¾%	n
1	1.0125 0000	1.0137 5000	1.0150 0000	1.0175 0000	1
2	0.5093 9441	0.5103 3597	0.5112 7792	0.5131 6285	2
3	0.3417 0117	0.3425 4173	0.3433 8296	0.3450 6746	3
4	0.2578 6102	0.2586 5243	0.2594 4478	0.2610 3237	4
5	0.2075 6211	0.2083 2510	0.2090 8932	0.2106 2142	5
6	0.1740 3381	0.1747 7877	0.1755 2521	0.1770 2256	6
7	0.1500 8872	0.1508 2157	0.1515 5616	0.1530 3059	7
8	0.1321 3314	0.1328 5758	0.1335 8402	0.1350 4292	8
9	0.1181 7055	0.1188 8906	0.1196 0982	0.1210 5813	9
10	0.1070 0307	0.1077 1737	0.1084 3418	0.1098 7534	10
11	0.0978 6839	0.0985 7973	0.0992 9384	0.1007 3038	11
12	0.0902 5831	0.0909 6764	0.0916 7999	0.0931 1377	12
13	0.0838 2100	0.0845 2903	0.0852 4036	0.0866 7283	13
14	0.0783 0515	0.0790 1246	0.0797 2332	0.0811 5562	14
15	0.0735 2646	0.0742 3351	0.0749 4436	0.0763 7739	15
16	0.0693 4672	0.0700 5388	0.0707 6508	0.0721 9958	16
17	0.0656 6023	0.0663 6780	0.0670 7966	0.0685 1623	17
18	0.0623 8479	0.0630 9301	0.0638 0578	0.0652 4492	18
19	0.0594 5548	0.0601 6457	0.0608 7847	0.0623 2061	19
20	0.0568 2039	0.0575 3054	0.0582 4574	0.0596 9122	20
21	0.0544 3748	0.0551 4884	0.0558 6550	0.0573 1464	21
22	0.0522 7238	0.0529 8507	0.0537 0331	0.0551 5638	22
23	0.0502 9666	0.0510 1080	0.0517 3075	0.0531 8796	23
24	0.0484 8665	0.0492 0235	0.0499 2410	0.0513 8565	24
25	0.0468 2247	0.0475 3981	0.0482 6345	0.0497 2952	25
26	0.0452 8729	0.0460 0635	0.0467 3196	0.0482 0269	26
27	0.0438 6677	0.0445 8763	0.0453 1527	0.0467 9079	27
28	0.0425 4863	0.0432 7134	0.0440 0108	0.0454 8151	28
29	0.0413 2228	0.0420 4689	0.0427 7878	0.0442 6424	29
30	0.0401 7854	0.0409 0511	0.0416 3919	0.0431 2975	30
31	0.0391 0942	0.0398 3798	0.0405 7430	0.0420 7005	31
32	0.0381 0791	0.0388 3850	0.0395 7710	0.0410 7812	32
33	0.0371 6786	0.0379 0053	0.0386 4144	0.0401 4779	33
34	0.0362 8387	0.0370 1864	0.0377 6189	0.0392 7363	34
35	0.0354 5111	0.0361 8801	0.0369 3363	0.0384 5082	35
36	0.0346 6533	0.0354 0438	0.0361 5240	0.0376 7507	36
37	0.0339 2270	0.0346 6394	0.0354 1437	0.0369 4257	37
38	0.0332 1983	0.0339 6327	0.0347 1613	0.0362 4990	38
39	0.0325 5365	0.0332 9931	0.0340 5463	0.0355 9399	39
40	0.0319 2141	0.0326 6931	0.0334 2710	0.0349 7209	40
41	0.0313 2063	0.0320 7078	0.0328 3106	0.0343 8170	41
42	0.0307 4906	0.0315 0148	0.0322 6426	0.0338 2057	42
43	0.0302 0466	0.0309 5936	0.0317 2465	0.0332 8666	43
44	0.0296 8557	0.0304 4257	0.0312 1038	0.0327 7810	44
45	0.0291 9012	0.0299 4941	0.0307 1976	0.0322 9321	45
46	0.0287 1675	0.0294 7836	0.0302 5125	0.0318 3043	46
47	0.0282 6406	0.0290 2799	0.0298 0342	0.0313 8836	47
48	0.0278 3075	0.0285 9701	0.0293 7500	0.0309 6569	48
49	0.0274 1563	0.0281 8424	0.0289 6478	0.0305 6124	49
50	0.0270 1763	0.0277 8857	0.0285 7168	0.0301 7391	50

TABLE C-9 PERIODIC RENT OF ANNUITY WHOSE PRESENT VALUE IS 1 (cont.)

$$\frac{1}{a_{\overline{n}|i}} = \frac{i}{1-(1+i)^{-n}} \qquad \left[\frac{1}{s_{\overline{n}|i}} = \frac{1}{a_{\overline{n}|i}} - i\right]$$

n	2%	2¼%	2½%	2¾%	n
1	1.0200 0000	1.0225 0000	1.0250 0000	1.0275 0000	1
2	0.5150 4950	0.5169 3758	0.5188 2716	0.5207 1825	2
3	0.3467 5467	0.3484 4458	0.3501 3717	0.3518 3243	3
4	0.2626 2375	0.2642 1893	0.2658 1788	0.2674 2059	4
5	0.2121 5839	0.2137 0021	0.2152 4686	0.2167 9832	5
6	0.1785 2581	0.1800 3496	0.1815 4997	0.1830 7083	6
7	0.1545 1196	0.1560 0025	0.1574 9543	0.1589 9747	7
8	0.1365 0980	0.1379 8462	0.1394 6735	0.1409 5795	8
9	0.1225 1544	0.1239 8170	0.1254 5689	0.1269 4095	9
10	0.1113 2653	0.1127 8768	0.1142 5876	0.1157 3972	10
11	0.1021 7794	0.1036 3649	0.1051 0596	0.1065 8629	11
12	0.0945 5960	0.0960 1740	0.0974 8713	0.0989 6871	12
13	0.0881 1835	0.0895 7686	0.0910 4827	0.0925 3252	13
14	0.0826 0197	0.0840 6230	0.0855 3653	0.0870 2457	14
15	0.0778 2547	0.0792 8852	0.0807 6646	0.0822 5917	15
16	0.0736 5013	0.0751 1663	0.0765 9899	0.0780 9710	16
17	0.0699 6984	0.0714 4039	0.0729 2777	0.0744 3186	17
18	0.0667 0210	0.0681 7720	0.0696 7008	0.0711 8063	18
19	0.0637 8177	0.0652 6182	0.0667 6062	0.0682 7802	19
20	0.0611 5672	0.0626 4207	0.0641 4713	0.0656 7173	20
21	0.0587 8477	0.0602 7572	0.0617 8733	0.0633 1941	21
22	0.0566 3140	0.0581 2821	0.0596 4661	0.0611 8640	22
23	0.0546 6810	0.0561 7097	0.0576 9638	0.0592 4410	23
24	0.0528 7110	0.0543 8023	0.0559 1282	0.0574 6863	24
25	0.0512 2044	0.0527 3599	0.0542 7592	0.0558 3997	25
26	0.0496 9923	0.0512 2134	0.0527 6875	0.0543 4116	26
27	0.0482 9309	0.0498 2188	0.0513 7687	0.0529 5776	27
28	0.0469 8967	0.0485 2525	0.0500 8793	0.0516 7738	28
29	0.0457 7836	0.0473 2081	0.0488 9127	0.0504 8935	29
30	0.0446 4992	0.0461 9934	0.0477 7764	0.0493 8442	30
31	0.0435 9635	0.0451 5280	0.0467 3900	0.0483 5453	31
32	0.0426 1061	0.0441 7415	0.0457 6831	0.0473 9263	32
33	0.0416 8653	0.0432 5722	0.0448 5938	0.0464 9253	33
34	0.0408 1867	0.0423 9655	0.0440 0675	0.0456 4875	34
35	0.0400 0221	0.0415 8731	0.0432 0558	0.0448 5645	35
36	0.0392 3285	0.0408 2522	0.0424 5158	0.0441 1132	36
37	0.0385 0678	0.0401 0643	0.0417 4090	0.0434 0953	37
38	0.0378 2057	0.0394 2753	0.0410 7012	0.0427 4764	38
39	0.0371 7114	0.0387 8543	0.0404 3615	0.0421 2256	39
40	0.0365 5575	0.0381 7738	0.0398 3623	0.0415 3151	40
41	0.0359 7188	0.0376 0087	0.0392 6786	0.0409 7200	41
42	0.0354 1729	0.0370 5364	0.0387 2876	0.0404 4175	42
43	0.0348 8993	0.0365 3364	0.0382 1688	0.0399 3871	43
44	0.0343 8794	0.0360 3901	0.0377 3037	0.0394 6100	44
45	0.0339 0962	0.0355 6805	0.0372 6752	0.0390 0693	45
46	0.0334 5342	0.0351 1921	0.0368 2676	0.0385 7493	46
47	0.0330 1792	0.0346 9107	0.0364 0669	0.0381 6358	47
48	0.0326 0184	0.0342 8233	0.0360 0599	0.0377 7158	48
49	0.0322 0396	0.0338 9179	0.0356 2348	0.0373 9773	49
50	0.0318 2321	0.0335 1836	0.0352 5806	0.0370 4092	50

TABLE C-9 PERIODIC RENT OF ANNUITY WHOSE PRESENT VALUE IS 1 (*cont.*)

$$\frac{1}{a_{\overline{n}|i}} = \frac{i}{1-(1+i)^{-n}} \qquad \left[\frac{1}{s_{\overline{n}|i}} = \frac{1}{a_{\overline{n}|i}} - i\right]$$

n	3%	3½%	4%	4½%	n
1	1.0300 0000	1.0350 0000	1.0400 0000	1.0450 0000	1
2	0.5226 1084	0.5264 0049	0.5301 9608	0.5339 9756	2
3	0.3535 3036	0.3569 3418	0.3603 4854	0.3637 7336	3
4	0.2690 2705	0.2722 5114	0.2754 9005	0.2787 4365	4
5	0.2183 5457	0.2214 8137	0.2246 2711	0.2277 9164	5
6	0.1845 9750	0.1876 6821	0.1907 6190	0.1938 7839	6
7	0.1605 0635	0.1635 4449	0.1666 0961	0.1697 0147	7
8	0.1424 5639	0.1454 7665	0.1485 2783	0.1516 0965	8
9	0.1284 3386	0.1314 4601	0.1344 9299	0.1375 7447	9
10	0.1172 3051	0.1202 4137	0.1232 9094	0.1263 7882	10
11	0.1080 7745	0.1110 9197	0.1141 4904	0.1172 4818	11
12	0.1004 6209	0.1034 8395	0.1065 5217	0.1096 6619	12
13	0.0940 2954	0.0970 6157	0.1001 4373	0.1032 7535	13
14	0.0885 2634	0.0915 7073	0.0946 6897	0.0978 2032	14
15	0.0837 6658	0.0868 2507	0.0899 4110	0.0931 1381	15
16	0.0796 1085	0.0826 8483	0.0858 2000	0.0890 1537	16
17	0.0759 5253	0.0790 4313	0.0821 9852	0.0854 1758	17
18	0.0727 0870	0.0758 1684	0.0789 9333	0.0822 3690	18
19	0.0698 1388	0.0729 4033	0.0761 3862	0.0794 0734	19
20	0.0672 1571	0.0703 6108	0.0735 8175	0.0768 7614	20
21	0.0648 7178	0.0680 3659	0.0712 8011	0.0746 0057	21
22	0.0627 4739	0.0659 3207	0.0691 9881	0.0725 4565	22
23	0.0608 1390	0.0640 1880	0.0673 0906	0.0706 8249	23
24	0.0590 4742	0.0622 7283	0.0655 8683	0.0689 8703	24
25	0.0574 2787	0.0606 7404	0.0640 1196	0.0674 3903	25
26	0.0559 3829	0.0592 0540	0.0625 6738	0.0660 2137	26
27	0.0545 6421	0.0578 5241	0.0612 3854	0.0647 1946	27
28	0.0532 9323	0.0566 0265	0.0600 1298	0.0635 2081	28
29	0.0521 1467	0.0554 4538	0.0588 7993	0.0624 1461	29
30	0.0510 1926	0.0543 7133	0.0578 3010	0.0613 9154	30
31	0.0499 9893	0.0533 7240	0.0568 5535	0.0604 4345	31
32	0.0490 4662	0.0524 4150	0.0559 4859	0.0595 6320	32
33	0.0481 5612	0.0515 7242	0.0551 0357	0.0587 4453	33
34	0.0473 2196	0.0507 5966	0.0543 1477	0.0579 8191	34
35	0.0465 3929	0.0499 9835	0.0535 7732	0.0572 7045	35
36	0.0458 0379	0.0492 8416	0.0528 8688	0.0566 0578	36
37	0.0451 1162	0.0486 1325	0.0522 3957	0.0559 8402	37
38	0.0444 5934	0.0479 8214	0.0516 3192	0.0554 0169	38
39	0.0438 4385	0.0473 8775	0.0510 6083	0.0548 5567	39
40	0.0432 6238	0.0468 2728	0.0505 2349	0.0543 4315	40
41	0.0427 1241	0.0462 9822	0.0500 1738	0.0538 6158	41
42	0.0421 9167	0.0457 9828	0.0495 4020	0.0534 0868	42
43	0.0416 9811	0.0453 2539	0.0490 8989	0.0529 8235	43
44	0.0412 2985	0.0448 7768	0.0486 6454	0.0525 8071	44
45	0.0407 8518	0.0444 5343	0.0482 6246	0.0522 0202	45
46	0.0403 6254	0.0440 5108	0.0478 8205	0.0518 4471	46
47	0.0399 6051	0.0436 6919	0.0475 2189	0.0515 0734	47
48	0.0395 7777	0.0433 0646	0.0471 8065	0.0511 8858	48
49	0.0392 1314	0.0429 6167	0.0468 5712	0.0508 8722	49
50	0.0388 6550	0.0426 3371	0.0465 5020	0.0506 0215	50

TABLE C-9 PERIODIC RENT OF ANNUITY WHOSE PRESENT VALUE IS 1 (*cont.*)

$$\frac{1}{a_{\overline{n}|i}} = \frac{i}{1-(1+i)^{-n}} \qquad \left[\frac{1}{s_{\overline{n}|i}} = \frac{1}{a_{\overline{n}|i}} - i\right]$$

n	5%	5½%	6%	6½%	n
1	1.0500 0000	1.0550 0000	1.0600 0000	1.0650 0000	1
2	0.5378 0488	0.5416 1800	0.5454 3689	0.5492 6150	2
3	0.3672 0856	0.3706 5407	0.3741 0981	0.3775 7570	3
4	0.2820 1183	0.2852 9449	0.2885 9149	0.2919 0274	4
5	0.2309 7480	0.2341 7644	0.2373 9640	0.2406 3454	5
6	0.1970 1747	0.2001 7895	0.2033 6263	0.2065 6831	6
7	0.1728 1982	0.1759 6442	0.1791 3502	0.1823 3137	7
8	0.1547 2181	0.1578 6401	0.1610 3594	0.1642 3730	8
9	0.1406 9008	0.1438 3946	0.1470 2224	0.1502 3803	9
10	0.1295 0458	0.1326 6777	0.1358 6796	0.1391 0469	10
11	0.1203 8889	0.1235 7065	0.1267 9294	0.1300 5521	11
12	0.1128 2541	0.1160 2923	0.1192 7703	0.1225 6817	12
13	0.1064 5577	0.1096 8426	0.1129 6011	0.1162 8256	13
14	0.1010 2397	0.1042 7912	0.1075 8491	0.1109 4048	14
15	0.0963 4229	0.0996 2560	0.1029 6276	0.1063 5278	15
16	0.0922 6991	0.0955 8254	0.0989 5214	0.1023 7757	16
17	0.0886 9914	0.0920 4797	0.0954 4480	0.0989 0633	17
18	0.0855 4622	0.0889 1992	0.0923 5654	0.0958 5461	18
19	0.0827 4501	0.0861 5006	0.0896 2086	0.0931 5575	19
20	0.0802 4259	0.0836 7933	0.0871 8456	0.0907 5640	20
21	0.0779 9611	0.0814 6478	0.0850 0455	0.0886 1333	21
22	0.0759 7051	0.0794 7123	0.0830 4557	0.0866 9120	22
23	0.0741 3682	0.0776 6965	0.0812 7848	0.0849 6078	23
24	0.0724 7090	0.0760 3580	0.0796 7900	0.0833 9770	24
25	0.0709 5246	0.0745 4935	0.0782 2672	0.0819 8148	25
26	0.0695 6432	0.0731 9307	0.0769 0435	0.0806 9480	26
27	0.0682 9186	0.0719 5228	0.0756 9717	0.0795 2288	27
28	0.0671 2253	0.0708 1440	0.0745 9255	0.0784 5305	28
29	0.0660 4551	0.0697 6857	0.0735 7961	0.0774 7440	29
30	0.0650 5144	0.0688 0539	0.0726 4891	0.0765 7744	30
31	0.0641 8212	0.0679 1665	0.0717 9222	0.0757 5393	31
32	0.0632 8042	0.0670 9519	0.0710 0234	0.0749 9665	32
33	0.0624 9004	0.0663 3469	0.0702 7293	0.0742 9924	33
34	0.0617 5545	0.0656 2958	0.0695 9843	0.0736 5610	34
35	0.0610 7171	0.0649 7493	0.0689 7386	0.0730 6226	35
36	0.0604 3446	0.0643 6635	0.0683 9483	0.0725 1332	36
37	0.0598 3979	0.0637 9993	0.0678 5743	0.0720 0534	37
38	0.0592 8423	0.0632 7217	0.0673 5812	0.0715 3480	38
39	0.0587 6462	0.0627 7991	0.0668 9377	0.0710 9854	39
40	0.0582 7816	0.0623 2034	0.0664 6154	0.0706 9373	40
41	0.0578 2229	0.0618 9090	0.0660 5886	0.0703 1779	41
42	0.0573 9471	0.0614 8927	0.0656 8342	0.0699 6842	42
43	0.0569 9333	0.0611 1337	0.0653 3312	0.0696 4352	43
44	0.0566 1625	0.0607 6128	0.0650 0606	0.0693 4119	44
45	0.0562 6173	0.0604 3127	0.0647 0050	0.0690 5968	45
46	0.0559 2820	0.0601 2175	0.0644 1485	0.0687 9743	46
47	0.0556 1421	0.0598 3129	0.0641 4768	0.0685 5300	47
48	0.0553 1843	0.0595 5854	0.0638 9766	0.0683 2506	48
49	0.0550 3965	0.0593 0230	0.0636 6356	0.0681 1240	49
50	0.0547 7674	0.0590 6145	0.0634 4429	0.0679 1393	50

TABLE C-9 PERIODIC RENT OF ANNUITY WHOSE PRESENT VALUE IS 1 (*cont.*)

$$\frac{1}{a_{\overline{n}|i}} = \frac{i}{1-(1+i)^{-n}} \qquad \left[\frac{1}{s_{\overline{n}|i}} = \frac{1}{a_{\overline{n}|i}} - i\right]$$

n	7%	7½%	8%	8½%	n
1	1.0700 0000	1.0750 0000	1.0800 0000	1.0850 0000	1
2	0.5530 9179	0.5569 2771	0.5607 6923	0.5646 1631	2
3	0.3810 5166	0.3845 3763	0.3880 3351	0.3915 3925	3
4	0.2952 2812	0.2985 6751	0.3019 2080	0.3052 8789	4
5	0.2438 9069	0.2471 6472	0.2504 5645	0.2537 6573	5
6	0.2097 9580	0.2130 4489	0.2163 1539	0.2196 0708	6
7	0.1855 5322	0.1888 0032	0.1920 7240	0.1953 6922	7
8	0.1674 6776	0.1707 2702	0.1740 1476	0.1773 3065	8
9	0.1534 8647	0.1567 6716	0.1600 7971	0.1634 2372	9
10	0.1423 7750	0.1456 8593	0.1490 2949	0.1524 0771	10
11	0.1333 5690	0.1366 9747	0.1400 7634	0.1434 9293	11
12	0.1259 0199	0.1292 7783	0.1326 9502	0.1361 5286	12
13	0.1196 5085	0.1230 6420	0.1265 2181	0.1300 2287	13
14	0.1143 4494	0.1177 9737	0.1212 9685	0.1248 4244	14
15	0.1097 9462	0.1132 8724	0.1168 2954	0.1204 2046	15
16	0.1058 5765	0.1093 9116	0.1129 7687	0.1166 1354	16
17	0.1024 2519	0.1060 0003	0.1096 2943	0.1133 1198	17
18	0.0994 1260	0.1030 2896	0.1067 0210	0.1104 3041	18
19	0.0967 5301	0.1004 1090	0.1041 2763	0.1079 0140	19
20	0.0943 9293	0.0980 9219	0.1018 5221	0.1056 7097	20
21	0.0922 8900	0.0960 2937	0.0998 3225	0.1036 9541	21
22	0.0904 0577	0.0941 8687	0.0980 3207	0.1019 3892	22
23	0.0887 1393	0.0925 3528	0.0964 2217	0.1003 7193	23
24	0.0871 8902	0.0910 5008	0.0949 7796	0.0989 6975	24
25	0.0858 1052	0.0897 1067	0.0936 7878	0.0977 1168	25
26	0.0845 6103	0.0884 9961	0.0925 0713	0.0965 8016	26
27	0.0834 2573	0.0874 0204	0.0914 4809	0.0955 6025	27
28	0.0823 9193	0.0864 0520	0.0904 8891	0.0946 3914	28
29	0.0814 4865	0.0854 9811	0.0896 1854	0.0938 0577	29
30	0.0805 8640	0.0846 7124	0.0888 2743	0.0930 5058	30
31	0.0797 9691	0.0839 1628	0.0881 0728	0.0923 6524	31
32	0.0790 7292	0.0832 2599	0.0874 5081	0.0917 4247	32
33	0.0784 0807	0.0825 9397	0.0868 5163	0.0911 7588	33
34	0.0777 9674	0.0820 1461	0.0863 0411	0.0906 5984	34
35	0.0772 3396	0.0814 8291	0.0858 0326	0.0901 8937	35
36	0.0767 1531	0.0809 9447	0.0853 4467	0.0897 6006	36
37	0.0762 3685	0.0805 4533	0.0849 2440	0.0893 6799	37
38	0.0757 9505	0.0801 3197	0.0845 3894	0.0890 0966	38
39	0.0753 8676	0.0797 5124	0.0841 8513	0.0886 8193	39
40	0.0750 0914	0.0794 0031	0.0838 6016	0.0883 8201	40
41	0.0746 5962	0.0790 7663	0.0835 6149	0.0881 0737	41
42	0.0743 3591	0.0787 7789	0.0832 8684	0.0878 5576	42
43	0.0740 3590	0.0785 0201	0.0830 3414	0.0876 2512	43
44	0.0737 5769	0.0782 4710	0.0828 0152	0 0874 1363	44
45	0.0734 9957	0.0780 1146	0.0825 8728	0.0872 1961	45
46	0.0732 5996	0.0777 9353	0.0823 8991	0.0870 4154	46
47	0.0730 3744	0.0775 9190	0.0822 0799	0.0868 7807	47
48	0.0728 3070	0.0774 0527	0.0820 4027	0.0867 2795	48
49	0.0726 3853	0.0772 3247	0.0818 8557	0.0865 9005	49
50	0.0724 5985	0.0770 7241	0.0817 4286	0.0864 6334	50

TABLE C-10 ACRS COST RECOVERY PERCENTAGES*

Year	3-Year Class	5-Year Class	10-Year Class	15-Year Realty	15-Year Utility
1	25%	15%	8%	12%	5%
2	38	22	14	10	10
3	37	21	12	9	9
4		21	10	8	8
5		21	10	7	7
6			10	6	7
7			9	6	6
8			9	6	6
9			9	6	6
10			9	5	6
11				5	6
12				5	6
13				5	6
14				5	6
15				5	6

*For property placed in service between 1981 and 1986.

Answers and Solutions to Try These Problems and Test Yourself Problems

CHAPTER 1

Answers to Try These Problems

Set I (page 5)

1. 6   **2.** 26   **3.** 7   **4.** 17   **5.** 30   **6.** 102   **7.** 21.8   **8.** 111.6   **9.** 4.75
10. 1.05 or $1\frac{1}{20}$ **11.** 0.96 or $\frac{24}{25}$ **12.** 7.728   **13.** 0.626   **14.** 5.050   **15.** $10.06
16. $0.33   **17.** $243.35   **18.** $62.00   **19.** 3.2%   **20.** 0.5%   **21.** 8.1%   **22.** 26.0%
23. 56.3   **24.** 8.9   **25.** 15.0   **26.** 123.5
     56.30   8.90   15.01   123.52
     56.299   8.896   15.006   123.523

Set II (page 9)

1. 3; 10; 12   **2.** 3; $\frac{3}{4}$; 5; −6   **3.** 4; $\frac{3}{5}$; −1; 1   **4.** $40   **5.** 15   **6.** 25   **7.** 16.6
8. $7\frac{3}{4}$ **9.** $13\frac{1}{2}$ **10.** $76\frac{1}{2}$; $\frac{5}{2}$; 1

Set III (page 13)

1. $8T$ **2.** $\frac{3}{2}Z$ or $1\frac{1}{2}Z$ **3.** $\frac{2}{3}A$ or $\frac{2A}{3}$ **4.** $3\frac{1}{2}Y$ or $\frac{7}{2}Y$ **5.** $8\frac{1}{3}C$ or $\frac{25}{3}C$
6. $6\frac{1}{3}P$ or $\frac{19P}{3}$ **7.** $1.4R - 1$ **8.** $1.8W$ **9.** $6b - 1 + c$ **10.** $1.5x + y$ **11.** $-9y$
12. $8\frac{1}{3}K - 12$ **13.** $8Z + \frac{1}{2}Y$ **14.** $\frac{1}{4}A + B + 3$

Set IV (page 19)

1. $R = 8$ **2.** $B = 19$ **3.** $A = 10$ **4.** $S = 14$ **5.** $L = 60$ **6.** $Y = 8.04$ **7.** $A = \frac{3}{40}$
8. $R = 50$ **9.** $P = \$81$ **10.** $Y = \frac{5}{3}$ or $1\frac{2}{3}$ **11.** $Q = 26$ **12.** $H = \frac{2}{9}$

Set V (page 25)

1. $y = 10$ 2. $d = \frac{23}{3}$ or $7\frac{2}{3}$ 3. $A = 27$ 4. $d = 10$ 5. $Y = 24$ 6. $T = \frac{21}{2}$ or $10\frac{1}{2}$
7. $Y = 90$ 8. $y = 3$ 9. $z = 3$ 10. $R = \frac{21}{2}$ or $10\frac{1}{2}$ 11. $c = 120$ 12. $X = 20$
13. $t = 10$ 14. $a = 12$ 15. $T = 3$ 16. $R = 5$

Set VI (page 29)

1. $\frac{2}{3}$ 2. $\frac{3}{1}$ 3. $\frac{2}{15}$ 4. $\frac{3}{2}$ 5. a. $\frac{11}{13}$ b. $\frac{2}{11}$ 6. a. $\frac{16}{5}$ b. $\frac{3}{2}$ c. $\frac{5}{84}$ 7. a. $\frac{564}{9,829} \cong \frac{3}{50}$
b. $\frac{564}{9,265} \cong \frac{1}{15}$ c. About 3 out of every 50 workers were unemployed. There was one unemployed worker for every 15 employed.   8. a. $\frac{5.7}{6.7} \cong \frac{6}{7}$ b. $\frac{1}{3}$ c. $\frac{5.7}{12.4} \cong \frac{1}{2}$ About half of total sales were exported.

Set VII (page 33)

1. 24   2. 20   3. 16   4. 4   5. 4   6. No, $\frac{5}{7}$ should be more than $\frac{1}{2}$ or 0.5   7. b
8. a   9. No, approximately $27 should be left.   10. $12

Solutions to Test Yourself (page 39)

1. $5 + 10 \times 3 - 16 \div 2 = 5 + 30 - 8 = 27$ 2. $625 \div 5 \div 5 = 125 \div 5 = 25$
3. $100(1 + 0.06 \times \frac{1}{2}) = 100(1 + 0.03) = 100(1.03) = 103$
4. $\frac{5}{4}(2) - 3(3)(2) + 25 = 2.5 - 18 + 25 = 9.5$ 5. $26.05   6. $339.80
7. 15.3, 15.27, 15.275   8. $3; \frac{1}{2}; 1; -5$
9. $7Z - \frac{Z}{2} + Y - 6 = 6\frac{1}{2}Z + Y - 6$ or $\frac{13}{2}Z + Y - 6$
10. $R(5 + 0.4) + (S - 0.1S) = R(5.4) + (1.0S - 0.1S) = 5.4R + 0.9S$

11. $\frac{4H}{5} = 64$ Check: $\frac{4}{5}(80) = 64$

$\frac{5}{4} \times \frac{4H}{5} = 64 \times \frac{5}{4}$ $64 = 64$

$H = 80$

12. $8a = a + 14$ Check: $8(2) = 2 + 14$

$7a = 14$ $16 = 16$

$a = 2$

13. $X + 5 = X\left(1 + 10 \times \frac{2}{5}\right)$ Check: $\frac{5}{4} + 5 = \frac{5}{4}\left(1 + 10 \times \frac{2}{5}\right)$

$ = X(1 + 4)$ $\frac{5}{4} + \frac{20}{4} = \frac{5}{4}(5)$

$X + 5 = 5X$ $\frac{25}{4} = \frac{25}{4}$

$5 = 5X - X$

$5 = 4X$

$X = \frac{5}{4}$

ANSWERS AND SOLUTIONS TO TRY THESE PROBLEMS AND TEST YOURSELF PROBLEMS **661**

14. $Z + 0.5Z = 60$ **Check:** $40 + 0.5(40) = 60$
$1.5Z = 60$ $40 + 20 = 60$
$Z = 40$ $60 = 60$

15. $21 = \dfrac{a}{3} + 2a$ **Check:** $21 = \dfrac{9}{3} + 2(9)$

$21 = \dfrac{7}{3}a$ $21 = 3 + 18$

$21 \times \dfrac{3}{7} = \dfrac{\cancel{3}}{\cancel{7}} \times \dfrac{\cancel{7}}{\cancel{3}} a$ $21 = 21$

$9 = a$

16. $8 \text{ to } 32 = \dfrac{8}{32} = \dfrac{1}{4}$ **17.** $(19.5 + 0.5) \text{ to } (2 + 8) = 20 \text{ to } 10 = \dfrac{20}{10} = \dfrac{2}{1}$ or 2

18. $2.1(8.9) - 0.994(4.99) \cong 2(9) - 1(5) \cong 13$ **19.** $\dfrac{3.1(8.06)}{11.97} \cong \dfrac{3(8)}{12} \cong 2$

20. c
$3.04(12.01) - 1.99(7.11) \cong 3(12) - 2(7) \cong 22$

21. a
$\dfrac{34.79}{6.9} \cong \dfrac{35}{7} \cong 5$

22. c
Account balance $\cong 450 - 250 + 210 \cong 410$ **23. a.** $\dfrac{20}{22} = \dfrac{10}{11}$ **b.** $\dfrac{20}{42} = \dfrac{10}{21}$

CHAPTER 2

Answers to Try These Problems

Set I (page 47)

1. $0.14, \dfrac{7}{50}$ **2.** $0.06, \dfrac{3}{50}$ **3.** $0.27, \dfrac{27}{100}$ **4.** $0.03, \dfrac{3}{100}$ **5.** $1.34, 1\dfrac{17}{50}$ **6.** $3.27, 3\dfrac{27}{100}$

7. $5.00, \dfrac{5}{1}$ **8.** $0.045, \dfrac{9}{200}$ **9.** $0.9944, \dfrac{1,243}{1,250}$ **10.** $0.0241, \dfrac{241}{10,000}$ **11.** $0.0081, \dfrac{81}{10,000}$

12. $0.0002, \dfrac{1}{5,000}$ **13.** $0.006, \dfrac{3}{500}$ **14.** $0.0175, \dfrac{7}{400}$ **15.** $0.055, \dfrac{11}{200}$ **16.** $0.3333, \dfrac{1}{3}$

17. $0.00375, \dfrac{3}{800}$ **18.** $0.004, \dfrac{1}{250}$ **19.** $0.0729, \dfrac{51}{700}$ **20.** $4.2225, 4\dfrac{89}{400}$

Set II (page 51)

1. 37%   **2.** 3%   **3.** 40%   **4.** 145%   **5.** 290%   **6.** 63.4%   **7.** 0.58%   **8.** 10%
9. 1%   **10.** 100%   **11.** 42.5% or $42\dfrac{1}{2}$%   **12.** 66.7% or $66\dfrac{2}{3}$%   **13.** 31.25%   **14.** 88.1%
15. 320%   **16.** 30%

Set III (page 57)

1. $672.81   **2.** 160   **3.** 30   **4.** 100   **5.** $98.23   **6.** $3.62   **7.** 48   **8.** 45%
9. 240%   **10.** $564.82   **11.** $P = 51.6$ **12.** $P = 5.46$ **13.** $P = 90$ **14.** $R = 6\%$
15. $R = 300\%$ **16.** $R = 4\%$ **17.** $R = 56\%$ **18.** $B = 350$ **19.** $B = 17{,}777.78$
20. $B = 6{,}000$

Set IV (page 63)

1. a. $426   **b.** $8,946   **2.** 44.7%   **3.** $40,250   **4.** 50.2%   **5.** $518.60   **6.** $1,500,000
7. 920   **8. a.** $123.50, $74.10   **b.** $296.40

Set V (page 69)

1. 12% increase   2. 9.7% decrease   3. 66.7% or $66\frac{2}{3}$% increase   4. 38.0% decrease
5. 33.3% or $33\frac{1}{3}$% increase   6. 31.8% decrease   7. 24.2% decrease   8. 100% increase
9. 15.2% decrease   10. 20% decrease

Solutions to Test Yourself (page 77)

1. a. $0.02 = \frac{2}{100} = \frac{1}{50}$ b. $0.046 = \frac{46}{1,000} = \frac{23}{500}$ c. $19.25\% = 0.1925 = \frac{1,925}{10,000} = \frac{77}{400}$
   d. $3.16 = 3\frac{16}{100} = 3\frac{4}{25}$ e. $0.005 = \frac{5}{1,000} = \frac{1}{200}$ f. $0.4\% = 0.004 = \frac{4}{1,000} = \frac{1}{250}$

2. a. 4%   b. 60%   c. 130%   d. 300%   e. 35%   f. 133.33% or $133\frac{1}{3}$%

3. $0.06(640) = P$
   $P = 38.4$

4. $0.045B = 9.72$
   $B = 216$

5. % change × original = amount of change
   $R \times 800 = 936 - 800$
   $800R = 136$
   $R = \frac{136}{800} = 0.17 = 17\%$ increase

6. % change × original = amount of change
   $R \times 41.2 = 41.2 - 10.3$
   $41.2R = 30.9$
   $R = \frac{30.9}{41.2} = 0.75 = 75\%$ decrease

7. $0.035 \times 82 = P$
   $P = 2.87$

8. $160R = 50$
   $R = \frac{50}{160} = 0.3125 = 31.25\%$

9. $90R = 4.5$
   $R = \frac{4.5}{90} = 0.05 = 5\%$

10. $0.03B = 33$
    $B = \frac{33}{0.03} = 1,100$

11. $0.0025 \times \$2,500 = P$
    $P = \$6.25$

12. $0.01B = \$2,500$
    $B = \frac{2,500}{0.01} = \$250,000$

13. $0.8 \times 465 = P$
    $P = 372$

14. $3,560R = 178$
    $R = \frac{178}{3,560} = 0.05 = 5\%$

15. $2,013 - 1,830 = \$183$
    $1,830R = 183$
    $R = \frac{183}{1,830} = 0.1 = 10\%$ increase

CHAPTER 3

ANSWERS TO TRY THESE PROBLEMS

Set I (page 87)

1. $4.26   2. $60.89   3. $25.74   4. a. $8.48   b. $3.48   5. a. $37.50   b. $11.25
6. a. $41.50   b. $17.50   7. 33%   8. 36%   9. 16%   10. $25.76   11. $3.50
12. $62.50   13. $36   14. $15.74   15. $125   16. $5.27   17. $960   18. $87
19. $568.18

Set II (page 93)

1. a. $90 b. 19.6% c. 16.4%   2. 33.3%, 25%   3. a. 25% b. 20%
4. No, the markup is only 40.2%   5. a. $22.50 b. 20%   6. a. $5.76 b. 56.3%
7. a. $6.75 b. 50%   8. a. $1.00 b. 28.2%   9. a. $21 b. 23.1%   10. a. $6.67
b. 33.3%   11. 42.9%   12. No, the markup of 37.5% is not enough.   13. 13.0%
14. a. 41.0% b. 37.1%

Set III (page 99)

1. a. $70 b. $103.60 c. 90 d. $1.15   2. $1.26   3. a. $250, $390.63 b. 47
c. $8.31   4. $1.39   5. $1.07   6. $2.37

Set IV (page 107)

1. a. $11,625 b. 55 c. $990 d. $10,635 e. 445 f. $23.90   2. a. $400 b. $36
c. $364 d. 85 e. $4.28   3. $1.05   4. $41.44   5. a. $1,125 b. 20 c. $76 d. 8
e. 172 f. $6.10   6. $6.44   7. $0.64   8. a. $215 b. Operating loss, $16 c. $145
9. a. $404.76 b. Below c. Absolute loss, $1 d. $22,806.40

SOLUTIONS TO TEST YOURSELF (page 121)

1. $M = OH + NP = 12 + 15.20 + 27.20
$C + M = S$
$32 + 27.20 = S$
$S = 59.20

2. $C + M = S$
$C + 480 = 960$
$C = 480
$M = OH + NP$
$480 = OH + 176$
$OH = 304

3. $C + M = S$
$40 + 25 = S$
$S = 65
$\%M \text{ on } C = \frac{M}{C} = \frac{25}{40} = 62.5\%$
$\%M \text{ on } S = \frac{M}{S} = \frac{25}{60} = 41.7\%$

4. $M = (\%M \text{ on } C) \times C$
$= 0.44 \times 89.50$
$M = 39.38
$C + M = S$
$89.50 + 39.38 = S$
$S = 128.88
$\%M \text{ on } S = \frac{M}{S}$
$= \frac{39.38}{128.88}$
$\%M \text{ on } S = 30.6\%$

5. $C + M = S$
$75 + 0.68(75) = S$
$S = 126

6. a. $M = S - C = 775 - 500 = 275
b. $OH = 0.4(500) = 200
$M = OH + NP$
$275 = 200 + NP$
$NP = 75

664 ANSWERS AND SOLUTIONS TO TRY THESE PROBLEMS AND TEST YOURSELF PROBLEMS

7. $C + M = S$
    $C + 0.39(75) = 75$
    $C + 29.25 = 75$
    $C = \$45.75$

8. $C + M = S$
    $36 + 0.4S = S$
    $36 = 0.6S$
    $S = \$60$
    $M = 60 - 36 = \$24$
    $\%M \text{ on } C = \dfrac{M}{C} = \dfrac{24}{36} = 66.7\%$

9. $C + M = S$
    $C + 0.4C = 56$
    $1.4C = 56$
    $C = \$40$

10. $M = S - C = 500 - 400 = \$100$
    a. $\%M \text{ on } C = \dfrac{M}{C} = \dfrac{100}{400} = 25\%$
    b. $\%M \text{ on } S = \dfrac{M}{S} = \dfrac{100}{500} = 20\%$

11. Total cost $= 200 \times 1.20 = \$240$
    $C + M = S$
    $240 + 0.45(240) = S$
    $240 + 108 = S$
    $S = \$348$
    10% of 200 = 20 lbs. spoil
    200 − 20 = 180 lbs. sold at original price
    Price per pound $= \dfrac{348}{180} = 1.933 = \$1.93$

12. Total cost $= 50 \times 1.10 = \$55$
    $C + M = S$
    $55 + 0.75S = S$
    $55 = 0.25S$
    $S = \$220$
    $8 \times 2.00 = \$16$ markdown sales
    $220 - 16 = \$204$ sales at original price
    Price per pound $= \dfrac{204}{42} = 4.857 = \$4.86$

13. Total cost $= 120 \times 1.30 = \$156$
    $C + M = S$
    $156 + 0.6(156) = S$
    $S = \$249.60$
    10% of 120 = 12 lbs. reduced
    $12 \times 1.59 = \$19.08$ reduced sales
    $249.60 - 19.08 = \$230.52$ sales at original price
    5% of 120 = 6 lbs. thrown out; therefore,
    $120 - 12 - 6 = 102$ lb. sold at original price
    Price per pound $= \dfrac{230.52}{102} = \$2.26$

14. a. Breakeven $= C + OH = 620 + 75 = \$695$
    b. $\%M \text{ on } C = \dfrac{M}{C} = \dfrac{75}{620} = 12.1\%$
    c. Markdown price $= \dfrac{1}{2} \times 695 = \$347.50$

    Absolute loss $=$ Cost $-$ Markdown price $= 620 - 347.50 = \$272.50$

CHAPTER 4 — ANSWERS TO TRY THESE PROBLEMS

Set I (page 131)

1. $D = \$3.95$; $N = \$11.85$ **2.** $D = \$30$; $N = \$45$ **3.** $D = \$110$; $N = \$220$ **4.** 20%; $75
5. $40; 30%   **6.** 66.7%; $32   **7.** 35%; $42.90   **8.** $100; $10   **9.** $580; $156.60
10. 25%; $54   **11.** $16; 15%   **12. a.** $105   **b.** $245   **13.** $127.50   **14. a.** $31.92
b. 38%   **15.** 37.5%   **16.** $13.80   **17.** 8%   **18.** $25   **19.** $100

Set II (page 141)

1. 0.792; $514.80; 20.8%   **2.** 0.648; $113.40; 35.2%   **3.** 0.64; $59.52; 36%
4. $150; 0.765; 23.5%   **5.** $4,000; 0.684; 31.6%   **6.** 30%; 0.7   **7.** $320; 0.839; 16.1%
8. 14.5%; 0.855   **9.** $279.53; 0.855; 14.5%   **10.** $12.96   **11.** 35/5   **12.** $40
13. $150   **14.** 23.1%   **15. a.** 20%   **b.** 40%   **16. a.** $240; $216   **b.** 10%   **17.** 3.4%
18. a. 0.729; 27.1%   **b.** 0.6885; 31.15%   **c.** 0.748125; 25.1875%   **19.** $900   **20.** $323.53

Set III (page 149)

1. a. October 3   **b.** October 23   **2. a.** July 10   **b.** July 20   **3. a.** March 1
b. March 16   **4.** April 27   **5.** August 15   **6.** December 10   **7.** $240.56
8. $2,009; No   **9. a.** $527.51   **b.** $538.28   **10.** $448.54; $2.46   **11.** $5,050
12. $623.90   **13.** $192.41   **14.** $49.39   **15.** No, he needs $12.48 more   **16.** $9.60
17. No, it should have been $287.80   **18. a.** $1,427.95   **b.** $1,213.76   **c.** $214.19
d. $24.28   **e.** $1,189.48

Set IV (page 155)

1. $470; $500   **2.** $1,530.61; $1,519.39   **3.** $3,880; $4,700   **4. a.** $3,673.47
b. $2,026.53   **c.** $73.47   **5.** $490   **6.** $17   **7.** $265.63   **8.** $155

Solutions to Test Yourself (page 167)

1.
$$L - \$D = N$$
$$L - 431.20 = \$1,108.80$$
$$L = \$1,540$$
$$\%D \times L = \$D$$
$$\%D \times 1,540 = 431.20$$
$$\%D = \frac{431.20}{1,540} = 28\%$$

2. $\%Pd \times L = N$
$$0.85L = 13.60$$
$$L = \frac{13.60}{0.85} = \$16$$

666 ANSWERS AND SOLUTIONS TO TRY THESE PROBLEMS AND TEST YOURSELF PROBLEMS

3. August 27 + 10 days = September 6
4. End of February + 10 days = March 10
5. October 13 + 15 days = October 28

6. $\%D \times Cr = \$D$ or $\%Pd \times Cr = \$Pd$
    $0.03 \times 14{,}550 = \$D$ $0.97 \times 14{,}550 = \$Pd$
    $\$D = \$436.50$ $\$Pd = \$14{,}113.50$
    $14{,}550 - 436.50 = \$Pd$
    $\$Pd = \$14{,}113.50$

7. $\%D \times Cr = \$D$ or $\%Pd \times Cr = \$Pd$
    $0.01 \times 742.38 = \$D$ $0.99 \times 742.38 = \$Pd$
    $\$D = \$7.42$ $\$Pd = \$734.96$
    $742.38 - 7.42 = \$Pd$
    $\$Pd = \$734.96$

8. $\%D \times L = \$D$ or $\%Pd \times L = N$
    $0.4 \times 210 = \$D$ $0.6 \times 210 = N$
    $\$84 = \$D$ $N = \$126$
    $L - \$D = N$ $L - N = \$D$
    $210 - 84 = N$ $210 - 126 = \$D$
    $N = \$126$ $\$D = \$84$

9. $\$D = L - N = 128 - 99.84 = \$28.16$
    $\%D \times L = \$D$
    $\%D \times 128 = 28.16$
    $\%D = \dfrac{28.16}{128} = 22\%$

10. $\%Pd \times L = N$
    $0.72L = 144$
    $L = \dfrac{144}{0.72} = \$200$

11. **a.** $\%Pd_1 \times \%Pd_2 \times L = N$
    $0.88 \times 0.90 \times 500 = N$
    $N = \$396$
    b. $\$D = L - N = 500 - 396 = \$104$
    c. Net cost rate factor = $\%Pd_1 \times \%Pd_2 = 0.88 \times 0.90 = 0.792$
    d. $\%Pd = 0.875 \times 0.90 = 0.792 = 79.2\%$
    $SED = 100\% - 79.2\% = 20.8\%$

12. $\%Pd_1 \times \%Pd_2 \times L = N$
    $0.80 \times 0.95 \times L = 68.40$
    $0.76L = 68.40$
    $L = \dfrac{68.40}{0.76} = \$90$

13. $\%Pd \times L = N$
    $0.80 \times 35 = N$
    $N = \$28$
    $\%D \times L = \$D$
    $\%D \times 28 = 3$ (28 − 25 = $3 additional discount needed)
    $\%D = \dfrac{3}{28} = 10.7\%$

ANSWERS AND SOLUTIONS TO TRY THESE PROBLEMS AND TEST YOURSELF PROBLEMS **667**

14.   %D × Cr = $D   or   %Pd × Cr = $Pd
    0.02 × 2,225 = $D   0.98 × 2,225 = $Pd
          $D = $44.50 $Pd = $2,180.50
    2,225 − 44.50 = $Pd
              $Pd = $2,180.50

15. Merchandise cost = 846 − 32 = $814
   a. %Pd × Cr = $Pd
        0.99 × 814 = $Pd
             $Pd = $805.86
      Total payment = $Pd on merchandise + freight = 805.86 + 32 = $837.86
   b. Amount Credited = $846
   c. Amount owed = $0

16. **a.** List price = $15,572.75
   b. $\%Pd_1 \times \%Pd_2 \times L = N$
        0.90 × 0.95 × 15,572.75 = N
                         N = $13,314.70
   c. $D = L − N
          = 15,572.75 − 13,314.70
      $D = $2,258.05
   d.   %Pd × Cr = $Pd
        0.98 × 13,314.70 = $Pd
                $Pd = $13,048.41

17. **a.** %Pd × Cr = $Pd
     0.98 × Cr = $109.76
          Cr = $112
   b. Amount owed = 219 − 112 = $107

CHAPTER 5

Answers to Try These Problems

Set I (page 177)

1.
END OF YEAR	ANNUAL DEPRECIATION	ACCUMULATED DEPRECIATION	BOOK VALUE (END OF YEAR)
—	—	—	$12,000 (cost)
1	$2,200	$ 2,200	9,800
2	2,200	4,400	7,600
3	2,200	6,600	5,400
4	2,200	8,800	3,200
5	2,200	11,000	1,000

2.
END OF YEAR	ANNUAL DEPRECIATION	ACCUMULATED DEPRECIATION	BOOK VALUE (END OF YEAR)
—	—	—	$2,500 (cost)
1	$525	$ 525	1,975
2	525	1,050	1,450
3	525	1,575	925
4	525	2,100	400

3.

END OF YEAR	ANNUAL DEPRECIATION	ACCUMULATED DEPRECIATION	BOOK VALUE (END OF YEAR)
—	—	—	$3,000 (cost)
1	$750	$ 750	2,250
2	750	1,500	1,500
3	750	2,250	750

4. a. $7,000   b. $14,000

5. $9,900

6. $1,280; $640

Set II (page 181)

1.

END OF YEAR	ANNUAL DEPRECIATION	ACCUMULATED DEPRECIATION	BOOK VALUE (END OF YEAR)
—	—	—	$1,800 (cost)
1	$480	$ 480	1,320
2	400	880	920
3	320	1,200	600
4	240	1,440	360
5	160	1,600	200
6	80	1,680	120

2.

END OF YEAR	ANNUAL DEPRECIATION	ACCUMULATED DEPRECIATION	BOOK VALUE (END OF YEAR)
—	—	—	$4,200 (cost)
1	$1,400	$1,400	2,800
2	1,050	2,450	1,750
3	700	3,150	1,050
4	350	3,500	700

3.

END OF YEAR	ANNUAL DEPRECIATION	ACCUMULATED DEPRECIATION	BOOK VALUE (END OF YEAR)
—	—	—	$3,000 (cost)
1	$1,200	$1,200	1,800
2	800	2,000	1,000
3	400	2,400	600

4. a. $1,750
   b. $6,300

5. a. $500
   b. $1,100
   c. $1,500

6. a. $9,600; $6,400; $3,200
   b. $6,000

7. $380

ANSWERS AND SOLUTIONS TO TRY THESE PROBLEMS AND TEST YOURSELF PROBLEMS 669

Set III (page 185)

1.
END OF YEAR	ANNUAL DEPRECIATION	ACCUMULATED DEPRECIATION	BOOK VALUE (END OF YEAR)
—	—	—	$1,400 (cost)
1	0.4 (1,400) = $560.00	$ 560.00	840.00
2	0.4 (840) = 336.00	896.00	504.00
3	0.4 (504) = 201.60	1,097.60	302.40
4	0.4 (302.40) = 120.96	1,218.56	181.44
5	0.4 (181.44) = 72.58	1,291.14	108.86

2.
END OF YEAR	ANNUAL DEPRECIATION	ACCUMULATED DEPRECIATION	BOOK VALUE (END OF YEAR)
—	—	—	$5,000 (cost)
1	0.5 (5,000) = $2,500.00	$2,500	2,500
2	0.5 (2,500) = 1,250.00	3,750	1,250
3	0.5 (1,250) = 625.00	4,375	625
4	0.5 (625) = ~~312.50~~ 225.00	4,600	400

3.
END OF YEAR	ANNUAL DEPRECIATION	ACCUMULATED DEPRECIATION	BOOK VALUE (END OF YEAR)
—	—	—	$2,000.00 (cost)
1	0.4(2,000) = $800.00	$ 800.00	1,200.00
2	0.4(1,200) = 480.00	1,280.00	720.00
3	0.4(720) = 288.00	1,568.00	432.00
4	0.4(432) = 172.80	1,740.80	259.20
5	0.4(259.20) = ~~103.68~~ 59.20	1,800.00	200.00

4. $4,642.86

5. $35,185.19

6. a. 20%
   b. $1,984

7. $3,555.55

Set IV (page 191)

1.
END OF YEAR	ANNUAL DEPRECIATION	ACCUMULATED DEPRECIATION	BOOK VALUE (END OF YEAR)
—	—	—	$22,000.00 (cost)
1	0.2 (22,000) = $4,400.00	$ 4,400.00	17,600.00
2	0.32 (22,000) = 7,040.00	11,440.00	10,560.00
3	0.192 (22,000) = 4,224.00	15,664.00	6,336.00
4	0.1152(22,000) = 2,534.40	18,198.40	3,801.60
5	0.1152(22,000) = 2,534.40	20,732.80	1,267.20
6	0.0576(22,000) = 1,267.20	22,000.00	0

2.

END OF YEAR	ANNUAL DEPRECIATION	ACCUMULATED DEPRECIATION	BOOK VALUE (END OF YEAR)
—	—	—	$8,000.00 (cost)
1	0.3333(8,000) = $2,666.40	$2,666.40	5,336.60
2	0.4445(8,000) = 3,556.00	6,222.40	1,777.60
3	0.1481(8,000) = 1,184.80	7,407.20	592.80
4	0.0741(8,000) = 592.80	8,000.00	0

3.

END OF YEAR	ANNUAL DEPRECIATION	ACCUMULATED DEPRECIATION	BOOK VALUE (END OF YEAR)
—	—	—	$5,000.00 (cost)
1	$ 714.50	$ 714.50	4,285.50
2	1,224.50	1,939.00	3,061.00
3	874.50	2,813.50	2,186.50
4	624.50	3,438.00	1,562.00
5	446.50	3,884.50	1,115.50
6	446.00	4,330.50	669.50
7	446.50	4,777.00	223.00
8	223.00	5,000.00	0

4. a. $28,800 **b.** $134,800 **c.** $115,200

5. 1999: $12,146.50
2000: $20,816.50
2001: $14,866.50

Set V (page 199)

1. FIFO, $220; LIFO, $204; Average cost, $213.60
2. FIFO, $62.00; LIFO, $73.60; Average cost, $64.00
3. FIFO, $1,605; LIFO, $1,330; Average cost, $1,459.90
4. FIFO, $420; LIFO, $338; Average cost, $373.66
5. FIFO, $990, LIFO, $924; Average cost, $959.79
6. **a.** FIFO, $640.
   b. LIFO, $510.
   c. Average cost, $587.40

Solutions to Test Yourself (page 209)

1. Annual depreciation $= \dfrac{23,000 - 5,000}{4} = \$4,500$

END OF YEAR	ANNUAL DEPRECIATION	ACCUMULATED DEPRECIATION	BOOK VALUE (END OF YEAR)
—	—	—	$23,000
1	$4,500	$ 4,500	18,500
2	4,500	9,000	14,000
3	4,500	13,500	9,500
4	4,500	18,000	5,000

2. Total depreciation = 31,300 − 8,500 = 22,800

Sum of the years' digits = $\dfrac{3 \times 4}{2} = 6$

END OF YEAR	ANNUAL DEPRECIATION	ACCUMULATED DEPRECIATION	BOOK VALUE (END OF YEAR)
—	—	—	$31,300
1	$\dfrac{1}{6} \times 22{,}800 = \$\ 3{,}800$	$ 3,800	27,500
2	$\dfrac{2}{6} \times 22{,}800 = \$\ 7{,}600$	11,400	19,900
3	$\dfrac{3}{6} \times 22{,}800 = \$11{,}400$	22,800	8,500

3. Double-declining-balance rate = $2\left(\dfrac{1}{5}\right) = \dfrac{2}{5} = 40\%$

END OF YEAR	ANNUAL DEPRECIATION	ACCUMULATED DEPRECIATION	BOOK VALUE (END OF YEAR)
—	—	—	$80,000
1	0.4(80,000) = $32,000	$32,000	48,000
2	0.4(48,000) = 19,200	51,200	28,800
3	0.4(28,800) = 11,520	62,720	17,280
4	0.4(17,280) = 6,912	69,632	10,368
5	0.4(10,368) = ~~4,147.20~~ 2,368	72,000	8,000

4.

END OF YEAR	ANNUAL DEPRECIATION	ACCUMULATED DEPRECIATION	BOOK VALUE (END OF YEAR)
—	—	—	$16,000.00
1	0.1429 × 16,000 = $2,286.40	$2,286.40	13,713.60
2	0.2449 × 16,000 = 3,918.40	6,204.80	9,795.20
3	0.1749 × 16,000 = 2,798.40	9,003.20	6,996.80

5. a.

NUMBER BOUGHT	COST	TOTAL COST
8	$2.50	$ 20
3	3.00	9
5	4.00	20
6	4.50	27
10	5.00	50
32		$126

Average cost = $\dfrac{126}{32} = 3.937 = \$3.94$

Inventory value = 9 × 3.94 = $35.46

b.

	NUMBER BOUGHT	COST	TOTAL COST
LIFO	8	$2.50	$20
	1	3.00	3
	9		$23

c. FIFO   9 $5.00 $45

672 ANSWERS AND SOLUTIONS TO TRY THESE PROBLEMS AND TEST YOURSELF PROBLEMS

CHAPTER 6 **Answers to Try these Problems**

Set 1 (page 217)

1. **a.** Joanne, $2,100; Maryann, $4,200; Kathryn, $2,100
    b. Joanne, $2,100; Maryann, $5,040; Kathryn, $1,260
    c. Joanne, $2,520; Maryann, $4,200; Kathryn, $1,680
    d. Joanne, $3,087.50; Maryann, $3,950.00; Kathryn, $1,362.50
    e. Joanne, $2,900; Maryann, $3,800; Kathryn, $1,700

2. $2,800 3. **a.** $10,500; $4,500; $6,000 **b.** $9,000; $3,857.14; $5,142.86

Set II (page 223)

1. Nat, $4,800; Oscar, $3,840 2. Nat, $4,420; Oscar, $4,220 3. Nat, $5,088; Oscar, $3,552
4. Nat, $11,176; Oscar, $5,124 5. Nat, $12,235; Oscar, $7,765
6. **a.** Harry, $54,000; Larry, $64,000 **b.** Harry, $80,600; Larry, $81,400

Set III (page 231)

1. **a.** $4,625 **b.** $2,312.50 2. **a.** $1,884.62 **b.** $942.31 3. $613.60 4. $285
5. $319.20 6. $890 7. $612.50 8. $7,797 9. $491.40 10. $527.38 11. $447.10
12. Able, 40, 2.5, $272, $25.50, $297.50
    Green, 40, 5, $376, $70.50, $446.50
    Lee, 36, 0, $306, 0, $306

Set IV (page 241)

1. $45.52; $10.65 2. $30.54; $7.14 3. $54.32; $12.70 4. $76.10; $17.80
5. $6.20; $8.58 6. $151.78; $40.60 7. $0.00; $52.81 8. $25.88; $6.05 9. $54
10. $141 11. $44 12. $126 13. $790.03; $377.78; $88.35; $1,256.16; $4,837.10
14. $19.11; $28.70; $6.71; $54.52; $408.39 15. $2,712.78; $555.41; $129.89; $3,398.08;
$5,560.15 16. $163.42; $76.36; $17.86; $257.64; $973.92 17. $1,028.60
18. $2,991.10

Solutions to Test Yourself (page 253)

1. **a.** $3 + 5 = 8$

    Sam: $\frac{3}{8} \times 24,000 = \$ 9,000$

    Tom: $\frac{5}{8} \times 24,000 = 15,000$

    b. Sam: $0.4 (24,000) = \$ 9,600$

    Tom: $0.6 (24,000) = 14,400$

ANSWERS AND SOLUTIONS TO TRY THESE PROBLEMS AND TEST YOURSELF PROBLEMS 673

    c. Total investment = $32,000 + $40,000 = $72,000

        Sam: $\dfrac{32}{72} \times 24{,}000 = \$10{,}666.67$

        Tom: $\dfrac{40}{72} \times 24{,}000 = \$13{,}333.33$

    d. $24,000 − 2 (2,000) = $20,000 to be divided

        3 + 7 = 10

        Sam: $\dfrac{3}{10} \times 20{,}000 = \$6{,}000 + \$2{,}000 = \$8{,}000$

        Tom: $\dfrac{7}{10} \times 20{,}000 = \$14{,}000 + \$2{,}000 = \$16{,}000$

2. a. Average investments:

    Jenny = $11,000

    Penny = $2(9{,}400) + 4(18{,}000) + 6(15{,}000) = \dfrac{\$180{,}800}{12}$

                                                                 = $15,066.67

    Total = $11,000 + 15,066.67 = $26,066.67

    Jenny: $\dfrac{11{,}000}{26{,}066.67} \times 10{,}000 = \$4{,}219.95$

    Penny: $\dfrac{\$15{,}066.67}{26{,}066.67} \times 10{,}000 = \$5{,}780.05$

  b. Jenny: (0.12)(11,000)   = $1,320

      Penny: (0.12)(15,066.67) = $1,808

    To be divided equally: $10,000 − $3,128 = $6,872

    $\dfrac{6{,}872}{2} = \$3{,}436$

    Jenny: 1,320 + 3,436 = $4,756

    Penny: 1,808 + 3,436 = $5,244

3. Biweekly gross = $\dfrac{37{,}350}{26} = \$1{,}436.54$

    Semimonthly gross = $\dfrac{37{,}350}{24} = \$1{,}556.25$

    Monthly gross = $\dfrac{37{,}350}{12} = \$3{,}112.50$

4. Hourly rate = $8.40

    Time and a half = 1.5 × 8.40 = $12.60

    Double time = 2 × 8.40 = $16.80

    Monday through Saturday = $8 + 8\tfrac{1}{2} + 9 + 8\tfrac{1}{2} + 8 + 5 = 47$ hours

    Regular pay: 40 × 8.40   = $336.00

    Overtime: Time and a half = 7 × 12.60 =   88.20

            Double time = $6\tfrac{1}{2} \times 16.80$ = 109.20

    Gross pay: $533.40

674 ANSWERS AND SOLUTIONS TO TRY THESE PROBLEMS AND TEST YOURSELF PROBLEMS

5. Gross at hourly rate: 40 × 12.75 = $510

   Total Production: 230 + 285 + 240 + 220 + 205 = 1,180 units

   Gross at piecework rate: 1,180 × 0.45 = $531

   Gross weekly earnings = $531

6. Regular earnings: 40 × 11.00   = $440.00

   Earnings above quota: 23 × 2.50 =   57.50

   Gross earnings: $497.50

7. Regular pay: $320

   Commission: 4% (14,000 − 5,000) = 0.04 (9,000) =   360

   Gross earnings: $680

8. Net sales = $12,300 − 1,400 = $10,900

   Commission: 0.05 × 3,000 = $150

               0.08 × 4,000 =   320

               0.11 × 3,900 =   429

                             $899

   Deductions: Draw   300

   Gross earnings: $599

9. Social security tax = 0.062 × 891.95 = $55.30

   Medicare tax = 0.0145 × 891.95 = $12.93

10. Social security tax = 0.062 × 443.17 = $27.48

    Medicare tax = 0.0145 × 443.17 = $6.43

11. Social security tax = 0.062 × (72,600 − 71,100) = $93

    Medicare tax = 0.0145 × 1,881.62 = $27.28

12. Social security tax = $0   (The maximum of $72,600 has been reached.)

    Medicare tax = 0.0145 × 3,450.81 = $50.04

13. $186   (Table C-1, single persons)

14. $91   (Table C-1, married persons)

15. 0.31 [6,093.26 − 5 (105.77) − 3,788] + 785.93 = $1,336.62   (Table C-2)

16. 0.28 (1,462.91 − 1,138) + 154.20 = $245.17   (Table C-2)

17. Deductions: Social security tax = 0.062 × 1,125 = $ 69.75

                       Medicare tax = 0.0145 × 1,125   =   16.31

                       Federal income tax (Table C-1)   = 133.00

                                                  $219.06

                   Net earnings = $1,125 − 219.06 = $905.94

18. Biweekly gross = $\frac{74{,}360}{26}$ = $2,860

Social security will be deducted because year-to-date earnings will not reach $72,600 for 25.38 biweekly pay periods (end of December).

$72,600 ÷ 2,860 = 25.38

Deductions:
Social security tax = 0.062 × 2,860	= $	177.32
Medicare tax = 0.0145 × 2,860	=	41.47
Federal income tax (Table C-2) = 0.31 [2,860 − 2 (105.77) − 2,250] + 478.20	=	601.72
State tax = 0.03 × 2,860	=	85.80
Pension = 0.0525 × 2,860	=	150.15
Health insurance	=	81.00
		$1,137.46

Net earnings = $2,860 − 1,137.46 = $1,722.54

CHAPTER 7

Answers to Try These Problems

Set I (page 267)

1. $I = \$312$; $S = \$4{,}212$ 2. $I = \$412.50$; $S = \$4{,}912.50$ 3. $r = 9.5\%$; $S = \$12{,}570$
4. $t = 5$ months; $S = \$245$ 5. $r = 6\%$; $I = \$150$ 6. $t = 3$ months; $I = \$31.25$
7. $P = \$5{,}000$; $S = \$5{,}675$ 8. $P = \$20{,}975$; $S = \$22{,}023.75$ 9. **a.** $590.63
b. $6,890.63   10. 8.5%   11. $425   12. 6 months

Set II (page 273)

1. 136 days   2. 142 days   3. 145 days   4. 106 days   5. 161 days
6. March 12 of the next year   7. May 2   8. April 30   9. February 27, 2001
10. May 14, 1999

Set III (page 279)

1. $41.33; $40.77   2. $62.50; $61.64   3. $39.05; $38.52
4. $t = 2$ years or 24 months or 720 days; $I = \$770$ 5. $I = \$308$; $S = \$4{,}928$
6. $I = \$142.50$; $S = \$6{,}142.50$ 7. $I = \$8.23$; $S = \$854.23$
8. $t = 15$ months or $1\frac{1}{4}$ years or 450 days; $I = \$220$ 9. $P = \$1{,}608$; $r = 11\%$
10. $P = \$2{,}520$; $S = \$2{,}709$ 11. $r = 6.5\%$; $I = \$35.49$ 12. $r = 10.75\%$; $S = \$4{,}425.75$
13. $t = 10.5$ months or 0.875 years or 315 days; $S = \$8{,}740.83$ 14. $22,500
15. **a.** June 16   **b.** $1,200   16. 11%   17. **a.** 4 months   **b.** March 5, 2001

676 ANSWERS AND SOLUTIONS TO TRY THESE PROBLEMS AND TEST YOURSELF PROBLEMS

Set IV (page 285)

1. Kate Kornell   2. Citrus Groves Inc.   3. $6,000   4. 60 days   5. May 4, 1999
6. $100   7. $6,100   8. $352.33   9. Maker, Sister; Payee, Fran Friendly
10. a. April 26   b. $1,270.26   11. a. Mr. Brown   b. Usury Inc.   c. $42.75
d. $2,742.75   12. $1,680   13. $2,000   14. $2,700   15. $960

Set V (page 291)

1. More   2. Equal   3. Less   4. $8,000   5. $4,582.34   6. a. $3,500   b. $3,492.39
7. $456.28   8. $2,216.30   9. $4,047.52   10. $5,242.63   11. $9,170.79
12. a. $747.60   b. $749.16

Solutions to Test Yourself (page 301)

1. $I = Prt$

$$I = 3{,}600 \times 0.07 \times \frac{4}{12} = \$84$$

$$S = P + I = 3{,}600 + 84 = \$3{,}684$$

2. $I = Prt$

$$544.50 = 13{,}200 \times r \times \frac{9}{12}$$

$$544.50 = 9{,}900r$$

$$r = \frac{544.50}{9{,}900} = 5.5\%$$

$$S = P + I = 13{,}200 + 544.50 = \$13{,}744.50$$

3. $I = S - P = 5{,}006.25 - 4{,}500 = \$506.25$

$I = Prt$

$$506.25 = 4{,}500 \times 0.075 \times \frac{t}{12}$$

$$506.25 = 28.125t$$

$$t = \frac{506.25}{28.125} = 18 \text{ months}$$

4. $P = S - I = 17{,}255 - 255 = \$17{,}000$

$I = Prt$

$$255 = 17{,}000 \times r \times \frac{3}{12}$$

$$255 = 4{,}250r$$

$$r = \frac{255}{4{,}250} = 6\%$$

5. July 12 to December 31 (193 to 365) = 172 days

This leaves 180 − 172 = 8 days in the following year. The maturity date is January 8 of the next year.

6. October 28

ANSWERS AND SOLUTIONS TO TRY THESE PROBLEMS AND TEST YOURSELF PROBLEMS 677

7. March 4 to August 1 (63 to 213) = 150 days

8. **a.** $I = Prt$

   $I = 3,200 \times 0.09 \times \dfrac{7}{12} = \$168$

   b. $S = P + I = 3,200 + 168 = \$3,368$

9. $I = S - P = 4,319 - 4,200 = \$119$

   $I = Prt$

   $119 = 4,200 \times r \times \dfrac{4}{12}$

   $119 = 1,400r$

   $r = \dfrac{119}{1,400} = 8.5\%$

10. $I = Prt$

    $5.10 = P \times 0.08 \times \dfrac{6}{12}$

    $5.10 = P(0.04)$

    $P = \$127.50$

11. $I = Prt$

    $150 = 2,500 \times 0.12 \times \dfrac{t}{12}$

    $150 = 25t$

    $t = \dfrac{150}{25} = 6$ months

12. ORDINARY   EXACT

    $I = 400 \times 0.10 \times \dfrac{80}{360}$ $I = 400 \times 0.10 \times \dfrac{80}{365}$

    $I = 8.888 = \$8.89$ $I = 8.767 = \$8.77$

13. **a.** Her boyfried

    b. Tiffany DuPont

    c. $I = Prt$

    $I = 800 \times 0.0525 \times \dfrac{120}{360} = \$14$

    d. $S = P + I = 800 + 14 = \$814$

14. March 2 to May 19 (61 to 139) = 78 days

    $P = \dfrac{S}{1 + rt}$

    $= \dfrac{1,325.35}{1 + 0.09 \times \dfrac{78}{360}}$

    $= \dfrac{1,325.35}{1 + 0.0195}$

    $= \dfrac{1,325.35}{1.0195}$

    $P = \$1,300$

15. $I = Prt = 900 \times 0.12 \times \dfrac{6}{12} = \$54$

$S = P + I = 900 + 54 = \$954$

$P = \dfrac{S}{1 + rt}$

$= \dfrac{954}{1 + 0.10 \times \dfrac{6}{12}}$

$= \dfrac{954}{1 + 0.05}$

$= \dfrac{954}{1.05}$

$P = \$908.57$

16. August 20 to October 31 (232 to 304) = 72 days

$I = Prt = 3{,}600 \times 0.11 \times \dfrac{72}{360} = \$79.20$

$S = P + I = 3{,}600 + 79.20 = \$3{,}679.20$

October 1 to October 31 = 30 days

$P = \dfrac{S}{1 + rt}$

$= \dfrac{3{,}679.20}{1 + 0.12 \times \dfrac{30}{360}}$

$= \dfrac{3{,}679.20}{1 + 0.01}$

$= \dfrac{3{,}679.20}{1.01}$

$P = \$3{,}642.77$

CHAPTER 8

Answers to Try These Problems

Set I (page 311)

1. $D = \$90$; $p = \$4{,}410$ **2.** $D = \$68.25$; $p = \$2{,}731.75$ **3.** $D = \$57.75$; $p = \$842.25$
4. $S = \$8{,}500$; $p = \$8{,}287.50$ **5.** $S = \$10{,}000$; $p = \$9{,}800$ **6.** $S = \$800$; $d = 8.5\%$
7. $S = \$3{,}000$; $d = 10\%$ **8.** $\$5{,}625$ **9.** $D = \$83.20$; $p = \$6{,}156.80$ **10.** 15% **11.** 14%
12. 4 months **13.** 1 month **14. a.** $882.70 **b.** $910 **c.** $27.30 **15.** $909.18

Set II (page 319)

1. $D = \$86.13$; $p = \$5{,}213.87$ **2.** $D = \$111.48$; $p = \$3{,}528.52$ **3.** $D = \$350$; $p = \$24{,}650$
4. $S = \$1{,}236$; $p = \$1{,}225.70$ **5.** $S = \$4{,}200$; $p = \$4{,}151$ **6.** $S = \$1{,}197$; $p = \$1{,}175.45$
7. June 23; 23 days **8.** February 19; 16 days **9.** January 29; 42 days
10. September 4; 63 days **11.** $S = \$3{,}765$; $p = \$3{,}647.34$ **12.** $S = \$43{,}000$;
$p = \$42{,}161.50$ **13.** $S = \$2{,}233$; $p = \$2{,}219.11$ **14.** $1,584 **15. a.** $118.13
b. $7,981.87 **16. a.** $458 **b.** $5.73 **c.** $452.27 **17. a.** $192 **b.** $164.80
c. $6,427.20 **d.** $27.20 **18.** $7,063.24

ANSWERS AND SOLUTIONS TO TRY THESE PROBLEMS AND TEST YOURSELF PROBLEMS 679

Set III (page 327)

1. $1,200   2. $1,355   3. $480   4. $1,948.72   5. $612.24   6. $3,215.17
7. a. $720   b. $720   8. $54.74   9. a. $3,360   b. $54.29   10. a. $600   b. $9.95
11. a. $648   b. $8.82   12. a. $127.87   b. $21.95
13. a. $65   b. Int. Note: $3,250   c. Int. Note: $3,315
              Disct. Note: $3,185   Disct. Note: $3,250
14. a. $P = \$613.21$ b. Int. Note: $613.21   c. Int. Note: $36.79
       $p = \$611$ Disct. Note: $611   Disct. Note: $39

Solutions to Test Yourself (page 337)

1. $D = Sdt$

   $D = 7{,}500 \times 0.08 \times \dfrac{30}{360} = \$50$

   $p = S - D = 7{,}500 - 50 = \$7{,}450$

2. $D = S - p = 510 - 471.75 = \$38.25$

   $D = Sdt$

   $38.25 = 510 \times d \times \dfrac{10}{12}$

   $38.25 = 425d$

   $d = \dfrac{38.25}{425} = 9\%$

3. $S = p + D = 4{,}541 + 259 = \$4{,}800$

   $D = Sdt$

   $259 = 4{,}800 \times 0.0925 \times \dfrac{t}{360}$

   $259 = 1.2\overline{3}\, t$

   $t = \dfrac{259}{1.2\overline{3}} = 210 \text{ days}$

4. Term of note: March 14 to June 2 (73 to 153) = 80 days

   Discount period: April 27 to June 2 (117 to 153) = 36 days

5. a. $D = Sdt$

      $D = 3{,}150 \times 0.07 \times \dfrac{40}{360} = \$24.50$

   b. $p = S - D = 3{,}150 - 24.50 = \$3{,}125.50$

   c. $S = \$3{,}150$

6. $D = S - p = \$1{,}000 - 962.50 = \$37.50$

   $D = Sdt$

   $\$37.50 = 1{,}000 \times d \times \dfrac{90}{360}$

   $\$37.50 = 250d$

   $d = \dfrac{37.50}{250} = 15\%$

7. $D = Sdt$

$$160 = 4{,}000 \times 0.12 \times \frac{t}{12}$$

$$160 = 40t$$

$$t = \frac{160}{40} = 4 \text{ months}$$

8. **a.** $D = Sdt$

$$425 = S \times 0.075 \times \frac{120}{360}$$

$$425 = S(0.025)$$

$$S = \frac{425}{0.025} = \$17{,}000$$

b. $p = S - D = 17{,}000 - 425 = \$16{,}575$

9. $I = Prt$

$$I = 11{,}000 \times 0.07 \times \frac{180}{360} = \$385$$

$$S = P + I = 11{,}000 + 385 = \$11{,}385$$

$$D = 11{,}385 \times 0.10 \times \frac{45}{360} = \$142.31$$

$$P = 11{,}385 - 142.31 = \$11{,}242.69$$

10. **a.** $D = Sdt$

$$= 3{,}250 \times 0.105 \times \frac{60}{360}$$

$$D = 56.875 = \$56.88$$

b. $p = S - D = 3{,}250 - 56.88 = \$3{,}193.12$

11. **a.** $I = Prt$

$$I = 2{,}000 \times 0.09 \times \frac{90}{360} = \$45$$

$$S = P + I = 2{,}000 + 45 = \$2{,}045$$

July 1 (182 + 90) = 272 = September 29

August 10 to September 29 (222 to 272) = 50 days

$D = Sdt$

$$D = 2{,}045 \times 0.12 \times \frac{50}{360} = 34.083 = \$34.08$$

$$p = S - D = 2{,}045 - 34.08 = \$2{,}010.92$$

b. The bank earned $34.08.

c. Amount earned = 2,010.92 − 2,000 = $10.92

12. $S = \dfrac{P}{1 - dt}$

$$= \frac{7{,}315}{1 - (0.09)\left(\frac{5}{12}\right)}$$

$$= \frac{7{,}315}{1 - 0.0375}$$

$$= \frac{7{,}315}{0.9625}$$

$$S = \$7{,}600$$

13. a. $\%Pd \times Cr = \$Pd$

$0.98(985.23) = 965.525 = \$965.53$

$$S = \frac{P}{1-dt}$$

$$= \frac{965.53}{1 - 0.12 \times \frac{20}{360}}$$

$$= \frac{965.53}{1 - 0.00\overline{6}}$$

$$= \frac{965.53}{0.99\overline{3}}$$

$$S = \$972.01$$

b. Amount saved = $985.23 - 972.01 = \$13.22$

14. $I = Prt$ or $D = Sdt = 6,000 \times 0.085 \times \frac{4}{12} = \$170$

15. Interest note: $6,000

Discount note: $p = S - D = 6,000 - 170 = \$5,830$

16. Interest note: $S = P + I = 6,000 + 170 = \$6,170$

Discount note: $6,000

CHAPTER 9

Answers to Try These Problems

Set I (page 349)

1. $300 2. $240 3. $294 4. $65 5. $10,500 6. $525 7. $42.50 8. $17.20
9. $30.40 10. $131.46 11. a. $114.75 b. $48.88 12. a. $269 b. $19.43

Set II (page 357)

1.

UNPAID BALANCE	+	FINANCE CHARGE	=	AMOUNT OWED	PAYMENT	TO PAY INTEREST	TO REDUCE DEBT	NEW BALANCE
$112.00		—		$112.00	$ 24.00	—	$24.00	$88.00
88.00		$1.32		89.32	24.00	$1.32	22.68	65.32
65.32		0.98		66.30	24.00	0.98	23.02	42.30
42.30		0.63		42.93	24.00	0.63	23.37	18.93
18.93		0.28		19.21	19.21	0.28	18.93	0.00
					$115.21	$3.21		

2.

UNPAID BALANCE	+	FINANCE CHARGE	=	AMOUNT OWED	PAYMENT	TO PAY INTEREST	TO REDUCE DEBT	NEW BALANCE
$90.00		—		$90.00	$15.00	—	$15.00	$75.00
75.00		$0.75		75.75	15.00	$0.75	14.25	60.75
60.75		0.61		61.36	15.00	0.61	14.39	46.36
46.36		0.46		46.82	15.00	0.46	14.54	31.82
31.82		0.32		32.14	15.00	0.32	14.68	17.14
17.14		0.17		17.31	17.31	0.17	17.14	0.00
					$92.31	$2.31		

3.

UNPAID BALANCE	+	FINANCE CHARGE	=	AMOUNT OWED	PAYMENT	TO PAY INTEREST	TO REDUCE DEBT	NEW BALANCE
$300.00		—		$300.00	$ 48.00	—	$48.00	$252.00
252.00		$3.78		255.78	48.00	$ 3.78	44.22	207.78
207.78		3.12		210.90	48.00	3.12	44.88	162.90
162.90		2.44		165.34	48.00	2.44	45.56	117.34
117.34		1.76		119.10	48.00	1.76	46.24	71.10
71.10		1.07		72.17	72.17	1.07	71.10	0.00
					$312.17	$12.17		

4. $129.97 **5.** $2.84 **6.** $169.84 **7.** $200.72 **8.** $8.84 **9.** $13.73 **10. a.** $35 **b.** $16.04 **11.** $0.50; $31.30

Set III (page 363)

1. 15.00% **2.** 14.25% **3.** 21.25% **4.** 13.75% **5.** 16.75% **6. a.** $192 **b.** 14.75% **7.** 18.00% **8.** 10.75% **9.** 13.00% **10.** 12.75% **11. a.** $49.11 **b.** $22.94 **c.** $477.12 **d.** 16.50% **12. a.** 12.5% **b.** 22.25%

Set IV (page 367)

1. $57.69 **2.** $151.79 **3.** $51.21 **4.** $258.89 **5. a.** $208.80 **b.** $31.32 **c.** $271.98 **6. a.** $972 **b.** $354 **c.** $119.37 **d.** $2,004.63

Solutions to Test Yourself (page 377)

1. Total of monthly payments = 12 × 465 = $5,580

   Finance charge = 5,580 − 5,000 = $580

2. Total of monthly payments = 36 × 61.50 = $2,214

   Finance charge = 2,214 − 1,725 = $489

3. Amount financed = Cash price − Down payment = 2,399 − 100 = $2,299

   Monthly payment = $\dfrac{\text{Total debt}}{\text{Number of payments}} = \dfrac{2,299 + 148}{10} = \$244.70$

4. Amount financed = Cash price − Down payment = 780 − 0.20 (780) = $624

   $I = Prt = 624 \times 0.09 \times \dfrac{24}{12} = \$112.32$

   Monthly payment = $\dfrac{\text{Total debt}}{\text{Number of payments}} = \dfrac{624 + 112.32}{24} = \$30.68$

5. Total of monthly payments = 12 × 50 = $600

   Finance charge = 600 − 500 = $100

6. a. Amount financed = Cash price − Down payment = 650 − 90 = $560

   $I = Prt = 560 \times 0.16 \times \dfrac{15}{12} = \$112$

   b. Total of payments = 560 + 112 = $672

   Monthly payment = $\dfrac{672}{15} = \$44.80$

   c. Total cost = 672 + 90 or 650 + 112 = $762

   d. Charge per $100 = $\dfrac{112}{560} \times 100 = \$20$

   Table C-4 value for 15 payments = 28.50%

7. Total on current bill = 918 + 13.77 = $931.77

Less payment	150.00
Unpaid balance on next bill	781.77
Finance charge ($1\tfrac{1}{2}\%$)	11.73
New purchases	205.00
Total on next bill	$998.50

8. $50 (There is no finance charge on the first bill.)

9. Finance charge = 0.015 × 172.29 = $2.58

   September total = 172.29 + 2.58 = $174.87

10. | | |
    |---|---|
    | September total | $174.87 |
    | Less payment | 25.00 |
    | October unpaid balance | 149.87 |
    | Finance charge ($1\tfrac{1}{2}\%$) | 2.25 |
    | New purchase | 26.50 |
    | October total | $178.62 |

11.

UNPAID BALANCE	+	FINANCE CHARGE	+	NEW PURCHASES	=	AMOUNT OWED	PAYMENT	TO PAY INTEREST	TO REDUCE DEBT	NEW BALANCE
$149.87		$2.25		$26.50		$178.62	$50.00	$2.25	$47.75	$128.62
128.62		1.93		—		130.55	50.00	1.93	48.07	80.55
80.55		1.21		—		81.76	50.00	1.21	48.79	31.76
31.76		0.50		—		32.26	32.26	0.50	31.76	0.00

12. a. Total of all payments = 199 + 20 = $219

Monthly payment = $\dfrac{219}{12}$ = $18.25

b. Charge per $100 = $\dfrac{20}{199} \times 100$ = $10.05

Table C-4 value = 18.00%

13. a. $I = Prt$

$36 = 300 \times r \times \dfrac{15}{12}$

$\dfrac{36}{300 \times 1.25} = r$

$9.6\% = r$

b. Charge per $100 = $0.096 \times \dfrac{15}{12} \times 100$ = $12

or $\dfrac{36}{300} \times 100$ = $12

Table C-4 value = 17.50%

14. a. Sum of remaining payments = $\dfrac{12(13)}{2}$ = 78

Sum of total number of payments = $\dfrac{48(49)}{2}$ = 1,176

Refund = $\dfrac{78}{1,176} \times 3,496$ = $231.88

b. Amount still due = 12 × 173.82 = $2,085.84
Refund   − 231.88
Total to pay loan $1,853.96

15. Finance charge = 65 × 18 − 899 = $271

Total months = $\dfrac{18 \times 19}{2}$ = 171

Remaining months = $\dfrac{9 \times 10}{2}$ = 45

a. Refund = $\dfrac{45}{171} \times 271$ = $71.32

b. Final payment = (65 × 9) − 71.32 = $513.68

CHAPTER 10 — Answers to Try these Problems

Set I (page 389)

1. 8%; 6   2. $3\frac{1}{2}$%; 30   3. 5%; 4   4. 3%; 16   5. $2\frac{1}{2}$%; 44   6. $4\frac{3}{8}$%; 24   7. $\frac{17}{24}$%; 108

8. $1\frac{5}{8}$%; 14   9. $\frac{11}{12}$%; 60   10. $1\frac{3}{4}$%; 26   11. 2%; 13   12. $\frac{3}{4}$%; 67

Set II (page 397)

1. $902.05; $377.05
2. $1,645.04; $1,025.04
3. $2,446.33; $1,496.33
4. $1,700.29; $500.29
5. $527.97; $127.97
6. $524.32; $124.32
7. $389.90
8. $2,323.92
9. $56,458.13
10. a. $2,209.88
    b. $209.88
11. a. $2,208.97
    b. $208.97
12. More frequent compounding gives more interest and a larger compound amount.
13. $91.13
14. $577.81
15. a. $1,211.28
    b. $311.28
16. $5,381.22

Set III (page 401)

1. $612.52; $112.52
2. $3,038.77; $538.77
3. $7,354.07; $4,354.07
4. $887.49; $47.49
5. $9,169.45; $2,169.45
6. $17,481.73; $5,481.73
7. $877.39
8. $2,875.10

686 ANSWERS AND SOLUTIONS TO TRY THESE PROBLEMS AND TEST YOURSELF PROBLEMS

9. $749.54
10. a. $3,900.46
    b. $900.46
11. a. $511.31
    b. $11.31
12. $1,667.90

Set IV (page 407)

1. $676.33; $176.33   2. $3,874.49; $874.49   3. $555.88; $155.88   4. $182.35; $82.35
5. $1,956.64; $856.64   6. $1,559.79; $259.79   7. $897.63; $447.63   8. 9.31%
9. 9.20%   10. 8.30%   11. 8%   12. They are equal.   13. a. $8,195.68   b. $4,195.68
c. 9.38%   14. a. $595.57   b. $195.57   c. 5.61%; 6.17%   15. $5,895.36

Set V (page 413)

1. $446.13; $153.87   2. $498.79; $901.21   3. $3,334.99; $3,865.01
4. $1,334.61; $1,165.39   5. $6,880.92; $3,119.08   6. $2,085.91; $1,914.09
7. $15,122.71; $4,877.29   8. $2,101.40   9. $9,946.53; $7,053.47   10. $37,712.98
11. a. $10,051.32   b. $9,948.68   12. No; Present value = $3,648.05 or Compound amount = $4,070.67   13. Buy it now; Present value = $86,730.26 or Compound amount = $92,216.95

Solutions to Test Yourself (page 423)

1. $i = \dfrac{6\%}{4} = 1\dfrac{1}{2}\%$ $n = 4 \times 15 = 60$

2. $i = \dfrac{3\dfrac{1}{2}\%}{12} = 3\dfrac{1}{2} \div 12$ $n = 12 \times 2\dfrac{10}{12} = 12 \times 2\dfrac{5}{6}$

   $= \dfrac{7}{2} \times \dfrac{1}{12} = \dfrac{7}{24}\%$ $= 12 \times \dfrac{17}{6} = 34$

3. $i = \dfrac{7\%}{1} = 7\%$ $n = 1 \times 3 = 3$

   Compound amount = 4,400 (1.2250430) = $5,390.19

   Compound interest = 5,390.19 − 4,400 = $990.19

4. $i = \dfrac{4\dfrac{1}{2}\%}{2} = 4\dfrac{1}{2} \div 2$ $n = 2 \times 8 = 16$

   $i = \dfrac{9}{2} \times \dfrac{1}{2} = \dfrac{9}{4} = 2\dfrac{1}{4}\%$

   Compound amount = 900 (1.427621) = $1,284.86

   Compound interest = 1,284.86 − 900 = $384.86

5. $i = \dfrac{9\%}{2} = 4\dfrac{1}{2}\%$ $\qquad n = 2 \times 8 = 16$

   Compound amount = 3,000 (2.0223701) = $6,067.11

   Compound interest = 6,067.11 − 3,000 = $3,067.11

6. $i = \dfrac{5\dfrac{1}{2}\%}{4} = 5\dfrac{1}{2} \div 4$ $\qquad n = 4 \times 9\dfrac{3}{12}$

   $ = \dfrac{11}{2} \times \dfrac{1}{4}$ $\qquad\qquad\qquad = 4 \times 9\dfrac{1}{4}$

   $ = \dfrac{11}{8} = 1\dfrac{3}{8}\%$ $\qquad\qquad = 4 \times \dfrac{37}{4} = 37$

   Compound amount = 600 (1.657456) = $994.47

   Compound interest = 994.47 − 600 = $394.47

7. $i = \dfrac{3\%}{1} = 3\%$ $\qquad n = 1 \times 4 = 4$

   Compound amount = 750 (1.125508) = $844.13

8. $i = \dfrac{4\%}{4} = 1\%$ $\qquad n = 4 \times 4 = 16$

   Compound amount = 2,000 (1.1725786) = $2,345.16

   New principal = 2,345.16 + 1,000 = $3,345.16 $n = 4 \times 2 = 8$

   Compound amount = 3,345.16 (1.0828567) = $3,622.33

9. $i = \dfrac{6\%}{4} = 1\dfrac{1}{2}\%$ $\qquad n = 4 \times 2 = 8$

   $i = \dfrac{7\%}{4} = 1\dfrac{3}{4}\%$ $\qquad n = 4 \times 3 = 12$

   Compound amount = 300 (1.126492) (1.231439) = $416.16

10. $i = \dfrac{6\%}{12} = \dfrac{1}{2}\%$ $\qquad N = 12$

    Effective rate = $(1 + i)^N - 1$ = 1.06167 − 1 = 0.06167 = 6.17%

11. Bank A is better.

    For Bank A: The effective rate is 4.25%, since annual compounding gives an effective rate equal to the nominal rate.

    For Bank B:

    $i = \dfrac{4\%}{12} = \dfrac{1}{3}\%$ $N = 12$

    Effective rate = $(1 + i)^N - 1$ = 1.04074 − 1 = 0.04074 = 4.07%

12. $i = \dfrac{8\%}{4} = 2\%$ $n = 4 \times 10 = 40$

    a. Present value = 20,000 (0.45289042) = $9,057.81

    b. Compound interest = 20,000 − 9,057.81 = $10,942.19

13. $i = \dfrac{7\%}{2} = 3\dfrac{1}{2}\%$ $\qquad n = 2 \times 1 = 2$

Present value = 170,000 (0.933510700) = $158,696.82

Buy in one year; the land is presently worth only $158,696.82 based on the future value and given rate.

or

Compound amount = 160,000 (1.071225000) = $171,396

If $160,000 is invested at the given rate for one year, it will be worth more than the future value of the property.

CHAPTER 11

Answers to Try these Problems

Set I (page 433)

1. $6,381.98; $4,800; $1,581.98   2. $16,499.94; $10,800; $5,699.94
3. $14,772.13; $8,800; $5,972.13   4. $18,366.78; $14,400; $3,966.78
5. $159,425.06; $9,425.06   6. $185,297.18; $81,297.18   7. $9,440.45; $1,740.45
8. $49,153.29; $7,153.29   9. $12,506.76   10. a. $11,784.82 b. $9,100 c. $2,684.82
11. a. $200,000 b. $229,227.59 c. $29,277.59   12. a. $20,979.51 b. $2,739.51
13. a. $10,435.47 b. $14,930.74 c. $5,930.74   14. a. $3,236.89 b. $836.89

Set II (page 441)

1. $4,026.05; $6,000; $1,973.95   2. $10,677.54; $20,000; $9,322.46
3. $5,554.63; $9,600; $4,045.37   4. $13,670.07; $16,800; $3,129.93
5. $2,110.03; $3,200; $1,089.97   6. $73,484; $108,000; $34,516
7. a. $8,192.38 b. $9,600 c. $1,407.62   8. $42,699.10   9. $1,025.78
10. a. $53,200 b. $46,166.30   11. $2,311.42
12. a. $12,633.35 b. $1,187.25 c. $22,812.75
13. a. $30,744.90 b. $9,533.01 c. $50,466.99

Set III (page 449)

1. $79; $3,792; $792   2. $995.91; $39,836.40; $14,836.40   3. $650.11; $8,451.43; $3,451.43   4. $2,331.93; $116,596.50; $56,596.50   5. $227.78; $6,833.40; $3,166.60
6. $134.86; $3,236.64; $1,263.36   7. $209.97; $9,448.65; $50,551.35
8. $142.41; $5,981.22; $1,018.78   9. $2,913.91   10. $636.89; $2,190.24
11. a. $260.78 b. $12,517.44 c. $2,482.56   12. a. $783.87 b. $18,645.20
13. a. $343.88 b. $5,130.72

ANSWERS AND SOLUTIONS TO TRY THESE PROBLEMS AND TEST YOURSELF PROBLEMS

Set IV (page 453)

1. **a.** $7,121.45 **b.** $11,059.39 **c.** $73.24 **d.** $113.74 **e.** $579.53 **f.** $1,397.67

   g. (a) $1,878.55 (b) $2,059.39 (c) $167.60 (d) $237.40 (e) $320.47 (f) $497.67

2. **a.** $25,870.70 **b.** (i) $7,730.75 (ii) $252.04 **c.** $19,838.38 **d.** $652.04

   e. $21,884.38

Solutions to Test Yourself (page 463)

1. $i = \dfrac{6\frac{1}{2}\%}{1} = 6\frac{1}{2}\%$ $n = 1 \times 9 = 9$

   $S_n = 500 \,(11.731852) = \$5,865.93$ (Table C-7)

   Interest earned $= 5,865.93 - (500 \times 9) = \$1,365.93$

2. $i = \dfrac{9\%}{12} = \dfrac{3}{4}\%$ $n = 12 \times 3 = 36$

   $A_n = 1,800 \,(31.4468052) = \$56,604.25$ (Table C-8)

   Interest earned $= (1,800 \times 36) - 56,604.25 = \$8,195.75$

3. $i = \dfrac{7\%}{4} = 1\dfrac{3}{4}\%$ $n = 4 \times 6 = 24$

   Rent $= 80,000 \,(0.05138565) = \$4,110.85$ (Table C-9)

   Interest earned $= (4,110.85 \times 24) - 80,000 = \$18,660.40$

4. $i = \dfrac{5\%}{2} = 2\dfrac{1}{2}\% = 0.025$ $n = 2 \times 5\dfrac{1}{2} = 11$

   Rent $= 27,000 \,(0.10510596 - 0.025)$ (Table C-9 $- i$)

   $= 27,000 \,(0.08010596) = \$2,162.86$

   Interest earned $= 27,000 - (2,162.86 \times 11) = \$3,208.54$

5. $i = \dfrac{5\%}{4} = 1\dfrac{1}{4}\%$ $n = 4 \times 11 = 44$

   Compound amount $= 1,000 \,(1.7273542) = \$1,727.35$ (Table C-5)

   Interest earned $= 1,727.35 - 1,000 = \$727.35$

6. $i = \dfrac{8\%}{1} = 8\%$ $n = 1 \times 15 = 15$

   a. $S_n = 20,000(27.1521139) = \$543,042.28$ (Table C-7)

   b. Total deposits $= 20,000 \times 15 = \$300,000$

   c. Interest earned $= 543,042.28 - 300,000 = \$243,042.28$

7. $i = \dfrac{10\%}{4} = 2\dfrac{1}{2}\%$ $n = 4 \times 5 = 20$

   a. $A_n = 5,000 \,(15.5891622) = \$77,945.81$ (Table C-8)

   b. Total withdrawals $= 5,000 \times 20 = \$100,000$

   c. Interest earned $= 100,000 - 77,945.81 = \$22,054.19$

690 ANSWERS AND SOLUTIONS TO TRY THESE PROBLEMS AND TEST YOURSELF PROBLEMS

8. $i = \dfrac{6\%}{2} = 3\% = 0.03 \qquad n = 2 \times 20 = 40$

   a. Rent = 900,000 (0.04326238 − 0.03) (Table C-9 − i)
      = 900,000 (0.01326238) = $11,936.14

   b. Total deposits = 11,936.14 × 40 = $477,445.60

   c. Interest earned = 900,000 − 477,445.60 = $422,554.40

9. $i = \dfrac{7\%}{4} = 1\dfrac{3}{4}\% \qquad n = 4 \times 7\dfrac{1}{2} = 30$

   a. Rent = 15,000 (0.04312975) = $646.95 (Table C-9)

   b. Interest earned = (646.95 × 30) − 15,000 = $4,408.50

10. $i = \dfrac{7\%}{4} = 1\dfrac{3}{4}\% \qquad n = 4 \times 4 = 16$

    Present value = 4,000 (0.7576163) = $3,030.47 (Table C-6)

    Interest earned = 4,000 − 3,030.47 = $969.53

11. a. $i = \dfrac{7\dfrac{1}{2}\%}{12} = \dfrac{5}{8}\% \qquad n = 12 \times 3\dfrac{1}{2} = 42$

    $S_n = 200(47.858363) = \$9,571.67$ (Table C-7)

    b. $i = \dfrac{5}{8}\% \qquad n = 12 \times 5 = 60$

    Compound amount = $9,571.67 (1.4532944) = $13,910.45 (Table C-5)

    c. Interest earned = 13,910.45 − (200 × 42) = $5,510.45

12. $i = \dfrac{8\dfrac{1}{2}\%}{1} = 8\dfrac{1}{2}\% \qquad n = 1 \times 10 = 10$

    a. $A_n = 15,000(6.56134806) = \$98,420.22$ (Table C-8)

    b. $i = \dfrac{8\dfrac{1}{2}\%}{1} = 8\dfrac{1}{2}\% \qquad n = 1 \times 15 = 15$

    Present value = 98,420.22 (0.29413989) = $28,949.31 (Table C-6)

    c. Interest earned = (15,000 × 10) − 28,949.31 = $121,050.69

CHAPTER 12 — Answers to Try these Problems

Set I (page 479)

1. Line

2. About 4%

3. About 2.5%

4. The first quarter of 1998

5. About 1.5%

6. Quarterly percent change fluctuated between 4% and 0.2% during the 3-year period. The overall trend was down.

ANSWERS AND SOLUTIONS TO TRY THESE PROBLEMS AND TEST YOURSELF PROBLEMS

7. Bar
8. About 810,000
9. 1997
10. About 70,000
11. The percent of working mothers is increasing.
12. Mothers with children under 6 years old.
13. Child care services

Set II (page 485)

1. 89%   2. 41%   3. Personal income tax   4. $535,000,000   5. $203,000,000
6. Component bar graph
7. Number of cars sold would have to be converted to percent of total.
8. 4.9 million   9. 3.4 million   10. 59.0%   11. 41.0%   12. 9.6%   13. 19.3%

Set III (page 495)

	MEAN	MEDIAN	MODE	RANGE	STANDARD DEVIATION
1.	7	3.5	4	27	9.5
2.	16.4	16	none	19	7.0
3.	$7	$7	none	$6	$2

4. $112   5. $112   6. $116   7. $26   8. $7.93   9. $0.00
10. The second student had more variation in salary.   11. 2.3   12. 2.6   13. $650
14. 80¢   15. 75¢   16. No, an unweighted mean implies equal importance for all items.

Set IV (page 501)

1. 104
2. Increase of 4% from 1996
3. 126
4. Increase of 26% from 1996
5. $I_{1992} = 120$; $I_{1994} = 130$; $I_{1996} = 150$; $I_{1998} = 156$
6. Compared with 1990, the Littons' entertainment expenses increased by 50% in 1996 and by 56% in 1998.
7. 64.6%
8. 40.8% increase
9. 146.3%
10. No, aspirin is not reported separately.

692 ANSWERS AND SOLUTIONS TO TRY THESE PROBLEMS AND TEST YOURSELF PROBLEMS

11. No, the index is for the entire country.
12. 16% decline from the base year 1982

Solutions to Test Yourself (page 509)

1. **a.** About 19% **b.** About 14%   2. Approximately 13%   3. 15–24 and 35–44
4. Approximately 15% of 267 million = 0.15 × 267 = 40.1 million
5. Approximately 10% of 227 million = 0.10 × 227 = 22.7 million
6. There will be fewer workers in this group to pay social security, medicare, federal income, state, local, and other taxes; lower numbers of high school graduates to enroll in colleges; fewer buyers for teenage and young-twenties merchandise and services; a smaller pool of workers to fill part-time and entry-level jobs; and so on.
7. 24.7%
8. Classical
9. 100% − 34.5% = 65.5%
10. Jazz sales = 13.9% of $825,000 = 0.139 × 825,000 = $114,675
11. Mean = $\frac{13 + 23 + 27 + 23 + 19}{5} = \frac{105}{5} = 21$ years
12. 13
    19
    23   Median = 23 years
    23
    27
13. Mode = 23 years
14. Range = 27 − 13 = 14 years
15.

VALUES	d	d²
13	−8	64
19	−2	4
23	2	4
23	2	4
27	6	36
		112

Standard deviation = $\sqrt{\frac{d^2}{N}} = \sqrt{\frac{112}{5}}$
= $\sqrt{22.4}$
= 4.7 years

16. 0 years
17. $ 75 × 2 = $ 150
    120 × 3 =   360
    150 × 4 =   600
    270 × 1 =   270
          10 $1,380

Weighted mean = $\frac{\text{Sum of (values × weights)}}{\text{Sum of weights}}$
= $\frac{1{,}380}{10}$ = $138

18. $I_{1985} = \dfrac{24.10}{21.60} \times 100 = 111.6$

$I_{1990} = \dfrac{20.00}{21.60} \times 100 = 92.6$

$I_{1995} = \dfrac{14.60}{21.60} \times 100 = 67.6$

19. In 1985, crude oil prices were 11.6% higher than in 1980.

20. In 1995, crude oil prices were 32.4% lower than in 1980.

CHAPTER 13

ANSWERS TO TRY THESE PROBLEMS

Set I (page 523)

1.
Cost of goods sold	59.7%
Gross profit	40.3
Operating expenses	21.8
Operating profit	18.5
Taxes	7.9
Net earnings	10.6

2. $66,900 3. 2.3 4. Working capital ratio 5. 1.7 6. Acid-test ratio 7. 17.1%

8. 8.5% 9. 15.3% 10. Yes: Current ratio is more than 2; quick ratio is more than 1.

Set II (page 533)

1. $390

2. 5.8%

3. $1,170

4. $104,000

5. $56,000

6. $3,000

7. 7.8%

8. 12.5

9. 7.3%

10. 7.6%

11. 5.7%

Set III (page 541)

1. 55\dfrac{1}{4}$ or $55.25

2. 1,960,600 shares

3. $1.45 per share

4. 1.7%

5. $17

6. +1\frac{1}{8}$ or + $1.125

7. $24,308

8. 7.5%

9. 2019

10. −$17.50

11. $81.25

12. $1,035

13. $10,275

SOLUTIONS TO TEST YOURSELF (page 549)

1. a. Working capital = 4,170 − 1,830 = $2,340

   b. Current ratio = $\frac{4,170}{1,830}$ = 2.3

   c. Quick ratio = $\frac{4,170 - 520}{1,830}$ = 2.0

   d. Return on investment = $\frac{1,420}{2,850}$ = 49.8%

   e. Return on total assets = $\frac{1,420}{4,680}$ = 30.3%

   f. Account receivable-net sales ratio = $\frac{1,250}{32,000}$ = 3.9%

   g. $\frac{\text{net income}}{\text{net sales}} = \frac{1,420}{32,000}$ = 4.4%

   h. Yes

2. Current yield = $\frac{0.82}{21}$ = 0.0390 = 3.9%

3. a. Dividend = (0.06 × 70) × 800 = $3,360

   b. % Yield = $\frac{4.20}{76.50}$ = 5.5%

   c. Dividend for 2000 = 3 × 3,360 = $10,080

4. a. ABC: Current yield = $\frac{60}{950}$ = 6.3%

   XYZ: Current yield = $\frac{70}{1,000}$ = 7.0%

   Buy XYZ bond.

   b. Approximate yield to maturity of XYZ bond equals current yield (7.0%) because the bond was bought at par.

5. Earnings per share = $\frac{210,000}{120,000}$ = $1.75

   PE ratio = $\frac{42.875}{1.75}$ = 24.5

ANSWERS AND SOLUTIONS TO TRY THESE PROBLEMS AND TEST YOURSELF PROBLEMS

6. a. Daily low = $55\frac{3}{4}$ or $55.75

   b. Sales volume = 333,800 shares

   c. Change from previous day = $-\$\frac{5}{8}$ or $-\$0.625$

   d. Current yield = 2.6%

7. Total amount paid = Trading cost + Commission

$$= (100 \times 115.25) + 0.03(100 \times 115.25)$$
$$= 11{,}525 + 345.75$$
$$= \$11{,}870.75$$

8. Total cost = $(400 \times 15.75) + 0.03(400 \times 15.75)$ = $6,489.00

Total received = $(400 \times 14.125) - 0.03(400 \times 14.125)$ = 5,480.50

Net loss	$1,008.50
Dividends ($0.50 per share)	200.00
Total loss	$ 808.50

Rate of return = $\dfrac{-808.50}{6{,}489} = -12.5\%$

9. a. Closing price = $87\frac{3}{4}\%$ of 1,000 = 0.8775 × 1,000 = $877.50

   b. Change from previous day = $+\frac{1}{2}\%$ of 1,000 = +0.005 × 1,000 = $5

   c. Dividend = $6\frac{1}{8}\%$ of $1,000 = 0.06125 × 1,000 = $61.25

   d. Sales volume = 16 bonds

10. Total purchase price = Cost of bonds + Commission

$$= (5 \times 1{,}125) + (5 \times 10)$$
$$= \$5{,}675$$

Answers to Odd-Numbered End-of-Chapter Problems

CHAPTER 1

(PAGE 35)

1. 5.2; 5.17; 5.170
3. 12.2; 12.25; 12.250
5. 79.9; 79.94; 79.945
7. $1,712.69
9. $12.40
11. $\frac{65}{70} = \frac{13}{14}$
13. $\frac{16}{96} = \frac{1}{6}$
15. $\frac{92}{8} = \frac{23}{2}$
17. $\frac{11}{7}$
19. $\frac{1}{3}$
21. 25.625
23. 34.204
25. 133.4
27. 1.5
29. 18
31. 6A
33. $1.68x + 5$
35. $\frac{5}{4}Z - 8Y + 3$
37. $37.8C$
39. $7N + 1.5H$
41. $Y = 3$
43. $R = 4$
45. $S = 5$
47. $C = 15$
49. $S = 6$
51. $C = 230$
53. $S = 40$
55. $P = 345$
57. $r = 0.202$
59. $S = 1,250$
61. 3
63. 4
65. c
67. b
69. b
71. $740
73. $\frac{1}{14}$
75. **a.** $\frac{2}{3}$ **b.** $\frac{3}{8}$ **c.** $\frac{5}{8}$
    d. Answers should add up to 1.

CHAPTER 2

(PAGE 71)

1. $0.22, \frac{11}{50}$
3. $2.26, 2\frac{13}{50}$ or $\frac{113}{50}$
5. $0.4, \frac{2}{5}$
7. $0.0275, \frac{11}{400}$
9. $0.0025, \frac{1}{400}$
11. $0.00321, \frac{321}{100,000}$
13. $0.0525, \frac{21}{400}$
15. $0.5, \frac{1}{2}$
17. $0.08125, \frac{13}{160}$
19. $0.008, \frac{1}{125}$
21. 37%
23. 20%

25. 6.18%
27. 240%
29. 300.18%
31. 30%
33. 27.27%
35. 209.09%
37. 40.38%
39. 60%
41. 24.43
43. $430.50
45. 14%
47. 105%
49. 100
51. 50% increase
53. 60% decrease
55. 11% increase
57. 51% decrease
59. 23% decrease
61. 170
63. 133.33% or $133\frac{1}{3}$%
65. 86
67. $3,612.50
69. 4.38%
71. $422,400
73. $82,000
75. 93.5%
77. $20.79
79. 400
81. 34.84%
83. 26%
85. 6.67% or $6\frac{2}{3}$%
87. 66.67% or $66\frac{2}{3}$% increase
89. 2,706
91. a. 15.69%   b. $100
93. 33.33% or $33\frac{1}{3}$%
95. 19.09%
97. $89, 80%
99. $174,300
101. $71,907
103. $5,890

CHAPTER 3

(PAGE 111)

1. $16.90; $51.90
3. $4.32; $10.50
5. $20; $6.75
7. $19.20; $79.20
9. $123; $175
11. $51; $136
13. $50; $8.50
15. $29.50; $37.50
17. $9; 16.1%; 13.8%
19. $75; 66.7%; 40%
21. $37.44; $34.56; 92.3%
23. $22.20; $59.20; 37.5%
25. $30.80; $9.20; 29.9%
27. $192; $48; 20%
29. $16; $40; 66.7%
31. $7.80; $12; 53.8%
33. $3.60; $12; 233.3%
35. $112.50; $163.13; 225 lbs; $0.73
37. a. $210   b. $273   c. $7.50   d. 285 lbs   e. $0.93 ea
39. a. $320   b. $486.40   c. $28   d. 36 dz   e. $12.73 dz
41. Loss of $5
43. $220
45. $17.97
47. a. 80.6%   b. 44.6%
49. $13.50
51. $144
53. $30.86
55. a. $58.87   b. 61.3%
57. a. $52   b. 53.8%
59. $5.29, 41.2%
61. a. $22   b. $33   c. 33.3% or $33\frac{1}{3}$%
63. $1.27 per pair
65. a. 30%   b. 12.5%
67. $0.99 per box

ANSWERS TO ODD-NUMBERED END-OF-CHAPTER PROBLEMS   699

69. $33.97 ea

71. $108.82

73. $70.42

75. $1.35 ea

77. $0.66 ea

79. $2.56 lb

81. 19.3%

83. a. $336.49   **b.** Below
   c. Operating loss, $4.78
   d. NP = $12,398.65

CHAPTER 4

(PAGE 157)

1. $450; $150; 0.75

3. $446.40; $173.60; 0.72; 28%

5. $1,124.55; $375.45; 0.7497; 25.03%

7. $20; $6.70; 0.665; 33.5%

9. $754.60; $770; 0

11. $3,920; $4,000; 0

13. $582; $129.79

15. $475.68; $480.48; 0

17. $147.50; $442.50

19. $0.90

21. 40%

23. a. $38.23   **b.** 35.2%

25. a. $25.54   **b.** 46.8%   **c.** 0.532

27. $85

29. $60

31. a. 15%   **b.** 32%

33. 12.3%

35. 15%

37. a. 20/15   **b.** $0.72

39. 20/4

41. $65.46

43. a. 7.4%   **b.** $25

45. $59.26

47. $826.83

49. a. $198.55   **b.** Feb. 6

51. a. Dec. 15   **b.** $617.60

53. $1,407.70

55. $1,496.76; $2,540

57. a. $483.84   **b.** $476.16

59. a. $1,500   **b.** $1,000

61. a. $1,200, larger discount   **b.** $525.77

CHAPTER 5

(PAGE 201)

1. Double-declining-balance rate $= 2\left(\dfrac{1}{4}\right) = \dfrac{1}{2} = 50\%$

End of Year	Annual Depreciation	Accumulated Depreciation	Book Value (End of Year)
—	—	—	$8,000 (cost)
1	0.5(8,000) = $4,000	$4,000	4,000
2	0.5(4,000) = 2,000	6,000	2,000
3	0.5(2,000) = 1,000	7,000	1,000
4	0.5(1,000) = 500	7,500	500

3. Annual depreciation = $1{,}842 - 75 = \dfrac{1{,}767}{6} = \$294.50$

End of Year	Annual Depreciation	Accumulated Depreciation	Book Value (End of Year)
—	—	—	$1,842.00 (cost)
1	$294.50	$ 294.50	1,547.50
2	294.50	589.00	1,253.00
3	294.50	883.50	958.50
4	294.50	1,178.00	664.00
5	294.50	1,472.50	369.50
6	294.50	1,767.00	75.00

5. Double-declining-balance rate = $2\left(\dfrac{1}{10}\right) = \dfrac{1}{5} = 20\%$

End of Year	Annual Depreciation	Accumulated Depreciation	Book Value (End of Year)
—	—	—	$40,000 (cost)
1	0.2 (40,000) = $8,000	$ 8,000	32,000
2	0.2 (32,000) = 6,400	14,400	25,600
3	0.2 (25,600) = 5,120	19,520	20,480

7. Annual depreciation = $2{,}500 - 300 = \dfrac{2{,}200}{4} = \$550$

End of Year	Annual Depreciation	Accumulated Depreciation	Book Value (End of Year)
—	—	—	$2,500 (cost)
1	$550	$ 550	1,950
2	550	1,100	1,400
3	550	1,650	850
4	550	2,200	300

9. $12,800

11. Total depreciation = $24{,}735 - 2{,}235 = \$22{,}500$

Sum of the years' digits = $1 + 2 + 3 + 4$ or $\dfrac{4 \times 5}{2} = 10$

End of Year	Annual Depreciation	Accumulated Depreciation	Book Value (End of Year)
—	—	—	$24,735 (cost)
1	0.4 (22,500) = $9,000	$ 9,000	15,735
2	0.3 (22,500) = 6,750	15,750	8,985
3	0.2 (22,500) = 4,500	20,250	4,485
4	0.1 (22,500) = 2,250	22,500	2,235

13. a.

End of Year	Annual Depreciation	Accumulated Depreciation	Book Value (End of Year)
—	—	—	$2,500 (cost)
1	0.2 (2,500) = $500	$ 500	2,000
2	0.32 (2,500) = 800	1,300	1,200
3	0.192 (2,500) = 480	1,780	720
4	0.1152 (2,500) = 288	2,068	432
5	0.1152 (2,500) = 288	2,356	144
6	0.0576 (2,500) = 144	2,500	0

b. Annual depreciation $= \dfrac{2,500 - 400}{5} = \dfrac{2,100}{5} = \$420$

End of Year	Annual Depreciation	Accumulated Depreciation	Book Value (End of Year)
—	—	—	$2,500 (cost)
1	$420	$ 420	2,080
2	420	840	1,660
3	420	1,260	1,240
4	420	1,680	820
5	420	2,100	400

c. Double-declining-balance rate $= 2\left(\dfrac{1}{5}\right) = 40\%$

End of Year	Annual Depreciation	Accumulated Depreciation	Book Value (End of Year)
—	—	—	$2,500 (cost)
1	0.4 (2,500) = $1,000	$1,000	1,500
2	0.4 (1,500) = 600	1,600	900
3	0.4 (900) = 360	1,960	540
4	0.4 (540) = 140*	2,100	400
5	0	2,100	400

*Not all of 0.4 can be taken

d. Total depreciation = 2,500 − 400 = $2,100

Sum of the years' digits $= 1 + 2 + 3 + 4 + 5$ or $\dfrac{5 \times 6}{2} = 15$

End of Year	Annual Depreciation	Accumulated Depreciation	Book Value (End of Year)
—	—	—	$2,500 (cost)
1	$\dfrac{5}{15} \times 2,100 = \$700$	$ 700	1,800
2	$\dfrac{4}{15} \times 2,100 = 560$	1,260	1,240
3	$\dfrac{3}{15} \times 2,100 = 420$	1,680	820
4	$\dfrac{2}{15} \times 2,100 = 280$	1,960	540
5	$\dfrac{1}{15} \times 2,100 = 140$	2,100	400

15. a. $300 **b.** $50

702 ANSWERS TO ODD-NUMBERED END-OF-CHAPTER PROBLEMS

17.
End of Year	Annual Depreciation	Accumulated Depreciation	Book Value (End of Year)
—	—	—	$18,000.00 (cost)
1	0.3333 (18,000) = $5,999.40	$ 5,999.40	12,000.60
2	0.4445 (18,000) = 8,001.00	14,000.40	3,999.60
3	0.1481 (18,000) = 2,665.80	16,666.20	1,333.80
4	0.0741 (18,000) = 1,333.80	18,000.00	0

19. a. $3,210 b. $19,260 c. $3,210

21. a. $\frac{1}{5}$ b. $2,080 c. $2,304 d. 40%

23. FIFO, because the most expensive bicycles (purchased last) would be in inventory.

25. a. $35.12 b. $33.60 c. $36.00

27. a. $173.00 b. $178.20 c. $168.60

CHAPTER 6

(PAGE 243)

1. Ruth, $2,285.71; Sue, $1,571.43; Ted, $2,142.86

3. Ruth, $1,500; Sue, $2,000; Ted, $2,500

5. Ruth, $2,190.48; Sue, $1,714.29; Ted, $2,095.24

7. Joe, $6,360; Roy, $7,395; Clare, $2,745

9. Dan, $6,600; Doreen, $3,300

11. 1st, $12,000; 2nd, $30,000; 3rd, $36,000; 4th, $42,000

13. a. Ursula, $2,610; Vera, $4,350; Will, $1,740
    b. Each would receive $2,900

15. Karen, $27,800; Keith, $11,200

17. Ellie, $10,666.67; Shelly, $5,333.33

19. a. Eve, $3,000; Enis, $2,000; Edna, $3,800
    b. Eve, $2,730; Enis, $2,100; Edna, $4,200

21. Elaine, $47,142.86; Bill, $32,857.14

23. Xavier, $6,350; Yolanda, $6,550; Zena, $7,100

25. a. $1,807.69
    b. $1,958.33

27. $333.63

29. $982

31. $832

33. $3,084

35. Duff: 40, 2, $340, $25.50, $365.50
    Hollis: 40, 5, $396, $74.25, $470.25
    Smith: 35, 0, $269.50, 0, $269.50

37. $4,840

39. Social security = $38.51; medicare = $9.01
41. Social security = $48.39; medicare = $11.32
43. $42.97
45. $0
47. $30
49. $93
51. $1,373.60; $385.18; $90.08; $1,848.86; $4,363.69
53. $146.63; $84.39; $19.74; $250.76; $1,110.41
55. $323.40
57. $22
59. First week in February, social security = $87.52; medicare = $20.47; federal income tax = $295.14

    Last week in December, social security = $37.91; medicare = $20.47; federal income tax = $295.14

CHAPTER 7

(PAGE 293)

1. 134 days
3. September 10
5. 72 days
7. Exact, $24; ordinary, $24.33
9. Exact, $36.99; ordinary, $37.50
11. $I = \$420$; $S = \$4,420$
13. $P = \$450$; $S = \$468$
15. $r = 11\%$; $S = \$513.75$
17. $t = 2$ months or 60 days; $S = \$1,620$
19. $P = \$4,000$; $I = \$420$
21. $563.37
23. **a.** $619 **b.** September 29
25. $16.76; $435.76
27. **a.** April 26 **b.** $1,494 **c.** $1,494.47
29. $828.22
31. 8%
33. 10.5%
35. 9%
37. 4 months or 120 days
39. 1.5 years or 18 months
41. 2 months or 60 days
43. $300
45. $6,000
47. $6,552.26
49. $1,400
51. **a.** 8.5% **b.** more
53. $2,341.46
55. $2,977.01
57. $883.07
59. **a.** It is less ($P = \$127.45$). **b.** $2.05

CHAPTER 8

(PAGE 329)

1. $p = \$838.31$; $D = \$35.69$
3. $d = 11\%$; $D = \$336.60$

5. $d = 14.5\%$; $p = \$1,484$

7. $S = \$360$; $D = \$12$

9. 180 days; 130 days

11. 102 days; 49 days

13. 69 days; 23 days

15. $1,524; $1,507.07

17. $625; $622.92

19. $956.72

21. a. $3,283   b. $67

23. Yes, he will have $1,176.

25. 4 months

27. 270 days

29. a. 45 days   b. September 29

31. $2,400

33. a. $75
    b. Interest note, $5,000; discount note, $4,925
    c. Interest note, $5,075; discount note, $5,000

35. a. $P = \$896.76$; $p = \$896.76$
    b. $896.76; $896.76
    c. $I = \$57.24$; $D = \$57.24$

37. a. $900   b. $13.50

39. $8,249.17

41. a. $7,243.02   b. $144   c. $100.98

43. $19.70

45. $17.35

CHAPTER 9

(PAGE 369)

1. $39.50

3. $55

5. $150

7. $71.83

9. $52.23

11. 16.75%

13. 21.75%

15. 19.5%

17. $130

19. $189

ANSWERS TO ODD-NUMBERED END-OF-CHAPTER PROBLEMS

21. a. $16.00 **b.** $289.99

23. a. $69.53 **b.** 18.25%

25. a. $47.36 **b.** 21.75%

27. a. $3,900 **b.** $900 **c.** 11.1% **d.** 19.75%

29. $40.15

31.

Unpaid Balance	+	Finance Charge	=	Amount Owed	Payment	To Pay Interest	To Reduce Debt	New Balance
$215.03		$2.15		$217.18	$55.00	$2.15	$52.85	$162.18
162.18		1.62		163.80	55.00	1.62	53.38	108.80
108.80		1.09		109.89	55.00	1.09	53.91	54.89
54.89		0.55		55.44	55.44	0.55	54.89	0.00
						$5.41		

33. a.

Unpaid Balance	+	Finance Charge	=	Amount Owed	Payment	To Pay Interest	To Reduce Debt	New Balance
$172.00		—		$172.00	$25.00	—	$25.00	$147.00
147.00		$1.47		148.47	25.00	$1.47	23.53	123.47
123.47		1.23		124.70	25.00	1.23	23.77	99.70
99.70		1.00		100.70	25.00	1.00	24.00	75.70
75.70		0.76		76.46	25.00	0.76	24.24	51.46
51.46		0.51		51.97	25.00	0.51	24.49	26.97
26.97		0.50		27.47	27.47	0.50	26.97	0.00
						$5.47		

b. $5.47

35.

Unpaid Balance	+	Finance Charge	+	New Purch.	=	Amount Owed	Payment	To Pay Interest	To Reduce Debt	New Balance
$32.14		$0.50		$83.42		$116.06	$22.00	$0.50	$21.50	$94.06
94.06		1.41		—		95.47	22.00	1.41	20.59	73.47
73.47		1.10		—		74.57	22.00	1.10	20.90	52.57
52.57		0.79		—		53.36	22.00	0.79	21.21	31.36
31.36		0.50		—		31.86	22.00	0.50	21.50	9.86
9.86		0.50		—		10.36	10.36	0.50	9.86	0.00

37. a. $75.74 **b.** $27.58 **c.** 21.75%

39. a. 7% **b.** 12.75%

41. a. $232 **b.** 10.75%

43. a. $108.92 **b.** $2,201.08 **c.** $6,821.08

CHAPTER 10

(PAGE 415)

1. 2%; 36
3. $5\frac{1}{2}$%; 4
5. $\frac{3}{8}$%; 45
7. 8.24%
9. 5.50%
11. 4.59%
13. $14,185.86; $5,885.86
15. $1,374.75; $674.75
17. $5,948.45; $2,051.55
19. $64.50; $235.50
21. $1,999.99; $753.66
23. $3,308.58; $2,908.58
25. $22,589.65
27. They are the same.
29. a. $9,939.21 b. $5,939.21 c. 11.30%
31. a. $1,227.45 b. $327.45
33. a. $10,914.37 b. $23,192.35 c. $16,192.35
35. a. $6,246.02 b. $3,347.75
37. a. $36,174.52 b. $21,330.81
39. $5,000 at 12% compounded semiannually; $1,264.52
41. a. $918.15 b. 10.2% c. 9.31%
43. Buy it in one year; P.V. = $46,192.27 or C.A. = $50,874.31
45. a. $4,859.47 b. $12,080.40 c. $38,754.39
47. $1,469.33
49. a. $9,303.91 b. $5,696.09 c. 12.68%
51. a. $2,440.31 b. $2,559.69
53. $1,231.47
55. $10,602.37
57. a. $10,568.27 b. $25,404.74

CHAPTER 11

(PAGE 455)

1. $12,720.34; $6,000.34
3. $21,730.64; $12,730.64

ANSWERS TO ODD-NUMBERED END-OF-CHAPTER PROBLEMS 707

5. $104.57; $54,771.50

7. $6,673.14; $526.86

9. $9,422.50; $4,977.50

11. $501.38; $2,049.68

13. a. $10,721.36 b. $2,321.36

15. a. $48,381.28 b. $29,181.28

17. a. $1,935,080.53 b. $1,135,080.53

19. a. $146,211.88 b. $96,211.88

21. a. $83,042.37 b. $150,000 c. $66,957.63

23. a. $20,621.87 b. $14,619.23 c. $9,380.77

25. $1,155,738.60

27. $2,334.17

29. $547.47

31. $9,346.36

33. a. $6,193.14 b. $337,270.72 c. $97,270.72

35. $412.79

37. $362.26, $2,611.52

39. a. $587.21 b. $11,744.20 c. $3,255.80

41. $20,702.81, $15,702.81

43. $3,464.91

45. a. $23,400.26 b. $6,599.74

47. Rent = $1,410.06

Payment No.	Amount of Payment	Interest for Period	Amount to Principal	Principal at End of Period
0	—	—	—	$5,000.00
1	$1,410.06	$250.00	$1,160.06	3,839.94
2	1,410.06	192.00	1,218.06	2,621.88
3	1,410.06	131.09	1,278.97	1,342.91
4	1,410.06	67.15	1,342.91	0
		$640.24	$5,000.00	

49. Rent = $337.53

Payment No.	Interest Earned	Amount of Payment	Yearly Increase	Total in Account
1	$ 0	$ 337.53	$337.53	$ 337.53
2	28.69	337.53	366.22	703.75
3	59.82	337.53	397.35	1,101.10
4	93.59	337.53	431.12	1,532.22
5	130.24	337.53	467.77	1,999.99
	$312.34	$1,687.65		

CHAPTER 12

(PAGE 503)

1. About $275 million
3. 1998
5. The scale is broken. (The part between 0 and 200 is different from the rest of the scale.)
7. About $550,000
9. 22%
11. $45.6 million for salaries and wages.
13. $597.88
15. **a.** 79.9 **b.** 80 **c.** 94
17. 86.55%
19. Mean = $32,142.86; median = $20,000; mode = $18,000. The median is most representative. The mean is higher than six out of the seven values. The mode is the lowest value.
21. **a.** 11, 4 **b.** 27, 9.5
23. **a.** 152 **b.** 171.4 **c.** 53.3

CHAPTER 13

(PAGE 543)

1. See the Chapter 13 glossary, page 715
3. $205,000
5. $11 million
7. 5.3%
9. 0.9; no, because it is less than 1.0.
11. 12.8%
13. Working capital, 1998 = $2,289,000
    Percent change = +9.3%; increase shows company is growing.
15. 1.3, better
17. $3.10 per share
19. $344,430
21. $1,087,000
23. 16.9
25. **a.** $60 **b.** 6.6% **c.** 7.0%
27. 23\frac{7}{8}$ or $23.875 per share
29. $+\frac{1}{4}$ or +$0.25
31. 6.3%
33. 11.3%
35. 47 bonds
37. 2006

APPENDIX A

EXERCISES

(PAGE 575)

1. 310,024	3. 6,015,600	5. 879,630	7. 276.39	9. 204
11. 0.63	13. 1.89	15. 20.33	17. 7.43	19. 0.90
21. $\frac{9}{10}$	23. $\frac{3}{50}$	25. $\frac{3}{2,500}$	27. $2\frac{1}{20}$	29. $\frac{53}{125}$
31. 0.367	33. 1,020	35. 0.001629	37. 0.85	39. 0.02659
41. $\frac{29}{36}$	43. $10\frac{101}{120}$	45. 9.125	47. 27.241	49. 14.517
51. 10.537	53. 0.7176	55. $2\frac{16}{35}$	57. $3\frac{11}{45}$	59. 6.1002
61. $\frac{3}{56}$	63. $\frac{1}{2,205}$	65. 45	67. $24\frac{2}{3}$	69. 0.00268
71. $\frac{27}{32}$	73. $\frac{7}{18}$	75. $4\frac{1}{6}$	77. 4,489.2	79. 11.2

APPENDIX B

Exercises (page 585)

1. 6.4 m	3. 12.7 cm	5. 68.58 cm	7. 0.86 in.
9. 7.02 in.	11. 328 ft	13. 141.75 g	15. 119.07 g
17. 1.93 kg	19. 0.11 oz	21. 3.3 short tons	23. 136.4 lb
25. 1.43 L	27. 17.81 L	29. 59.16 mL	31. 0.6 tsp
33. 5 tsp	35. 18.72 gal	37. 0°C	39. 38°C
41. 29°C	43. 32°F	45. 95°F	47. 122°F

Problems (page 587)

1. quart	3. mile	5. yard	7. 77°F
9. 425.25 g	11. 39°C	13. 64.43 L	15. 106.68 mm

Glossary

CHAPTER 1

An **ALGEBRAIC EXPRESSION** is a combination of signs and symbols that represent one number or quantity. $\left(5x + 3 \text{ or } \frac{y}{b} - 1.\right)$

COEFFICIENTS are numbers that variables are multiplied by. [4 is the coefficient of $4y$ and of $4(a - b)$.]

CONSTANTS are definite numbers whose value does not change.

An **EQUATION** is a statement that two algebraic expressions are equal. ($2x + 5 = 11$.)

GENERAL NUMBERS (VARIABLES), usually represented by letters, stand for quantities for which the description is known but not the value. (c = cost or p = profit.)

A **RATIO** is a comparison between two similar numbers. It may be written in different forms. ($\frac{3}{5}$, 3 to 5, 3:5.)

A **RECIPROCAL** is the result of inverting a number. (The reciprocal of $\frac{2}{3}$ is $\frac{3}{2}$.)

SIMILAR TERMS are terms that include the same variable. ($2a - b + 4a$ has two similar terms.)

TERMS of an algebraic expression are the parts that are separated by plus and minus signs. ($4xy + 3x$ contains two terms.)

VARIABLES (See General Numbers.)

CHAPTER 2

ASSESSED VALUE is the value of residential or commercial property used by the local government for purposes of taxation.

BASE is the whole quantity to which the rate (percent) is related. It usually follows the word *of* or *on*.

COMMISSION is a form of salary paid to some salespersons. It can vary from week to week because it depends on sales that are made.

PERCENT means hundredths. One percent is the fraction $\frac{1}{100}$ or the decimal 0.01.

PRODUCT is the result of multiplying the rate and base.

RATE is the percent that multiplies any base.

CHAPTER 3

ABSOLUTE (or GROSS) LOSS occurs if the selling price is less than the cost of merchandise. It is the difference between cost and selling price.

BREAKEVEN POINT is the selling price that equals the cost plus overhead. There is neither profit nor loss when an item is sold at the breakeven point.

MARGIN is the same as markup.

MARKDOWN is the amount deducted from the original selling price to determine the reduced price.

MARKUP is the amount added to the cost to determine the selling price. It covers all expenses and net profit.

NET PROFIT is the amount left after all expenses have been deducted from the markup.

OPERATING LOSS occurs if the selling price is below the breakeven point. It is the difference between the selling price and total cost (wholesale cost plus overhead).

OVERHEAD includes all expenses such as rent, taxes, utilities, equipment maintenance, and the like.

REDUCED PRICE is the selling price after a discount or deduction off the regular price.

REGULAR PRICE is the selling price at which merchandise would normally be marked. This price usually covers all overhead expenses and net profit.

CHAPTER 4

CASH DISCOUNT is a reduction in price given for payment of a bill for merchandise before the payment is due.

CONSUMER is one who buys a product for his or her own use.

CREDIT is the amount entered as the payment received.

E.O.M. (end of month) is a sales term indicating that the discount period starts at the beginning of the following month.

EXTRA (−X) is a sales term in which the discount period extends for an additional number of days. (2/10–40X indicates that a discount can be taken for 50 days.)

An **INVOICE** is a bill displaying the amount that must be paid.

LIST PRICE is a suggested retail price or a stated price before a trade discount.

NET COST RATE FACTOR is the product of the "percents paid" in a trade discount problem. It is multiplied by the list price to obtain the net price.

NET PRICE is the price after all discounts are taken.

ORDINARY DATING is a sales term indicating that the discount period starts at the invoice date.

A **PRODUCER (MANUFACTURER)** is a person or business that makes a product and resells it to a wholesaler or retailer.

PROX. (Proximo) See E.O.M.

A **RETAILER** is a person or business that sells merchandise directly to the consumer.

R.O.G. (receipt of goods) is a sales term indicating that the cash discount period begins on the day the merchandise arrives.

SERIES DISCOUNTS are two or more trade discounts given on the same purchase.

A **SINGLE EQUIVALENT DISCOUNT** is the one trade discount that will give the same net price as a series of discounts.

A **TRADE DISCOUNT** is a reduction from a quoted or list price which may be given for quantity orders or for other special reasons.

A **WHOLESALER** is a person or business that buys merchandise from a manufacturer and resells it to retail stores.

CHAPTER 5

The **ACCELERATED COST-RECOVERY SYSTEM (ACRS)** is a depreciation method created as part of the Economic Tax Recovery Act of 1981. Annual depreciation is a predetermined percent of the cost of an asset, and the entire cost is eventually recovered.

ACCUMULATED DEPRECIATION is the sum of the annual depreciations to date.

ASSETS are items of value owned by a business.

AVERAGE COST is a method of inventory valuation that assumes that all items bought during a specific time period have the same cost, equal to the average cost of the items during that period.

BOOK VALUE is the value of an asset in the company's records. An asset's book value does not imply that the asset can be sold for that amount.

COST RECOVERY is the concept, used in recent tax laws, of the decline in the value of an asset.

DECLINING-BALANCE is a method of depreciation in which the maximum annual depreciation occurs in the first year. A fixed percent, up to twice the straight-line rate, may be used as the multiplier of the book value.

DEPRECIATION is the loss in value of an asset as it is used to produce revenue.

The **DEPRECIATION SCHEDULE** shows the annual depreciation, book value, and accumulated depreciation for each year of an asset's useful life.

DOUBLE-DECLINING-BALANCE is an accelerated method of depreciation that uses twice the straight-line rate multiplied times the book value.

FIFO (first in-first out) is a method of inventory valuation that assumes that merchandise bought first is sold first.

A **HALF-YEAR CONVENTION** is an assumption under MACRS whereby an asset is allowed one-half year of depreciation for the first year the asset is placed in service, regardless of when the asset is placed in service during the year.

INVENTORY is the merchandise at hand at a specific time.

LIFO (last in–first out) is a method of inventory valuation that assumes that merchandise bought last is sold first. This assumption is used only for accounting purposes.

The **MODIFIED ACCELERATED COST-RECOVERY SYSTEM (MACRS)** is a depreciation method introduced as part of the Tax Reform Act of 1986. It retained many of the features of the ACRS method while returning to a recovery period that more nearly approaches a true life.

A **PERIODIC INVENTORY SYSTEM** is one in which a physical count of inventory is taken at regular intervals.

A **PERPETUAL INVENTORY SYSTEM** is an inventory system in which a computer is normally used to continuously update inventory records.

SALVAGE (SCRAP) VALUE is the value of an asset at the end of its useful life.

The **SPECIFIC IDENTIFICATION METHOD** is an inventory valuation method that matches each item with its actual cost.

STRAIGHT-LINE is a method of depreciation in which total depreciation is evenly divided over the asset's life. Annual depreciation is the same for each year.

SUM-OF-YEARS'-DIGITS is an accelerated method of depreciation in which the largest annual depreciation occurs in the first year. Annual depreciation decreases each year of an asset's useful life.

TOTAL DEPRECIATION is the original cost less the resale or salvage value.

USEFUL LIFE is the number of years during which an asset is used and depreciated.

CHAPTER 6

COMMISSIONS are employee earnings that represent a certain percentage of the net sales of the employee.

In a **DIFFERENTIAL PIECE-RATE PLAN** the rate paid per item depends on the number of items produced.

DOUBLE TIME is twice the regular hourly rate.

A **DRAW** is an advance against future commissions received by a salesperson. Once commissions are earned, the draw reduces the amount of earnings.

An **EXEMPTION** is an amount of gross earnings that is not subject to tax.

The **FAIR LABOR STANDARDS ACT** sets minimum wage standards and overtime regulations for all those workers covered by this law.

The **FEDERAL INSURANCE CONTRIBUTIONS ACT (FICA)** provides for deductions of a certain percent of a base amount of most workers' yearly wages. This amount is matched by employers and is used by the federal government to pay for retirement, disabled workers, survivor's benefits, medicare, and so on. FICA is now separated into social security and medicare.

GROSS EARNINGS are employee earnings before FICA, income tax, or other deductions.

An **HOURLY WAGE** is a rate of compensation expressed as a certain amount of dollars per hour.

INCOME TAX WITHHOLDING is the federal tax deducted from each employee's pay. It is determined by gross earnings, marital status, and number of dependents.

The **MEDICARE TAX** is a percentage deduction based on an employee's gross earnings, which is matched by the employer. The money is used to fund health insurance benefits.

OVERTIME is the number of hours an employee works in excess of 40 per week.

A **PARTNERSHIP** is a business having two or more people as co-owners.

The **PERCENTAGE METHOD** is a way of computing federal income tax withholding using tax rates and taxable earnings to calculate the amount to be withheld. This method does not require the pages of tables needed with the wage bracket method and is therefore more adaptable to computerized processing of payroll.

PIECEWORK wages are compensation based on units produced or completed.

A **QUOTA** is an expected level of production. A bonus may be paid for exceeding a quota.

A **SALARY** is a fixed amount of compensation per pay period.

The **SLIDING-SCALE COMMISSION** is a method of pay that uses different commission rates for different levels of net sales.

SOCIAL SECURITY. See Federal Insurance Contributions Act.

STRAIGHT COMMISSION is a method of compensation in which earnings are a percent of net sales.

A **TIME-AND-A-HALF RATE** is one and one-half times the normal rate of pay for any hours worked in excess of 40 per week.

WAGE BRACKET TABLES are used by employers to determine federal income tax withholding. The Internal Revenue Service publishes many tables encompassing different pay periods and single or married status.

WITHHOLDING is the deduction made from an employee's gross earnings for social security tax, medicare tax, income tax, pensions, health insurance, union dues, etc.

WITHHOLDING ALLOWANCE. See Exemption.

CHAPTER 7

ADD-ON INTEREST is simple interest that is added to the amount borrowed and received, along with the amount borrowed, as repayment at the end of the period.

The **AMOUNT** of a note is the maturity value of the note.

The **BANKER'S RULE** is the use of ordinary interest.

COMPOUND INTEREST is interest charged or received on both principal and interest.

The **DUE DATE** of a note is the maturity date of the note.

EXACT INTEREST is a method of calculating interest in which the time is expressed as a fraction such as

$$t = \frac{\text{Exact number of days in the loan}}{365}$$

EXACT TIME is the time used in calculating exact interest, namely, a period in which the year has 365 days.

The **FACE VALUE** of a simple interest note is the principal of the loan.

INTEREST is money paid for the use of money.

The **MAKER** of a note borrows the money and signs the note.

The **MATURITY DATE** of a note is the date on which payment of the note is due.

The **MATURITY VALUE** of a note is the amount repaid.

The **MONEY MARKET RATE** is the average rate of interest being paid by banks and other financial institutions.

A note is **NEGOTIABLE** if it can be sold or legally transferred to another person or institution.

ORDINARY INTEREST is a method of calculating interest in which the time is expressed as a fraction such as

$$t = \frac{\text{Exact number of days in the loan}}{360}$$

ORDINARY TIME is the time used in calculating ordinary interest, namely, a period in which the year has 360 days.

The **PAYEE** of a loan lends the money and receives payment on the due date.

The **PRESENT VALUE** is the amount that would have to be invested today in order to have a given value in the future.

The **PRINCIPAL** of a loan is the amount borrowed.

A **PROMISSORY NOTE** is a written promise to pay a debt.

The **RATE** of a loan is the percent interest charged per year.

SIMPLE INTEREST is interest calculated on the principal only.

TERM is the time between the making of a note and the due date of the note.

CHAPTER 8

BANK DISCOUNT (or **DISCOUNT**) is the interest deducted from the maturity value at the time a loan is granted.

The **DISCOUNT PERIOD** is the time from the date of discount to the maturity date.

DISCOUNTING A NOTE is receiving cash from selling a note to a bank before the maturity date of the note.

FACE VALUE is the amount shown on any promissory note. For a simple discount note, the face value is the maturity value.

MATURITY VALUE is the total amount of a note that must be repaid. On a discount note, it is the amount on which the discount is computed and is equal to proceeds plus bank discount.

PROCEEDS are the amount the borrower receives when a loan is discounted, after the interest has been deducted. It is the amount the borrower has for his use.

A SIMPLE DISCOUNT NOTE is a note whose interest is deducted from the face value of the note. The borrower receives the amount borrowed minus the interest.

The **TRUTH IN LENDING ACT** is a federal law that requires lenders to reveal their total finance charge and the annual percentage rate based on the amount actually received by a borrower.

WITH RECOURSE is an agreement that the seller of a note is responsible for payment if the original maker does not pay.

CHAPTER 9

An **AMORTIZATION SCHEDULE** is a table that gives in detail the steps by which a debt is reduced and paid off.

The **AMOUNT FINANCED** is the amount actually borrowed, found by subtracting the down payment from the cash price.

The **ANNUAL PERCENTAGE RATE** is the true annual interest rate charged by lenders—which must be quoted to a borrower according to the Truth in Lending Act. It is calculated on the balance due at the time of each installment payment.

The **FINANCE CHARGE** is the extra money paid for buying on credit. It is often called a service charge, carrying charge, or interest.

An **INSTALLMENT PLAN** is a payment plan whereby a borrower makes regular periodic payments to pay off a debt, usually with a high rate of interest.

The **NOMINAL INTEREST RATE** is the simple interest rate calculated on the original amount of the debt for the entire period during which payments are made. It is not legal for the lender to quote this rate to the borrower.

The **REFUND FRACTION** is the portion of the finance charge that is refunded according to the rule of 78. Its numerator is the sum of the numbers of the remaining months and its denominator is the sum of the numbers of the total months.

A REVOLVING CHARGE ACCOUNT is an open-ended charge account in which charges are allowed up to a specified maximum, a minimum payment is required, and interest is charged on the outstanding balance.

The **RULE OF 78** is a method for finding the interest saved when an installment plan is paid in full before the last payment.

The **TRUTH IN LENDING ACT** is a federal law enacted in 1969 that requires that borrowers be informed in writing of the finance charge and the annual percentage rate paid on any installment purchase, charge account, or any other loan.

The **UNITED STATES RULE** states that each payment on a loan or installment debt must first be used to pay any interest owed. The balance of the payment is then used to decrease the debt and obtain a new balance on which future interest will be calculated.

CHAPTER 10

A **CERTIFICATE OF DEPOSIT (CD)** is a savings investment where the depositor agrees to leave a minimum amount of money for a specific period of time. Rates on CDs are higher than those on passbook accounts.

COMPOUND AMOUNT is the final sum at the end of the term.

COMPOUND INTEREST is the interest that is computed periodically and added to previous principal. This total then becomes the principal for the next interest period.

COMPOUNDING is the process of adding interest to principal at regular intervals.

The **CONVERSION PERIOD** or **COMPOUNDING PERIOD** is the regular interval for which interest is computed and added to the principal.

The **EFFECTIVE RATE** or **EFFECTIVE ANNUAL YIELD** is the simple interest rate that would give the same return in 1 year as the compound rate.

A **MONEY-MARKET ACCOUNT** pays interest comparable to CDs and offers penalty-free withdrawal of funds as well as limited check writing privileges.

The **NOMINAL RATE** is the stated annual rate of interest without regard to compounding.

A **PASSBOOK ACCOUNT** is a savings account where money can be deposited or withdrawn anytime with no penalty.

PRESENT VALUE is the principal that must be invested at a given rate to accumulate to a given amount at the end of a definite period.

CHAPTER 11

AMORTIZATION is the process of repaying a loan by a series of equal payments that cover principal and interest.

An **AMORTIZATION SCHEDULE** is a listing of the principal and interest included in each loan payment, and the balance still owed.

The **AMOUNT OF AN ANNUITY** is the final value at the end of the term, the sum of all payments and compound interest earned.

An **ANNUITY** is a series of payments, usually equal, made at regular intervals.

The **CASH EQUIVALENT** is the sum that can be invested today that would have the same maturity value as would the down payment and periodic payment of an annuity.

An **ORDINARY, SIMPLE ANNUITY CERTAIN** is an annuity in which the payments are made at the end of the period, the interest conversion periods and the payment interval are the same, and there are definite dates for the beginning and end of the payments.

The **PAYMENT INTERVAL** is the time between two successive payments.

The **PRESENT VALUE OF AN ANNUITY** is the value at the beginning of the term in order to receive given payments from it.

RENT is the periodic payment of the annuity.

A **SINKING FUND** is an annuity established to repay a debt at a future date or to replace equipment at the end of the depreciation period.

A **SINKING FUND TABLE** is a listing of the current balance and the interest earned for any sinking fund.

The **TERM OF THE ANNUITY** is the length of time the annuity continues.

CHAPTER 12

ARITHMETIC MEAN (see **Mean**).

The **AVERAGE** is one number that can be used to summarize a group of numbers. The mean, median, and mode are different kinds of averages.

A **BAR GRAPH** is a visual display of values in which vertical or horizontal bars are used to make comparisons or show relationships.

A **BASE YEAR** is a year in the past that is used to compare values in later years.

The **CONSUMER PRICE INDEX** is a measure of the average change in prices in a fixed "market basket" of goods and services purchased for day-to-day living.

DISPERSION is the scatter or spread of the numbers in any group. The range and standard deviation are measures of dispersion.

An **INDEX NUMBER** is a special form of ratio used to show percent changes over a period of time.

A **LINE GRAPH** is a visual display of values that uses a line to show changes over a period of time.

The **MEAN** is the average found by dividing the sum of the values by the number of values.

MEASURE OF CENTRAL TENDENCY is the statistical term for average.

The **MEDIAN** is the middle value in a series of values that are arranged in order of magnitude.

The **MODE** is the value that occurs most frequently.

A **PIE CHART** (or **CIRCLE GRAPH**) is a visual display of values in which a circle is used to show the breakdown of a whole unit into parts.

The **RANGE** is the difference between the largest and smallest numbers.

A **SAMPLE** is a part of the whole.

SAMPLING is a statistical process in which a conclusion about a whole population is made on the basis of a sample.

The **STANDARD DEVIATION** is a measure of dispersion that is based on the distance of each number from the mean.

The **STATISTICAL MAP** uses shading or colors and a key to show how a particular characteristic is distributed over a geographical area.

STATISTICS is the science of collecting, organizing, analyzing, presenting, and interpreting numerical data.

The **WEIGHTED MEAN** is used when the values to be averaged are not equally important. Each value is given its relative importance.

CHAPTER 13

ACCOUNTS PAYABLE is the total money owed to creditors by the company for goods and services it has bought on credit.

ACCOUNTS RECEIVABLE is the total money owed to the company by customers for goods and services purchased on credit.

ACCRUED EXPENSES are those which must be paid within a short period of time.

ASSETS are items of value owned by the business.

A **BALANCE SHEET** shows what the company owes and owns at a particular time.

BONDS are direct obligations of a corporation or government with a fixed percent return and a definite maturity date.

CALLABLE BONDS may be paid off by the issuer before maturity under specified conditions.

CALLABLE PREFERRED STOCK gives the company the right to buy back the stock under specified conditions.

CAPITAL STOCK is the value of the stock certificates issued by the company.

COMMON STOCK gives the owner the right to share in profits and to vote on company policy. The dividend is determined each year.

CONVERTIBLE BONDS may be exchanged for other securities, usually common stock of the same company, under specified conditions.

CONVERTIBLE PREFERRED STOCK gives the owner the right to exchange shares for common stock under specified conditions.

CUMULATIVE PREFERRED STOCK is stock such that if a dividend is not paid one year, it remains as an obligation for the next year.

CURRENT ASSETS include cash and those assets that will be converted to cash within one year (or one operating cycle) of the balance sheet date.

CURRENT LIABILITIES are debts that must be paid within a short time, usually 1 year.

DEBENTURES are bonds backed only by the general credit of the issuer.

A **DISCOUNT** is the amount below face value that is paid for a bond.

A **DIVIDEND** is a stockholder's return on investment. It is paid by a company to the holders of stock.

The **DOW JONES INDUSTRIAL AVERAGE** is an average of the prices of thirty industrial stocks.

FIXED ASSETS (or **PLANT ASSETS** or **PLANT AND EQUIPMENT**) include land, buildings, machinery, furniture, fixtures, trucks, and the like, bought for the company's use, not for sale.

GOVERNMENT BONDS are debts of the U.S. government.

An **INCOME STATEMENT** summarizes a company's operations during a period of time and shows whether there has been a profit or loss.

INVENTORIES include raw materials, in-process work, and finished stock.

LIABILITIES are amounts owed by the company to its employees and creditors.

LONG-TERM LIABILITIES are debts that must be paid more than 1 year after the balance sheet date.

MUNICIPAL BONDS are issued by a state, political subdivision of a state, or by a state agency or authority.

A **MUTUAL FUND** is an investment where individuals pool their money to buy stocks, bonds, and other securities selected by the professional managers of the fund.

NET INCOME is the amount of net sales remaining after deducting cost of goods sold, operating expenses, and taxes.

NOTES PAYABLE are short-term notes that a company owes to its suppliers and/or bank.

NOTES RECEIVABLE are promissory notes owed to the company.

An **ODD LOT** is fewer than 100 shares of stock.

PAR VALUE (of a bond) is the face value or redemption value of the bond. It is also the value upon which the annual interest is computed.

PAR VALUE (of a stock) is the original price set by the company when the stock is issued. A stock may have no par value.

PARTICIPATING PREFERRED STOCK gives the owner the possibility of sharing in the company profits in addition to the fixed-percent dividend.

PREFERRED STOCK has a fixed percent of its par value as dividends.

A **PREMIUM** is the amount above face value that is paid for a bond.

QUICK ASSETS are cash and assets that can be converted to cash quickly.

RETAINED EARNINGS are the accumulated profits that have not been distributed to the shareholders but have been kept for future use in the business.

A **ROUND LOT** is a multiple of 100 shares of stock.

SECURITIES is another name for stocks and bonds.

The **STANDARD & POOR'S 500 INDEX** is an unmanaged index of 500 large publicly traded stocks representing a variety of industries.

A **STOCK EXCHANGE** is a place where stocks that are registered with the exchange are bought and sold.

A **STOCKBROKER** is a person who acts as an agent in the buying and selling of stock.

STOCKHOLDERS' EQUITY (or **NET WORTH**) is the owners' share of a company's assets; the difference between total assets and total liabilities.

WORKING CAPITAL is the excess of current assets over current liabilities.

APPENDIX A

A **COMMON FRACTION** has integers in both the numerator and denominator.

A **COMMON MULTIPLE** is a multiple of two or more numbers.

A **COMPLEX DECIMAL** contains a common fraction.

A **COMPLEX FRACTION** has a fraction or an indicated operation in one or both terms.

A **DECIMAL**, or **DECIMAL FRACTION**, has a denominator equal to 10, 100, 1,000, or any number of tens multiplied together. When written in the short form with a decimal point instead of a denominator, it is called a decimal rather than a decimal fraction.

A **FACTOR** of a number can be divided into the number without a remainder.

An **IMPROPER FRACTION** is a fraction in which the numerator is larger than the denominator.

The **LEAST COMMON DENOMINATOR (LCD)** is the smallest number that is a multiple of all the denominators.

LIKE FRACTIONS have the same denominators.

A **MIXED DECIMAL** is a whole number with a fraction in decimal form.

A **MIXED NUMBER** is an integer plus a proper fraction.

A **MULTIPLE** of a number is exactly divisible by the number.

A **PRIME NUMBER** has no factors except itself and 1.

A **PROPER FRACTION** is a fraction in which the numerator is less than the denominator.

UNLIKE FRACTIONS have different denominators.

Index

A

Absolute loss, 104–106, 711
Accelerated cost recovery system (ACRS), 187, 522, 712
    percentage table, 658
Accounts payable, 519, 715
Accounts receivable, 519, 715
Accounts receivable-net sales ration, 522
Accrued (accumulated) expenses, 519, 715
Accrued (accumulated) taxes, 519
Accumulated depreciation, 174, 712
Acid-test ration, 520
ACRS, 187, 712
    percentage table, 658
Add-on interest, 262, 713
Addition, 1–2
    of decimals, 570
    of mixed numbers, 560–561
    of terms, 11, 32
Algebraic expressions, 7–11, 711
Amortization, 443–446, 714
Amortization schedule, 351–352, 445–446, 714
Amount:
    of an annuity, 428–431, 714
    at compound interest, 386, 391–400, 714
    of discount, 305–309, 714
    financed, 345, 714
    at simple interest, 262–263, 283–284, 713
Annual percentage rate (APR), 344, 359–362, 714
Annual percentage rate (APR) table, 597–601
Annuity, 427–452, 714
    amount of, 428–431, 714
    cash equivalent price of, 436–438
    ordinary, simple, certain, 427, 715
    present value of, 435–437, 715
    rent for, 428, 443–448
Annuity tables:
    amount, 634–641
    periodic rent, 642–649
    present value, 650–657
Approximate yield to maturity, 531–532
Arithmetic and algebraic expressions, 7–8
Arithmetic mean, 487, 715
Articles of copartnership, 213
Assessed value, 59–60, 711
Asset depreciation, 173–174, 179, 187–188, 712
Assets, 173, 518–519, 712, 715
    current, 518–519, 716
    in financial report, 518–519
    fixed, 519, 716
    quick, 500, 716
    return on total, 522
    useful life for, 174, 712
Average-cost inventory valuation, 194–196, 712
Average investment per month, partners agreement, 219
Averages, 487–491, 715

B

Balance sheet, 517–520, 715
Bank discount, 305–309, 714
Bank loan, cash discount savings advantage, 323–325
Banker's rule, 275–276, 713
Bar graph, 475–477, 715
    component, 483
Base, 53–61, 65, 711
Base year, 497, 715
Basic equation, 53–54, 59–61
Billing, merchandise, 143–148, 153–154
Bimodal, 488
Bond quotations, 537–539
Bond trading, 530
Bond yields, 530–531
Bonds, 525, 530–532, 715
    callable, 530, 715
    convertible, 530, 715
    debenture, 530, 715
    face value of, 530, 715
    government, 530, 716
    municipal, 530, 716
    par value of, 530, 716
    yields of, 530–532
Book value, 168–171, 174–175, 183, 189, 712
Borrowing, 261–262
    to anticipate invoice, 323–325
    Rule of 78's for, 365–366, 714
Breakeven point, 104, 711

C

Calculators, problem solving hints, 4, 43, 61, 68, 84, 135, 180, 189, 195, 214, 215, 263, 264, 282, 322, 399, 400, 410, 430, 491
Calendar table, 596
Callable bonds, 530, 715
Callable preferred stock, 526, 715
Capital stock, 519, 715
Carrying charge (*see* Finance charge)
Cash discount, 143–148, 711
    short-term loans to save, 323–325
Cash equivalent price of annuity, 436–438
Celsius, 582–583
Centi, 580
Certificates of deposit (CD's) weekly yields, 391, 470–471, 714
Charge accounts, 344–345, 352–355
Circle graph (*see* Pie chart)
Coefficient, 8, 711
Commission, 59, 228–230, 711–712
    basic equation applied to, 59
    payroll and, 228–229
    with quota, 229
    plus salary, 228–230
    straight, 228, 713
    sliding scale, 215, 229, 713
Common factor, fraction, 557
Common fraction, 556, 716
Common multiple, 555, 716
Common stock, 526–529, 715
Complex decimal, 568, 716
Complex fraction, 565–566, 716
Component bar graph, 483
Compound amount, 386, 391–400, 714
    for additional deposit, 397–417
    formula for, 399–400
    present value of, 409–411
    for rate change, 403
Compound interest, 247, 383–427
    investment annuity including, 427–452
    simple interest vs., 383–386
Compound interest tables:
    for amount, 602–617
    for present value, 618–633
Compounding, 359–370, 714
Compounding (conversion) period, 383–384, 399–400, 403–404, 714
    for additional deposit, 403–404
    for annuity, 427–428
    for rate change, 403
    n larger than last table entry, 404–405
Computer spreadsheet functions (*see* Functions, computer spreadsheet)
Constants, 7, 711
Consumer, 125, 711
Consumer Price Index (CPI), 499

717

Conversion (compounding) period, 386–387, 399–400, 403–404, 714
   for annuity, 427–428
   for additional deposit, 403–404
   for rate change, 403
   n larger than last table entry, 404–405
Convertible bonds, 530, 715
Convertible preferred stock, 526, 715
Cost, 81–85
   of credit, 340–342
   inventory value and, 193–196
   markup on, 82–85
   net, 126
   percent markup on, 89–92
Cost recovery, 173–174, 712
Credit (commercial), 143–148, 711
Credit (consumers), 343–382
Credit cards, 352–355
Cumulative preferred stock, 526, 715
Current assets, 518–519, 715
   accounts receivable, 519, 715
   inventories, 519, 716
   notes receivable, 519, 716
Current liabilities, 519, 716
   accounts payable, 519, 715
   accrued expenses, 519, 715
   accrued taxes, 519
   notes payable, 519, 716
Current ratio, 520
Current yield:
   bond, 530–531
   stock, 528–529

D

Date, calendar, and time exact, 269–271
Date of loan, 283
Date format, 171
Day-of-the-year number table, 596
Debentures, 530, 716
Debt amortization, 351–352
Deci, 580
Decimal fraction, 566–568, 716
Decimal places for compounding interest, 391
Decimal point, 566–567
Decimals, 566–573
   addition/subtraction of, 570–571
   complex, 568, 716
   conversion to/from fractions, 566–570
   conversion to/from percents, 41–50
   division of, 571–572
   division/multiplication of by powers of ten, 567–568
   mixed, 568, 716
   multiplication of, 571
   rounding of, 3–4, 32
Declining-balance depreciation, 183–184, 712
   maximum rate for, 183
Denominator, 555–563, 565–570
Depreciation, 173–190, 712
   ACRS, 187, 712
   declining-balance, 183–184, 712
   MACRS, 187–190, 712
   straight-line, 174–176, 712
   sum-of-years' digits, 179–180, 712
Depreciation schedule, 175–184, 188–189, 712
Differential piece-rate plan, 227, 712

Digits, life of asset, 179
Disclosing finance charges, 344
Discount, on a bond, 537, 716
Discount note,
   simple, 305–309, 714
   vs. promissory note, 326
Discount period, 313–317, 714
Discount loan, maturity value of, 305–306, 321–322, 714
Discounting a promissory note, 313–317, 714
Discounts:
   bank, 305–309, 714
   cash, 143–148, 711
   commercial, 125–171
   percent, 125–148
   series, 133–137, 712
   single equivalent, 137–139, 712
   trade, 125–137, 712
Dispersion, 491–493, 715
   range, 491, 715
   standard deviation, 491–493, 715
Distance (metric measure), 580–581
Dividends, stock, 525–530, 716
Division, 2, 32
   of decimals, 571–572
Double time, 226, 712
Double-declining-balance depreciation, 183–184, 712
Dow Jones Industrial Average, 527, 716
Draw, 229, 712
Due date, 270–271, 713

E

E.O.M. (end of month), 145–146, 711
Earnings per share, 529
Earnings, gross, 225–230, 233, 235, 713
Economic Recovery Tax Act, 173
Effective rate, 405–406, 714
Equal shares, partners agreement, 214
Equation, 15–17, 32, 711
Estimated useful life, 174, 712
Exact interest, 275–276, 713
Exact time, 269–271, 275, 713
Exemption, 235, 712
Extra(X), 146, 711

F

Face value:
   of bond, 530
   of discounted note, 305–306, 326, 714
   of promissory note, 283–284, 713
Factor, 555, 716
Fair Labor Standards Act, 225, 713
Fahrenheit, 582–583
Federal income tax, 235–240
   withholding tables, 590–595
Federal Insurance Contribution Act, 233, 713
Federal Reserve Board, 359
FICA (Medicare and Social Security) tax, 233–235
FIFO inventory valuation, 193–194, 712
Finance charge, 344–366, 714
   disclosure of, 344
Financial ratios, 520–522
   current, 520
   quick, 520–521

   return on investment, 521–522
   return on total assets, 522
   accounts receivable-net sales ratio, 522
   price-earnings, 529
Fixed assets, 519, 716
Fixed percents, partners agreement, 214
Fixed ratios, partners agreement, 214–215
Fractions, 41, 555–566
   common, 556, 716
   complex, 565–566, 716
   conversion to/from decimal, 566–570
   conversion to/from percents, 41–46
   decimal, 566–569, 716
   division of, 564–565
   improper and proper, 556–558, 716
   like and unlike, 558–560, 716
   multiplication of, 562–563
Functions, computer spreadsheet:
   AVERAGE, 514
   COUNT, 515
   DDB, 212
   FV, 465
   IF, 171, 257–259, 303–304, 381, 425, 515, 554
   MAX, 515
   MEDIAN, 515
   MIN, 515
   MOD, 554
   NPER, 426
   OR, 515
   PMT, 465
   PV, 466
   SLN, 211
   STDEVP, 515
   SYD, 212

G

General numbers, 7, 711
Government bonds, 530, 716
Gram, 580–581
Graphs, 444, 473–484
   bar, 475–477, 715
   component bar, 483
   line, 473–475, 715
   pictograph, 476–477
   pie chart, 481–482, 715
   statistical map, 483–484, 715
Gross earnings, 225–230, 233, 235, 713
Gross profit, 521

H

Half-year convention, 188, 712
Hourly wages, 226–227, 713
   double time, 226, 712
   overtime, 226, 713
   time-and-a-half, 226, 713

I

Improper fractions, 556, 716
   mixed numbers and, 556–558, 716
Income statement, 517, 521, 716
Income tax withholding, 235–240, 713
   wage bracket tables for, 590–593
   percentage method tables for, 594–595
Index number, 497–499, 715
Industrial Production Index, 499

INDEX

Installment plan, 343–352, 365–366, 714
   at annual percentage rate, 351–352, 359–362
   cash equivalent price, 438–439
   cost of credit, 344–346
   periodic payment, 346–347
   rebate using rule of 78, 365–366, 714
Interest, 261–263, 713
   add-on, 262, 713
   compound, 261, 383–426, 713–714
   exact, 275–276, 713
   ordinary, 275–276, 713
   simple, 261–290, 713
Interest on investment partners agreement, 220
Interest rate:
   annual percentage rate, 344, 359–362, 714
   compounding, with change of, 403
   effective, 405–406, 714
   nominal, 346, 360–362, 386, 393–394, 714
   simple, 262, 264, 281, 283, 713
Internal accounting, 174
Inventory, 193, 712
Inventories, 519–520, 716
Inventory valuation, 193–197
   average-cost method of, 194–195, 712
   FIFO method of, 193–194, 712
   LIFO method of, 194, 712
Investment, securities, 517
Investment returns, 521, 525, 537
Invoice, 143–148, 711
   borrowing to anticipate a cash discount, 323–325

K
Kilo, 580

L
Least common denominator (LCD), 558–559, 716
Least common multiple, 555
Liabilities, 519, 716
   current, 519, 716
   long-term, 519, 716
Like and unlike fractions, 558–560, 716
Line, fraction, 555
Line graph, 473–475, 715
LIFO inventory valuation, 194, 712
List price, 126–129, 143, 712
Liter, 580–581
Loans:
   amortization of, 443–446, 714
   time for interest on, 269–271
Loss, absolute (gross), 104–106, 711

M
Maker, 283, 713
Manufacturer, 125, 712
Margin (*see* markup)
Markdown, 101–106
Markup, 81–104, 711
   on cost, 83–85
   on selling price, 83–85
   percent, 89–92
   on perishables, 97–98
MasterCard, 343, 345
Maturity date:
   of loan, 270–271
   of note, 270–271, 713
Maturity value:
   of annuity, 428–431
   at compound interest, 386–411
   of discounted loan, 305–306, 321–322, 714
   finding principal for, 281–282
   of promissory note, 283–284, 714
   finding simple interest on, 262–263
Mean, 487–490, 715
   arithmetic, 487, 715
   weighted, 489–491, 715
Measure, metric units of, 579–583
Measures of central tendency, 487–490, 715
Median, 487–488, 715
Medicare tax, 233–235, 713
Meter, 580–581
Metric conversion, 98, 579–583
Metric Conversion Act, 579
Metric system (measure), 579–583
Milli, 580
Minimum wage, 225
Mixed decimal, 568–569, 716
Mixed numbers, 556–558, 560–562, 716
   addition of, 560–561
   improper fractions and, 556–558
   subtraction of, 561–562
Mode, 488, 715
Modified Accelerated Cost Recovery System (MACRS), 187–190, 712
Money market account, 391, 714
Money market rate, 287, 713
Mortgages:
   amortization of, 443–446, 714
   investment/securities and, 519, 530
   U.S. Rule and, 351–352, 714
Multiple, 555, 716
Multiplication, 1–2
   algebraic expression and, 7
   of decimals, 571
   of fractions, 562–563
Municipal bonds, 530, 716
Mutual fund, 529–530, 716

N
NASDAQ, 526
NASDAQ Composite Index, 527
Negotiable promissory notes, 283, 313, 713
Net cost, 126
Net cost rate factor, 134, 138, 712
Net income, 521–522, 716
Net price, 126–129, 133–137, 712
Net profit, 81–85, 104–106, 711
Net worth, 519, 716
New York Stock Exchange (NYSE), 526
Nominal interest rate, 346, 360–362, 386, 393–394, 714
Notes:
   bank discount of, 305–309, 714
   interest-bearing, 283–290
   discounting of, 313–317, 714
   non-interest bearing, 313–314
   payable, 519, 716
   receivable, 519, 716
   promissory, 283–290, 313–317, 713
Numerator, fraction, 555

O
Odd lots, 526, 716
Omnibus Trade and Competitiveness Act, 579
Operating loss, 105–106, 711
Ordinary annuity certain, 427, 715
Ordinary dating, 145, 712
Ordinary interest, 275–276, 713
Ordinary time, 275, 713
Original investment, partners agreement, 215
Outstanding balance, 351–352
Overhead, 81–85, 104–106, 711
Overtime, 226–227, 713
Owner's equity, 518–519, 716
   capital stock, 519, 715
   retained earnings, 519, 716

P
Par value, 526, 530, 716
Parentheses, 2–3
Partial payments, 153–154
Participating preferred stock, 526, 716
Partnership, articles of co-, 213
Partnership agreements, 213–221, 713
   average investment per month, 219
   combination of methods, 220–221
   equal shares, 214
   fixed percents, 214
   fixed ratios, 214–215
   interest on investment, 220
   original investment, 215
Passbook account, 391, 714
Payee, 283, 713
Payment interval, annuity, 428, 715
Payments, periodic, 346–347
Payroll, 225–240
   for commissioned employee, 228–229
   deductions, 233–240
   for hourly employee, 226–227
   for piecework wage earner, 227–228
   for salaried employee, 225
Percent, 41–68, 711
   change measured, 65–69
   conversions from decimal/fraction, 49–50
   conversions to decimal/fraction, 41–46
   relativity of, 53
Percentage method, 236, 239–240, 713
   percentage method tables, 594–595
Periodic inventory system, 193, 712
Perpetual inventory system, 193, 712
Pictograph, 466–477
Pie chart (circle graph), 481–482, 715
Piecework wages, 227–228, 713
Plant & equipment, 519
Pound (measure), 580–581
Preferred stock, 526, 716
   callable, 526, 715
   convertible, 526, 715
   cumulative, 526, 715
   participating, 526, 716
Premium, on a bond, 537, 716
Present value:
   of annuity, 435–437, 715
   of compound amount, 409–411
   defined, 281–282, 287, 714
   at simple interest, 281–282
   of promissory note, 287–290, 713

Price:
　list, 126–129, 143, 712
　net, 126–129, 133–137, 712
Price-earnings ratio, 529
Pricing, 81–106
　for markdown, 101–106
　perishables, 97–98
Prime number, 555, 716
Principal:
　defined, 261–262, 281, 713
　present value (P) and, 281–282
　simple interest on, 261–263
Proceeds, 305–308, 314, 321–326, 714
　calculation of, 306–309
Producer, 125, 712
Producer Price Index, 499
Product, 53–61, 65, 677
Product marketing (merchandising) path, 125, 133
Product pricing, 81–106
Profits and losses in partnership agreement, 213–221
Promissory note, 283–290, 313–317, 713
　discounting of, 313–317, 714
　face value of, 283–284, 713
　present value of, 287–290, 713
Proper fraction, 556, 716
Property tax, 59–60
PROX, 145–146, 712

Q

Quick assets, 520, 716
Quick ratio, 520
Quota, 229, 713
Quotations:
　bond, 535–536
　stock, 537–539

R

R.O.G. (receipt of goods), 146, 712
Range, 491, 715
Rate,
　annual percentage, 344, 359–362, 714
　at compound interest, 386–387
　effective, 405–406, 714
　money market, 287, 713
　nominal, 346, 360–362, 386, 393–394, 714
　in percent formula, 49–57, 61, 711
　at simple interest, 262, 264, 281, 283, 713
　solving for, 53–67
Rate of return on stock investment, 537
Ratios, 27–28, 711
　financial, 520–522, 529
　in partnership profits, 214–215
Reciprocal, 17, 711
Recourse, 313, 714
Reduced price, 101, 105–106, 711
Refund fraction, 365, 714
Regular price, 101–106, 711
Rent of an annuity, 428, 715
　solving for: amount known, 447–448
　present value known, 443–446
Retailer, 125–126, 133, 712
Retailer-wholesaler discounting, 125–127, 133
Retained earnings, 519, 716
Return on investment, 521–522

Return on total assets, 522
Revolving credit, 353–355, 714
Round lots, 526, 716
Rule of 78, 365–366, 714

S

Salary, 225–226, 713
Sales, net, income statement, 521, 716
Salvage (scrap) value, 174, 712
Sample, 497, 715
Sampling, 497, 715
Savings account, 391
Securities, 517, 525–532, 716
Selling price, 81–106
Series discounts, 133–137, 712
Service charge (*see* Finance charge)
Similar terms, 11, 21–22, 711
Simple-discount note, 305–309, 714
　vs. promissory note, 326
Simple interest, 261–290, 713
　amount at maturity, 262–263, 283–284, 713
　vs. compound interest, 383–386
　on installment loan, 344, 351–352
　present value at, 281–282, 713
　promissory notes, 283–290, 313–317, 713
　vs. simple discount, 326
Single equivalent discount, 137–139, 712
Sinking fund, annuity, 447–448, 715
Sinking fund table, 448, 715
Sliding scale commission, 229, 713
Social Security tax, 233–235, 713
Specific identification method, 193, 712
Square root, 491–492
Standard and Poor's 500 Index, 527, 716
Standard deviation, 491–493, 715
Statistics, 469–515, 715
Statistical Abstract of the United States, 470
Statistical map, 483–484, 715
Stock, 525–530
　certificate, 525
　common, 526, 715
　dividend, 525–530, 716
　exchange, 526, 716
　lots, odd or round, 526, 716
　market, 526
　par value of, 526, 716
　preferred, 526, 716
　quotations, 535–536
　rate of return on, 537
　trading, 526–527
　yields, 528–529
Stockbrokers, 526, 716
Stockholders' equity, 519, 716
　capital stock, 519, 715
　retained earnings, 519, 716
Straight commission, 228, 713
Straight piecework, 227
Straight-line depreciation, 174–176, 712
Substitution, algebraic, 8
Subtraction, 2–3
　of decimals, 570–571
　of mixed numbers, 561–562
Sum-of-years' digits depreciation, 179–180, 712

T

Tax Reform Act, 173

Tax:
　exemption, 235
　federal income, 235–240
　FICA, 233, 713
　medicare, 233–235, 713
　property, 59–60
　social security, 233–235, 713
　state and city income, 236
Temperature (metric units), 582–583
Term:
　of an annuity, 428, 715
　of a note, 283, 713
Terms, algebraic, 8–11, 711
Time:
　by calendar table, 269–271
　exact, 269–271, 275, 713
　fraction, 275
　at simple interest, 262, 264, 281, 283, 713
　ordinary, 275, 713
Time-and-a-half, 226, 713
Total depreciation, 174, 712
Trade discount, 125–137, 712
Trade-in value, 174
Trading:
　bonds, 530
　stocks, 526–527
Truth-in-Lending Act, 308, 344, 351, 359, 714

U

United States rule, 351–352, 714
Useful life, 174, 712

V

Valuation of inventory, 193–196
　by average cost, 194–196, 712
　by FIFO, 193–194, 712
　by LIFO, 194, 712
Variables, 7, 711
Volume (metric measure), 581–582

W

Wage bracket method, 236–238
Wage bracket tables, 590–593, 713
Wages, 216–218
　hourly, 216–217, 713
　piecework, 227–228, 713
Wall Street Journal, 535–538
Weight (metric measure), 581–582
Weighted mean, 489–491, 715
Wholesaler, 125, 133, 712
Wholesaler-retailer trade discounting, 125–127, 133
With recourse, 313, 714
Withholding, federal income tax, 235–240, 713
　percentage method, 236, 239–240, 713
　wage bracket method, 236–238
Withholding allowance, 235, 713
Withholding tables:
　percentage method, 594–595
　wage bracket, 590–593
Working capital, 520, 716
Working capital ratio, 520
W-4 form, 235

Z

Zero:
　in decimals, 567–568